Sensor Array Signal Processing

SECOND EDITION

SECOND EDITION

SENSOR ARRAY SIGNAL PROCESSING

PRABHAKAR S. NAIDU

CRC Press
Taylor & Francis Group
Boca Raton London New York

CRC Press is an imprint of the
Taylor & Francis Group, an **informa** business

CRC Press
Taylor & Francis Group
6000 Broken Sound Parkway NW, Suite 300
Boca Raton, FL 33487-2742

First issued in paperback 2017

ISBN 13: 978-1-138-11397-8 (pbk)
ISBN 13: 978-1-4200-7190-0 (hbk)

Visit the Taylor & Francis Web site at
http://www.taylorandfrancis.com

and the CRC Press Web site at
http://www.crcpress.com

This work is dedicated to the memory of the great visionary, J. R. D. Tata, who shaped the Indian Institute of Science for many decades.

Contents

Prologue from First Edition

An array of sensors is often used in many diverse fields of science and engineering, particularly where the goal is to study propagating wavefields. Some examples are astronomy (radio astronomy), medical diagnosis, radar, communication, sonar, nonrestrictive testing, seismology, and seismic exploration (see Ref. [1] for different applications of the array signal processing). The main goal of array signal processing is to deduce the following information through an analysis of wavefields:

(a) Source localization as in radar, sonar, astronomy, and seismology, etc.
(b) Source waveform estimation as in communication, etc.
(c) Source characterization as in seismology
(d) Imaging of the scattering medium as in medical diagnosis, seismic exploration, etc.

The tools of array signal processing remain the same, cutting across the boundaries of different disciplines. For example, the basic tool of beamformation is used in many areas mentioned above. The present book aims at unraveling the underlying basic principles of array signal processing without a reference to any particular application. However, an attempt is made to include as many tools as possible from different disciplines in an order that reflects the underlying principle.

In the real world, different types of wavefields are used in different applications, for example, acoustic waves in sonar, mechanical waves in seismic exploration, electromagnetic waves in radar and radio astronomy. Fortunately, all wavefields can be characterized under an identical mathematical framework. This common mathematical framework is briefly summarized in Chapter 1. Here we have described the basic equations underlying different wavefields and the structure of array signals and the background noise when the noise sources follow some simple geometrical distribution. The topics covered are wavefield in open space, bounded space including multipath propagation and layered medium. Also covered is the weak scattering phenomenon, which is the basis for tomographic imaging. In Chapter 2, we study different types of sensor configurations. The emphasis is, however, on the commonly used uniform linear array (ULA), uniform circular array (UCA). Many practical sensor array systems can be studied in terms of the basic ULA and UCA systems (cylindrical array in radar and sonar, cross-array in astronomy and seismology). Like sensors, the sources can also be configured in the form of an array. The source array is useful in synthesizing a desired wavefront and/or waveform. In Chapter 3, we examine the issues connected with the design of 2D digital filters for wavefield analysis.

Since the propagating wavefields possess some interesting spectral characteristics in the frequency wavenumber domain, for example, the spectrum of a propagating wavefront is always on a radial line, it is natural to take into account these features in the design of digital filters for separation of interfering wavefields. Specifically, we cover in detail the design of a fan filter and quadrant filter. Also, the classical Wiener filter as an optimum least squares filter is covered in this chapter.

The theme in Chapters 4 and 5 is localization of a source. In Chapter 4, we describe the classical methods based on the frequency wavenumber spectrum of the observed array output. We start with the Blackman Tukey type frequency wavenumber spectrum and then go on to modern nonlinear high-resolution spectrum analysis methods, such as Capon's maximum likelihood spectrum, which is also known as the minimum variance distortionless response (MVDR) beamformer and maximum entropy spectrum. Localization essentially involves estimation of parameters pertaining to the source position, for example, azimuth and elevation angles, range, speed if the source is moving, etc. In the last two decades, a host of new methods of source localization have been invented. We elaborate these new approaches in Chapter 5. These include subspace-based methods, use of man-made signals such as in communication, and finally multipath environment. Quite often, localization must be done in the real time and it may be necessary to track a moving source. Adaptive techniques are best suited for such tasks. A brief discussion on adaptive approach is included. In Chapter 6, we look into methods for source waveform separation and estimation. The direction of arrival (DOA) is assumed to be known or has been estimated. We shall describe a Wiener filter that minimizes the mean square error in the estimation of the desired signal coming from a known direction and a Capon filter that, while minimizing the power, ensures that the desired signal is not distorted. We also talk about the estimation of the direction of arrival in a multipath environment encountered in wireless communication.

The next two chapters are devoted to array processing for imaging purposes. In Chapter 7 we look at different types of tomographic-imaging systems: nondiffracting, diffracting, and reflection tomography. The received wavefield is inverted under the assumption of weak scattering to map any one or more physical properties of the medium, for example, sound speed variations in a medium. For objects of regular shape, scattering points play an important role in geometrical diffraction theory. Estimation of these scattering points for the determination of shape is also discussed. In Chapter 8, we study the method of wavefield extrapolation for imaging, extensively used in seismic exploration. The raw seismic traces are stacked in order to produce an output trace from a hypothetical sensor kept close to the source (with zero-offset). A suite of such stacked traces may be modeled as a wavefield recorded in an imaginary experiment wherein small charges are placed on the reflector and exploded at the same time. The zero-offset wavefield is used for imaging of reflectors. The imaging process may be looked upon as a downward continuation of the wavefield, or inverse source problem, or propagation backward in time, i.e., depropagation to the reflector. All three view-points are very briefly described.

The book is based on a course entitled "Digital Array Processing" offered to the graduate students who had already taken a course on digital signal processing (DSP) and a course on modern spectrum analysis (MSA). It has been my conviction that a student should be exposed to all basic concepts cutting across the different disciplines without being burdened with the questions of practical applications that are usually dealt with in specialty courses. The most satisfying experience is that there is a common thread that connects seemingly different tools used in different disciplines. An example is beamformation, a commonly used tool in radar/sonar, which has a close similarity with stacking used in seismic exploration. In this exposition, I have tried to bring out the common thread that exists in the analysis of wavefields used in a wide variety of application areas. The proposed book has a significantly different flavor, both in coverage and depth in comparison with books on the market [1–5]. The first book, edited by Haykin, is a collection of chapters, each devoted to an application. It rapidly surveys the state-of-art in respective application areas, but does not go deep enough and describe the basic mathematical theory required for the understanding of array processing. The second book by Ziomek is entirely devoted to array signal processing in underwater acoustics. It covers in great depth the topic of beamformation by linear and planar arrays, but confines to linear methods. Modern array-processing tools do not find a place in this book. The third book by Pillai [3] has a very narrow scope as it only deals with the subspace-based methods in great detail. The fourth book by Bouvet and Bienvenu (Eds) is again a collection of papers largely devoted to modern subspace techniques. It is not suitable as a text. Finally, the present book has some similarities with a book by Johnson and Dudgeon [3], but differs in one important respect, namely, it does not cover the application of arrays to imaging, though a brief mention of tomography is made. Also, the present book covers newer material that was not available at the time of the publication of the book by Johnson and Dudgeon. During the last two decades, there has been intense research activity in the area of array signal processing. There have been at least two review papers summarizing the new results obtained during this period. The present book is not a research monograph, but it is an advanced level text, which focuses on the important developments that, the author believes, should be taught to give a broad "picture" of array signal processing.

I have adopted the following plan of teaching. As the entire book cannot be covered in one semester (about 35 hours), I preferred to cover it in two parts in alternate semesters. In the first part, I covered Chapter 1 (exclude Section 1.6), Chapter 2, Chapters 4, 5, and 6. In the second part, I covered Chapter 1, Chapter 2 (exclude Section 2.3), Chapter 3 (exclude Section 3.5), Chapters 7 and 8. Exercises are given at the end of each chapter. (The solution guide may be obtained from the publisher.)

References

1. S. Haykin (Ed.), *Array Signal Processing*, Prentice Hall, Englewood Cliffs, 1985.
2. L. J. Ziomek, *Underwater Acoustics, A linear Systems Theory*, Academic Press, Orlando, FL, 1985.
3. S. U. Pillai, *Array Signal Processing*, Springer-Verlag, New York, 1989.
4. M. Bouvet and G. Bienvenu, *High Resolution Methods in Underwater Acoustics*, Springer-Verlag, Berlin, 1991.
5. D. H. Johnson and D. E. Dudgeon, *Array Signal Processing*, Prentice Hall, Englewood Cliffs, NJ, 1993.
6. H. Krim and M. Viberg, Two decades of array signal processing, *IEEE Signal Process. Mag.*, July, pp. 67–94, 1996.
7. T. Chen (Ed.), Highlights of statistical signal and array processing, *IEEE Signal Process. Mag.*, September, pp. 21–64, 1998.

Prologue for Second Edition

It has been my endeavor to provide a broad "picture" of array signal processing. The second edition has been planned with this as a guiding principle. Accordingly, six new sections and two new chapters have been added to the original book. The new sections deal with the following topics: distributed sensor array, multicomponent sensor array, Doppler-azimuth processing, communication signals, azimuth/elevation estimation, wideband adaptive beamforming, and frequency invariant beamformer. These topics have been widely researched in the last ten to fifteen years. I believe that it is now time that these topics found a place in a graduate course.

Wireless communication has become the most popular application technology. With growing numbers of mobile users, there is a definite need for increasing the capacity of networks. Use of antenna array seems to be a natural choice. I have, keeping this in mind, included two new chapters (Chapters 7 and 8). Chapter 7 deals with multipath channel, which forms a core issue in any effort to increase the capacity, and Chapter 8 deals with bit stream estimation in multipath channels. The coverage is introductory in nature. It does not cover all aspects of wireless communication, but only the role played by an antenna array. A recent book, entitled *Introduction to Space-Time Wireless Communication* by A. Paulraj, R. Nabar and D. Gore (Published by Cambridge University Press, 2003) is a comprehensive work on the use of antenna array in wireless communication.

I have maintained the same style of presentation as in the first edition. All ideas are presented in mathematical language with illustrative examples wherever possible. Exercises are included at the end of each chapter.

Acknowledgments

The thought of formalizing the lecture notes into a text occurred to me when I was visiting the Rurh Universitaet, Bochum, Germany in 1996 as a Humboldt Fellow. Much of the groundwork was done during this period. I am grateful to AvH Foundation who supported my stay. Professor Dr. J. F. Boehme was my host. I am grateful to him for the hospitality extended to me. Many of my students, who credited the course on Array Signal Processing, have contributed by way of working out the exercises cited in the text. I am particularly grateful to the following: S. Jena, S. S. Arun, P. Sexena, P. D. Pradeep, G. Viswanath, K. Ganesh Kumar, J. Joseph, V. Krishnagiri, N. B. Barkar. A. Buveneswari and A. Vasuki, who have significantly contributed to Chapter 7 through their dissertation work. Dr. K. V. S. Hari read the manuscript at an early stage and made many constructive suggestions. Parts of the present second edition were prepared at Maharaj Vijayram Gajapathi Raj (MVGR) College of Engineering, Vizianagaram, where I have been visiting for last few years. I must make a special mention of Professor K. V. L. Raju who extended generous support. I wish to thank the CRC Press Inc., in particular Nora Konopka and Jill Jurgensen for their promptness and patience.

Finally, I owe deep gratitude to my family—my wife, Madhumati and sons Srikanth, Sridhar, and Srinath for their forbearance. I must specially thank my son, Srinath, who carefully scrutinized parts of the manuscript.

1

An Overview of Wavefields

A sensor array is used to measure wavefields and extract information about the sources and the medium through which the wavefield propagates. It is therefore imperative that some background in different types of wavefields and the basic equations governing the wavefield must be acquired for complete understanding of the principles of array signal processing (ASP). In an idealistic environment of open space, homogeneous medium and high frequency (where ray approach is valid), a thorough understanding of the wave phenomenon may not be necessary (those who are lucky enough to work in such an idealistic environment may skip this chapter). But in a bounded inhomogeneous medium and at low frequencies where diffraction phenomenon is dominating, the physics of the waves plays a significant role in ASP algorithms. In this chapter, our aim is essentially to provide the basics of the physics of the waves, which will enable us to understand the complexities of the ASP problems in a more realistic situation. The subject of wave physics is vast and, naturally, no attempt is made to cover all its complexities.

1.1 Types of Wavefields and Governing Equations

The most commonly encountered wavefields are: (i) acoustic waves including sound waves, (ii) mechanical waves in solids including vibrations, and (iii) electromagnetic (EM) waves including light. The wavefields may be classified into two types, namely, scalar and vector waves. In the scalar wavefield we have a scalar physical quantity that propagates through the space, for example, hydrostatic pressure is the physical quantity in acoustic scalar wavefields. In a vector wavefield, the physical quantity involved is a vector, for example, the displacement vector in mechanical waves, electric and magnetic vectors in EM waves. A vector has three components, all of which travel independently in a homogeneous medium without any exchange of energy. But at an interface separating two different media, the components do interact. For example, at an interface separating two solids, a pressure wave will produce a shear wave and vice versa. In a homogeneous medium without any reflecting boundaries, there is no energy transfer among components. Each component of a vector field then behaves as if it is a scalar field, like an acoustic pressure field.

1.1.1 Acoustic Field

An acoustic field is a pressure (hydrostatic) field. The energy is transmitted by means of propagation of compression and rarefaction waves. The governing equation in a homogeneous medium is given by

$$\nabla^2 \phi = \frac{\rho}{\gamma \phi_0} \frac{d^2 \phi}{dt^2}, \tag{1.1a}$$

where ϕ_0 is ambient pressure, γ is the ratio of specific heats at constant pressure and volume, and ρ is density. The wave propagation speed is given by

$$c = \sqrt{\frac{\gamma \phi_0}{\rho}} = \sqrt{\frac{\kappa}{\rho}}, \tag{1.1b}$$

where κ is the compressibility modulus and the wave propagates radially away from the source. In an inhomogeneous medium, the wave equation is given by

$$\begin{aligned} \frac{1}{c^2(\mathbf{r})} \frac{d^2 \phi}{dt^2} &= \rho(\mathbf{r}) \nabla \cdot \left(\frac{1}{\rho(\mathbf{r})} \nabla \phi \right) \\ &= \rho(\mathbf{r}) \nabla \left(\frac{1}{\rho(\mathbf{r})} \right) \cdot \nabla \phi + \nabla^2 \phi . \end{aligned} \tag{1.2a}$$

After rearranging the terms in Equation 1.2a we obtain

$$\nabla^2 \phi - \frac{1}{c^2(\mathbf{r})} \frac{d^2 \phi}{dt^2} = \frac{\nabla \rho(\mathbf{r})}{\rho(\mathbf{r})} \cdot \nabla \phi, \tag{1.2b}$$

where $\mathbf{r}$ stands for position vector. The acoustic impedance is equal to the product of density and propagation speed ρc and the admittance is given by the inverse of the impedance or it is also defined in terms of the fluid speed and the pressure,

$$\text{Acoustic admittance} = \frac{\text{fluid speed}}{\text{pressure}} = \frac{\nabla \phi}{j \omega \phi}.$$

Note that the acoustic impedance in air is 42, but in water it is 1.53×10^5.

1.1.2 Mechanical Waves in Solids

The physical quantity that propagates is the displacement vector, that is, particle displacement with respect to its stationary position. Let $\mathbf{d}$ stand for the displacement vector. The wave equation in a homogeneous medium is given by [1, p. 142]

$$\rho\frac{\partial^2 \mathbf{d}}{\partial t^2} = (2\mu + \lambda)\,grad\,div\,\mathbf{d} - \mu\,curl\,curl\,\mathbf{d},$$

where μ is shear constant and λ is Young's modulus. In terms of these two basic lame constants, we define other more familiar parameters:

$$\text{Pressure wave speed: } \alpha = \sqrt{\frac{(2\mu + \lambda)}{\rho}},$$

$$\text{Shear wave speed: } \beta = \sqrt{\frac{\mu}{\rho}},$$

$$\text{Poisson ratio: } \sigma = \frac{\lambda}{2(\mu + \lambda)},$$

$$\text{Bulk modules } \kappa = \left(\frac{2}{3}\mu + \lambda\right).$$

The above parameters are observable from experimental data.

A displacement vector can be expressed as a sum of gradient of a scalar function ϕ and curl of a vector function ψ (Helmholtz theorem)

$$\mathbf{d} = \nabla\phi + \nabla \times \psi, \tag{1.3a}$$

ϕ and ψ satisfy two different wave equations:

$$\nabla^2\phi = \frac{1}{\alpha^2}\frac{\partial^2\phi}{\partial t^2}$$

$$\nabla \times \nabla \times \psi = -\frac{1}{\beta^2}\frac{\partial^2\psi}{\partial t^2} \qquad \nabla\cdot\psi = 0, \tag{1.3b}$$

where $\nabla\times$ is a curl operator on a vector. The operator is defined as follows:

$$\nabla \times \mathbf{F} = \det\begin{bmatrix} \mathbf{e}_x & \mathbf{e}_y & \mathbf{e}_z \\ \frac{\partial}{\partial x} & \frac{\partial}{\partial y} & \frac{\partial}{\partial z} \\ f_x & f_y & f_z \end{bmatrix}$$

$$= \left(\frac{\partial f_z}{\partial y} - \frac{\partial f_y}{\partial z}\right)\mathbf{e}_x + \left(\frac{\partial f_x}{\partial z} - \frac{\partial f_z}{\partial x}\right)\mathbf{e}_y + \left(\frac{\partial f_y}{\partial x} - \frac{\partial f_x}{\partial y}\right)\mathbf{e}_z,$$

where $\mathbf{e}_x$, $\mathbf{e}_y$, and $\mathbf{e}_z$ are unit vectors in the direction of *x*, *y*, and *z*, respectively, and f_x, f_y, and f_z are components of vector **F**. The scalar potential gives rise to longitudinal waves or pressure waves (p-waves) and the vector potential gives rise to transverse waves or shear waves (s-waves). The p-waves travel with speed α and the s-waves travel with speed β. The components of displacement vector can be expressed in terms of ϕ and ψ. From Equation 1.3a we obtain

$$\mathbf{d}=(d_x, d_y, d_z)$$

$$\begin{aligned} d_x &= \frac{\partial\phi}{\partial x}+\frac{\partial\psi_z}{\partial y}-\frac{\partial\psi_y}{\partial z} \\ d_y &= \frac{\partial\phi}{\partial y}+\frac{\partial\psi_x}{\partial z}-\frac{\partial\psi_z}{\partial x}, \\ d_z &= \frac{\partial\phi}{\partial z}+\frac{\partial\psi_y}{\partial x}-\frac{\partial\psi_x}{\partial y} \end{aligned} \tag{1.4}$$

where $\psi=(\psi_x, \psi_y, \psi_z)$. In solids, we must speak of stress and strain tensors. An element of solid is not only compressed but also twisted, while an element of fluid is only capable of being compressed but not twisted. We have to use tensors for characterizing the phenomenon of twisting. We shall define the stress and strain tensors and relate them through Hooke's law. A stress tensor is a matrix of nine components

$$\underline{\underline{\mathbf{s}}} = \begin{bmatrix} s_{xx} & s_{yx} & s_{zx} \\ s_{xy} & s_{yy} & s_{zy} \\ s_{xz} & s_{yz} & s_{zz} \end{bmatrix}. \tag{1.5a}$$

The components of the stress tensor represent stresses on different faces of a cube (see Figure 1.1). A strain tensor is given by

$$\underline{\underline{\varepsilon}} = \begin{bmatrix} \varepsilon_{xx} & \varepsilon_{yx} & \varepsilon_{zx} \\ \varepsilon_{xy} & \varepsilon_{yy} & \varepsilon_{zy} \\ \varepsilon_{xz} & \varepsilon_{yz} & \varepsilon_{zz} \end{bmatrix}. \tag{1.5b}$$

The first subscript refers to the plane perpendicular to the axis denoted by the subscript and the second subscript denotes the direction in which the vector is pointing. For example, s_{xx} is a stress in a plane perpendicular to the x-axis (i.e., y-z plane) and pointing along the x-axis.

The stress components on different faces of a cuboid are shown in Figure 1.1. The torque on the cuboid should not cause any rotation. For this, we must have $s_{xy}=s_{yx}$ and similarly all other nondiagonal elements in the stress matrix. Thus, $\underline{\underline{\mathbf{s}}}$ must be a symmetric matrix. The components of a strain matrix are related to the displacement vector

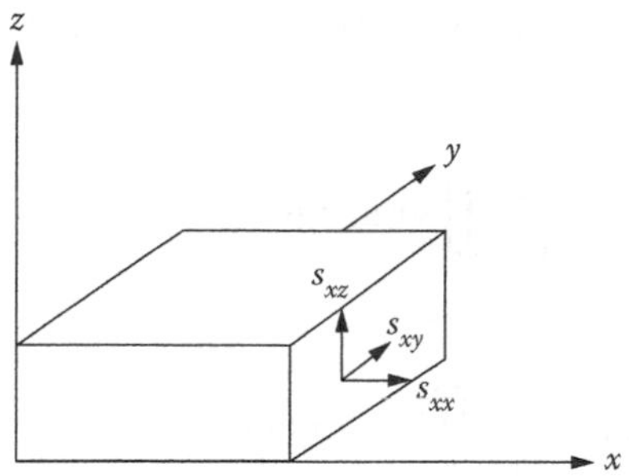

FIGURE 1.1
An element of volume (cuboid) and stresses are shown on a face perpendicular to the x-axis.

$$\varepsilon_{xx} = \frac{\partial d_x}{\partial x}, \ \varepsilon_{yy} = \frac{\partial d_y}{\partial y}, \ \varepsilon_{zz} = \frac{\partial d_z}{\partial z},$$

$$\begin{aligned} \varepsilon_{xy} = \varepsilon_{yx} &= \frac{\partial d_x}{\partial y} + \frac{\partial d_y}{\partial x}; \\ \varepsilon_{yz} = \varepsilon_{zy} &= \frac{\partial d_y}{\partial z} + \frac{\partial d_z}{\partial y}; \\ \varepsilon_{zx} = \varepsilon_{xz} &= \frac{\partial d_z}{\partial x} + \frac{\partial d_x}{\partial z}. \end{aligned} \tag{1.6}$$

Finally, the stress and strain components are related through Hooke's Law:

$$\begin{aligned} s_{xx} &= \rho\alpha^2\varepsilon_{xx} + \rho(\alpha^2 - 2\beta^2)\,(\varepsilon_{yy} + \varepsilon_{zz}), \\ s_{yy} &= \rho\alpha^2\varepsilon_{yy} + \rho(\alpha^2 - 2\beta^2)\,(\varepsilon_{xx} + \varepsilon_{zz}), \\ s_{zz} &= \rho\alpha^2\varepsilon_{zz} + \rho(\alpha^2 - 2\beta^2)\,(\varepsilon_{xx} + \varepsilon_{yy}), \\ s_{xy} &= s_{yx} = \rho\beta^2\varepsilon_{xy}, \\ s_{yz} &= s_{zy} = \rho\beta^2\varepsilon_{yz}, \\ s_{zx} &= s_{xz} = \rho\beta^2\varepsilon_{zx}. \end{aligned} \tag{1.7}$$

Using Equations 1.4, 1.6, and 1.7, we can express all nine stress components in terms of the scalar and vector potential functions, ϕ and ψ. For example, it is possible to show that s_{xx} is given by

$$s_{xx} = \rho\alpha^2 \frac{\partial^2\phi}{\partial x^2} + \rho(\alpha^2 - 2\beta^2)\left(\frac{\partial^2\phi}{\partial y^2} + \frac{\partial^2\phi}{\partial z^2}\right) + 2\rho\beta^2\left(\frac{\partial^2\psi_z}{\partial x\partial y} - \frac{\partial^2\psi_y}{\partial x\partial z}\right).$$

A general solution of Equation 1.3b may be given by

$$\phi(x,y,z,\omega)=\frac{1}{4\pi^2}\iint_{-\infty}^{\infty}\Phi(u,v,\omega)e^{-\sqrt{u^2+v^2-k_\alpha^2}z}e^{j(ux+vy)}du\,dv, \tag{1.8a}$$

$$\psi(x,y,z,\omega)=\frac{1}{4\pi^2}\iint_{-\infty}^{\infty}\Psi(u,v,\omega)e^{-\sqrt{u^2+v^2-k_\beta^2}z}e^{j(ux+vy)}du\,dv, \tag{1.8b}$$

where $k_\alpha=\omega/\alpha$, $k_\beta=\omega/\beta$. $\Phi(u, v, \omega)$ and $\Psi(u, v, \omega)$ are, respectively, the Fourier transforms of the displacement potentials ϕ and ψ evaluated on the surface $z=0$. Furthermore, ψ must satisfy zero divergence condition (Equation 1.3b). This will place additional constraints on $\Psi(u, v, \omega)$, namely,

$$ju\Psi_x(u,v,\omega)+jv\Psi_y(u,v,\omega)-\sqrt{u^2+v^2-k_\beta^2}\,\Psi_z(u,v,\omega)=0. \tag{1.8c}$$

Recall that the pressure waves (p-waves) travel at speed α and the shear waves travel at speed β, where α is generally greater than β. The displacement vector is in the direction of the gradient of the scalar potential, but it is in the direction of curl of the vector potential (Equation 1.4), that is, normal to the vector potential. Thus, there is a fundamental difference in the nature of propagation of the shear and the pressure waves. The shear waves are polarized; the displacement vector is always perpendicular to the direction of wave propagation. The displacement vector executes a motion depending upon the phase difference between the components of the displacement vector; a line when the phase difference is zero, a circular path when the phase difference is 90°, or a random path when the phase difference is randomly varying. These factors play an important role in the design of sensor array systems and processing of vector potential signals.

1.1.3 EM Fields

In EM fields there are two vectors, namely, electric vector **E** and magnetic vector **H**, each with three components, thus a six-component vector field. The basic equations governing the EM fields are the Maxwell's equations (in mks units),

$$\left.\begin{aligned}\nabla\times\mathbf{E}&=-\frac{\partial\mathbf{B}}{\partial t}\\ \nabla\cdot\mathbf{D}&=\rho\end{aligned}\right\}\quad\text{Faraday's law,} \tag{1.9a}$$

$$\left.\begin{aligned}\nabla\times\mathbf{H}=\mathbf{J}+\frac{\partial\mathbf{D}}{\partial t}\\ \nabla\cdot\mathbf{B}=0\end{aligned}\right\}\text{Ampere's law,} \tag{1.9b}$$

where

$$\mathbf{D}=\varepsilon\mathbf{E}\quad \varepsilon\text{: dielectric constant,}$$

$$\mathbf{B}=\mu\mathbf{H}\quad \mu\text{: magnetic susceptibility,} \tag{1.9c}$$

$$\mathbf{J}=\sigma\mathbf{E}\quad \sigma\text{: conductivity.}$$

We shall introduce two potential functions ϕ, a scalar potential, and ψ, a vector potential, in terms of which the components of the EM field are expressed. The electric and magnetic fields are defined in terms of ϕ and ψ as follows:

$$\begin{aligned}\mathbf{H}&=\nabla\times\psi\\ \mathbf{E}&=-\nabla\phi-\frac{\partial\psi}{\partial t}.\end{aligned} \tag{1.10a}$$

The vector potential is further required to satisfy (Lorentz gauge)

$$\nabla\cdot\psi+\varepsilon\mu\frac{\partial\phi}{\partial t}=0. \tag{1.10b}$$

Using Equation 1.10 in Equation 1.9, we obtain two decoupled equations,

$$\nabla^2\phi-\varepsilon\mu\frac{\partial^2\phi}{\partial t^2}=-\frac{1}{\varepsilon}\rho, \tag{1.11a}$$

$$\nabla\times\nabla\times\psi-\varepsilon\mu\frac{\partial^2\psi}{\partial t^2}=-\mu\mathbf{J}. \tag{1.11b}$$

When there is no free charge, that is $\rho=0$, the scalar potential may be canceled by a suitable choice, using the principle of gauge transformation [1, p. 207]. Then, equations in Equation 1.10a reduce to

$$\begin{aligned}\mathbf{H}&=\nabla\times\psi\\ \mathbf{E}&=-\frac{\partial\psi}{\partial t},\end{aligned} \tag{1.12}$$

and the vector potential satisfies

$$\nabla \times \nabla \times \psi - \varepsilon\mu \frac{\partial^2 \psi}{\partial t^2} = -\mu \mathbf{J}$$
$$\nabla \cdot \psi = 0 \tag{1.13}$$

Both electric and magnetic fields travel with the same speed unlike p and s waves in solids. The electric and magnetic vectors lie in a plane perpendicular to the direction of propagation. The tip of a field vector will execute a smooth curve known as a polarization ellipse. For example, in a vertically polarized electric field, the tip of the electric vector lies on a vertical line while the magnetic field lies on a horizontal line. The minor and major axes of the polarization ellipse are a and b, respectively. Define an angle $\chi = \tan^{-1}(a/b)$. The major axis of the polarization ellipse is inclined at an angle ζ (see Figure 1.2). Consider a plane wave with its wave vector in y-z plane making an angle θ with z axis. The electric vector will lie in a plane perpendicular to the direction of propagation. This plane is known as the x-θ plane. The electric vector can be split into two components, E_x and E_θ, lying in x-θ plane, that is,

$$\mathbf{E} = -E_x \mathbf{e}_x + E_\theta \mathbf{e}_\theta$$

where,

$E_x = E \cos\gamma$ horizontal component,

$E_\theta = E \sin\gamma\, e^{j\kappa}$ vertical component,

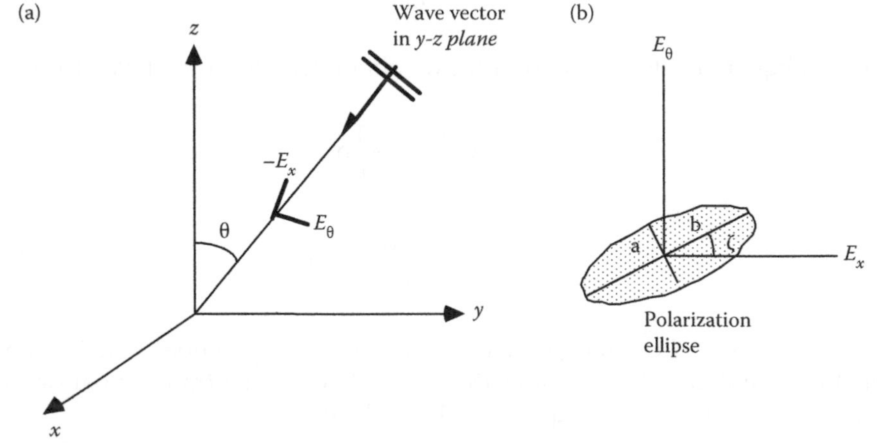

FIGURE 1.2
(a) EM wave propagation vector is in y-z plane. Electric and magnetic vectors lie in a plane perpendicular to the propagation vector and it contains the x-axis. (b) The tip of the electric vector executes a smooth curve, for example, an ellipse. The electric vector can be split into two components, E_x and E_θ.

where γ and κ are related to χ and ζ [2],

$$\cos(2\gamma)=\cos(2\chi)\cos(2\zeta) \quad 0\le\gamma\le\frac{\pi}{2},$$
$$\tan(\kappa)=\tan(2\chi)\csc(2\zeta) \quad -\pi\le\kappa\le\pi.$$

A plane EM wave is characterized by four parameters, namely, E (amplitude), θ (direction), χ, and ζ (polarization parameters).

When there is no free charge current (free moving charge, i.e., $\rho=0$), the vector potential for a plane wave EM field is given by

$$\psi=\mathbf{a}_p A\exp(j(\mathbf{k}\cdot\mathbf{r}-\omega t))$$

where $\mathbf{k}=(\omega/c)\mathbf{a}_k$ and A is a complex constant; $\mathbf{a}_k$ and $\mathbf{a}_p$ $(=-\cos\gamma\mathbf{e}_x+\sin\gamma\cos\theta\mathbf{e}_y-\sin\gamma\sin\theta\mathbf{e}_z)$ are mutually perpendicular unit vectors. The electric and magnetic field vectors are given by

$$\begin{aligned}\mathbf{E}&=j\omega\mathbf{a}_p A\exp(j(\mathbf{k}\cdot\mathbf{r}-\omega t))\\ \mathbf{H}&=j\omega(\mathbf{a}_k\times\mathbf{a}_p)A\exp(j(\mathbf{k}\cdot\mathbf{r}-\omega t))=\mathbf{a}_k\times\mathbf{E}.\end{aligned}\tag{1.14}$$

1.1.4 Properties of Wavefields

We shall briefly state some properties of the wavefields. (a) Radiation condition: A solution to a wave equation must satisfy the radiation condition or Sommerfeld condition given by [24, p. 499]

$$\frac{\partial\phi}{\partial r}\xrightarrow[r\to\infty]{}O\left(\frac{1}{r}\right)$$
$$\frac{\partial\phi}{\partial r}+jk\phi\xrightarrow[r\to\infty]{}O\left(\frac{1}{r}\right).$$

(b) Time reversal: All wave equations without time varying coefficients or time derivatives of odd order share an interesting property of time reversal. If $\phi(x, y, z, t)$ is a solution of the wave equation, it is easy to verify that $\phi(x, y, z, t_0-t)$ is also a solution of the same wave equation for any constant t_0 [3]. This fact has been used for reverse time propagation of seismic wavefields. The seismic wavefield observed on a surface is time-reversed and fed into a loudspeaker broadcasting the recorded wavefield.

1.1.5 Sensing of Wavefields

A device is required to convert a physical wavefield, say, pressure wave, into an electrical signal, which is then sampled and digitized. The resulting numbers are stored in computer memory for further processing. Such a device is known as a sensor. While the mechanism of conversion of a physical wavefield into an electrical signal is not important from the point of ASP, the speed of conversion and the dynamic range are very relevant; accordingly, the sampling rate and the number of bits per sample are fixed. The speed of conversion controls the bandwidth of a sensor, faster is the conversion larger is its bandwidth. As we have two types of wavefields, scalar wavefield and vector wavefield, there are also two types of sensors. A scalar sensor is used to sense a scalar wavefield such as pressure or any one component of the EM field. The most common example of the scalar sensor is a microphone or a hydrophone. A vector sensor will measure all components of the vector wavefield, three components of mechanical waves in solid, or six components of EM field. Three-component seismometers are sometimes used in seismic exploration. Six-component EM sensors are likely to be available very soon off-the-shelf [4]. Modern sensor arrays consist of several tens or hundreds of sensors. One major problem is the lack of uniformity in sensor response. Ideally, it is assumed that all sensors are omnidirectional with unit response. But in practice, the response of a sensor may depend upon the direction of incident wavefront and it may be complex. This behavior in itself is not so disturbing as the variation of this behavior from sensor to sensor. It is often required to carefully estimate the response of each sensor in a large array. This process is known as array calibration. Commonly used sensors measure just one component of the wavefield. This seems to be adequate, as all components travel with the same speed (except p and s waves in solids), yielding the same delay information on which many array-processing algorithms are based. Additional information, such as the polarization and the particle motion, has also recently been used in array processing. Some interesting developments in sensor technology, which will have considerable bearing on the future of sensor array processing, are taking place. For example, the vector sensors capable of measuring all six components of the EM field and four components of the mechanical field (pressure and three particle velocity components) have become commercially available [4].

In addition to the above wavefield sensors, we have chemical sensors capable of detecting a very small quantity of chemicals in the vapor state. The distribution of vapor is governed by a diffusion equation in place of a wave equation. Extremely sensitive detectors of magnetic fields based on the principle of superconducting quantum interference have also appeared and have been used in magneto-encephalography.

1.2 Wavefield in Open Space

In a homogenous space without any reflecting boundaries, the wave equation is easily solved. The propagation of wavefields may be described within the framework of filter theory. The propagating and nonpropagating (transient) waves are well demarcated in the frequency wavenumber domain. Continuation of a wavefield from one plane to another plane is easily achieved through a filtering operation. We shall introduce some of these basic concepts in this section. Another important constraint imposed by practical considerations is that the wavefield is measured on a set of discrete points with discrete sensors. Thus, the spatial sampling is always implied; in contrast, a temporal sampling of the analog sensor output is required only when digital processing is desired.

1.2.1 Fourier Representation of Wavefield

The generic form of a wave equation in a homogeneous medium is given by

$$\nabla^2 f = \frac{1}{c^2}\frac{\partial^2 f}{\partial t^2}, \tag{1.15}$$

where $f(\mathbf{r}, t)$ stands for any one of the wave types, for example, pressure or one of the components of a vector field. We introduce the Fourier integral representation of a wavefield,

$$f(\mathbf{r},t) = \frac{1}{8\pi^3}\iiint_{-\infty}^{\infty} F(u,v,\omega)H(u,v,z)e^{j(\omega t-ux-vy)}dudvd\omega, \tag{1.16}$$

in wave Equation 1.15. We observe that $H(u, v, z)$ must satisfy an ordinary differential equation given by

$$\frac{d^2H(u,v,z)}{dz^2} = \left(u^2+v^2-\frac{\omega^2}{c^2}\right)H(u,v,z), \tag{1.17}$$

whose solution is

$$H(u,v,z) = \exp(\pm\sqrt{(u^2+v^2-k^2z)}, \tag{1.18}$$

where $k=\omega/c$ is known as a wavenumber. When $\sqrt{(u^2+v^2)} > k$, we choose (−) sign for $z > 0$ and (+) sign for $z < 0$ so that the wavefield does not diverge. In both cases, the field will rapidly decay as $|z|\rightarrow\infty$. These are known as evanescent waves. When $\sqrt{(u^2+v^2)} < k$, we get propagating waves whose integral representation reduces to

$$f(x,y,t) = \frac{1}{8\pi^3}\iiint_{-\infty}^{\infty} F(u,v,\omega)e^{\pm j(\sqrt{k^2-u^2-v^2}z)}e^{j(\omega t-ux-vy)}dudvd\omega' \tag{1.19}$$

where the sign in $e^{\pm j(\sqrt{k^2-u^2-v^2}z)}$ is selected depending on whether the waves are diverging or converging. The convention is (−) sign for diverging waves and (+) sign for converging waves (see Figure 1.3). Note that in a bounded space, both diverging and converging waves can coexist and hence it would be necessary to use both signs in describing wavefields in a bounded space.

Equation 1.19 suggests an interesting possibility, that is, if a wavefield is observed on an x-y plane, it is possible to extrapolate it into the space above or below the plane of observation. Further, Equation 1.19 may be looked upon as a sum of an infinitely large number of plane waves of the type $e^{j(\omega t - ux - vy - \sqrt{k^2-u^2-v^2}z)}$ with complex amplitude $F(u, v, \omega)$. The direction of propagation of an individual wave is prescribed by the spatial frequencies u and v,

$$u = k \sin\theta \cos\varphi$$

$$v = k \sin\theta \sin\varphi,$$

where φ and θ are, respectively, azimuth and elevation angles of a plane wave. The elevation angle is an angle between the z-axis and the wave vector and the azimuth angle is an angle between the x-axis and the projection of the wave vector on the x-y plane. The representation given in Equation 1.19 is also known as a plane wave decomposition (PWD) of a wavefield. In (u, v, ω) space, a single frequency plane wave is represented by a point, and a wideband wave by a line passing through the center of the coordinate system (see Figure 1.4). The slope of the line is inversely proportional to the direction cosines (defined on page 16) of the vector perpendicular to the plane wavefront and directly proportional to the speed of propagation.

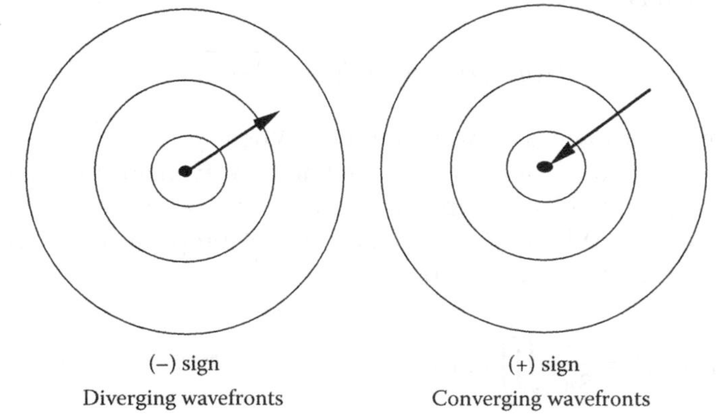

FIGURE 1.3
Sign convention in diverging and converging waves.

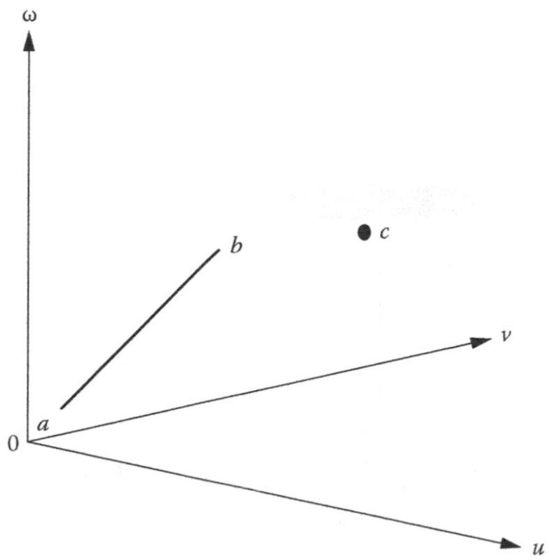

FIGURE 1.4
Representation of a plane wave in (u, v, ω) space. Point c represents a single frequency (narrowband) plane wave and line ab represents a wideband plane wave.

1.2.2 Domain of Propagation

We have noted earlier that propagation of waves is possible only when $\sqrt{(u^2+v^2)} < k$. For a fixed ω the spatial frequencies u and v must lie within a circle of radius $k(=\omega/c)$, that is, a disc defined by

$$(u^2+v^2) \le \left(\frac{\omega}{c}\right)^2. \tag{1.20}$$

Equation 1.20 represents, as a function of ω a conical surface in (u, v, ω) space. It is a vertical cone with an apex at the center of the coordinates and the angle of the cone is inversely proportional to the speed of propagation. For a real signal, plane wave representation in (u, v, ω) space (see Figure 1.4) extends below the (u,v) plane. The domain of wave propagation is obtained by reflecting the conical surface below the (u,v) plane. This results in an hourglass figure shown in Figure 1.5.

1.2.3 Apparent Propagation Speed

Apparent propagation speed refers to the speed with which a wave appears to travel across an array of sensors placed on a line. For example, consider an array of sensors on the x-axis and a plane wavefront incident at angle θ and φ as shown in Figure 1.6. Travel time from p to o is $dt = po\cos(\theta)/c$. Hence, the speed of propagation along the vertical axis is

$$\frac{po}{dt} = c_z = \frac{c}{\cos(\theta)}. \tag{1.21a}$$

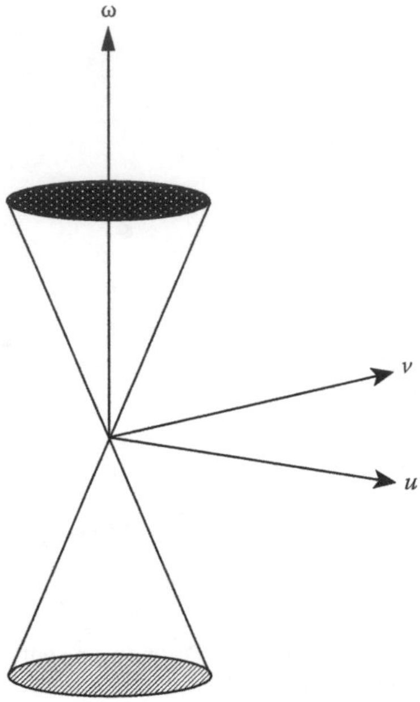

FIGURE 1.5
Domain of propagation in (u, v, ω) space (an hourglass figure).

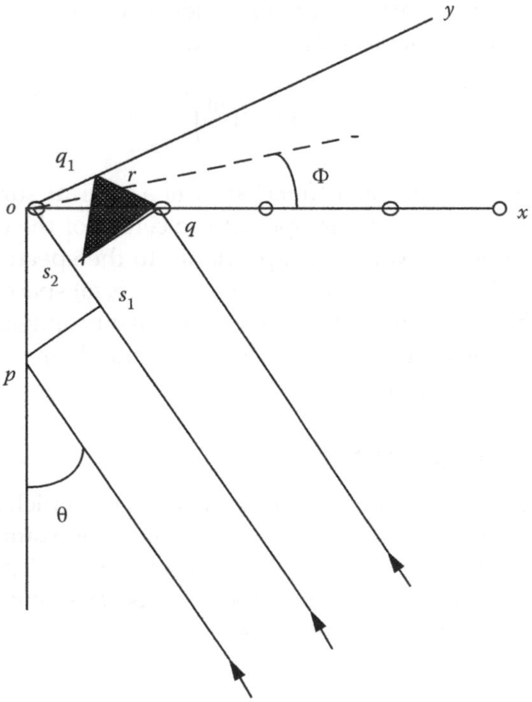

FIGURE 1.6
Geometrical derivation of apparent speeds.

To compute travel time from q to o, do the following construction:

(i) Draw a wavefront qs_2q_1 perpendicular to incident ray.
(ii) Draw a vertical plane through os_2. Since qq_1 is in the horizontal plane xoy and or is in the vertical plane, $\angle orq=\pi/2$.
(iii) Since rs_2 is in the wavefront plane, $\angle rs_2o=\pi/2$.

From (i), (ii), and (iii) we obtain the following result:

$$or=oq\cos\varphi$$

$$os_2=or\sin\theta.$$

$$=oq\sin\theta\cos\varphi$$

The travel time from s_2 to o is $dt=(os_2/c)=(oq/c)\sin\theta\cos\varphi$. Hence, the apparent speed along the x-axis is given by

$$c_x=\frac{oq}{dt}=\frac{c}{\sin\theta\cos\varphi}. \tag{1.21b}$$

Similarly, we can show that

$$c_y=\frac{c}{\sin\theta\sin\varphi}. \tag{1.21c}$$

Equations 1.21b and 1.21c can be rewritten in terms of spatial frequencies

$$c_x=\frac{\omega}{k\sin\theta\cos\varphi}=\frac{\omega}{u}$$

$$c_y=\frac{\omega}{k\sin\theta\sin\varphi}=\frac{\omega}{v} \tag{1.22}$$

$$c_z=\frac{\omega}{k\cos\theta}=\frac{\omega}{\sqrt{k^2-u^2-v^2}}.$$

From Equation 1.22 it follows that

$$\frac{1}{c_x^2}+\frac{1}{c_y^2}+\frac{1}{c_z^2}=\frac{1}{c^2}, \tag{1.23}$$

and that the apparent speeds are always greater than or equal to the wave speed. Closely related to the apparent speed are the so-called ray parameters,

$$p_x=\frac{\sin\theta\cos\varphi}{c},\ p_y=\frac{\sin\theta\sin\varphi}{c},\ p_z=\frac{\cos\theta}{c}. \tag{1.24}$$

The significance of ray parameters is that as a ray propagates through media of different wave speeds, the angles of incidence and emergence will change such that the ray parameters remain fixed, equal to that at the start of the ray. From Equation 1.22 it is easy to show that the ray parameters are related to the apparent speeds,

$$p_x=\frac{1}{c_x},\ p_y=\frac{1}{c_y},\ p_z=\frac{1}{c_z}. \tag{1.25a}$$

The direction cosines of a ray are defined as

$$\alpha=\sin\theta\cos\varphi,\ \beta=\sin\theta\sin\varphi,\ \gamma=\cos\theta. \tag{1.25b}$$

From Equations 1.24 and 1.25, the ray parameters can be expressed in terms of direction cosines

$$\alpha=p_x c,\ \beta=p_y c,\ \gamma=p_z c.$$

1.2.4 Continuation of Wavefield

Consider a thin layer of sources on the $z=0$ plane. Let $f_0(x, y, t)$ be the wavefield observed close to the $z=0$ plane. The wavefield on a horizontal plane $z=z_1$ can be expressed in terms of that on the $z=0$ plane. For this we make use of Equation 1.19,

$$f(x,y,z_1,t)=\frac{1}{8\pi^3}\iiint_{-\infty}^{\infty} F_0(u,v,\omega)e^{\pm j\left(\sqrt{k^2-u^2-v^2 z_1}\right)}e^{j(\omega t-ux-vy)}du\,dv\,d\omega, \tag{1.26}$$

where $F_0(u, v, \omega)$ is the Fourier transform of the wavefield observed on the $z=0$ surface. Thus, it is possible to extrapolate a wavefield observed on one plane to another plane provided the intervening space is source free. Note that the Fourier transforms of the wavefield observed on two parallel surfaces differ by a phase factor only. In the case of horizontal surfaces, the phase factor is simply given by $e^{\pm j(\sqrt{k^2-u^2+v^2\Delta z})}$ where Δz is vertical separation. Each component of the PWD is subjected to a phase shift whose magnitude depends upon the spatial frequencies, u and v, or on the direction of the wave vector. For example, for $\Delta z=\lambda$ the phase shift applied to different plane waves is shown in Figure 1.7 as a function of u (keeping $v=0$). It may be recalled that the Fourier transform of the wavefield for $(u^2+v^2)>k^2$ is rapidly vanishing and hence the phase shift is set to zero in this range.

1.2.5 Point Source

A point source is often used as a source of illumination, but it will only generate spherical waves that produce a much more complex (mathematically

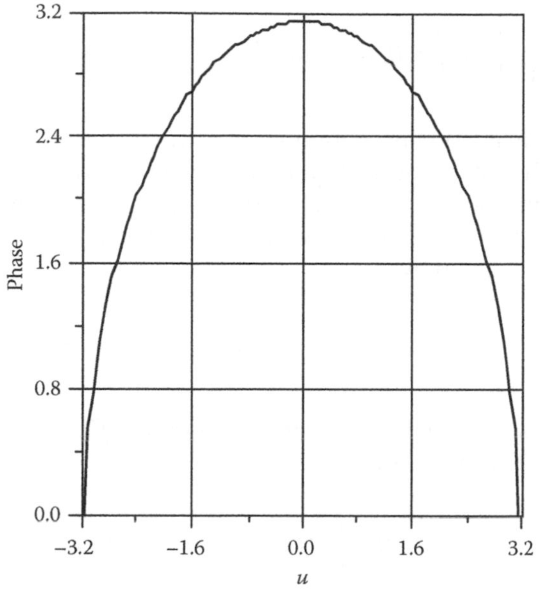

FIGURE 1.7
Phase shift as a function of u (keeping $v=0$) for vertical separation equal to wavelength (assumed to one).

speaking) response from a target compared with a plane wave. Fortunately, a point source wavefield may be written as a sum of infinitely many plane waves (a PWD is described in Section 1.2.1). Additionally, a response of a target may be obtained as a sum of plane wave responses. To obtain the Fourier integral representation of a point source wavefield, we go back to Equation 1.19 where we shall assume that the Fourier transform of the wavefield has radial symmetry,

$$f(x,y,z,t)=\frac{1}{8\pi^3}\iiint_{-\infty}^{\infty}F_0(u,v,\omega)e^{\pm j(\sqrt{k^2-u^2-v^2}z)}e^{j(\omega t-ux-vy)}dudvd\omega,$$
$$=\frac{1}{4\pi^2}\int F_0(s,\omega)e^{j\omega t}d\omega\int_0^{\infty}se^{\pm j(\sqrt{k^2-s^2}z)}ds\frac{1}{2\pi}\int_0^{2\pi}e^{-j(sr\cos(\theta-\varphi))}d\theta \quad (1.27a)$$

where $F_0(u, v, \omega)$ is the Fourier transform of $f(x, y, z=0, t)$. We can further simplify Equation 1.27a as

$$f(r,z,t)=\frac{1}{4\pi^2}\int_{-\infty}^{\infty}F_0(s,\omega)e^{j\omega t}d\omega\int_0^{\infty}sJ_0(sr)e^{\pm j(\sqrt{k^2-s^2}z)}ds, \quad (1.27b)$$

where $s=\sqrt{u^2+v^2}=k\sin\gamma$ where $0\le\gamma\le\pi/2$. In Equation 1.27b, we replace s by $k\sin\gamma$ and rewrite Equation 1.27b as

$$f(r,z,t)=\frac{1}{8\pi^2}\int_{-\infty}^{\infty}d\omega\int_{0}^{\pi/2}F_0(k\sin\gamma,\omega)e^{j\omega\left(t\pm\frac{z}{c}\cos\gamma\right)}k^2\sin(2\gamma)J_0(kr\sin\gamma)d\gamma. \quad (1.28)$$

Equation 1.28 is a PWD of a point source wavefield. In an r-z plane, for a fixed ω and γ, the integrand in Equation 1.28 represents a plane wave component, traveling with an angle of incidence γ, as shown in Figure 1.8, and $\Delta t=(z/c)\cos\gamma$ is the propagation time from surface to a depth z. The inverse of Equation 1.28 is given by

$$F_0(s,\omega)e^{\pm j(\sqrt{k^2-s^2}z)}=\int_{0}^{\infty}F(r,z,\omega)rJ_0(sr)dr, \quad (1.29)$$

where $F(r, z, \omega)$ is a Fourier transform (temporal) of $f(r, z, t)$. On the $z=0$ surface, Equation 1.29 reduces to

$$F_0(k\sin\gamma,\omega)=\int_{0}^{\infty}F(r,z=0,\omega)rJ_0(kr\sin\gamma)dr, \quad (1.30)$$

where we have substituted $s=k\sin\gamma$. Equation 1.30 in the time domain may be expressed as

$$f(t,\sin\gamma)=\int_{0}^{\infty}\left\{f(r,z=0,t)^{*}\frac{2}{\sqrt{\left(\frac{r}{c}\sin\gamma\right)^2-t^2}}\right\}rdr. \quad (1.31)$$

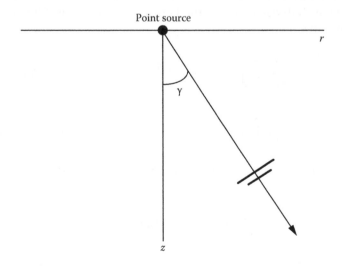

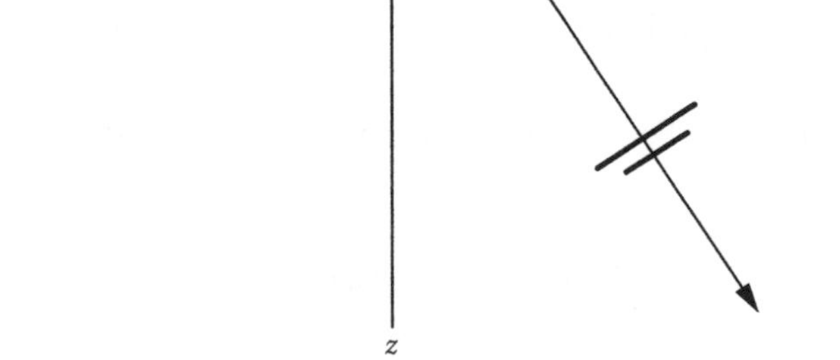

FIGURE 1.8
In a homogeneous medium, the wavefield due to a point source may be expressed as a sum of plane waves traveling with an angle of incidence γ, where $0\le\gamma\le(\pi/2)$.

Equation 1.31 enables us to convert the wavefield due to a point source into that due to a plane wave source. A linear array of sensors is used in a radial direction with respect to point source. The output of each sensor is first filtered with a filter having response

$$h(t)=\frac{2}{\sqrt{\left(\frac{r}{c}\sin\gamma\right)^2-t^2}},$$

and then summed over all sensors. The impulse response function may be expressed as an integral involving a delta function [5],

$$h(t)=\frac{2}{\sqrt{\left(\frac{r}{c}\sin\gamma\right)^2-t^2}}=\int_{-\pi}^{\pi}\delta(t-t_0\cos\beta)d\beta,$$

where $t_0=((r/c)\sin\gamma)$. Using the above integral representation of the impulse response function in Equation 1.31, we obtain

$$\begin{aligned}
f(t,\sin\gamma)&=\int_0^\infty\left\{\int_{-t_0}^{t_0} f(r,z=0,t-t')\frac{2}{\sqrt{t_0^2-(t')^2}}dt'\right\}rdr\\
&=\int_0^\infty\left\{\int_{-t_0}^{t_0} f(r,z=0,t-t')dt'\int_{-\pi}^{\pi}\delta(t'-t_0\cos\beta)d\beta\right\}rdr\\
&=\int_0^\infty\left\{\int_{-\pi}^{\pi} f(r,z=0,t-t_0\cos\beta)d\beta\right\}rdr.
\end{aligned}\tag{1.32}$$

For a discrete array, the integral is replaced by a sum,

$$f(n\Delta t_0,\sin\gamma)=\int_{-\pi}^{\pi}\sum_{n=0}^{\infty}n\Delta r^2 f(n\Delta r,z=0,t-n\Delta t_0\cos\beta)d\beta\,,\tag{1.33}$$

where $\Delta t_0=\Delta r/c\sin\gamma$. The inner sum in Equation 1.33 is the sum-after-delay operation, commonly known as slant stacking in seismic exploration. In Chapter 2, we will show how this operation is related to the radon transform.

1.2.6 Spatial Sampling and Aliasing

A wavefield is measured on a set of discrete points with discrete sensors. Thus, spatial sampling is always implied; in contrast, a temporal sampling of an analog sensor output is used only when digital processing is desired. Often the temporal sampling is preceded by some kind of low pass filtering, which determines the maximum frequency and the maximum sampling rate according to the well-known sampling theorem. It is thus possible to avoid the temporal aliasing, but it is a different story with spatial aliasing, which is intimately related to propagation speed. For simplicity of analysis, let us assume only spatial sampling. Consider a broadband plane wave incident on an infinitely long array of sensors,

$$f(x,t)=f\left(t-\frac{x}{c_x}\right)\sum_{i=-\infty}^{\infty}\delta(x-i\Delta x)$$

$$=\frac{1}{2\pi}\int_{-\infty}^{\infty}F(\omega)e^{j(t-(x/c_x))\omega}d\omega\sum_{i=-\infty}^{\infty}\delta(x-i\Delta x).$$

Note that c_x stands for wave speed in the direction of the x-axis or apparent speed. Taking the 2D Fourier transform of $f(x,t)$

$$F(u,\omega)=F(u,\omega)=\int\int_{-\infty}^{+\infty}f(x,t)e^{j(\omega t-ux)}dxdt$$

$$=\int_{-\infty}^{\infty}F(\omega')d\omega'\sum_{i=-\infty}^{\infty}\int\int_{-\infty}^{+\infty}e^{-j(\omega t-ux)}e^{j(t-(x/c))\omega'}\delta(x-i\Delta x)dxdt$$

$$=\int_{-\infty}^{\infty}F(\omega')\delta(\omega'-\omega)d\omega'\sum_{i=-\infty}^{\infty}\int_{-\infty}^{\infty}e^{j(u-(\omega'/c_x))x}\delta(x-i\Delta x)dx \tag{1.34}$$

$$=\int_{-\infty}^{\infty}F(\omega')\delta(\omega'-\omega)d\omega'\sum_{i=-\infty}^{\infty}e^{j(u-(\omega'/c_x))i\Delta x}$$

$$=\int_{-\infty}^{\infty}F(\omega')\delta(\omega'-\omega)d\omega'\sum_{k=-\infty}^{\infty}\delta\left(u-\frac{\omega'}{c_x}-\frac{2\pi}{\Delta x}k\right).$$

The spatial Fourier transform of the array output is sketched in Figure 1.9. The signal spectrum is concentrated on sloping lines in the u-ω plane, as seen in Figure 1.9.

Depending upon the signal bandwidth and the apparent speed, an alias will show up within the principal band. In Figure 1.9, there is no aliasing when the signal bandwidth is as shown by the dark square, which in this case corresponds to a sensor spacing $\Delta x=\lambda_{min}/2$ or $\Delta x=(\pi/\omega_{max})c_x$. Note that, since the sampling interval $\Delta t=\pi/\omega_{max}$, $\Delta x=\Delta t\ c_x$. There is no aliasing whenever the sensor spacing and the time sampling interval are such that $(\Delta x/\Delta t)\le c_x$. For the vertical angle of incidence, since $c_x=\infty$, there is no aliasing effect for any Δx.

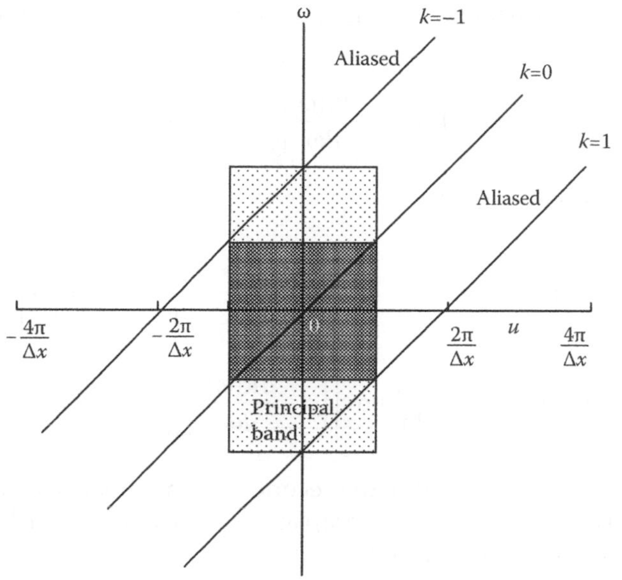

FIGURE 1.9
The Fourier transform of spatially sampled wavefield (plane wave) lies on a set of sloping lines (with slope = $1/c_x$) corresponding to different values of k in Equation 1.34.

1.2.7 Dispersion

A medium is said to be dispersive when a wavefront travels at a speed that is a function of frequency. Consider a wavefront at a fixed temporal frequency, ω_0. Let us track a particular point, say, the crest of a sinusoidal waveform. The speed with which a point on the waveform travels is said to be phase speed, which is equal to λ_0/T_0, where λ_0 is the wavelength and T_0 is the period of a single frequency waveform. Thus, phase speed, by definition, is given by

$$c_{ph} = \frac{\lambda_0}{T_0} = \frac{\omega_0}{k_0} = c,$$

and it is equal to the propagation speed.

Now consider a group of waves whose temporal spectrum is centered at frequency ω_0 and it is spread over a frequency interval $\Delta\omega$. The wavefield may be expressed as

$$f(\mathbf{r},t) = \frac{1}{2\pi} \int_{\omega_0-(\Delta\omega/2)}^{\omega_0+(\Delta\omega/2)} F(\omega)\exp(j(\omega t - \mathbf{k}\cdot\mathbf{r})d\omega, \tag{1.35a}$$

where $\mathbf{k} = (u, \upsilon, \sqrt{(\omega/c)^2 - u^2 - \upsilon^2})$ is the wave vector. We assume that the propagation speed is a function of frequency, and hence the wave vector is also a function of frequency. Using a Taylor's series expansion of the wave vector

$$\mathbf{k}(\omega)=\mathbf{k}(\omega_0)+\left.\frac{d\mathbf{k}(\omega)}{d\omega}\right|_{\omega=\omega_0}(\omega-\omega_0)+\dots,$$

in Equation 1.35a we obtain

$$\begin{aligned} f(\mathbf{r},t) &= F(\omega_0)e^{j(\omega_0 t-\mathbf{k}(\omega_0)\cdot\mathbf{r})}\frac{1}{2\pi}\int_{-(\Delta\omega/2)}^{\Delta\omega/2}\exp\left[j\left(t-\left.\frac{d\mathbf{k}(\omega)}{d\omega}\right|_{\omega=\omega_0}\cdot\mathbf{r}\right)\tilde{\omega}\right]d\tilde{\omega} \\ &= F(\omega_0)e^{j(\omega_0 t-\mathbf{k}(\omega_0)\cdot\mathbf{r})}\frac{\Delta\omega}{2\pi}\operatorname{sinc}\left[\left(t-\left.\frac{d\mathbf{k}(\omega)}{d\omega}\right|_{\omega=\omega_0}\cdot\mathbf{r}\right)\frac{\Delta\omega}{2}\right], \end{aligned} \tag{1.35b}$$

where $\tilde{\omega}=(\omega-\omega_0)$. Observe that the second term modulates the carrier wave (first term). The second term is a propagating waveform, which travels at a speed of c_{gp} which satisfies the equation

$$\left|\left.\frac{d\mathbf{k}(\omega)}{d\omega}\right|_{\omega=\omega_0}\right|c_{gp}=1, \tag{1.36a}$$

or

$$c_{gp}=\frac{1}{\left|\left.\frac{d\mathbf{k}(\omega)}{d\omega}\right|_{\omega=\omega_0}\right|}. \tag{1.36b}$$

The group speed differs from the phase speed only when the medium is dispersive. In a nondispersive medium both speeds are equal. The modulating waveform (second term in Equation 1.35b) travels at the same speed as the carrier wave (first term). An example of propagation of a narrowband signal is shown in Figure 1.10, where we have assumed that $c_{ph}=c_{gp}=1$. The arrival of the signal is denoted by the arrival of the crest of the modulating wave that travels with a speed equal to c_{gp}. Hence, the energy is transmitted at a speed equal to the group speed.

1.3 Wavefield in Bounded Space

Wavefield in a space bounded by a plane reflecting boundaries is of interest in many practical problems. Examples of bounded space are: (i) acoustic field in a room; (ii) acoustic field in a shallow water channel; and (iii) mechanical

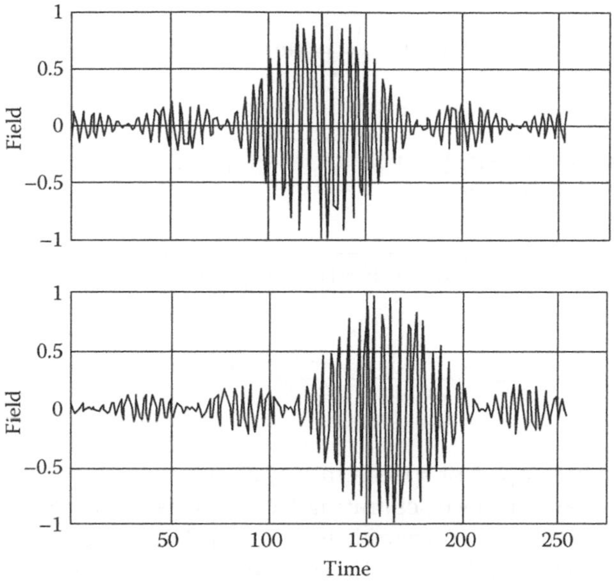

FIGURE 1.10
A narrowband signal with a center frequency at 0.23 Hz and a bandwidth equal to 0.02 Hz. The envelope travels at the group speed, which, in this case, is equal to the phase speed (assumed to be one). The signal in the lower panel arrives 32 time units later.

waves in a layered medium, e.g., in shallow earth, etc. The wavefield in a bounded space is normally studied under two different approaches, namely rays and wavefronts, and the exact solution of the wave equation leading to modes. While the former is more versatile and easy to use, it is less accurate. The second approach is mathematically accurate but difficult to use except in simple geometries.

1.3.1 Ray Propagation

A point source emits a spherical wavefront, a surface of equal phase. Lines perpendicular to the wavefront are rays along which the wave energy is transmitted. A spherical wavefront at a large distance from a point source may be approximated by a plane

$$f(\mathbf{r},t)=\frac{1}{2\pi}\int_{-\infty}^{\infty}F(\omega)e^{j\left[\omega t-u_0(x-x_0)-v_0(y-y_0)-\sqrt{k^2-s_0^2}(z-z_0)\right]}d\omega,$$

where (x_0, y_0, z_0) is the position of a point source and $\mathbf{r}=((x-x_0), (y-y_0), (z-z_0))$ is the position vector of a point on a plane wavefront. Note that the direction cosines are related to the spatial frequencies $u_0=k\alpha_0$ and $v_0=k\beta_0$, where $\alpha_0, \beta_0, \sqrt{1-\alpha_0^2-\beta_0^2}$ are direction cosines of a ray. If a sensor is placed at point (x_1, y_1, z_1), the wavefield received at (x_1, y_1, z_1) is given by

$$f(\mathbf{r}_1,t)=\frac{1}{2\pi}\int_{-\infty}^{\infty}F(\omega)e^{j[\omega t-\phi_1]}d\omega,$$

where $\phi_1=u_0(x_1-x_0)+v_0(y_1-y_0)+\sqrt{k^2-s_0^2}(z_1-z_0)$ is a constant phase. We can also write $\phi_1=\omega t_1$ where

$$\begin{aligned}t_1&=\frac{\sqrt{(x_1-x_0)^2+(y_1-y_0)^2+(z_1-z_0)^2}}{c}\\&=\frac{(x_1-x_0)}{c_x}+\frac{(y_1-y_0)}{c_y}+\frac{(z_1-z_0)}{c_z},\end{aligned}$$

where c_x, c_y, c_z are apparent speeds in x, y, z directions, respectively. When a ray encounters an interface separating two contrasting media, it splits itself into two rays, a reflected and a refracted ray. The laws of reflection and refraction are summarized in Figure 1.11. The reflection and refraction coefficients are related to the impedance of the media on both sides of the interface. For example, the reflection and transmission coefficients at the interface separating two fluid media are given by

$$\hat{r}=\frac{\frac{\rho_2}{\rho_1}\cos\theta_i-\sqrt{\left(\frac{c_1}{c_2}\right)^2-\sin^2\theta_i}}{\frac{\rho_2}{\rho_1}\cos\theta_i+\sqrt{\left(\frac{c_1}{c_2}\right)^2-\sin^2\theta_i}},\tag{1.37a}$$

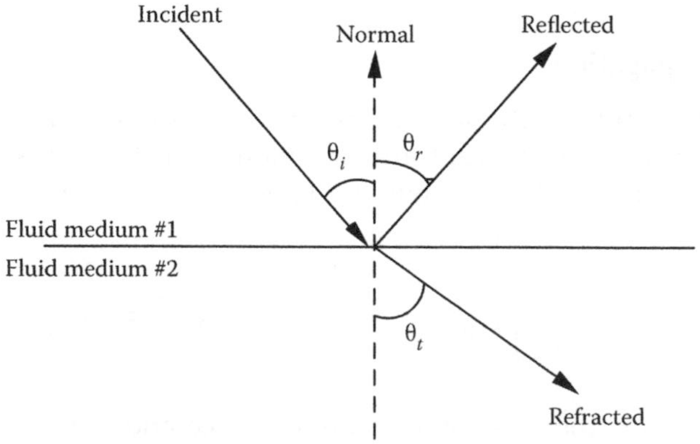

FIGURE 1.11
Laws of reflection and refraction: (1) Incident wave, reflected wave, refracted wave, and the normal to the interface at the point of incidence lie in the same plane. (2) $\theta_i=\theta_r$. (3) $(\sin\theta_i/c_1)=(\sin\theta_t/c_2)=p$ (ray parameter).

$$\hat{t} = \frac{2\cos\theta_i}{\left(\frac{\rho_2}{\rho_1}\right)\cos\theta_i + \sqrt{\left(\frac{c_1}{c_2}\right)^2 - \sin^2\theta_i}}. \tag{1.37b}$$

Reflection and refraction at an interface separating two elastic media are more complex. An incident longitudinal wave will give rise to two reflected and two refracted rays. The first ray is a p-wave and the second ray is a s-wave. The laws of reflection and refraction for a p-wave incident at the interface are summarized in Figure 1.12. The coefficients of reflection and refraction, $(\hat{r}_p, \hat{r}_s, \hat{t}_p, \hat{t}_s)$, are obtained by solving the following system of four linear equations [6, p3–101]:

$$\begin{aligned} &\hat{r}_p\cos\theta_i - \hat{r}_s\sin\theta_r' + \hat{t}_p\cos\theta_t - \hat{t}_s\sin\theta_t' = \cos\theta_i \\ &-\hat{r}_p\sin\theta_i - \hat{r}_s\cos\theta_r' + \hat{t}_p\sin\theta_t + \hat{t}_s\cos\theta_t' = \sin\theta_i, \end{aligned} \tag{1.38a}$$

$$\begin{aligned} &-\hat{r}_p\frac{(\lambda_1 + 2\mu_1 - 2\mu_1\sin^2\theta_r)}{c_p^I} + \hat{r}_s\frac{\mu_1}{c_s^I}\sin 2\theta_r' \\ &+\hat{t}_p\frac{(\lambda_2 + 2\mu_2 - 2\mu_2\sin^2\theta_t)}{c_p^{II}} - \hat{t}_s\frac{\mu_2}{c_s^{II}}\sin 2\theta_t' = \frac{(\lambda_1 - 2\mu_1 - 2\mu_1\sin^2\theta_r)}{c_p^I}, \\ &\hat{r}_p\frac{\mu_1}{c_p^I}\sin 2\theta_r + \hat{r}_s\frac{\mu_1}{c_s^I}\cos 2\theta_r' + \hat{t}_p\frac{\mu_2}{c_p^{II}}\sin 2\theta_t - \hat{t}_s\frac{\mu_2}{c_s^{II}}\cos 2\theta_t' = \frac{\mu_1}{c_p^I}\sin 2\theta_r \end{aligned} \tag{1.38b}$$

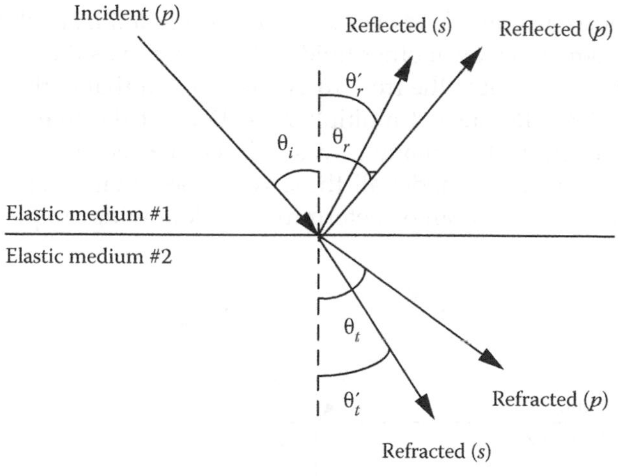

FIGURE 1.12
Reflection and refraction of a p-wave at an interface. Four rays as shown are generated. All rays, including the incident ray and the normal to the interface, lie in a plane. The angle of incidence and the angles of reflection and refraction are related. $\theta_i = \theta_r$ and $(\sin\theta_i/c_p^I) = (\sin\theta_r'/c_s^I) = (\sin\theta_t/c_p^{II}) = (\sin\theta_t'/c_s^{II}) = \text{const}$.

where (λ_1, μ_1) are the elastic constants and (c_p^I, c_s^I) are, respectively, the *p*-wave and the *s*-wave speed in the upper medium. Other parameters refer to the lower medium. For vertical incidence, that is, θ_i=0 from Equation 1.38, it follows that $\hat{r}_s = \hat{t}_s = 0$. This is a consequence of the fact that there is no *s*-wave generation for vertical incidence.

1.3.1.1 Polarization in the Plane of Incidence

Finally, an EM wave at an interface between two different media undergoes reflection and refraction. However, there is only one reflected and one transmitted wave, which travel at the same speed. The process of reflection and refraction is strongly influenced by the polarization angle. We shall consider one simple case of polarization in the plane of incidence (Figure 1.13). The coefficient of reflection and transmission is given by (Fresnel's equations)

$$\hat{r}_E = \frac{\chi - \zeta}{\chi + \zeta}, \quad \hat{t}_E = \frac{2}{\chi + \zeta}, \tag{1.39}$$

where $\chi = \cos\theta_t / \cos\theta_i$ and $\zeta = \mu_1 c_1 / \mu_2 c_2$. Note, when $\chi = \zeta$, $\hat{r}_E = 0$ and $\hat{t}_E = 1$; that is, there is no reflected energy. This phenomenon takes place at a specific angle of incidence known as Brewster angle given by

$$\sin^2\theta_B = \frac{1 - \zeta^2}{(c_2/c_1)^2 - \zeta^2}. \tag{1.40}$$

1.3.2 Propagation in Channel: Ray Theory

Consider a source inside a fluid channel that is bounded from above by a free surface and from below by another fluid (this is known as the Pekeris model). The reflection coefficient at the free surface is (–1) and that at the bottom is $\hat{r}_b$ (see Equation 1.37). Because of multiple reflections at the free surface and at the bottom (see Figure 1.14), many waves will reach a sensor array at different times. It is convenient to model all these waves as emanating from a series of images whose position can be determined by following simple geometrical

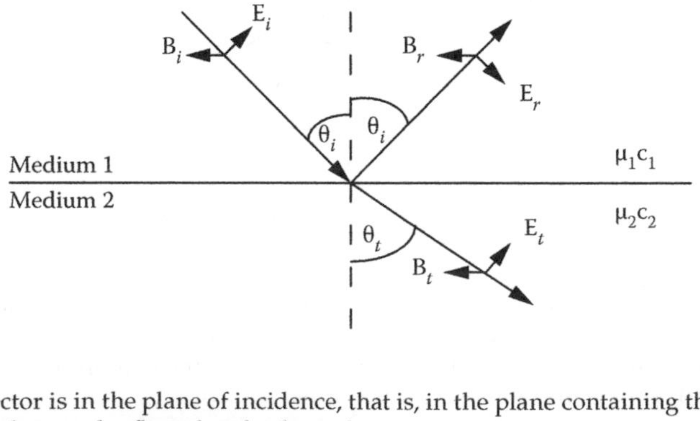

FIGURE 1.13
The electric vector is in the plane of incidence, that is, in the plane containing the incident ray, normal to interface, and reflected and refracted rays.

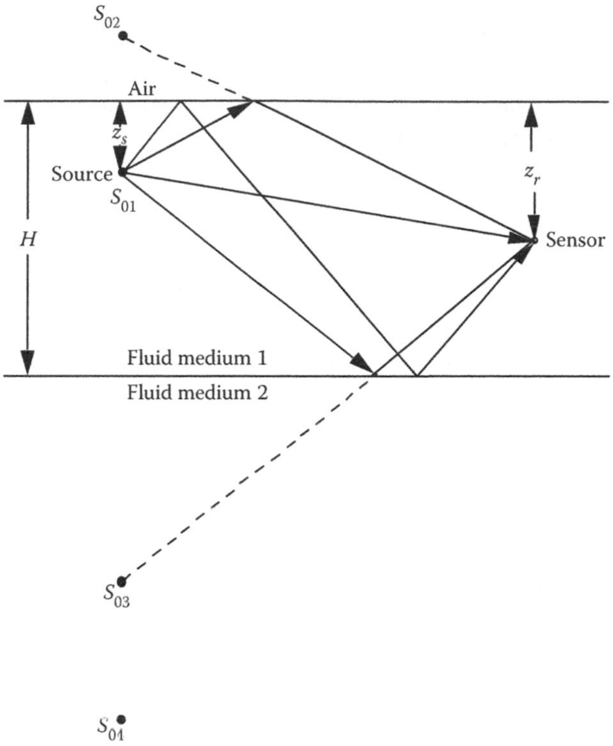

FIGURE 1.14
A shallow water channel with a sound source (•)at depth z_S and a sensor at depth z_r. Rays starting from the source are reflected at the surface and bottom and finally reach the sensor after one or more reflections. Some of the images (•)are also shown.

optics rules. An infinite set of images is formed between two parallel reflecting surfaces. We index these images with two integers (i,k); the first integer represents a group and the second integer represents an image within a group. There are four images in each group. For example, in the $i=0$ group the images are: s_{01} and s_{02}, which are equal but opposite in strength, that is, a surface dipole, and s_{03} and s_{04}, which are caused by the reflection of the surface dipole onto the bottom. The surface dipole and its image are separated by a distance of $2H$, where H is the depth of the fluid channel. The next group of images, that is for $i=1$, is obtained by sliding the surface dipole two depth units above and the image dipole two depth units below the bottom. The vertical distance with respect to the top sensor in the array to different images in the ith group is given by

$$\begin{aligned}
H_{i1} &= 2iH + z_r - z_s \\
H_{i2} &= 2iH + z_r + z_s \\
H_{i3} &= 2(i+1)\,H - z_r - z_s. \\
H_{i4} &= 2(i+1)\,H - z_r + z_s \qquad i=0,1,2,\ldots
\end{aligned} \tag{1.41}$$

An image whose second index is 1 or 2 lies above the bottom and an image whose second image is 3 or 4 lies below the bottom. The strength of an image depends upon the number of bounces the ray has undergone before reaching the sensor and also, on account of geometrical spreading and absorption, on the actual distance traveled, which is given by

$$\begin{aligned} l_{i0} &= \sqrt{d^2 + H_{i0}^2} \quad l_{i1} = \sqrt{d^2 + H_{i1}^2} \\ l_{i2} &= \sqrt{d^2 + H_{i2}^2} \quad l_{i3} = \sqrt{d^2 + H_{i3}^2}, \end{aligned} \tag{1.42}$$

where d is the horizontal distance to the sensor from the source. The strength of the images is given by

$$\begin{aligned} \alpha_{i0} &= (-1)^i \hat{r}_b^i \frac{e^{-\beta l_{i0}}}{l_{i0}} \quad \alpha_{i1} = (-1)^{i+1} \hat{r}_b^i \frac{e^{-\beta l_{i1}}}{l_{i1}} \\ \alpha_{i2} &= (-1)^i \hat{r}_b^{i+1} \frac{e^{-\beta l_{i2}}}{l_{i2}} \quad \alpha_{i3} = (-1)^{i+1} \hat{r}_b^{i+1} \frac{e^{-\beta l_{i3}}}{l_{i3}}, \end{aligned} \tag{1.43}$$

where β is the attenuation coefficient in the top liquid layer. The signal reaching the sensor (Figure 1.14) may be expressed as

$$p(t) = \frac{1}{2\pi} \sum_{i=0}^{\infty} \sum_{m=0}^{3} \int_{-\infty}^{\infty} \alpha_{im} P(\omega) e^{j(t-\tau_{im})\omega} d\omega, \tag{1.44}$$

where $p(t)$ is the pressure field received by a sensor and $P(\omega)$ is the Fourier transform of the waveform emitted by the source; τ_{im} is propagation delay $\tau_{im}=l_{im}/c$. The pressure field given by Equation 1.44 is a complex field. The magnitude and phase are found to vary rapidly with the depth of the sensor. For example, for a channel with the following parameters:

$H=100$ m,

$d=2000$ m,

$z_s=50$ m,

$\rho_1=1$, $c_1=1500$

$\rho_2=2$, $c_2=1600$,

the magnitude and phase variations as a function of the depth of a sensor are shown in Figure 1.15. Eighty images were taken into account in computing the pressure field.

As the range increases, while the position of image sources remains unchanged, the strength of the images may increase on account of the phenomenon of total internal reflection. Recalling Equation 1.37, it may be noticed that whenever $\sin(\theta_i)=c_1/c_2$ $(c_1 < c_2)$, $|r_b|=1$. This angle of incidence is known as the *critical angle*, θ_c. For $\theta_i > \theta_c$, r_b becomes purely imaginary. The wave energy travels along the interface and as it travels some energy is re-radiated back into the channel at the critical angle. It is possible to account for long-distance transmission of acoustic energy using the above ray model even at frequencies as low as 200 Hz [7]. By introducing certain corrections (i.e., beam displacement), the ray model has been improved in the low frequency region (60–140 Hz) [8]. The ray model has been used for source localization in shallow water [9].

1.3.3 Propagation in Channel: Normal Mode Theory

The pressure field due to a point source in a fluid channel, as depicted in Figure 1.14, is given by

$$p(r,z,t)=\frac{j}{2}\int_{-\infty}^{\infty}\sum_{n=0}^{\infty}\phi_n(z)\phi_n(z_s)H_0^1(k_n r)P(\omega)e^{j\omega t}d\omega, \tag{1.45}$$

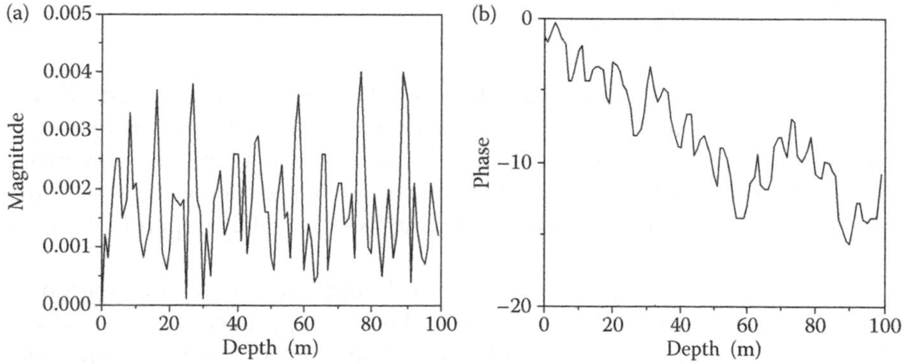

FIGURE 1.15
(a) Magnitude and (b) phase of the computed pressure field in a shallow water channel. Eighty images were taken into consideration for evaluating the net field at a desired point. Note the highly complex structure of the field caused by interference.

where $\phi_n(z)$ is an eigenfunction and k_n is its corresponding eigenvalue obtained by solving a differential equation with homogeneous boundary conditions

$$\frac{d^2\phi}{dz^2}+\gamma_n^2\phi=0$$

$$\phi(0)=0, \tag{1.46}$$

$$\left.\frac{d\phi}{dz}\right|_{z=H}=0 \quad (\textit{rigid bottom})$$

where $k_n^2+\gamma_n^2=\omega^2/c^2(z)$; k_n is a horizontal wavenumber and γ_n is a vertical wavenumber. The source is assumed to be at a depth z_s. Note that both wavenumbers are functions of frequency and also of depth when the sound speed is a function of depth. Consider a special case of isospeed channel and large r. Firstly, the Hankel function can be approximated as $H_0^1(k_n r)\approx\sqrt{2/(k_n\pi r)}e^{j(k_n r+(\pi/4))}$; secondly, both wavenumbers are independent of depth and therefore the solution of Equation 1.46 is given by

$$\phi_n(z)=\sin(\gamma_n z)=\frac{e^{j\gamma_n z}-e^{-j\gamma_n z}}{2j},$$

where $\gamma_n=(n-(1/2)\pi)/H$. For this special case, the pressure field in the fluid channel with the rigid bottom is given by

$$p(r,z,t)=\frac{e^{j(\pi/4)}}{2}\int_{-\infty}^{\infty}\sum_{n=0}^{\infty}\sqrt{\frac{1}{2k_n\pi r}}\sin(\gamma_n z_s)\{e^{j(\gamma_n z+k_n r+\omega t)}-e^{j(-\gamma_n z+k_n r+\omega t)}\}P(\omega)d\omega. \tag{1.47}$$

From Equation 1.47 it is possible to infer that each mode is a sum of two plane wavefronts traveling in the vertical plane at angle $\pm\theta_n$, where $\tan\theta_n=\gamma_n/k_n$ with respect to the horizontal plane. In three dimensions, the wavefront is a conical wavefront (see Figure 1.16); the angle of the cone is $\pi/2-\theta_n$. Note that the direction of propagation of the wavefronts is solely dependent on the channel characteristics.

The comparison of ray and normal mode approaches to propagation in a shallow water channel is instructive. The ray approach is essentially a high frequency approximation to the exact solution of the wave equation. The normal mode approach, on the other hand, is a series approximation to the exact solution of the wave equation. The conical wavefronts obtained by decomposing each mode may be looked upon as a result of constructive interference of spherical wavefronts from the source and its images [8]. At low frequency, the accuracy of the ray approach is enhanced if one were to use the concept of beam displacement. At a point of reflection, the beam is found to be laterally displaced, a phenomenon first observed in optics [10].

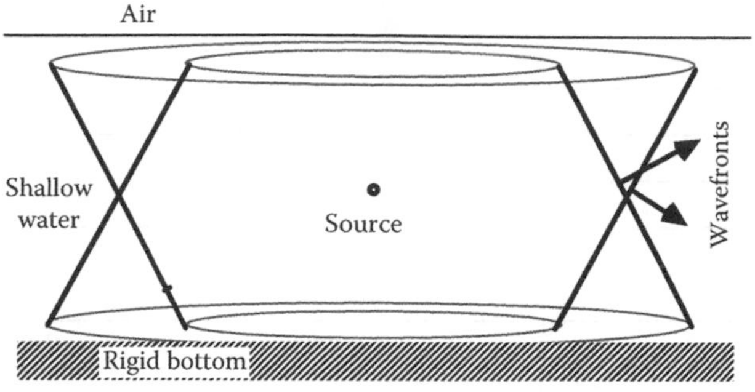

FIGURE 1.16
A mode can be decomposed into two conical wavefronts. In the vertical plane, the wavefronts travel at the angle $\pm\theta_n$, where $\tan\theta_n = \gamma_n / k_n$.

Both numerical and experimental evidence in support of the fact that beam displacement does help to increase the accuracy of the ray approach in relation to the exact solution has been widely reported [11, 12].

1.3.4 Propagation through Layered Medium

In seismic exploration, a horizontally layered medium is often used as a model. Vertically incident longitudinal waves (*p*-waves) are preferred as there is no loss of energy through conversion into *s*-waves at each interface. Also, as *s* waves arrive slightly later, they would interfere with late arriving *p*-wave signals. A layered medium is modeled as uniform horizontal layers stacked one above the other (see Figure 1.17). A vertically propagating plane wavefront is repeatedly reflected and transmitted at each interface, thus producing a complex reverberation pattern that may be conveniently described within the framework of filter theory [13].

A plane wave of unit amplitude is vertically incident on an interface. It is split into two waves, the first wave traveling upwards (reflected wavefront) and the second wave traveling downwards (refracted wavefront). Similarly, a plane wave incident from below is split into two plane waves (see Figure 1.18). The amplitude of the reflected wave is equal to $-\hat{r}$ (reflection coefficient) and that of the refracted wave is equal to $\hat{t}$ (transmission coefficient). Now consider a layer sandwiched between two semi-infinite layers. The thickness of the layer is measured in units of return travel time Δt, that is, one unit corresponds to a physical thickness Δh where $\Delta h = c/2\,\Delta t$. Figure 1.19 shows repeated reflections and refractions at two faces of the layer along with their respective amplitudes; r_0 and t_0 are, respectively, reflection and refraction coefficients at the top face of the layer; similarly r_1 and t_1 are those at the bottom face of the layer.

Let $f_{refl}(t)$ be the reflected waveform, which consists of a sum of all successively reflected wave components,

$$f_{refl}(t) = \frac{1}{2\pi}\int_{-\infty}^{\infty} F_0(\omega)\frac{r_0 + r_1 e^{-j\omega}}{1 + r_0 r_1 e^{-j\omega}} e^{j\omega t}, \tag{1.48a}$$

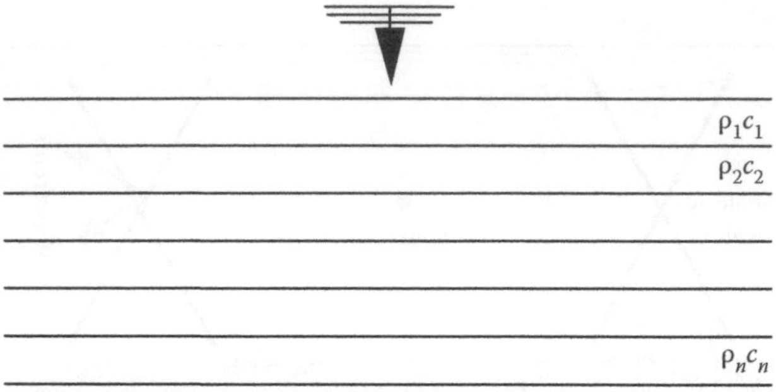

FIGURE 1.17
A stack of uniform layers. All layers are of the same thickness but with different impedances.

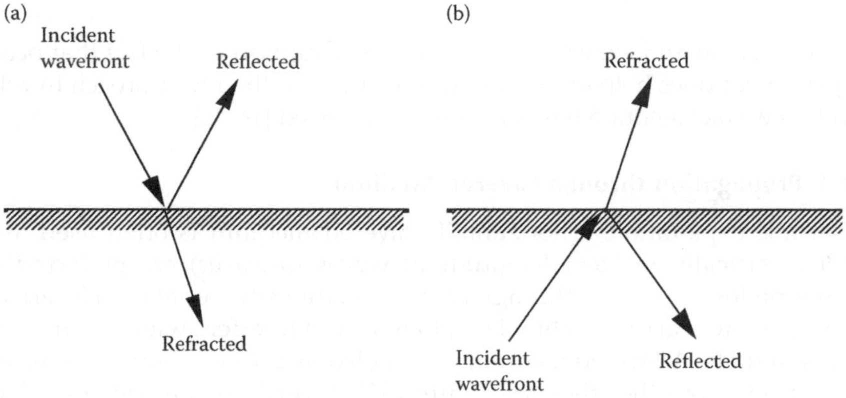

FIGURE 1.18
Reflected and transmitted waves at an interface. (a) Incident from above and (b) incident from below.

and similarly the transmitted waveform is given by

$$f_{trans}(t) = \frac{1}{2\pi}\int_{-\infty}^{\infty} F_0(\omega)\frac{t_0 t_1 e^{-j(\omega/2)}}{1+r_0 r_1 e^{-j\omega}} e^{j\omega t}. \tag{1.48b}$$

Let $z = e^{j\omega}$ and define the reflection and transmission response functions as

$$R(z) = \frac{r_0 + r_1 z^{-1}}{1 + r_0 r_1 z^{-1}}, \quad T(z) = \frac{t_0 t_1}{1 + r_0 r_1 z^{-1}} z^{-(1/2)}. \tag{1.49}$$

We can now express the *z*-transforms of the reflected and the transmitted waveforms as

$$F_{refl}(z) = F_0(z)R(z)$$
$$F_{trans}(z) = F_0(z)T(z).$$

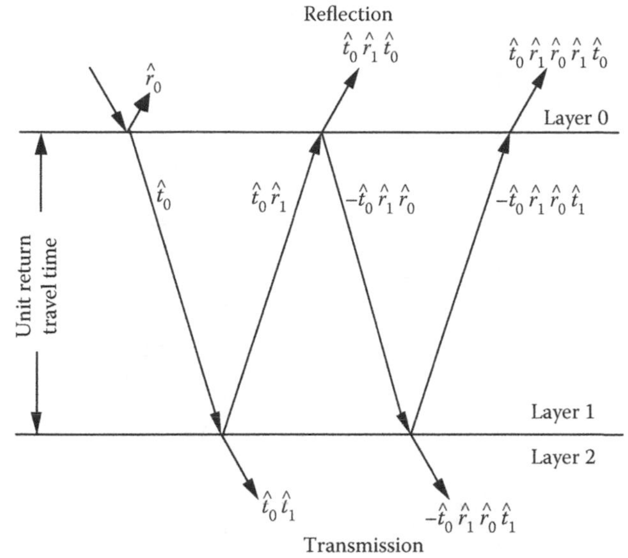

FIGURE 1.19
Repeated reflections at the two faces of a layer produce a sequence of reflected and transmitted waves. Some amount of energy is also trapped inside the layer. All waves travel vertically, though for the sake of clarity the waves are shown as inclined.

It is interesting to note that $T(z)$ has the form of a first order autoregressive (AR) process filter and $R(z)$ has the form of a autoregressive moving average (ARMA) process filter of order (1,1) [14]. Since $r_0\, r_1 \leq 1$, the only pole of the transfer function (Equation 1.49) lies within the unit circle, making $R(z)$ and $T(z)$ stable. The signal flow diagrams of transmission and reflection filters are shown in Figure 1.20.

The above characterization of a single layer model has been extended to a multilayer model. The basic structure of the transmission and reflection filters, however, remains unchanged. Further exposition of this approach may be found in Ref. [13].

1.4 Stochastic Wavefield

A signal emitted by a source, even if it is a simple sinusoid, may be modeled as a stochastic process because the signal may be controlled by a set of unknown random parameters, for example, a random phase in a sinusoid. The noise in the array output may consist of thermal noise, the wavefield emitted by numerous sources, either natural or man-made, and transmitted signal that has been scattered by numerous random scatterers or reflectors. In all these models there is an element of stochastic nature that imparts a stochastic character to the wavefield. In this and in the next section we

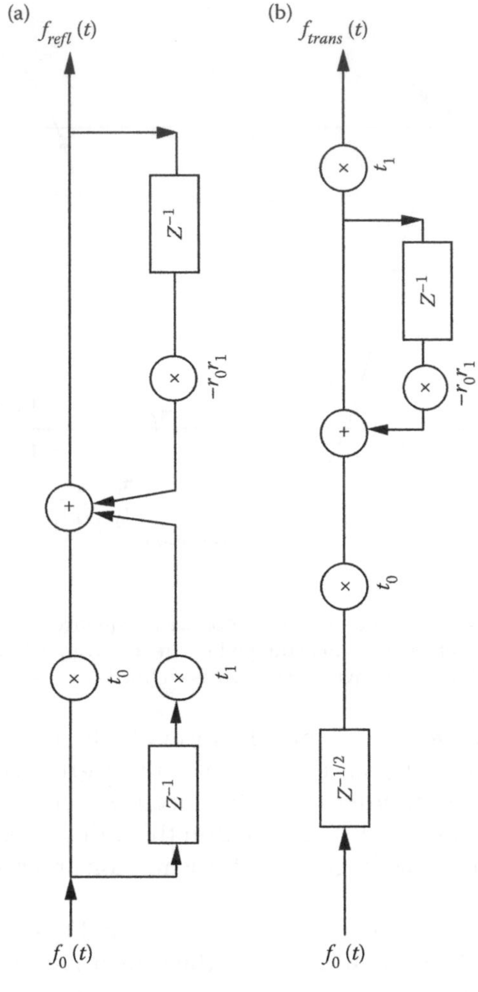

FIGURE 1.20
The response of a layer to an incident wave is described within the framework of filter theory. (a) Reflection filter and (b) transmission filter.

shall study some of these stochastic models of the wavefield from the point of ASP. First, we consider the wavefield emitted by a large number of random sources that are distributed in two or three-dimensional space, both in open space and bounded space. Basically, we shall study the correlation and the spectral characteristics of the output of an elementary array of two sensors. In Section 1.5, we shall study scattering or reflections produced by many random scatterers that act as obstacles in radio communication or microvariations of speed and density in underwater detection and communication.

1.4.1 Frequency Wavenumber Spectrum

A stochastic wavefield observed over an infinite line or plane is best characterized in terms of a frequency wavenumber spectrum. For a stochastic wavefield, in place of Fourier representation (Equation 1.19), we have spectral representation (on $z=0$ plane)

$$f(x,y,t)=\frac{1}{8\pi^3}\iiint\limits_{-\infty}^{\infty} dF(u,v,\omega)e^{j(\omega t-ux-vy)},$$

which is analogous to the spectral representation of a 1D stationary stochastic process [15]. A differential of the generalized Fourier transform of the stochastic wavefield is $dF(u, v, \omega)$, having the property that $E\{|dF(u, v, \omega)|^2\} \propto S_f(u, v, \omega)$, which we shall use to relate the covariance function to the frequency wavenumber spectrum.

$$C_f(\Delta x,\Delta y,\tau)=\frac{1}{8\pi^3}\iiint\limits_{-\infty}^{\infty} S_f(u,v,\omega)e^{j(\omega\tau-u\Delta x-v\Delta y)}du\,dv\,d\omega, \tag{1.50}$$

where $C_f(\Delta x, \Delta y, \tau)=E\{f(x, y, t)\, f(x+\Delta x, y+\Delta y, t+\tau)\}$ is the covariance function and $S_f(u, v, \omega)$ is the frequency wavenumber spectrum. In the spectrum analysis of time series (Equation 1.50) is known as the Wiener–Khinchin relation of great significance [14]. $S_f(u, v, \omega)$ represents power received at a given temporal frequency ω and spatial frequencies u and v. Since u and v are related to the direction cosines, $S_f(u, v, \omega)$ may be looked upon at a fixed temporal frequency as a function of the direction cosines.

1.4.2 Open Space

Consider noise sources in the far field region, distributed over a sphere or a circle (in two dimensions). Each point source emits a stationary stochastic waveform uncorrelated with all other sources. Let $f_i(t)$ be the stochastic waveform emitted by the ith source at angular distance $\varphi_0+\theta_i$, where θ_i is a random variable uniformly distributed over an interval $\pm\theta_0$ (see Figure 1.21). The signal received at the upper sensor is given by $f_i(t-d/2c\sin(\varphi_0+\theta_i))$ and that at the lower sensor is $f_i(t+d/2c\sin(\varphi_0+\theta_i))$, where d is the sensor separation. The total signal obtained by summing over all sources is given by

$$\begin{aligned} g_1(t)&=\sum_i f_i\left(t-\frac{d}{2c}\sin(\varphi_0+\theta_i)\right)\\ g_2(t)&=\sum_i f_i\left(t+\frac{d}{2c}\sin(\varphi_0+\theta_i)\right). \end{aligned} \tag{1.51a}$$

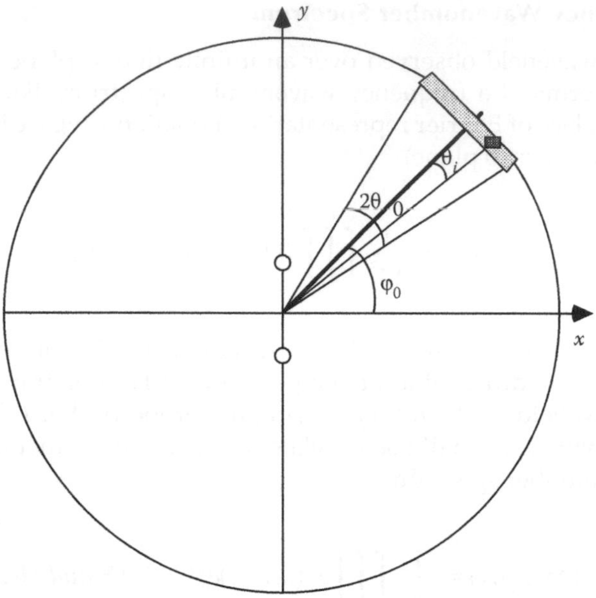

FIGURE 1.21
Point sources are uniformly distributed over an arc of a large circle in the x, y plane. An elementary array of two sensors is on the y-axis.

We shall now replace the random function in Equation 1.51a with its spectral representation [14] and rewrite it as

$$
\begin{aligned}
g_1(t) &= \frac{1}{2\pi}\int_{-\infty}^{\infty}\sum_i dF_i(\omega)e^{j(\omega(t-(d/2c)\sin(\varphi_0+\theta_i))},\\
g_2(t) &= \frac{1}{2\pi}\int_{-\infty}^{\infty}\sum_i dF_i(\omega)e^{j(\omega(t+(d/2c)\sin(\varphi_0+\theta_i)))}.
\end{aligned}
\tag{1.51b}
$$

We will now compute the cross-covariance function between the two outputs,

$$
\begin{aligned}
E\{g_1(t)g_2(t+\tau)\} &= \frac{1}{4\pi^2}\int_{-\infty}^{\infty}\int_{-\infty}^{\infty}\sum_i\sum_i E\{dF_i^*(\omega)dF_k(\omega')\}\\
&\quad\times E\left\{e^{-j(\omega(t-(d/2c)\sin(\varphi_0+\theta_i)))}\times e^{j(\omega'(t+\tau+(d/2c)\sin(\varphi_0+\theta_k)))}\right\}d\omega d\omega'.
\end{aligned}
\tag{1.52a}
$$

Noting the properties of the generalized Fourier transforms of stationary processes [14], Equation 1.52a may be simplified to yield the following result:

$$c_{12}(\tau)=\frac{1}{2\pi}\int_{-\infty}^{\infty}\sum_{i} S_{ii}(\omega)E\{e^{j(\omega(\tau+(d/c)\sin(\varphi_0+\theta_i))}\}d\omega, \tag{1.52b}$$

where we have assumed that the sources are uncorrelated with an identical spectrum. Since θ_i is a uniformly distributed random variable, the expected value in Equation 1.52b may be replaced by an integral

$$c_{12}(\tau)=c_{12}(\tau)=\frac{1}{2\pi}\int_{-\infty}^{\infty}S_0(\omega)d\omega\frac{1}{2\theta_0}\int_{-\theta_0}^{\theta_0}e^{j(\omega(\tau+(d/c)\sin(\varphi_0+\theta)))}d\theta. \tag{1.53a}$$

The integral over θ can be evaluated in the form of a series [16],

$$\int_{-\theta_0}^{\theta_0}e^{j(\omega(\tau+(d/c)\frac{d}{c}\sin(\varphi_0+\theta)))}d\theta$$

$$=2\left[\begin{array}{l}\left\{\begin{array}{l}\cos(\omega\tau)\sum\limits_{n=0,2,4\ldots}\delta_n J_n\left(\frac{\omega d}{c}\right)\frac{\sin(n\theta_0)\cos(n\varphi_0)}{n}\\+\sin(\omega\tau)\sum\limits_{n=1,3,5\ldots}2J_n\left(\frac{\omega d}{c}\right)\frac{\sin(n\theta_0)\sin(n\varphi_0)}{n}\end{array}\right\}\\ j\left\{\begin{array}{l}\sin(\omega\tau)\sum\limits_{n=0,2,4\ldots}\delta_n J_n\left(\frac{\omega d}{c}\right)\frac{\sin(n\theta_0)\cos(n\varphi_0)}{n}\\-\cos(\omega\tau)\sum\limits_{n=1,3,5\ldots}2J_n\left(\frac{\omega d}{c}\right)\frac{\sin(n\theta_0)\sin(n\varphi_0)}{n}\end{array}\right\}\end{array}\right], \tag{1.53b}$$

where $\delta_0=1$ and $\delta_n=2$ for all n. In the limiting case of circularly distributed noise sources, $\theta_0=\pi$, Equation 1.53b reduces to $2\pi J_0(\omega d/c)$.

$$c_{12}(\tau)=\frac{1}{\pi}\int_{0}^{\infty}S_0(\omega)J_0\left(\frac{\omega d}{c}\right)\cos(\omega\tau)d\omega. \tag{1.53c}$$

The normalized spatial covariance function at zero lag as a function of sensor separation is shown in Figure 1.22. Notice that the sensor outputs become increasingly uncorrelated as the angular extent of the noise sources increases. In the limiting case of circularly distributed noise sources, the correlation

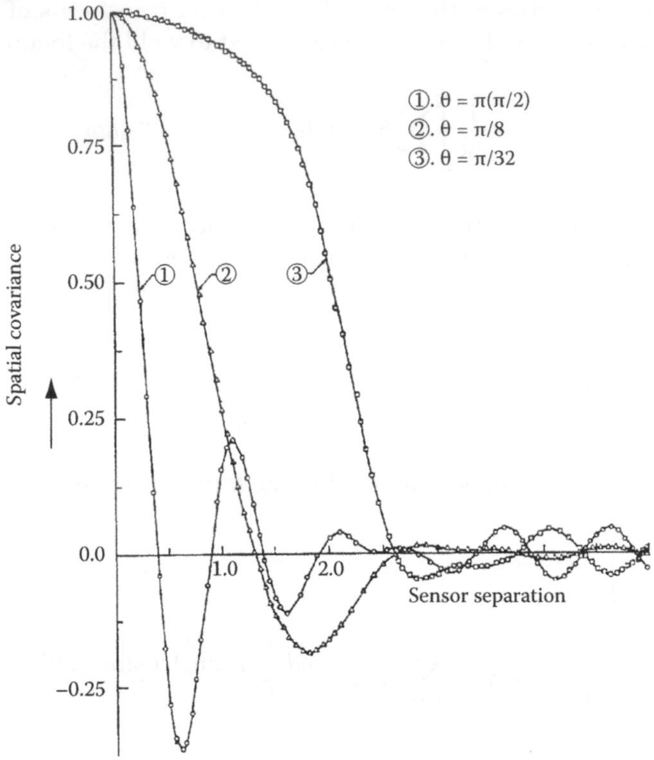

FIGURE 1.22
Spatial covariance function (at zero lag) as a function of sensor spacing, (d/λ). The covariance function is shown for three different angular widths of the distant noise sources. (1) 180°, (2) 22.5°, (3) 5.6°.

becomes negligible even for a separation of the order of one wavelength. When the sources are uniformly distributed on a sphere of large radius, the spatial covariance function is given by $c_{12}(0)=\sin(kd)/kd$ [17].

1.4.3 Channel

In this model, we shall consider a situation where the noise sources are uniformly distributed on one of the faces of a channel. This situation is close to an ocean channel where all noise sources are on or close to the ocean surface (Figure 1.23) [18].

Let $\psi(r, \varphi, t)$, a homogeneous random function, represent the noise sources on the top face of the channel, and $S_\psi(s, \alpha, \omega)$ be its spectrum where s is the radial frequency and α is the azimuth angle. The acoustic power radiated per unit area at a fixed temporal frequency is given by

$$S_\psi(\omega)=\frac{1}{4\pi^2}\int_0^{\infty}\int_0^{2\pi} sS_\psi(s,\alpha,\omega)d\alpha ds.$$

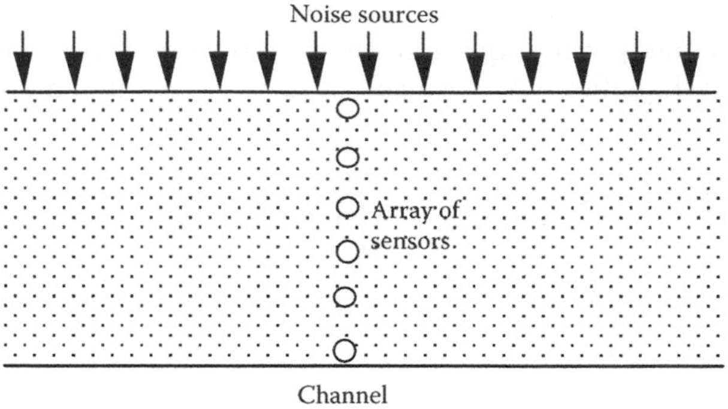

FIGURE 1.23
The noise sources on the surface of an ocean channel. Correlation structure of wavefield in the vertical direction is investigated.

For wind-generated noise on the ocean surface the frequency wavenumber spectrum is approximately modeled as

$$S_\psi(s,\alpha,\omega)=\frac{2p}{k^2}\Gamma(m)\left(1-\frac{s^2}{k^2}\right)^{m-1}S_0(\omega)\quad s<k$$
$$=0\qquad\qquad s\geq k,$$

where p and m are constants ($2\leq m\leq 3$) [17, 19]. Note the spectral representation of noise sources, at a fixed point on the plane,

$$\psi(r,\varphi,t)=\frac{1}{2\pi}\int_{-\infty}^{\infty}d\Psi(r,\varphi,\omega)e^{j\omega t},\tag{1.54a}$$

and over the entire plane

$$\psi(r,\varphi,t)=\frac{1}{(2\pi)^3}\iiint_{-\infty}^{+\infty}d\Psi(u,v,\omega)e^{j(\omega t-ux-vy)}.\tag{1.54b}$$

The pressure field due to a small element of noise sources at a sensor at a depth z_1 is given by Equation 1.45,

$$p_1(t)=\frac{1}{2\pi}\int_{-\infty}^{\infty}\frac{\omega}{2H}d\Psi(r,\varphi,\omega)e^{j\omega t}\sum_m\phi_m(z_1)\phi_m^*(z_s)H_0^1(k_m r)rdrd\varphi,\tag{1.55}$$

where $d\Psi(r, \varphi, \omega)\, rdrd\varphi$ is the strength of the noise sources. The total pressure is obtained by integrating Equation 1.55 over the entire plane. The spectral representation (Equation 1.54b) is used in carrying out the integration.

$$p_1(t) = \frac{1}{(2\pi)^2} \int_{-\infty}^{\infty} \frac{\omega}{2H} e^{j\omega t} \sum_m \phi_m(z_1)\phi_m^*(z_s) \int\int_{-\infty}^{+\infty} d\Psi(u,v,\omega) \int_0^{\infty} H_0^1(k_m r) J_0(sr) r dr. \quad (1.56)$$

Finally, we note that [17]

$$\int_0^{\infty} H_0^1(k_m r) J_0(sr) r dr = \frac{2}{\pi} \frac{1}{k_m^2 - s^2},$$

and obtain

$$p_1(t) = \frac{1}{2\pi^3} \int_{-\infty}^{\infty} \frac{\omega}{2H} e^{j\omega t} \int\int_{-\infty}^{+\infty} d\Psi(u,v,\omega) \sum_m \frac{\phi_m(z_1)\phi_m^*(z_s)}{k_m^2 - s^2}. \quad (1.57a)$$

The pressure field at another sensor placed at a depth z_2 is given by

$$p_2(t) = \frac{1}{2\pi^3} \int_{-\infty}^{\infty} \frac{\omega}{2H} e^{j\omega t} \int\int_{-\infty}^{+\infty} d\Psi(u,v,\omega) \sum_m \frac{\phi_m(z_2)\phi_m^*(z_s)}{k_m^2 - s^2}. \quad (1.57b)$$

The cross-spectrum between these two sensor outputs may be evaluated as follows

$$\begin{aligned} \frac{1}{2\pi} S_{12}(\omega) d\omega &= E\left\{ \frac{1}{2\pi} dP_1(\omega) \frac{1}{2\pi} dP_2^*(\omega) \right\} \\ &= \frac{\omega^2}{2\pi^3 H^2} \frac{1}{2\pi} d\omega \sum_m \sum_n \phi_m(z_1)\phi_m^*(z_s)\phi_n^*(z_2)\phi_n(z_s) \int_0^{\infty} \frac{sS_\psi(s,\omega)}{(k_m^2 - s^2)(k_n^{*2} - s^2)} ds, \end{aligned} \quad (1.58)$$

where

$$p_1(t) = \frac{1}{2\pi} \int_{-\infty}^{\infty} dP_1(\omega) e^{j\omega t} \text{ and } p_2(t) = \frac{1}{2\pi} \int_{-\infty}^{\infty} dP_2(\omega) e^{j\omega t}.$$

Simplifying Equation 1.58 we obtain

$$S_{12}(\omega)=\frac{\omega^2}{2\pi^3H^2}\sum_m\sum_n\phi_m(z_1)\phi_m^*(z_s)\phi_n^*(z_2)\phi_n(z_s)\int_0^\infty\frac{sS_\psi(s,\omega)}{(k_m^2-s^2)(k_n^{*2}-s^2)}ds \quad (1.59)$$

For spatially white noise sources, $S_\psi(s,\omega)=S_0(\omega)$, and the integral in Equation 1.59 reduces to

$$\int_0^\infty\frac{sS_\psi(s,\omega)}{(k_m^2-s^2)(k_n^{*2}-s^2)}ds=\frac{4}{\pi^2}\ln\left(\frac{k_m}{k_n^*}\right)\frac{S_0(\omega)}{(k_m^2-k_n^{*2})}.$$

For details on the derivation and numerical results, the reader is urged to see Refs. [20] and [21] where a more general model is dealt with. As an illustration of the general variation of the spectrum and the coherence, the numerical results for a channel where $H=4\lambda$ are shown in Figure 1.24. The noise sources are located on an annular ring with inner radius=100λ and outer radius=1,000λ. The noise is presumed to have been generated by wind ($m=1$). For coherence calculation, one sensor is kept fixed at 2λ and the other sensor is moved along the depth axis. We considered two types of bottoms, rigid and soft bottoms ($\rho_1=1$ g/cc, $c_1=1{,}500$ m/sec, $\rho_2=2.0$ g/cc, $c_2=1{,}600$ m/sec). The results are shown in Figure 1.24. The y-axis in Figure 1.24a represents a ratio $S_z(\omega)/S_\psi(\omega)$, where $S_z(\omega)$ is the spectrum of the pressure field at z. It is interesting to note that in the hard bottom channel, because of the trapping of energy, the spectrum is always greater than that in the soft bottom channel and it increases with depth. The coherence as a function of the sensor separation (Figure 1.24b) is highly oscillatory, roughly following the interference pattern of vertically traveling modes.

The stochastic wavefield in a bounded medium is extremely complex largely due to the interference of multiple reflected waves. What we have described above is a simple example involving just two reflectors. Imagine the complexity of wavefield in a room with six reflectors! Often, in such a complex situation, we model the wavefield as a diffused wavefield, which is briefly discussed in Section 1.4.2.

1.5 Multipath Propagation

When wave energy travels from point A to point B along more than one path, the propagation is said to be multipath propagation. Such a propagation regime is the result of local micro-inhomogeneities or point scatterers. Since the local micro-inhomogeneities or point scatterers are stochastic in

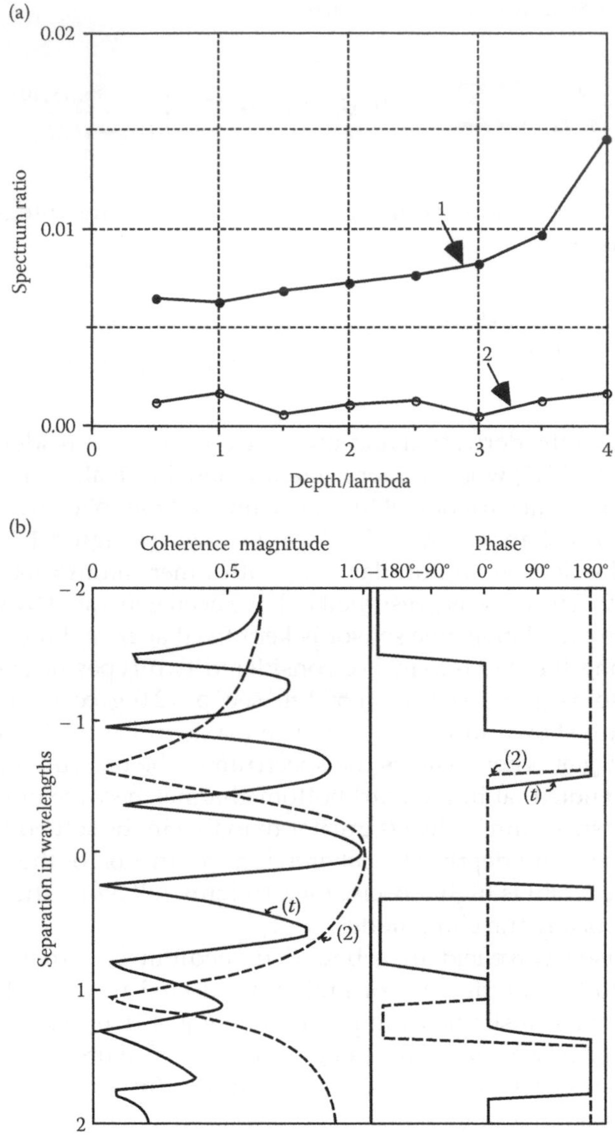

FIGURE 1.24
Spectrum and coherence as a function of depth. (a) Spectrum. Empty circles represent soft bottom and the filled circles represent hard bottom. (b) Coherence as a function of sensor separation. (1) Hard bottom. (2) Soft bottom. The magnitude is shown in the left panel and the phase in the right panel.

nature, it is appropriate to characterize the resulting multipath propagation under a stochastic framework. It often happens that the local inhomogeneities as well as the point scatterers are time varying, consequently multipath propagation is also time varying, sometimes very rapidly as in

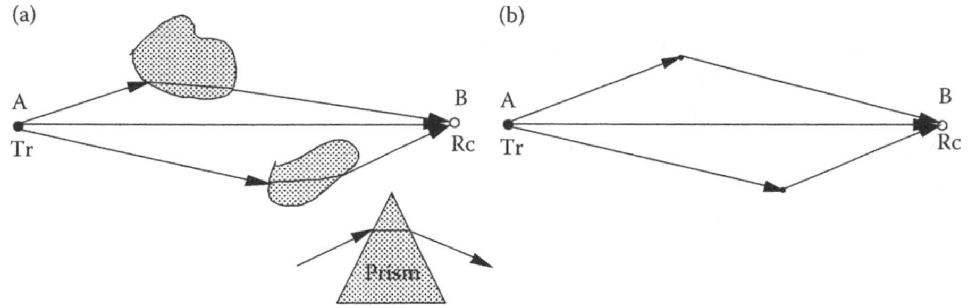

FIGURE 1.25
Phenomenon of a ray bending as it passes through a blob of inhomogeneity. For comparison, a ray is shown to bend as it passes through a prism. Scattering by a point scatterer produces a similar effect.

mobile wireless communication. The phenomenon of bending of a ray as it passes through a blob of inhomogeneity is illustrated in Figure 1.25. Scattering from a point scatterer produces a similar effect, but is more pronounced as the wave energy is scattered in all directions.

The dominant effect of the multipath propagation is a large collection of rays, known as microrays [22], impinging on an array from different directions. In a weakly scattering medium, a microray follows a path very close to that of an unperturbed ray. Hence, all microrays would have gone through similar macrovariations but with slight path variations. A sensor will then receive a train of coherent signals arriving at different time intervals, possibly Doppler shifted or scaled versions of the transmitted signal, when the scatterers are in motion. The amplitude variation among the microrays is likely to be less and of lower significance. Multipath propagation can cause severe loss of signal fidelity, particularly in those problems where the wave propagation is through natural channels, such as in underwater, solid earth, and urban environment, for wireless communication. We shall consider two types of multipath environments. First, we shall consider a random distribution of blobs of inhomogeneity as encountered in underwater channels and in subsurface channels. Next, we shall consider a collection of scattering points around a transmitter typically encountered in wireless propagation. There are other possibilities, but only the above two environments are well understood. The inhomogeneities in the medium are either stationary or very slowly moving; hence the Doppler effect can be ignored in the first instance.

1.5.1 Random Filter Model

The conceptual picture of many microrays closely following an unperturbed ray and reaching a sensor is mathematically described through the random filter model. The output of a sensor may be expressed as follows:

$$f_n(t) = \frac{1}{N_{rays}} \sum_k a_k g_0(t - t_n - \tau_k - \delta\tau_k(t)), \tag{1.60}$$

where

$g_0(t)$: waveform transmitted by a source,
t_n: arrival time of unperturbed ray at nth sensor,
τ_k: relative arrival time of kth microray,
$\delta\tau_k(t)$: relative delay due to time varying scattering effect,
a_k: coefficient of attenuation for kth ray,
N_{rays}: number of rays.

In the frequency domain, Equation 1.60 may be expressed as

$$\begin{aligned} f_n(t) &= \frac{1}{2\pi}\int_{-\infty}^{\infty} dG_0(\omega)\left[\frac{1}{N_{rays}}\sum_k a_k e^{-j\omega(t_n+\tau_k+\delta\tau_k(t))}\right]e^{j\omega t} \\ &= \frac{1}{2\pi}\int_{-\infty}^{\infty} dG_0(\omega)H(\omega,t)e^{j\omega t}, \end{aligned} \tag{1.61}$$

where

$$H(\omega,t) = \frac{1}{N_{rays}}\sum_k a_k e^{-j\omega(t_n+\tau_k+\delta\tau_k(t))},$$

is the time-varying random filter representation of the multipath propagation effect. Here, both τ_k and $\delta\tau_k(t)$ are modeled as random variables; hence the filter is known as a random time-varying filter. Sea experimental results as reported in Ref. [23] indicate that the delays may be approximated as a uniformly distributed random variable and that they are uncorrelated when the sensor separation is of the order of a few hundred feet. Using the above model of fluctuations, we can compute the mean and the variance of the time-varying filter transfer function. The mean is given by

$$\begin{aligned} E\{H_n(\omega,t)\} &= E\left\{\frac{1}{N_{rays}}\sum_k a_k e^{-j\omega(t_n+\tau_k+\delta\tau_k(t))}\right\} \\ &= \frac{1}{N_{rays}} e^{-j\omega t_n}\sum_k E\{a_k\}E\{e^{-j\omega\tau_k}\}E\{e^{-j\omega\delta\tau_k(t)}\} \\ &= A_n e^{-j\omega t_n}\Phi_0(\omega)\Phi_1(\omega,t), \end{aligned} \tag{1.62}$$

where $A_n = (1/N_{rays})\sum_k E\{a_k\}$, $\Phi_0(\omega) = E\{e^{-j\omega\tau_k}\}$ and $\Phi_1(\omega, t) = E\{e^{-j\omega\delta\tau_k(t)}\}$. For a uniformly distributed random variable in the interval $\pm\Delta t/2$,

$$\Phi_0(\omega) = \frac{\sin\left(\omega\frac{\Delta t}{2}\right)}{\omega\frac{\Delta t}{2}}.$$

1.5.2 Point Scatterers

In radio communication, the EM waves traveling along straight-line ray paths may encounter obstacles that would reflect or scatter the incident wavefield.

Since the frequency used is very high (800–1000 MHz, $\lambda \approx 1$ *meter*) most of the obstacles are likely to be much larger than the wavelength. There will be significant reflections and corner diffractions. The air medium is, however, assumed to be homogeneous and free from any scattering. Furthermore, a transceiver used in a modern wireless communication system is likely to be in motion, causing a significant Doppler shift. One consequence of the reflection or the scattering of waves is the possibility of more than one ray path connecting a transmitter and a receiver. Such multipath propagation in wireless communication is quite common. As a result of this, the different components of the wavefield reach a receiver at slightly different time instants, at different angles, and with different Doppler shifts, but coherently. Thus, multipaths are characterized by the following attributes: (i) delay diversity (0–10 μsec), (ii) angular diversity (5–6 degrees), and (iii) Doppler shift (0±50 Hz).

1.5.2.1 Delay Diversity

The signals arrive at a receiver (usually a single sensor) at different time intervals, the delay being due to different path lengths. A long delay implies a weak signal due to multiple reflections, attenuation in the air, and also by geometrical spreading. The quantity of great interest is the power received at a sensor as a function of delay, known as the power delay profile. A typical power delay profile is sketched in Figure 1.26. Referring to this figure, we define an excess delay spread τ_e as the delay within which 90% of the total

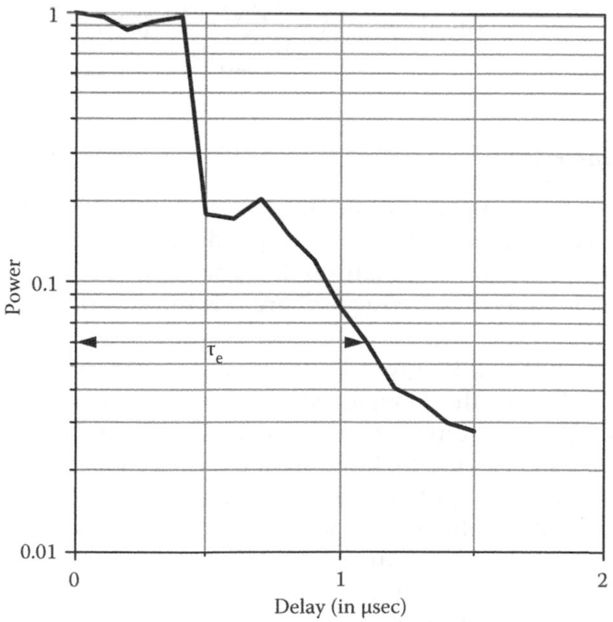

FIGURE 1.26
A sketch of the power delay profile. Excess delay parameter τ_e is shown.

power reaches the sensor. There are two other parameters commonly used to characterize a power delay profile,

$$\text{Mean delay } (\bar{\tau}) = \frac{\sum p_k \tau_k}{\sum p_k},$$

$$\text{rms delay spread} = \sqrt{\frac{\sum p_k \tau_k^2}{\sum p_k} - (\bar{\tau})^2}, \tag{1.63}$$

where p_k is the power in the kth path arriving with a delay τ_k. A cumulative plot of power, a plot of $\sum_{k=1}^{m} p_k$ vs. $\sum_{k=1}^{m} \tau_k$, is useful in deciphering different groups of multipaths.

1.5.2.2 Angular Diversity

The angle of arrival of different multipaths is a random variable. We define mean angle and root mean square (rms) angle spread as follows:

$$\text{Mean angle } (\bar{\varphi}) = \frac{\sum p_k \varphi_k}{\sum p_k},$$

$$\text{rms angle spread} = \sqrt{\frac{\sum p_k \varphi_k^2}{\sum p_k} - (\bar{\varphi})^2}. \tag{1.64}$$

1.5.2.3 Doppler Shift

It is common that a transceiver is moving, often at high speed, while the base station antenna is stationary. Consider a speeding car carrying a transceiver past a stationary scatterer (see Figure 1.27). The component of car velocity in the direction of scatterer is $v\cos(\varphi)$, where v is the speed of the car moving along the road.

The Doppler shift will be given by $\Delta\omega = v/c\ \omega_c \cos(\varphi)$, where ω_c is the carrier frequency. Plot of the Doppler shift as a function angle is shown in Figure 1.27 (inset). As a numerical example, let $v=120$ km/h, i.e., 100/3 m/sec, $c=3\times 10^8$ m/sec, and $\omega_c=18\pi\ 10^8$ the Doppler shift is equal to $\Delta\omega|_{\max}=200\pi$.

It is interesting to compute a power delay profile for a simple model of a street lined with buildings that act as obstacles. For simplicity we have assumed regularly spaced building (spacing=15 m and gap=5 m) on either sides of a street, 20 m wide (see Figure 1.28a). A stationary source in the middle of the street emits a spike waveform. A suite of rays starting from the source and ending at the base station antenna, which is 200 m away from

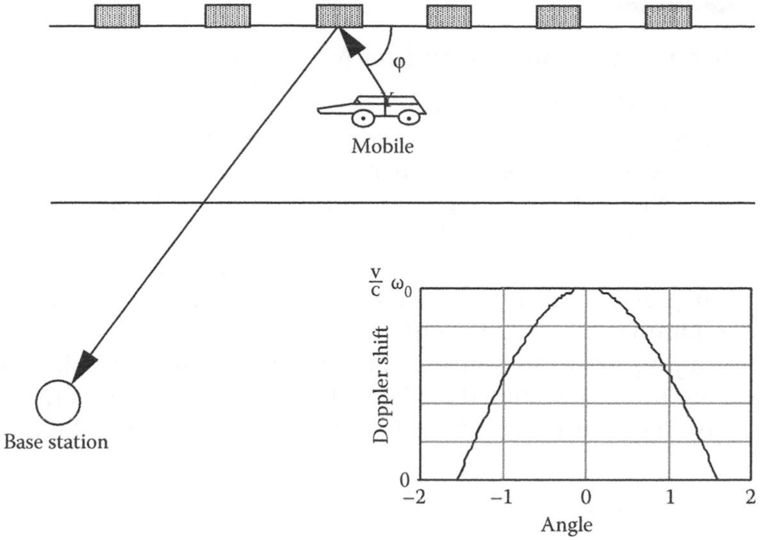

FIGURE 1.27
A moving car and stationary scatterer will produce a Doppler shift at the base station antenna. Inset shows the Doppler shift as a function of angle.

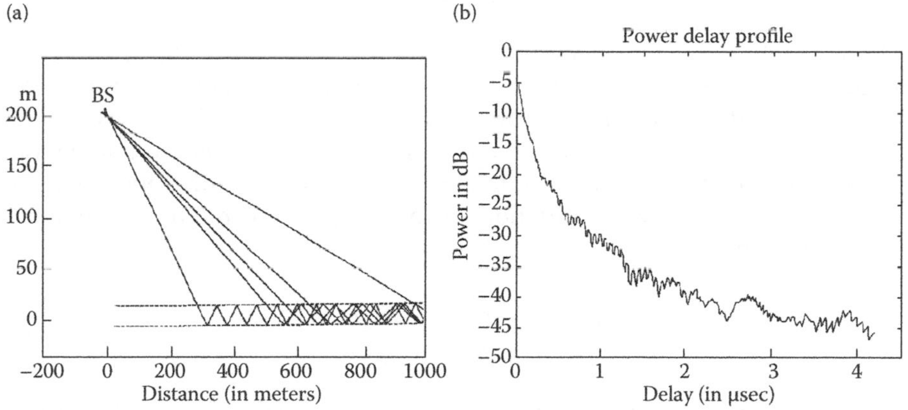

FIGURE 1.28
A simple model of a street lined with buildings that act as obstacles. (a) A sample of five rays is shown as they propagate through the street. (b) Computed power delay profile.

the street, were traced taking into effect all possible reflections at different buildings. A sample of five rays thus traced is shown in Figure 1.28a. Perfect reflection was assumed (reflection coefficient = 1), but geometrical spreading was taken into account. The power delay profile was computed by noting the power received at computed delay. The computed power delay profile is shown in Figure 1.28b. Other parameters are mean delay = 0.042 μsec, rms delay = 0.149 μsec, and excess delay spread = 0.05 μsec.

1.6 Propagation through Random Medium

Imaging of a medium through which a wave is propagating is one of the important applications of the sensor arrays, in particular, in medical tomography and in seismic exploration where reflected or scattered field is used for imaging. When the inhomogeneous medium is made of simple layers, as in subsurface imaging, the reflected field is all that is required for imaging purposes. This is, by and large, true in seismic imaging, which we shall cover in Chapter 10. But, when the medium is highly inhomogeneous, the scattering dominates. Currently, only weak scattering has been extensively used for imaging. In this section, we would like to review the weak scattering and point out how the scattered field enables imaging of an inhomogeneous medium. In Chapter 9, we will describe the use of sensor arrays for imaging.

1.6.1 Acoustic Field

The acoustic pressure field in an inhomogeneous medium satisfies the following wave equation,

$$\nabla^2 f - \frac{1}{c(x,y,z)^2}\frac{\partial^2 f}{\partial t^2} = 0, \tag{1.65}$$

where c is the acoustic wave speed, which is a function of the space coordinates. We shall assume that the density remains constant. (Equation 1.2b deals with both speed and density variations.) Let $c = c_0 + \delta c\ (x, y, z)$, where c_0 is the mean wave speed in the medium and $\delta c(x, y, z)$ is the fluctuation around the mean value. A medium is said to be weakly inhomogeneous if $|\delta c\ (x, y, z)| \ll c_0$. The wave equation for inhomogeneous medium (Equation 1.65) reduces to

$$\nabla^2 f - \frac{1}{c_0^2}\frac{\partial^2 f}{\partial t^2} = -\frac{2\delta c}{c_0^3}\frac{\partial^2 f}{\partial t^2}, \tag{1.66}$$

where the term on the right-hand side of Equation 1.66 represents the contribution due to speed fluctuations in the medium. Let us represent $(\delta c / c_0) = \varepsilon \delta \tilde{c}$, where $\delta \tilde{c}$ is a normalized function with unit rms magnitude and ε ($\ll 1$) is a constant. Equation 1.66 may be expressed as

$$\nabla^2 f - \frac{1}{c_0^2}\frac{\partial^2 f}{\partial t^2} = -\frac{2\varepsilon}{c_0^2}\delta \tilde{c}\frac{\partial^2 f}{\partial t^2}. \tag{1.67}$$

We shall now try to find a series solution of Equation 1.67. Let the series solution be given by

$$f = f_0 + \varepsilon f_1 + \varepsilon^2 f_2 + \cdots + \varepsilon^p f_p. \tag{1.68}$$

On substituting Equation 1.68 in Equation 1.67 and equating the coefficients of the powers of ε to zero, we obtain the following system of partial differential equations,

$$
\begin{aligned}
&\nabla^2 f_0 - \frac{1}{c_0^2}\frac{\partial^2 f_0}{\partial t^2} = 0, \\
&\nabla^2 f_1 - \frac{1}{c_0^2}\frac{\partial^2 f_1}{\partial t^2} = -\frac{2}{c_0^2}\delta\tilde{c}\frac{\partial^2 f_0}{\partial t^2} \\
&\cdots \\
&\nabla^2 f_p - \frac{1}{c_0^2}\frac{\partial^2 f_p}{\partial t^2} = -\frac{2}{c_0^2}\delta\tilde{c}\frac{\partial^2 f_{p-1}}{\partial t^2}.
\end{aligned}
\tag{1.69}
$$

We like to solve these equations in an unbounded medium. Note that $f_i(\mathbf{r})$, for $i=1, 2,\ldots,$ must satisfy the radiation condition. Note that the Green's function is the same for all equations in Equation 1.69; only the driving function on the right-hand side differs. The solution of ith equation may be expressed as

$$
f_i(\mathbf{r},t) = \frac{1}{4\pi}\iiint_{-\infty}^{+\infty} \frac{-\frac{2}{c_0^2}\delta\tilde{c}(x',y',z')\frac{\partial^2 f_{i-1}}{\partial t^2}}{|\mathbf{r}-\mathbf{r}'|} e^{j\mathbf{k}\cdot(\mathbf{r}-\mathbf{r}')}dx'dy'dz'. \tag{1.70}
$$

Consider the case of plane wave illumination, $f_0(\mathbf{r},t) = A_0 e^{j(\omega t - u_0 x - v_0 y - w_0 z)}$, where u_0, v_0, and w_0 are related to the direction cosines of the traveling plane wave (see Section 1.2). Using the above illumination function in Equation 1.70, we obtain the following expression for the first order term, also known as Born approximation,

$$
f_1(\mathbf{r},t) = \frac{1}{4\pi}\iiint_{-\infty}^{+\infty} \frac{2k_0^2\delta\tilde{c}e^{j(k_0|\mathbf{r}-\mathbf{r}'|)}}{|\mathbf{r}-\mathbf{r}'|} f_0(\mathbf{r}',t)dx'dy'dz'. \tag{1.71}
$$

We shall now use the Sommerfeld formula [26]

$$
\begin{aligned}
\frac{e^{j(k_0|\mathbf{r}-\mathbf{r}'|)}}{|\mathbf{r}-\mathbf{r}'|} &= \frac{e^{j(k_0\sqrt{\rho^2+z^2})}}{\sqrt{\rho^2+z^2}} \\
&= \frac{1}{2\pi}\int_0^{\infty}\int_0^{2\pi} \frac{\lambda d\lambda}{\sqrt{\lambda^2-k_0^2}} e^{j(\lambda(x-x')\cos\theta+\lambda(y-y')\sin\theta}e^{-\sqrt{\lambda^2-k_0^2}|z-z'|}d\theta,
\end{aligned}
\tag{1.72}
$$

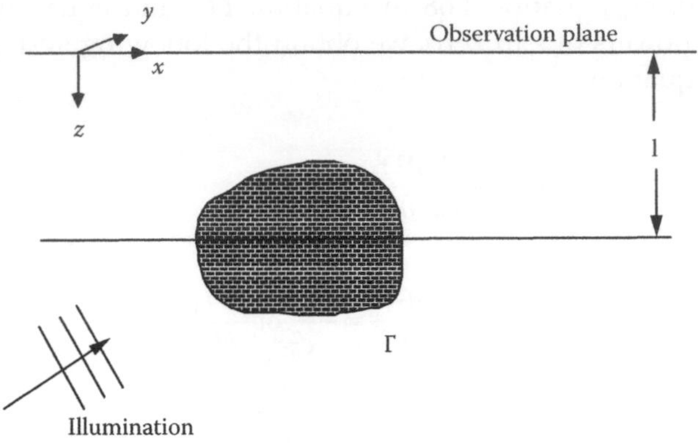

FIGURE 1.29
Γ represents the space occupied by the scattering medium. $\delta c = 0$ outside Γ. The scattered field is evaluated on observation plane l units above Γ.

to simplify Equation 1.71

$$f_1(\mathbf{r},t)=\frac{k_0^2}{4\pi^2}\int_{-\infty}^{\infty}\int_{0}^{2\pi}\frac{\lambda d\lambda d\theta}{\sqrt{\lambda^2-k_0^2}}$$

$$\times\left[\int_{\Gamma}\delta\tilde{c}(x',y',z')f_0(\mathbf{r}',t)e^{j(\lambda(x-x')\cos\theta+\lambda(y-y')\sin\theta)}e^{-\sqrt{\lambda^2-k_0^2}|z-z'|}dx'dy'dz'\right], \tag{1.73}$$

where Γ represents the space occupied by the scattering medium (see Figure 1.29).

Define $u'=\lambda\cos\theta$ and $v'=\lambda\sin\theta$. It follows that $du'dv'=\lambda d\lambda d\theta$. Equation 1.73 can be expressed as

$$f_1(x,y,\omega)$$

$$=\frac{-jk_0^2}{4\pi^2}\int_{-\infty}^{+\infty}\int\int_{\Gamma}\delta\tilde{c}(x',y',z')e^{-j((u'-u_0)x'+(v'-v_0)y'+(\sqrt{k_0^2-u'^2-v'^2}-w_0)z'}dx'dy'dz'. \tag{1.74}$$

$$\frac{e^{-j(\sqrt{k_0^2-(u'^2+v'^2)}-w_0)l}}{\sqrt{k_0^2-(u'^2+v'^2)}}e^{j(u'x+v'y)}du'dv'$$

Notice that the inner integral over Γ represents a three-dimensional Fourier transform of $\delta\tilde{c}(x',y',z')$. Hence, Equation 1.74 can be written in the frequency domain as follows:

$$f_1(x,y,\omega)$$

$$=\frac{-jk_0^2}{4\pi^2}\int_{-\infty}^{+\infty}\int\Delta\tilde{c}(u'-u_0,v'-v_0,\sqrt{k_0^2-u'^2-v'^2}-w_0)\frac{e^{-j(\sqrt{k_0^2-(u'^2+v'^2)}-w_0)l}}{\sqrt{k_0^2-(u'^2+v'^2)}}e^{j(u'x+v'y)}du'dv'. \tag{1.75}$$

Note that the factor $e^{j\sqrt{k_0^2-(u'^2+v'^2)}l}$ rapidly decays for $(u'^2+v'^2)>k_0^2$ and $l>0$. Such waves correspond to evanescent waves, which will be significant only in the immediate neighborhood of the scattering object. The presence of density fluctuation merely introduces an extra term in Equation 1.75. We will not go into the details, but see Ref. [25] where the density fluctuation is accounted for. For a two-dimensional object, the scattered field has been obtained by Kak [26].

$$f_1(x,\omega)=\frac{jk_0^2}{4\pi}\int_{-\infty}^{+\infty}\Delta\tilde{c}(u'-u_0,\sqrt{k_0^2-u'^2}-v_0)\frac{e^{j(\sqrt{k_0^2-u'^2}-v_0)l}}{\sqrt{k_0^2-u'^2}}e^{ju'x}du'. \tag{1.76}$$

1.6.2 Far-Field Approximation

When the scattering object is finite and the array is placed quite far from the object, we shall apply the far field approximation (in Equation 1.71), namely $(e^{jk_0|\mathbf{r}-\mathbf{r}'|}/|\mathbf{r}-\mathbf{r}'|)\approx(e^{jk_0(r-\mathbf{r}'\cdot\hat{\mathbf{r}})}/r)$, where $\hat{\mathbf{r}}$ is the unit vector in the direction of $\mathbf{r}$. The error due to this approximation, especially in the numerator, is illustrated in Figure 1.30. To assess the quantitative effect, consider the binomial expansion of

$$|\mathbf{r}-\mathbf{r}'|=\sqrt{r^2+r'^2-2rr'\cos\theta}\approx r\left(1+\frac{\left(\frac{r'}{r}\right)^2-2\frac{r'}{r}\cos\theta}{2}+\cdots\right)$$

$$=\left(r+\frac{r'^2}{2r}-r'\cos\theta+\cdots\right)\approx(r-r'\cos\theta).$$

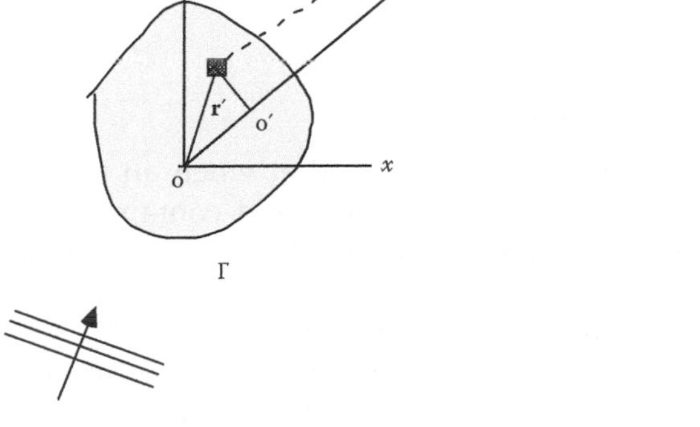

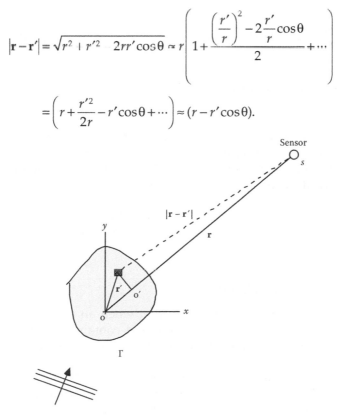

FIGURE 1.30
In far field approximation $|\mathbf{r}-\mathbf{r}'|$ is approximated by o's.

The error is of the order of $r'^2/2r$ in the first term of binomial expansion. This error will introduce a phase error, $\pi r'^2/\lambda r$. For the phase error to be small we must have $r'^2 \ll \lambda r$ or the largest dimension of the object must be much smaller than $\sqrt{\lambda r}$. Using this as the far field approximation in Equation 1.71, we obtain

$$f_1(\mathbf{r},t) = \frac{e^{-j(\omega t - k_0 r)}}{4\pi r} \int_\Gamma 2k_0^2 \delta\tilde{c}(x',y',z') e^{-jk_0(\mathbf{r}'\cdot\hat{\mathbf{r}})} e^{j(u_0 x' + v_0 y' + w_0 z')} dx'dy'dz'. \quad (1.77)$$

In Equation 1.77 we note that

$$\begin{aligned} e^{-jk_0(\mathbf{r}'\cdot\hat{\mathbf{r}})} &= e^{-jk_0(\alpha x' + \beta y' + \gamma z')} \\ &= e^{-j(ux' + vy' + wz')}, \end{aligned}$$

where $u = k_0\alpha$, $v = k_0\beta$, $w = k_0\gamma$ and (α, β, γ) are the direction cosines of unit vector $\hat{\mathbf{r}}$. Using this simplification in Equation 1.77 we obtain

$$\begin{aligned} f_1(\mathbf{r},t) &= \frac{e^{-j(\omega t - k_0 r)}}{4\pi r} \int_\Gamma 2k_0^2 \delta\tilde{c}(x',y',z') e^{-j[(u-u_0)x' + (v-v_0)y' + (w-w_0)z']} dx'dy'dz' \\ &= \frac{e^{-j(\omega t - k_0 r)}}{4\pi r} 2k_0^2 \Delta\tilde{c}(u-u_0, v-v_0, w-w_0), \end{aligned} \quad (1.78)$$

where $\Delta\tilde{c}(.)$ is the Fourier transform of $\delta\tilde{c}$. The result derived in Equation 1.78 has some significance. In the far field region the first order scattered field has the form of a point scatterer (term outside the square brackets), that is, spherical waves. The term inside the square brackets depends on the Fourier transform of the speed fluctuations, evaluated at spatial frequencies determined by the direction of illumination and the direction of sensor. In Chapter 9, we shall exploit this result for reconstruction of speed fluctuations.

1.6.3 Multisource Illumination

The basic fact used in tomography is that when an object is illuminated from different directions, the scattered field contains useful information for three-dimensional reconstruction. This property of the wavefield is elaborated in this simple example. Consider a source and sensor array on opposite sides of an object to be imaged (see Figure 1.31). Let the mth source be fired and the scattered field be sensed by the sensor array. Let $\mathbf{r}_m^a$ be a vector to the mth source and $\mathbf{r}_n^b$ be a vector to the nth sensor. $\mathbf{r}'$ is vector to

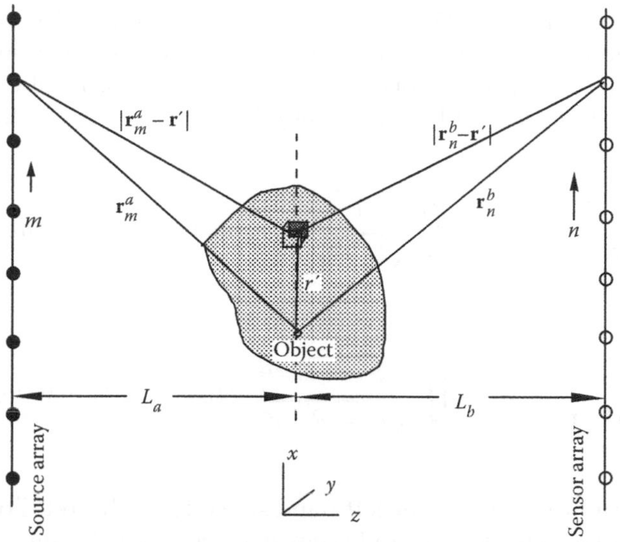

FIGURE 1.31
An array of sources is used to illuminate an object and the scattered wavefield is received by another array of sensors. The object Fourier transform can be directly related to the 2D Fourier transform of the received wavefield.

a scattering element. The scattered field due to the scattering element at nth sensor is given by

$$\Delta f_1(m,n) = \frac{1}{(4\pi)^2}\frac{2k_0^2\delta\tilde{c}e^{jk_0|\mathbf{r}_n^b-\mathbf{r}'|}}{|\mathbf{r}'-\mathbf{r}_n^b|}\frac{e^{jk_0|\mathbf{r}'-\mathbf{r}_m^a|}}{|\mathbf{r}'-\mathbf{r}_m^a|}dx'dy'dz'. \tag{1.79}$$

Note that the source and sensor arrays are equispaced linear arrays in the $y=0$ plane. Therefore, the tips of the vectors $\mathbf{r}_m^a$ and $\mathbf{r}_n^b$ will lie at $[md, 0, -L_a, m=0, \pm 1, \pm 2, \ldots]$ and $[nd, 0, -L_b, n=0, \pm 1, \pm 2, \ldots]$, respectively. Using Sommerfeld formula (Section 1.6.1) to express

$$\frac{e^{jk_0|\mathbf{r}_m^a-\mathbf{r}'|}}{|\mathbf{r}_m^a-\mathbf{r}'|} = \frac{-j}{2\pi}\int\int_{-\infty}^{+\infty}\frac{e^{-j\sqrt{k_0^2-u^2-v^2}(z'+L_a)}}{\sqrt{k_0^2-u^2-v^2}}e^{j[(md-x')u-y'v]}dudv, \tag{1.80a}$$

$$|z'| < L_a$$

$$\frac{e^{jk_0|\mathbf{r}_n^b-\mathbf{r}'|}}{|\mathbf{r}'-\mathbf{r}_n^b|} = \frac{-j}{2\pi}\int\int_{-\infty}^{+\infty}\frac{e^{-j\sqrt{k_0^2-u^2-v^2}(-z'+L_b)}}{\sqrt{k_0^2-u^2-v^2}}e^{j[(nd-x')u-y'v]}dudv. \tag{1.80b}$$

$$|z'| < L_b$$

Next we compute the 2D discrete Fourier transform (DFT) of Δf_1 (m, n). From Equation 1.79 it may be seen that the discrete Fourier transform of Δf_1 (m, n) is equal to the product of discrete Fourier transforms of Green's functions (i.e., left-hand side of Equation 1.80). These are given as follows:

$$DFT\left\{\frac{e^{jk_0|\mathbf{r}_m^a-\mathbf{r}'|}}{|\mathbf{r}_m^a-\mathbf{r}'|}\right\}_m = \frac{-j}{2\pi}\iint_{-\infty}^{+\infty}\frac{e^{-j\sqrt{k_0^2-u^2-v^2}|z'+L_a|}}{\sqrt{k_0^2-u^2-v^2}}\delta(ud-u_1)e^{-j[x'u+y'v]}\,du\,dv, \quad (1.81a)$$

$$DFT\left\{\frac{e^{jk_0|\mathbf{r}_n^a-\mathbf{r}'|}}{|\mathbf{r}'-\mathbf{r}_n^a|}\right\}_n = \frac{-j}{2\pi}\iint_{-\infty}^{+\infty}\frac{e^{-j\sqrt{k_0^2-u^2-v^2}|-z'+L_a|}}{\sqrt{k_0^2-u^2-v^2}}\delta(ud-u_2)e^{-j[x'u+y'v]}\,du\,dv. \quad (1.81b)$$

The subscripts m and n on the left-hand side refer to discrete Fourier transforms with respect to index m and index n, respectively. Using Equation 1.81 in Equation 1.79, we obtain the 2D Fourier transform of the response of a scattering element

$$\Delta F_1(u_1,u_2) = \frac{-k_0^2}{2\pi^2}\iint_{-\infty}^{+\infty}\left[\begin{array}{c}\dfrac{e^{-j\sqrt{k_0^2-(u_1/d)^2-v'^2}\,(z'+L_a)}}{\sqrt{k_0^2-(u_1/d)^2-v'^2}}\dfrac{e^{-j\sqrt{k_0^2-(u_2/d)^2-v^2}\,(-z'+L_b)}}{\sqrt{k_0^2-(u_2/d)^2-v^2}}\\ \times\delta\tilde{c}e^{-j[((u_1/d)+(u_2/d))x'+(v'+v)y']}dv'dv\end{array}\right]dx'dy'dz'. \quad (1.82)$$

Equation 1.82 is now summed over all scattering elements covering the entire object. We obtain

$$\begin{aligned}F_1(u_1,u_2) &= \frac{-k_0^2}{2\pi^2}\iint_{-\infty}^{+\infty}\Delta\tilde{c}\left(\frac{u_1}{d}+\frac{u_2}{d},\,v+v',\left(\sqrt{k_0^2-\left(\frac{u_1}{d}\right)^2-v^2}-\sqrt{k_0^2-\left(\frac{u_2}{d}\right)^2-v'^2}\right)\right)\\ &\quad\times\frac{e^{-j\sqrt{k_0^2-(u_2/d)^2-v'^2}L_a}}{\sqrt{k_0^2-(u_1/d)^2-v'^2}}\frac{e^{-j\sqrt{k_0^2-(u_1/d)^2-v^2}L_b}}{\sqrt{k_0^2-(u_2/d)^2-v^2}}dv'dv. \quad (1.83)\end{aligned}$$

As a special case, we consider an object having a slow variation in the y direction. Then,
$\Delta\tilde{c}(u, v, w) \approx \Delta\tilde{c}(u, w)\,\delta(v)$ and Equation 1.83 reduces to

$$F_1(u_1,u_2)=\frac{k_0^2}{\pi}\int_{-\infty}^{\infty}\Delta\tilde{c}\left(\frac{u_1}{d}+\frac{u_2}{d},\left(\sqrt{k_0^2-\left(\frac{u_1}{d}\right)^2-v^2}-\sqrt{k_0^2-\left(\frac{u_2}{d}\right)^2-v^2}\right)\right)$$
$$\times\frac{e^{-j\sqrt{k_0^2-(u_1/d)^2-v^2}L_a}}{\sqrt{k_0^2-(u_1/d)^2-v^2}}\frac{e^{-j\sqrt{k_0^2-(u_2/d)^2-v^2}L_b}}{\sqrt{k_0^2-(u_2/d)^2-v^2}}dv \tag{1.84}$$

If we were to use a line source in place of a point source, Equation 1.84 reduces to, by letting $v=0$,

$$F_1(u_1,u_2)=2k_0^2\Delta\tilde{c}\left(\frac{u_1}{d}+\frac{u_2}{d},\sqrt{k_0^2-\left(\frac{u_1}{d}\right)^2}-\sqrt{k_0^2-\left(\frac{u_2}{d}\right)^2}\right).$$
$$\times\frac{e^{-j\sqrt{k_0^2-(u_1/d)^2}L_a}}{\sqrt{k_0^2-(u_1/d)^2}}\frac{e^{-j\sqrt{k_0^2-(u_2/d)^2}L_b}}{\sqrt{k_0^2-(u_2/d)^2}} \tag{1.85}$$

The above result is identical to that given in Ref. [27].

1.6.4 Scattering of EM Field

The electric properties of a medium are dielectric permitivity (ε), magnetic permeability (μ), and conductivity (σ). The external EM wavefield will induce electric current in inhomogeneous medium. The induced current will, in turn, create EM wavefields outside the medium. The induced current density at a point is given by

$$\mathbf{J}_1(x,y,z)=(\kappa_1^2(x,y,z)-\kappa_0^2)\mathbf{E}(x,y,z), \tag{1.86}$$

where $\kappa_1^2(x,y,z)=\omega^2\mu_0\varepsilon_1(x,y,z)-j\omega\mu_0\sigma_1(x,y,z)$, $\kappa_0^2=\omega^2\mu_0\varepsilon_0$ and $\mathbf{E}(x, y, z)$ is an electric field which induces the electric current [25, 28]. It is assumed that the space outside the inhomogeneous medium is air, and hence $\sigma_0=0$. The electric field at a point outside the inhomogeneous medium is given by

$$\mathbf{E}_1(x,y,z,\omega)=j\omega\int_{\Gamma}\mathbf{J}_1(x',y',z',\omega)\frac{\exp(jk_0|\mathbf{r}-\mathbf{r}'|)}{4\pi|\mathbf{r}-\mathbf{r}'|}dx'dy'dz', \tag{1.87a}$$

and the magnetic field is given by

$$\mathbf{H}_1(x,y,z,\omega)=\int_{\Gamma}\nabla\times\mathbf{J}_1(x',y',z',\omega)\frac{\exp(jk_0|\mathbf{r}-\mathbf{r}'|)}{4\pi|\mathbf{r}-\mathbf{r}'|}dx'dy'dz'. \tag{1.87b}$$

Under Born approximation $\mathbf{E}(x,y,z) \approx \mathbf{E}_0 e^{jk_0(\alpha x+\beta y+\gamma z)}$, where (α, β, γ) are direction cosines of the wave vector and $\mathbf{E}_0$ is the incident electric field. Note that $\mathbf{E}_1$ is in the same direction as $\mathbf{E}_0$, but $\mathbf{H}_1$ is perpendicular to the incident vector. Let the incident electric field be in the z direction, then the scattered magnetic field will be in the (x,y) plane. The x and y components of the magnetic field are given by

$$H_{1x}(x,y,z,\omega) = (1 + jk_0)E_0 \times \int_{\Gamma} (\kappa_1^2(x',y',z') - \kappa_0^2)(y - y') \frac{\exp(jk_0|\mathbf{r} - \mathbf{r}'|)}{4\pi|\mathbf{r} - \mathbf{r}'|^2} dx'dy'dz', \tag{1.88a}$$

and

$$H_{1y}(x,y,z,\omega) = (1 + jk_0)E_0 \int_{\Gamma} (\kappa_1^2(x',y',z') - \kappa_0^2)(x - x') \frac{\exp(jk_0|\mathbf{r} - \mathbf{r}'|)}{4\pi|\mathbf{r} - \mathbf{r}'|^2} dx'dy'dz'. \tag{1.88b}$$

Analogous to Equation 1.75 we can write, in the frequency domain, an expression for the scattered electric and magnetic fields as

$$H_{1x}(x,y,l,\omega) = \frac{E_0}{4\pi^2} \iint_{-\infty}^{+\infty} \left[\begin{array}{l} K_1(u' - u_0, v' - v_0, \sqrt{k_0^2 - u'^2 - v'^2} - w_0) \\ \times \dfrac{e^{j(\sqrt{k_0^2 - (u'^2 + v'^2)} - w_0)l}}{\sqrt{k_0^2 - (u'^2 + v'^2)}} (jv') e^{j(u'x + v'y)} \end{array} \right] du'dv', \tag{1.89a}$$

$$H_{1y}(x,y,l,\omega) = \frac{E_0}{4\pi^2} \iint_{-\infty}^{+\infty} \left[\begin{array}{l} K_1(u' - u_0, v' - v_0, \sqrt{k_0^2 - u'^2 - v'^2} - w_0) \\ \times \dfrac{e^{j(\sqrt{k_0^2 - (u'^2 + v'^2)} - w_0)l}}{\sqrt{k_0^2 - (u'^2 + v'^2)}} (ju') e^{j(u'x + v'y)} \end{array} \right] du'dv', \tag{1.89b}$$

where K_1 stands for the 3D Fourier transform of $(\omega^2 \mu_0 \varepsilon_1 (x, y, z) - j\omega \mu_0 \sigma_1 (x, y, z) - \omega^2 \mu_0 \varepsilon_0)$. From Equations 1.75 and 1.89 it may be inferred that a strong similarity between the scattered acoustic field and the scattered EM field exists.

1.7 Exercises

(1) In Equation 1.53c, let the noise be a band-limited process with a spectrum limited to $\omega_0 - (\delta/2) \le \omega \le \omega_0 + (\delta/2)$. Show that

$$c_{12}(\tau) = c_{11}(\tau) J_0\left(\omega_0 \frac{d}{c}\right),$$

under the condition that $(d/c)\,\delta << 1$. What is the implication of this condition? [29].

(2) Consider a diffused field in 3D space. One possible model is that point sources are assumed to be uniformly distributed over a large sphere of radius R. Each point source emits a stochastic waveform that is uncorrelated with the waveforms emitted by all other point sources. Show that the field at a sensor placed on the z-axis at a distance $d/2$ units from the origin of the coordinates is given by

$$f_1(t) \approx \frac{1}{2\pi} \frac{e^{j(\omega/c)R}}{R} \int_{-\infty}^{\infty} dF(\omega,\theta,\varphi) e^{j\omega(t-(d/2c)\cos\theta)},$$

where θ, φ are, respectively, elevation and azimuth of the point source. Place another sensor, also on the z-axis, at a distance $-d/2$. Show that the cross correlation between the outputs is given by

$$c_{12}(\tau) \approx \frac{1}{2\pi} \int_{-\infty}^{\infty} S_f(\omega) \operatorname{sinc}\left(\frac{\omega d}{c}\right) e^{j\omega\tau} d\omega.$$

(3) Derive the reflection and transmission response of a single layer (shown in Figure 1.19) when it is illuminated from below. Derive the reflection response of two layers illuminated from the top.

(4) In a local scattering model, show that the covariance matrix of the array output becomes diagonal, that is, the wavefields become spatially uncorrelated when the angular width subtended by the cloud of scatterers at the array is equal to $(\lambda/(d\cos\theta_0))$. Consequently, it is not possible to estimate the DOA of the transceivers.

References

1. P. M. Morse and H. Feshbach, *Methods of Theoretical Physics*, vol. I, McGraw-Hill, New York, 1953.

2. J. Li and R. T. Compton, Angle and polarization estimation using ESPRIT with polarization sensitive array, *IEEE Trans., AP-39*, pp. 1376–1383, 1991.
3. S. A. Levin, Principle of reverse time migration, *Geophys.* vol. 49, No. 5, pp. 581–583, 1984.
4. A. Nehorai, Advanced-sensor signal processing *in Highlights of Statistical Signal and Array Processing, IEEE Signal Proc. Mag.*, pp. 43–45, 1998.
5. S. Treitel, P. R. Gutowski and D. Wagner, Plane wave decomposition of seismograms, *Geophys.* vol. 47, pp. 1375–1401, 1982.
6. P. M. Morse, Vibrations of elastic bodies; Wave propagation in elastic solids, in *Handbook of Physics*, (Eds.) E. U. Condon and H. Odishaw, McGraw-Hill, New York, 1958.
7. K. V. Mackenzie, Long range shallow water transmission, *J. Acoust. Soc. Am.*, vol. 33, pp. 1505–1514, 1961.
8. C. T. Tindle and G. E. J. Bold, Improved ray calculations in shallow water, *J. Acoust. Soc. Am.*, vol. 70, pp. 813–819, 1981.
9. P. S. Naidu and P. G. Krishna Mohan, Signal subspace approach in localization of sound source in shallow water, *Signal Proc.*, vol. 24, pp. 31–42, 1991.
10. Von F. Goos and H. Hanchen, Ein neuer und fundamentaler Versuch zur Totalreflexion, *Ann. Phys.*, vol. 1, pp. 333–346, 1947.
11. C. T. Tindle, Ray calculations with beam displacement, *J. Acoust. Soc. Am.*, vol. 73, pp. 1581–1586, 1983.
12. E. K. Westwood and C. T. Tindle, Shallow water time series simulation using ray theory, *J. Acoust. Soc. Am.*, vol. 81, pp. 1752–1761, 1987.
13. E. A. Robinson, Iterative least squares procedure for ARMA spectral estimation, in *Nonlinear Methods of Spectral Analysis*, Ed. S. Haykin, Springer Verlag, Berlin, pp. 127–154, 1979.
14. P. S. Naidu, *Modern Spectrum Analysis of Time Series*, CRC Press, Boca Raton, FL, 1996.
15. A. H. Yaglom, *Introduction to Theory of Stationary Random Functions*, Prentice-Hall, Englewood Cliffs, NY, 1962.
16. M. Subbarayudu, Performance of the eigenvector (EV) method in the presence of coloured noise, PhD Thesis, Indian Institute of Science, Bangalore, 1985.
17. B. F. Cron and C. H. Sherman, Spatial-correlation functions for various noise models, *J. Acoust. Soc. Am.*, vol. 34, pp. 1732–1736, 1962.
18. W. A. Kuperman and F. Ingenito, Spatial correlation of surface generated noise field in a stratified ocean, *J. Acoust. Soc. Am.*, vol. 67, pp. 1988–1996, 1980.
19. R. M. Hampson, The theoretical response of vertical and horizontal line arrays to wind generated noise in shallow water, *J. Acoust Soc. Am.*, vol. 78, pp. 1702–1712, 1985.
20. P. S. Naidu and P. G. Krishna Mohan, A study of the spectrum of an acoustic field in shallow water due to noise sources at the surface, *J. Acoust Soc. Am.*, vol. 85, pp. 716–725, 1989.
21. P. G. Krishna Mohan, Source localization by signal subspace approach and ambient noise modelling in shallow water, PhD Thesis, Indian Institute of Science, Bangalore, 1988.
22. S. M. Flatte, Wave propagation through random media: Contribution from ocean acoustics, *Proc. IEEE*, vol. 71, pp. 1267–1294, 1983.

23. H. W. Broek, Temporal and spatial fluctuations in single-path underwater acoustic wave fronts, I–III, *J. Acoust. Soc. Am.*, vol. 72, pp. 1527–1543, 1982.
24. A. N. Tychonoff and A. A. Samarski, *Differentialgleichungen der Mathematischen Physik*, VEB Verlag der Wissenschaften, Berlin, 1959.
25. S. J. Norton, Generation of separate density and compressibility images in tissue, *Ultrason. Imag.*, vol. 5, pp. 240–252, 1983.
26. A. C. Kak, Tomographic imaging with diffracting and nondiffracting sources, in *Array Processing*, Ed. S. Haykin, Springer-Verlag, New York, pp. 351–433, 1985.
27. R-S. Wu and M. N. Toksoz, Diffraction tomography and multisource holography applied to seismic imaging, *Geophysics*, vol. 52, pp. 11–25, 1987.
28. C. Pichot, L. Jofre, G. Peronnet, and J. C. Bolomey, Active microwave imaging of inhomogeneous bodies, *IEEE Trans.*, AP-32, pp. 416–424, 1985.
29. N. T. Gaarder, The design of point detector arrays, I, *IEEE Trans.*, IT-13, pp. 42–53, 1967.

2

Sensor Array Systems

An array of sensors distributed over a horizontal plane surface is used to receive a propagating wavefield with the following objectives:

(1) To localize a source

(2) To receive a message from a distant source

(3) To image a medium through which the wavefield is propagating

In this chapter we shall study the basic structure of a sensor array system, and in the sequel learn how the above objectives are achieved. The most commonly used array geometries are uniform linear array (ULA) and uniform circular array (UCA). A uniform planar array (UPA), where sensors are placed on an equispaced rectangular grid, is more common in large military-phased array systems. A wavefront that propagates across the array of sensors is picked up by all sensors. Thus, we have not one but many outputs that constitute an array signal. In the simplest case, all components of the array signals are simply delayed replicas of a basic signal waveform. In the worst case, individual sensor outputs are strongly corrupted with noise and other interference, leaving very little resemblance among them. Array processing now involves combining all sensor outputs in some optimal manner so that the coherent signal emitted by the source is received and all other inputs are maximally discarded. The aperture of an array, that is, the spatial extent of the sensor distribution is a limiting factor on resolution. Use of multicomponent or vector sensors has been proposed to effectively increase the array aperture. The aperture can also be synthetically increased by moving a source or sensor. The synthetic aperture concepts are extensively used in mapping radars and sonars. In this chapter, we concentrate on sensor array systems that will form the basic material for the subsequent chapters. Two new sections on distributed sensor array and multicomponent sensor array have been added to widen the scope of the chapter.

2.1 Uniform Linear Array

2.1.1 Array Response

Consider a plane wavefront, having a temporal waveform $f(t)$ incident on a ULA of sensors (see Figure 2.1) at an angle θ. In signal-processing literature,

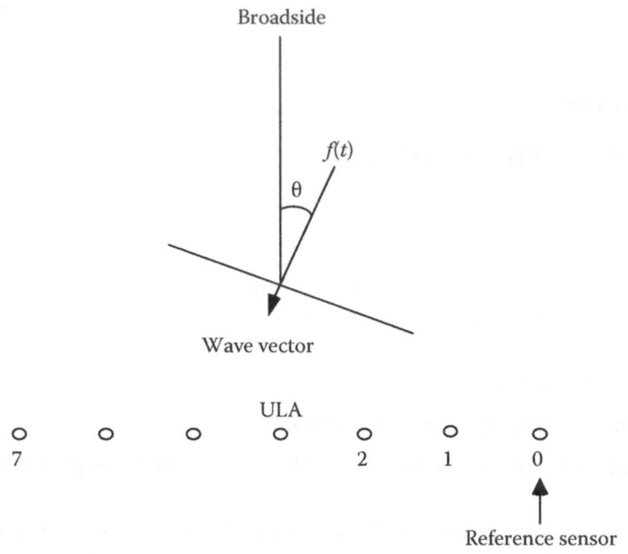

FIGURE 2.1
Uniform linear array of sensors. Note the convention of sensor indexing. The left most sensor is the reference sensor with respect to which all time delays are measured.

the angle of incidence is also known as the direction of arrival (DOA). Note that the DOA is always measured with respect to the normal to array aperture, while another related quantity azimuth, which was introduced in Chapter 1, is measured with respect to the x-axis, independent of array orientation. In this work, θ stands for DOA and φ stands for azimuth. We shall assume that a source emits a stationary stochastic signal $f(t)$. Let $f_m(t)$, $m=0, 1, 2,\ldots, M-1$ be the outputs of the sensors. The signal arrives at successive sensors with an incremental delay. The output of the first sensor is $f_0(t)=f(t)$, the output of the second sensor is $f_1(t)=f(t-\Delta t)$, and so on. Thus, the output of the mth sensor is $f_m(t)=f(t-m\Delta t)$. Sometimes it is convenient to represent the sensor output in the frequency domain,

$$f_m(t) = \frac{1}{2\pi}\int_{-\infty}^{\infty} dF(\omega)e^{j\omega(t-(md/c)\sin\theta)}, \tag{2.1}$$

where we have used the spectral representation of a stationary stochastic process [1]. The simplest form of array signal processing is to sum all sensor outputs without any delay.

$$g(t) = \frac{1}{M}\sum_{m=0}^{M-1} f_m(t) = \frac{1}{2\pi}\int_{-\infty}^{\infty} dF(\omega)e^{j\omega t}\frac{1}{M}\sum_{m=0}^{M-1} e^{-j\omega(md/c)\sin\theta}$$

$$= \frac{1}{2\pi} \int_{-\infty}^{\infty} dF(\omega) H(\omega\tau) e^{j\omega t}, \tag{2.2}$$

where $H\,(\omega\tau)$ is array response function, $\tau = (d/c)\sin\theta$, and d is sensor spacing. The array response function for a ULA is given by

$$H(\omega\tau) = \frac{1}{M} \sum_{m=0}^{M-1} e^{-j\omega(md/c)\sin\theta} = \frac{\sin\left(\frac{M}{2}\omega\tau\right)}{M \sin\frac{\omega\tau}{2}} e^{-j(M-1/2)\omega\tau}. \tag{2.3a}$$

When the sensor output is weighted with complex coefficients, a_m, $m = 0, 1, \ldots, M-1$ the array response becomes

$$H(\omega\tau) = \frac{1}{M} \sum_{m=0}^{M-1} a_m e^{-j\omega(md/c)\sin\theta}. \tag{2.3b}$$

A few samples of the frequency response function (magnitude only) are shown in Figure 2.2 for different values of M, that is, array size. The response function is periodic with a period 2π. The maximum occurs at $\omega\tau = 2n\pi$. The peak at $n=0$ is known as the main lobe and other peaks at $n = \pm 1, \pm 2, \ldots$ are known as grating lobes. Since the magnitude of the array response is plotted, the period becomes p as seen in Figure 2.2. The grating lobes can be avoided if we restrict the range of $\omega\pi$ to $\pm\pi$, that is, at a fixed frequency the DOA must satisfy the relation $(d/\lambda)\sin\theta \le (1/2)$. For θ in the interval $\pm(\pi/2)$ this requirement is satisfied if $(d/\lambda) \le (1/2)$. If the range of θ is reduced it is possible to increase the sensor spacing, for example, for $-(\pi/4) \le \theta + (\pi/4)$ the sensor spacing need satisfy the constraint $(d/\lambda) \le (1/\sqrt{2})$. The phase of the frequency response is a linear function of $\omega\tau$. This useful property of a ULA is lost when the sensors are nonuniformly spaced (see Section 2.1.5).

The array response is a function of the product of frequency ω and delay τ; or more explicitly, $\omega\,(d/c)\sin\theta$. The implication of this dependence is that two wavefronts whose waveform is a simple sinusoid but with different frequencies (ω_1, ω_2) arriving at different angles (θ_1, θ_2) will produce an identical array response if $\omega_1 \sin\theta_1 = \omega_2 \sin\theta_2$. Later, we shall discuss such ambiguous issues when we look into the broadband beamformation. The response function has a main lobe that is surrounded by many sidelobes of decreasing magnitude, just as we find in spectral windows. The first zero is at

$$\theta_{zero} = \sin^{-1}\frac{\lambda}{Md}, \tag{2.4}$$

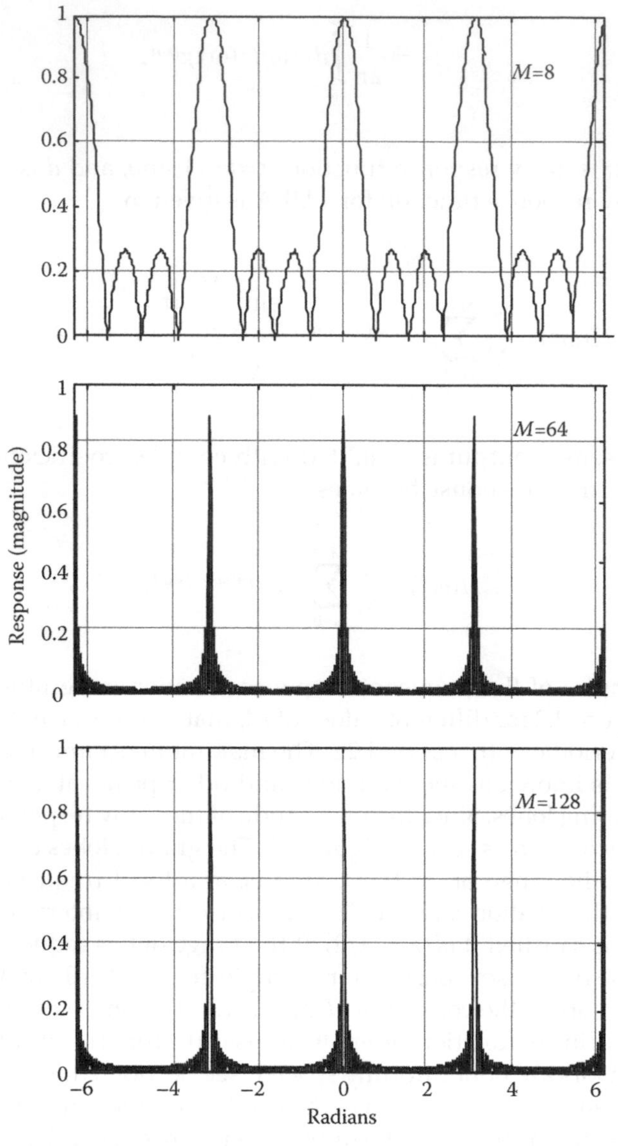

FIGURE 2.2
Array response function (magnitude) for different values of M. Notice that the main lobe becomes sharper as the array size is increased.

which for large M, becomes inversely proportional to the array length expressed in terms of wavelength. The first sidelobe is 13.5 dB below the main lobe. It is well known that both the width of the main lobe and the magnitude of the sidelobes can be controlled by using a suitable weight function as in spectrum analysis [1].

2.1.2 Array Steering

We have seen that the array response is maximum when the DOA is on the broadside ($\theta=0$). The maximum, however, can be changed to any direction through the simple act of introducing a time delay to each sensor output before summation. This is known as array steering. Let an incremental delay of τ per channel be introduced. The sum output of the array is now given by

$$
\begin{aligned}
g(t) &= \frac{1}{M}\sum_{m=0}^{M-1} f_m(t+m\tau) \\
&= \frac{1}{2\pi}\int_{-\infty}^{\infty} dF(\omega)e^{j\omega t}\frac{1}{M}\sum_{m=0}^{M-1} e^{j(\tau-(d/c)\sin\theta_0)\omega m} \\
&= \frac{1}{2\pi}\int_{-\infty}^{\infty} dF(\omega)H\left(\left(\tau-\frac{d}{c}\sin\theta_0\right)\omega\right)e^{j\omega t},
\end{aligned}
\tag{2.5}
$$

where we have assumed that the DOA is θ_0. Let $\tau=(d/c)\sin\theta$. Then the array response is maximum whenever $\theta=\theta_0$. We say that the array is steered in the direction θ_0, that is, in the DOA of the incident wavefront. The array response is now a function of DOA. This is demonstrated in Figure 2.3. It is interesting to note that the width of the main lobe increases with increasing DOA. To further understand this broadening effect we shall study the array response function around its maximum, that is, at $\tau=(d/c)\sin\theta_0$. The first zero will occur at

$$
\frac{M}{2}\left[\omega\frac{d}{c}\sin\theta_0-\omega\frac{d}{c}\sin(\theta_0-\Delta\theta)\right]=\pi. \tag{2.6a}
$$

Upon simplifying Equation 2.6a, we get

$$
\sin\theta_0-\sin(\theta_0-\Delta\theta)=\frac{\lambda}{Md}, \tag{2.6b}
$$

whose solution is given by

$$
\Delta\theta=\theta_0-\sin^{-1}\left[\sin\theta_0-\frac{\lambda}{Md}\right]. \tag{2.6c}
$$

The dependence of $\Delta\theta$ on the DOA for different array sizes is illustrated in Figure 2.4. The broadening of the main lobe is due to the reduction in the

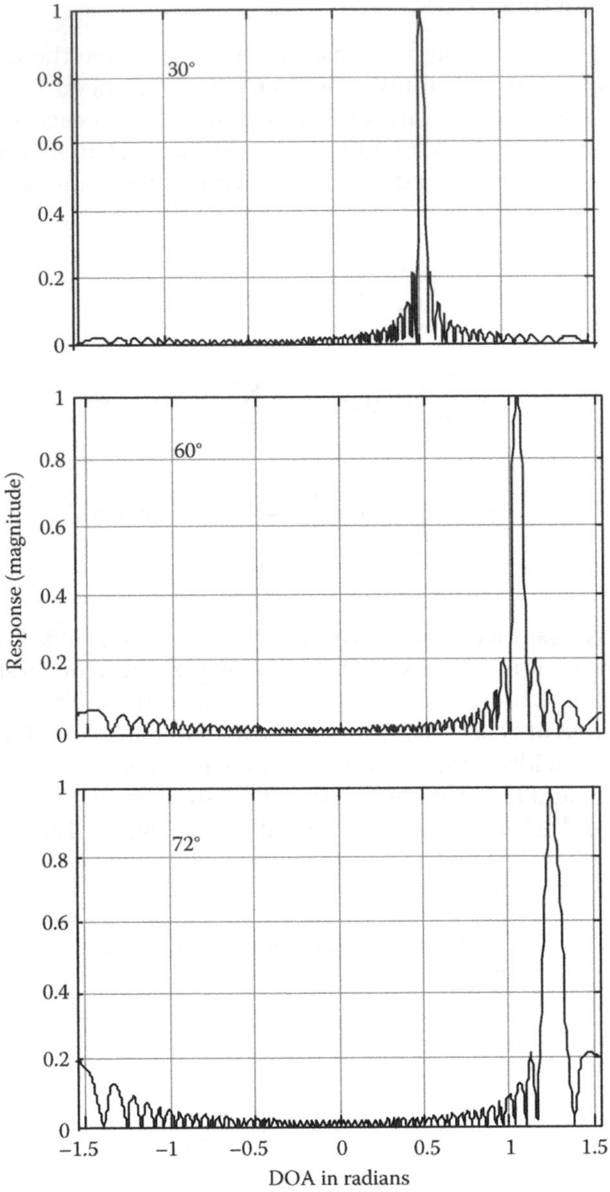

FIGURE 2.3
The effect of angle of arrival of a wavefront on the array response. The main lobe broadens and the side lobes become asymmetric.

array aperture for a wavefront that is incident away from the broadside. The response is maximum whenever

$$\omega \frac{d}{c}(\sin\theta_0 - \sin\theta) = 2\pi n$$

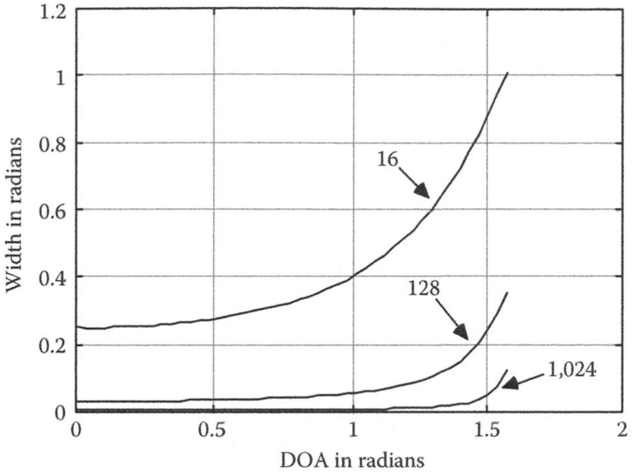

FIGURE 2.4
Width of the main lobe as a function of DOA for three different sizes of the sensor array ($M=16$, 128, 1,024). The sensor spacing is assumed to be $\lambda/2$.

or

$$\frac{d}{\lambda}(\sin\theta_0 - \sin\theta) = n.$$

For $(d/\lambda) \leq (1/2)$, the acceptable solution is $\theta=\theta_0$ for which $n=0$. For $(d/\lambda) > (1/2)$ there are more than one solution, one for each grating lobe. For example, let $(d/\lambda)=1$, a solution of $\theta=\sin^{-1}(\sin\theta_0 - n)$ exists only for $n=0$ and ± 1.

Now, let $\tau=(d/c)\,(k/M)$. The array response function can be written as a discrete Fourier transform (DFT) of a complex sinusoid, $\exp\left(-j\,(2\pi d/\lambda)\sin\theta_0 m\right)$,

$$H(k) = \frac{1}{M}\sum_{m=0}^{M-1} e^{-j(2\pi d/\lambda)\sin\theta_0 m} e^{j(2\pi km/M)(d/\lambda)}. \tag{2.7}$$

Now $H(k)$ is the kth inverse DFT coefficient, which should correspond to the array response at a steering angle, $\sin^{-1}(k/M)$. The array response is thus computed only at a set of discrete angles. Since M is finite, usually a few tens, the angular range of $\pm(\pi/2)$ is coarsely sampled. To overcome this limitation it may be necessary to pad zeros to the complex sinusoid before computing the DFT. However, it must be remembered that this step does not enhance the resolution, but only improves the sampling of the otherwise windowed Fourier transform (FT). Use of DFT for beamformation was first suggested in Ref. [2].

2.1.3 Broadband Source

Often a remote source such as broadband radar, engine noise, or earthquake, etc. emits a broadband stochastic waveform. The simplest approach to DOA estimation in such a situation is to compute the spectrum of the sum of the sensor outputs. From Equation 2.2 we have

$$g(t) = \frac{1}{M}\sum_{m=0}^{M-1} f_m(t) = \frac{1}{2\pi}\int_{-\infty}^{\infty} dF(\omega)H(\omega\tau)e^{j\omega t}, \tag{2.8}$$

which may be considered as a spectral representation of the sum of the sensor outputs. Hence, its spectrum is given by $S_f(\omega) = S_f(\omega)|H(\omega\tau)|^2$ [1], where $S_f(\omega)$ is the spectrum of the waveform emitted by the source. We can approximate $S_f(\omega)$ by the spectrum of the output of any one sensor. Thus, we obtain

$$|H(\omega\tau)|^2 = \frac{S_g(\omega)}{S_f(\omega)}. \tag{2.9}$$

Now consider a plot of $|H(\omega\tau)^2|$ as a function of ω. There is always one peak at $\omega=0$ and a stream of peaks caused by the incident wavefront [3], at positions given by

$$\omega_{peak} = 2\pi n \frac{c}{d\sin\theta}, \tag{2.10a}$$

where $n=0, \pm 1, \pm 2, \dots$. We introduce a quantity called minimum array sampling frequency $\omega_{\min}=2\pi(c/d)$. An array of sensors may be considered a waveform sampler that samples the waveform as it propagates across the array. The sampling interval is $(d/c)\sin\theta$ and the maximum interval or minimum sampling frequency occurs when $\theta=(\pi/2)$. In terms of the minimum array sampling frequency, the peak may be written as

$$\omega_{peak} = \frac{\omega_{\min} n}{\sin\theta_0}. \tag{2.10b}$$

Evidently, ω_{peak} must be in the range $\omega_{\min} n \le \omega_{\text{peak}} \le \infty$. For the sake of illustration, let the signal spectrum be of infinite width. Now, a plot of $|H(\omega\tau)|^3$ will show an infinite set of peaks spaced at an interval $(\omega_{\min}/\sin\theta)$. For example, for $\theta=45°$ an idealized plot of $|H(\omega\tau)|^2$ is shown in Figure 2.5a. A numerical example is shown in Figure 2.5b, where we have assumed a 16-sensor ULA with spacing $d=15$ m. A broadband signal with bandwidth=(±200 Hz) is incident at DOA angle equal to 45°. The average spacing of peaks is 42.0 Hz against the theoretical value of 42.43 Hz.

2.1.3.1 Angular Spectrum

Recall that the frequency wavenumber spectrum of a plane wave is a line passing through the origin with a slope inversely proportional to the direction cosines of the wave vector, in particular, in Chapter 1, Section 1.2 ..., we have

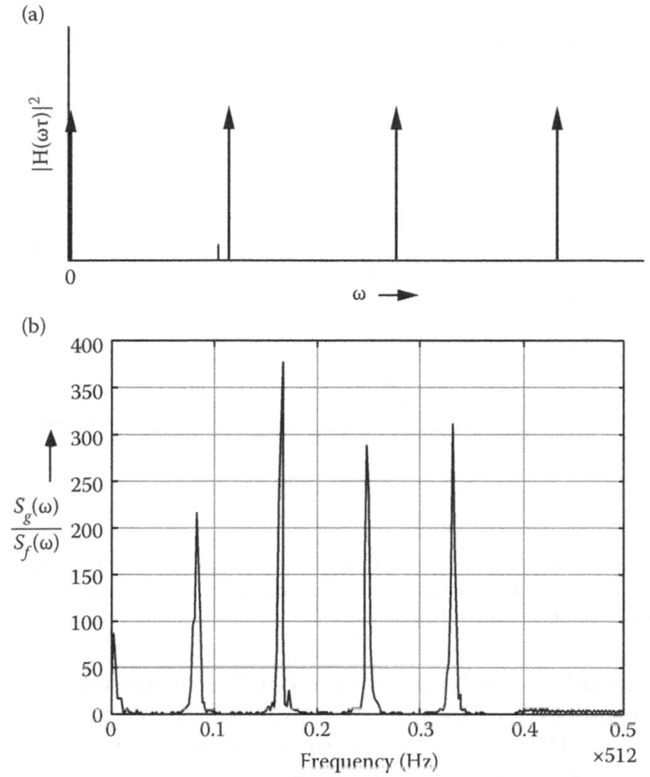

FIGURE 2.5
(a) A plot of $|H(\omega\tau)|^2$ for $\theta_0=41.81°$. The minimum array sampling frequency is ω_{min}. The position of the first peak (excluding the one at $\omega=0$) or the interval between the peaks can be used for estimation of DOA. (b) The ratio of the spectrum of the sum of all sensor outputs divided by the spectrum of the first sensor output. There are four peaks of $|H(\omega\tau)|^2$ within the available bandwidth. A 16 sensor ULA ($d=15$ m) is illuminated by a broadband signal (±200 Hz) incident at 45° DOA. Wave speed is 1,500 m/sec.

$$\omega = \frac{c}{\alpha}u, \quad \omega = \frac{c}{\beta}v, \tag{2.11}$$

where α and β are direction cosines. Conversely, the spectrum on a line joining the origin with a point in the frequency wavenumber space corresponds to the power of a plane wave propagating with the direction cosines, as in Equation 2.11. Consider the frequency wavenumber spectrum of the output of a ULA and integrate the power along a series of radial lines. The integrated power thus obtained is plotted as a function of DOA θ. Such a plot will reveal the presence of plane waves incident on the ULA. The background incoherent noise will tend to average out, giving constant or slowly varying power. As an example, we consider a ULA of 16 sensors along the x-axis and two incoherent broadband (80–120 Hz) sources at DOAs 40° and 50°. The angular spectrum obtained by averaging over different radial lines clearly resolves

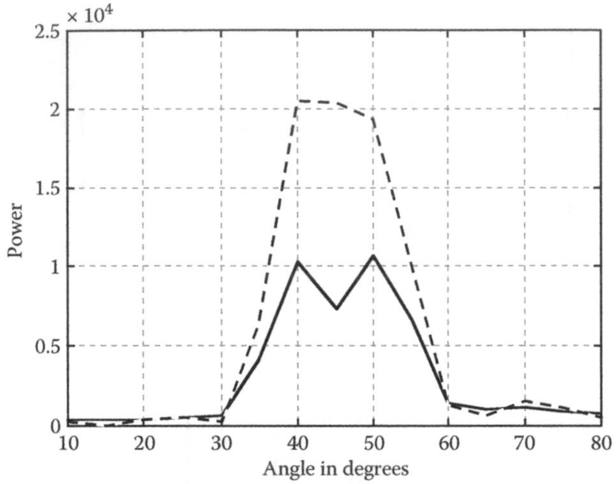

FIGURE 2.6
Angular spectrum obtained by averaging along the radial lines. Two broadband sources (solid line: 80–120 Hz) and two narrowband sources (dashed line: 98–102 Hz). Sixteen sensor array, 128 samples with sampling interval 0.005 sec. SNR=0 dB.

two peaks, but fails to resolve the narrowband signals (see Figure 2.6). Based on the above property of the frequency wavenumber spectrum, a method of estimating the DOA by projecting the spectrum along a radial line onto $\omega=\pi$ line has been proposed in Ref. [4]. In Chapter 1, Section 1.4, it was mentioned that the frequency wavenumber spectrum may be considered as a directional spectrum. Indeed, it is possible to estimate the DOAs by integrating over the temporal frequency band

$$\tilde{S}_f(u,v)=\int_{freq\ band} S_f(u,v,\omega)d\omega. \tag{2.12}$$

Such a method was in fact proposed in Ref. [3] where the integrated wavenumber spectrum $\tilde{S}_f(u,v)$ was called zero-delay wavenumber spectrum.

2.1.3.2 Slant Stacking

Closely related to the idea of array steering (or beamforming) is slant stacking, which is used extensively in seismic exploration. Stacking is also related to the radon transform [5]. Consider a wavefield $f(t, x)$, where we shall replace t by $\tau+p_x x$, where $p_x=(1/c_x)=(u/\omega)$. The stacking operation is defined as

$$\hat{f}(\tau,p_x)=\int_{-\infty}^{\infty} f(\tau+p_x x,x)dx. \tag{2.13}$$

Let $u=p_x\omega$ for a fixed ω

$$\hat{f}(\tau,p_x)=\frac{1}{4\pi^2}\iint_{-\infty}^{+\infty}\omega F(\omega,p\omega)e^{j\omega\tau}d\omega dp\int_{-\infty}^{\infty}e^{j(p_x\omega x-p\omega x)}dx$$

$$=\frac{1}{2\pi}\int_{-\infty}^{+\infty}F(\omega,p_x\omega)e^{j\omega\tau}d\omega. \tag{2.14a}$$

Taking the inverse FT, we obtain from Equation 2.14a

$$\hat{F}(\omega,p_x)=F(\omega,p_x\omega). \tag{2.14b}$$

Thus, 1D FT of the stacked output is equal to 2D FT of the wavefield. Beam steering assumes a plane wave model, but stacking does not require this assumption. As shown in Equation 2.14, the stacked output is directly proportional to the spatial FT of the wavefield, which is equal to the array response function (Equation 2.7). When the incident wavefield is nonplanar the right thing to do is plane wave decomposition, which is achieved through stacking (see Equation 1.33). Such a situation arises in seismic exploration where a low frequency (wavelengths on the order of a few hundred meters are common) source is used to energize rock strata. The wavefield observed on the surface is a function of two spatial coordinates and time. We shall first derive a result similar to that in Equation 2.14, but for a two-dimensional wavefield.

The stacked output of a planar array is defined as

$$\hat{f}(\tau,p_x,p_y)=\iint_{-\infty}^{+\infty}f(\tau+p_x x+p_y y,x,y)dxdy. \tag{2.15}$$

In Equation 2.15 we shall replace the integrand by its Fourier representation,

$$f(t,x,y)=\frac{1}{8\pi^3}\iint_{-\infty}^{+\infty}\omega^2F(\omega,p_x\omega,p_y\omega)e^{j(\omega t-p_x\omega x-p_y\omega y)}d\omega dp_x dp_y,$$

where we have used the relations $u=p_x\omega$ and $v=p_y\omega$ and obtain

$$\hat{f}(\tau,p_x,p_y)$$

$$=\iint_{-\infty}^{+\infty}dxdy\frac{1}{8\pi^3}\iint_{-\infty}^{+\infty}\omega^2F(\omega,p_x\omega,p_y\omega)e^{j(\omega\tau+(p_x'-p_x)\omega x+(p_y'-p_y)\omega y)}d\omega dp_x'dp_y'.$$

$$=\frac{1}{2\pi}\int_{-\infty}^{\infty}F(\omega,p_x'\omega,p_y'\omega)e^{j\omega\tau}d\omega$$

Hence,

$$\hat{F}(\omega, p_x, p_y) = F(\omega, p_x\omega, p_y\omega). \tag{2.16a}$$

When there is axial symmetry, as happens in horizontally stratified rocks, Equation 2.16a takes a different form [6]

$$\hat{F}(\omega, p) = \int_0^\infty r dr J_0(\omega p r) F(\omega, r). \tag{2.16b}$$

2.1.4 Matrix Formulation

When the incident signal is a narrowband signal, the output of an array, in particular a ULA, may be conveniently represented in a matrix format, which reveals some interesting properties. This is also true of a broadband signal, but the processing has to be in the frequency domain.

2.1.4.1 Representation of Narrowband Signals

A narrowband signal $f_{nb}(t)$ may be represented as

$$f_{nb}(t) = s_0(t)\cos(\omega_c t + \varphi_0(t)), \tag{2.17a}$$

where $s_0(t)$ is a slowly varying waveform, often called envelop and $\varphi_0(t)$ is also a slowly varying phase; $\cos(\omega_c t)$ is a rapidly varying sinusoid, often known as a carrier, and ω_c is known as carrier frequency. Many active array systems radiate narrowband signals, for example, a phased array radar. Equation 2.17a may be expressed as

$$f_{nb}(t) = f_i(t)\cos(\omega_c t) - f_q(t)\sin(\omega_c t)$$

where

$$f_i(t) = s_0(t)\cos(\varphi_0(t)),$$

$$f_q(t) = s_0(t)\sin(\varphi_0(t)).$$

$f_i(t)$ is known as an inphase component and $f_q(t)$ is a quadrature component. The inphase and quadrature components are uncorrelated. They have the same spectral density function. The inphase and quadrature can be uniquely recovered from a narrowband signal by a process known as mixing, which involves multiplication with $2\cos(\omega_c t)$ and $-2\sin(\omega_c t)$ and low pass filtering [7]. A complex analytical signal is defined as $f_c(t) = f_i(t) + jf_q(t)$. Consider a narrowband signal delayed by one quarter period. Assuming that both inphase and quadrature components are slowly varying signals, we get the following approximate result:

$$
\begin{aligned}
f_{nb}\left(\left(t-\frac{\tau_0}{4}\right)\right) &= f_i(t)\cos\left(\omega_c\left(t-\frac{\tau_0}{4}\right)\right) - f_q(t)\sin\left(\omega_c\left(t-\frac{\tau_0}{4}\right)\right) \\
&\approx f_i(t)\sin(\omega_c t) + f_q(t)\cos(\omega_c t) \\
&= f_{nb}^{Hilb}(t).
\end{aligned}
$$

We define a complex analytical signal as

$$
\begin{aligned}
f_{nb}(t) + jf_{nb}\left(t-\frac{\tau_0}{4}\right) &= f_i(t)e^{j\omega_c t} + jf_q(t)e^{j\omega_c t} \\
&= f_c(t)e^{j\omega_c t}.
\end{aligned} \tag{2.17b}
$$

The representation given by Equation 2.17b is useful in narrowband beamformation. The process described in Equation 2.17b is often referred to as quadrature filtering, which is illustrated in Figure 2.7. Note that the input to quadrature filter is real, but the output is complex.

Consider the mth sensor of a ULA. The complex output of the quadrature filter is

$$
\begin{aligned}
f_m(t) &= f_{nb}\left(t-m\frac{d}{c_x}\right) + jf_{nb}\left(t-m\frac{d}{c_x}-\frac{\tau_0}{4}\right) \\
&= f_c(t)e^{j\omega_c t - jm\omega_c(d/c_x)}.
\end{aligned} \tag{2.17c}
$$

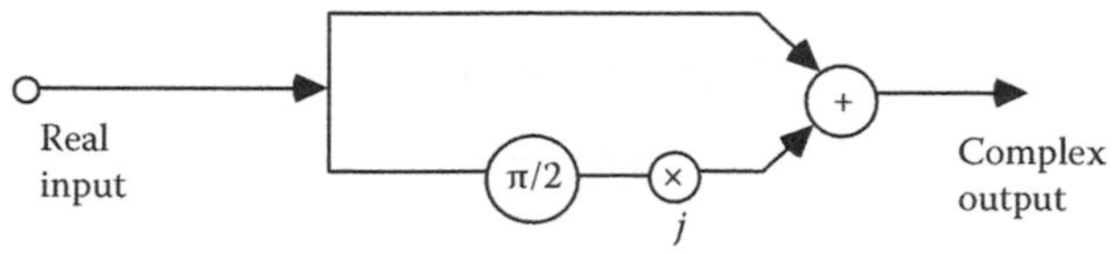

FIGURE 2.7
Quadrature filter structure. Since the phase change due to propagation appears in the complex sinusoid at the output, it is easy to introduce phase adjustments for beamformation.

The term representing propagation delay now appears in the complex sinusoid. Naturally, in order to form a beam, it is convenient to introduce phase adjustments. Let $w_0, w_1, \ldots, w_{M-1}$ be a set of complex weight coefficients for beamformation. The beam output will be given by

$$\text{output} = f_c(t)\sum_{m=0}^{M-1} w_m e^{j\omega_c t - jm\omega_c(d/c_x)}.$$

Through complex weight coefficients it is possible to adjust both amplitude and phase so that the resulting response is closest to any desired response.

2.1.4.2 Matrix

A snapshot is a vector representing the outputs of all sensors taken at the same time instant t. Let $\mathbf{f}(t)=col\{f_0(t), f_1(t), \ldots, f_{M-1}(t)\}$ be a snapshot, where $f_0(t), f_1(t), \ldots, f_{M-1}(t)$ stand for the sensor outputs at time t. When the incident signal is narrowband, the signal varies slowly with time (assume that the carrier has been removed). In a noise-free case, a single time shot is adequate as it contains all available information. A snapshot vector for the narrowband signal may be expressed using Equation 2.17c as

$$\begin{aligned}\mathbf{f}(t) &= f_c(t)col\{1, e^{-j\omega_c(d/c_x)}, \ldots, e^{-j(M-1)\omega_c(d/c_x)}\} \\ &= f_c(t)\phi(\theta_0),\end{aligned} \tag{2.17d}$$

where, it may be recalled, the apparent speed $c_x=(c/\sin\theta_0)$. Further, let the sensor response matrix be $\alpha(\theta_0)=\text{diag}\{\alpha_0(\theta_0), \alpha_1(\theta_0), \ldots, \alpha_{M-1}(\theta_0)\}$, in which each element represents the response of a sensor as a function of the angle of incidence of the wavefront. The propagation effect of the medium on a wavefront propagating across the array is represented by $\phi(\theta_0)$. Together $\phi(\theta_0)$ and $\alpha(\theta_0)$ form a direction vector $\mathbf{a}(\theta_0)=\alpha(\theta_0)\phi(\theta_0)$, representing the response of an array to a wavefront incident at angle θ_0 (DOA). Finally, the array output may be expressed as follows:

$$\begin{aligned}\mathbf{f}(t) &= f_c(t)\alpha(\theta_0)\phi(\theta_0) \\ &= f_c(t)\mathbf{a}(\theta_0).\end{aligned} \tag{2.17e}$$

When there are P narrowband sources radiating simultaneously, the array output may be expressed as a linear combination of P terms of the type shown in Equation 2.17e

$$\mathbf{f}(t) = \begin{bmatrix} \alpha(\theta_0)\phi(\theta_0), \alpha(\theta_1)\phi(\theta_1), \ldots, \alpha(\theta_{P-1})\phi(\theta_{P-1}) \\ (M \times P) \end{bmatrix} \begin{bmatrix} f_{c_0}(t) \\ f_{c_1}(t) \\ \cdot \\ \cdot \\ \cdot \\ f_{c_{P-1}}(t) \\ (P \times 1) \end{bmatrix} + \eta(t), \quad (2.18a)$$

where $\eta(t)$ is the noise vector assumed to be uncorrelated with the signal terms. Equation 2.18a may be written in a more compact form where P columns

$$\mathbf{f}(t) = \mathbf{A}\mathbf{s} + \eta(t), \quad (2.18b)$$

of $\mathbf{A}$ matrix are P-direction vectors pointing to P sources. The matrix representation of the array output model, as in Equation 2.18b, plays a very crucial role in the development of high-resolution methods for DOA estimation.

The array steering can also be represented in terms of a matrix operation. To steer an array to a desired direction, θ, we form an inner product of the steering vector and the array snapshot

$$\mathbf{a}^H(\theta)\mathbf{f}(t) = f_c(t)\mathbf{a}^H(\theta)\mathbf{a}(\theta_0). \quad (2.19a)$$

The output power is given by

$$\begin{aligned} \mathbf{a}^H(\theta)\mathbf{C}_f\mathbf{a}(\theta) &= \sigma_{s_0}^2 E\left\{\left|\mathbf{a}^H(\theta)\mathbf{a}(\theta_0)\right|^2\right\} \\ &= \sigma_{s_0}^2 M^2 \left|H\left(\omega\frac{d}{c}(\sin\theta - \sin\theta_0)\right)\right|^2, \end{aligned} \quad (2.19b)$$

where $\mathbf{C}_f = E\{\mathbf{f}(t)\mathbf{f}^H(t)\}$ is the spatial covariance matrix (SCM). Whenever $\theta = \theta_0$, that is, when the steering angle is equal to the DOA, the left-hand side of Equation 2.19b equals $\sigma_{s_0}^2 M^2$, giving the power of the source.

The M dimensional steering vector will span an M-dimensional space known as the array manifold. The tip of the steering vector traces a closed curve in the array manifold or a closed surface when the steering vector is a function of two variables, for example, azimuth and elevation. Consider the case of identical sensors, that is,

$$\alpha_0(\theta_0) = \alpha_1(\theta_0) = \cdots = \alpha_{M-1}(\theta_0) = \alpha(\theta_0).$$

In this case, the direction vector is given by

$$\mathbf{a}(\theta_0) = \alpha(\theta_0)\text{col}\{1, e^{-j\omega(d/c)\sin\theta_0}, e^{-j\omega(2d/c)\sin\theta_0}, \ldots, e^{-j\omega((M-1)d/c)\sin\theta_0}\}.$$

In the event of sensors being omnidirectional, that is, $\alpha(\theta_0)$=constant, the array manifold becomes a closed curve on a sphere (in M-dimensional space). For uniqueness the array manifold must not intersect, otherwise, at the point of intersection the steering vector will point to two different directions, θ_1 and θ_2 such that $\mathbf{a}(\theta_1)=\mathbf{a}(\theta_2)$. Such a possibility exists only when $(d/\lambda) > 0.5$. To show this, consider the steering vector for omnidirectional sensors. Let θ_1 and θ_2 be two such directions for which $\mathbf{a}(\theta_1)=\mathbf{a}(\theta_2)$, that is, for all m

$$e^{j2\pi(d/\lambda)m\sin\theta_1} = e^{j2\pi(d/\lambda)m\sin\theta_2}.$$

This is possible when

$$\frac{d}{\lambda}[\sin\theta_1 - \sin\theta_2] = 1,$$

or

$$\sin\theta_1 = \frac{\lambda}{d} + \sin\theta_2, \tag{2.20}$$

A solution of Equation 2.20 exists only when $(\lambda/d) < 2$; for example, when $(\lambda/d)=1.2$ the following pairs of directions are the possible solutions: (36.87°, –36.87°), (23.58°, –53.13°), and (11.54°, –90°).

The steering vector satisfies the following properties:

(a) $\mathbf{a}(\theta)=\mathbf{a}(\pi-\theta)$.
(b) $\mathbf{a}^*(\theta)=\mathbf{a}(-\theta)$.
(c) $\mathbf{a}(\theta)$ is periodic with a period $\pm(\pi/2)$ only if $d=(\lambda/2)$.

Property (a) implies a wavefront coming from the north and another symmetrically opposite direction from the south (a and b in Figure 2.8) cannot be distinguished. This is known as north–south ambiguity. Property (b) implies a wavefront coming from the east, and another symmetrically opposite direction from the west (a and c in Figure 2.8) can be distinguished only if the signal is complex. This is known as east–west ambiguity. To show this, recall Equation 2.17e and compare the outputs of a ULA for a real input signal incident at angle θ and $-\theta$. Let $\mathbf{f}_\theta(t)$ be the output of a ULA for an incident angle θ, and $\mathbf{f}_{-\theta}(t)$ be the output for an incident angle, $-\theta$. For a real signal $\mathbf{f}_\theta(t)=\mathbf{f}^*_{-\theta}(t)$, but for a complex signal $\mathbf{f}_\theta(t)\neq\mathbf{f}^*_{-\theta}(t)$. Property (c) implies that there is no grating lobe in the range $\pm(\pi/2)$ when the sensor spacing is $d\leq(\lambda/2)$.

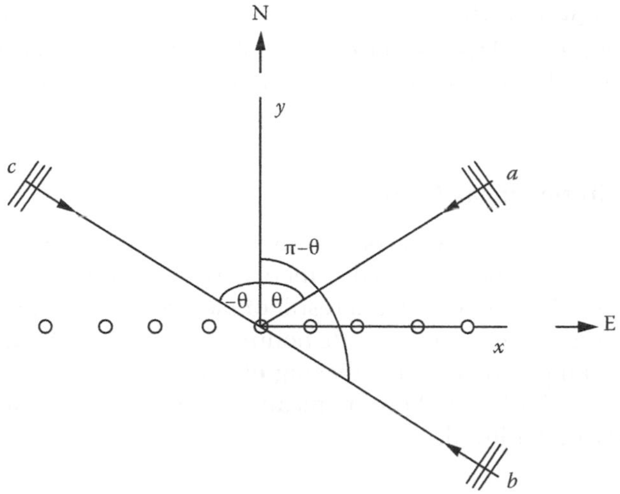

FIGURE 2.8
A ULA cannot distinguish wavefronts *a* and *b* (north-south ambiguity). However, it can distinguish wavefronts *a* and *c* if the signal is complex (east-west ambiguity).

The steering vector is closely related to the array response function. To show this we define a unit vector, $\mathbf{1}=col\{1, 1, 1,\ldots, 1\}$ and consider a dot product,

$$\mathbf{a}(\theta)^H\mathbf{1} = \sum_{m=0}^{M-1} e^{j\omega m(d/c)\sin(\theta)}, \tag{2.21a}$$

which follows from Equation 2.3. For the sake of simplicity, we have assumed that all sensors are identical and omnidirectional. In real array, the individual sensor response is likely to be directional and varying. Then, the array response is given by

$$\mathbf{a}(\theta)^H\mathbf{1} = \sum_{m=0}^{M-1} \alpha_m(\theta)e^{j\omega m(d/c)\sin(\theta)}. \tag{2.21b}$$

An array is steered to a desired direction by introducing delays to each sensor output. The response of such a steered array is given by (from Equation 2.5).

$$\begin{aligned} H\left(\tau-\omega\frac{d}{c}\sin\theta_0\right) &= \frac{1}{M}\sum_{m=0}^{M-1} e^{j(\tau-(d/c)\sin\theta_0)\omega m} \\ &= \frac{1}{M}\mathbf{a}^H(\theta)\mathbf{a}(\theta_0), \end{aligned} \tag{2.22}$$

where θ is the desired direction to which the array is steered and θ_0 is the DOA of a wavefront. The response of the steered array is expressed as an inner product of the steering vector and direction vector, as shown in Equation 2.22.

2.1.5 Nonuniform Linear Arrays

There are reasons for having to consider non-ULAs. These are: (a) redundant sensors are removed and employed to increase the array aperture; (b) certain sensors in a long ULA may fail as a result of factors beyond our control; and (c) array spacing is intentionally made nonuniform in order to derive certain benefits, for example, there is no aliasing effect if the periodicity of a ULA is destroyed. We shall show how, by means of nonuniform array, the above objectives may be achieved.

2.1.5.1 Redundant Sensors

Let us first consider the case of redundant sensors. We shall rewrite Equation 2.22 as

$$H\left(\left(\tau-\frac{d}{c}\sin\varphi_0\right)\omega\right)=\frac{1}{M}\mathbf{b}^H\phi(r), \tag{2.23}$$

where $\mathbf{b}=col\{\alpha_0(\theta_0)\alpha_0^*(\theta),\alpha_1(\theta_0)\alpha_1^*(\theta),\ldots\alpha_{P-1}(\theta_0)\alpha_{P-1}^*(\theta)\}$ and $r=(\sin\theta-\sin\theta_0)\omega(d/c)$. The power output of the array is simply proportional to the square of the transfer function.

$$power\propto|H|^2=\frac{1}{M^2}\mathbf{b}^H\phi(r)\phi^H(r)\mathbf{b}. \tag{2.24}$$

Let us expand matrix $\phi(r)\phi^{H}(r)$

$$\mathbf{H}=\phi(r)\phi^H(r)=\begin{bmatrix}1 & e^{jr} & e^{j2r} & e^{j3r} & \ldots & e^{j(p-1)r}\\ e^{-jr} & 1 & e^{jr} & e^{j2r} & \ldots & e^{j(p-2)r}\\ e^{-j2r} & e^{-jr} & 1 & e^{jr} & \ldots & e^{j(p-3)r}\\ & & & & \ldots & \\ e^{-j(p-1)r} & e^{-j(p-2)r} & & & \ldots & 1\end{bmatrix}. \tag{2.25}$$

It may be observed that $\mathbf{H}$ is a Toeplitz matrix, that is, along any diagonal the entries are repeated even though they refer to different sensors in the array. For example, consider the second upper diagonal where the entries refer to a pair of sensors whose indices are m and m–2, where m=2, 3,…, M–2; explicitly, the pairs of sensors involved in creating the terms on this

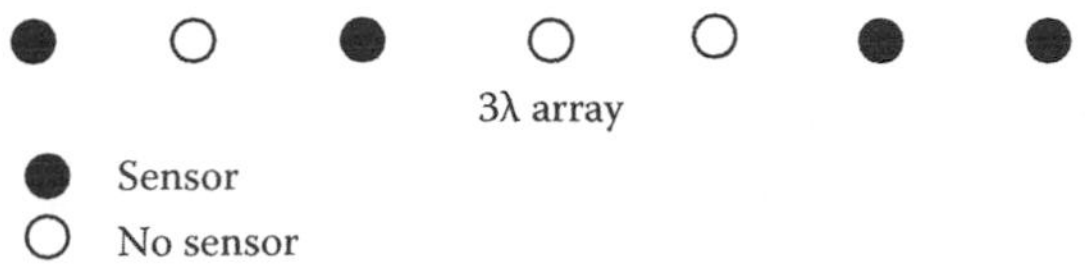

FIGURE 2.9
A four-sensor array spread over 3λ aperture will produce all entries of the matrix in Equation 2.25.

diagonal are (2, 0), (3, 1), (4, 2), etc. Thus, insofar as the second diagonal is concerned, there are several redundant pairs of sensors. This redundancy can be removed by selectively removing sensors [8]. For example, consider a seven sensor uniform array shown in Figure 2.9. All entries in the **H** matrix (Equation 2.25) can be obtained from just four sensors shown by filled circles. For example, the first diagonal may be obtained from sensors at position 5 and 6, the second diagonal from sensors at position 0 and 2, the third diagonal from sensors 2 and 5, the fourth diagonal from sensors 2 and 6, the fifth diagonal from sensors 0 and 5, and the last diagonal from sensors 0 and 6. Thus, we are able to generate a complete 7×7 matrix just from a four-sensor nonuniformly spaced array by removing all redundant sensors. In general, it is possible to arrive at a distribution of the minimum number of sensors required to fill all entries in the matrix in Equation 2.25 [9]. For a given array length, the minimum number of sensors required to generate the **H** matrix is found through exhaustive search. As the array size increases, there exists more than one distribution of a zero redundancy array, that is, an array having sensors just enough to generate all diagonals of the **H** matrix. For example, for an array length of 15 (16 sensors spaced at $(\lambda/2)$), the minimum redundant array has seven sensors and there are 77 different distributions. For an array length of 31 (32 sensors spaced at $(\lambda/2)$), the minimum redundant array has ten sensors and there are 888 different distributions. A few sample distributions are shown in Table 2.1.

The theoretical response of a minimum redundancy array is identical with that of a normal array (redundant array), but in the presence of background noise the response of a redundant array is considerably inferior. To demonstrate this, we have considered an array of length 15 but having only seven sensors in the redundant array. Broadband array transfer was computed with background noise variance equal to one. First, the **H** matrix was computed. To simulate a ULA, a complex noise of unit variance was added to each element of the matrix. But to simulate a minimum redundancy array, a complex noise of unit variance was added to each diagonal (same noise element to all elements in a given diagonal). The transfer functions are shown in Figure 2.10. The transfer functions show the main lobe and the first grating lobe at correct frequency in both cases; however, the noisy sidelobes are about ten times greater in the case of the minimum redundancy array. The upper and lower bounds on the signal to noise ratio (SNR) gain were theoretically derived in Ref. [10].

TABLE 2.1

Sensor locations for minimum redundancy

Array Length in Units of (λ/2)	Sensor Locations for Minimum Redundancy
6	0, 2, 5, 6
15	0, 1, 2, 3, 7, 11, 15
	0, 1, 3, 6, 10, 14, 15
	0, 1, 4, 8, 13, 14, 15
	0, 2, 4, 5, 8, 14, 15
31	0, 1, 2, 3, 4, 5, 12, 18, 25, 31
	0, 1, 2, 3, 6, 11, 18, 25, 27, 31
	0, 1, 2, 3, 8, 14, 18, 22, 27, 31
	0, 1, 2, 3, 12, 18, 22, 26, 29, 31

Note: For an array length of six there is only one arrangement, but for array lengths 15 and 31 there are 77 and 888 arrangements, respectively.

2.1.5.2 Missing Sensors

In a long chain of sensors, such as in a towed array, there may be a few sensors that have either failed or are malfunctioning. Such sensors are often skipped, leading to loss of fidelity in the array steering response. Here, we shall analyze the effect of a missing sensor on the array performance. Let p be the probability that a sensor will be malfunctioning and it is independent of all other sensors. Let x_l, $l=0,1,2,...N-1$ be the locations of live sensors. Note that x_1 is a multiple of sensor spacing. Let $\Delta x_l = x_{l+1} - x_l$ be the spacing between the $l+1$st and lth sensors. It may be expressed as $k \cdot d$, where d is a nominal sensor spacing and $k-1$ is a Bernoulli random variable so that

$$prob\{\Delta x_l = k \cdot d\} = (1-p)p^{k-1}. \tag{2.26}$$

The output of the nth sensor may be expressed as

$$\begin{aligned} f_n(t) &= \frac{1}{2\pi}\int_{-\infty}^{+\infty} dF(\omega)e^{j(\omega t-(\omega/c)\sin\theta_0 \Sigma_{l=0}^{n}\Delta x_l)} \\ &= \frac{1}{2\pi}\int_{-\infty}^{+\infty} dF(\omega)e^{j(\omega t)}\prod_{l=0}^{n} e^{-j(\omega/c)\sin\theta_0\Delta x_l}, \end{aligned} \tag{2.27}$$

where θ_0 is the DOA of a wavefront. We further assume that $x_0=0$. We consider a delay and sum type of beam steering. Each sensor output is delayed

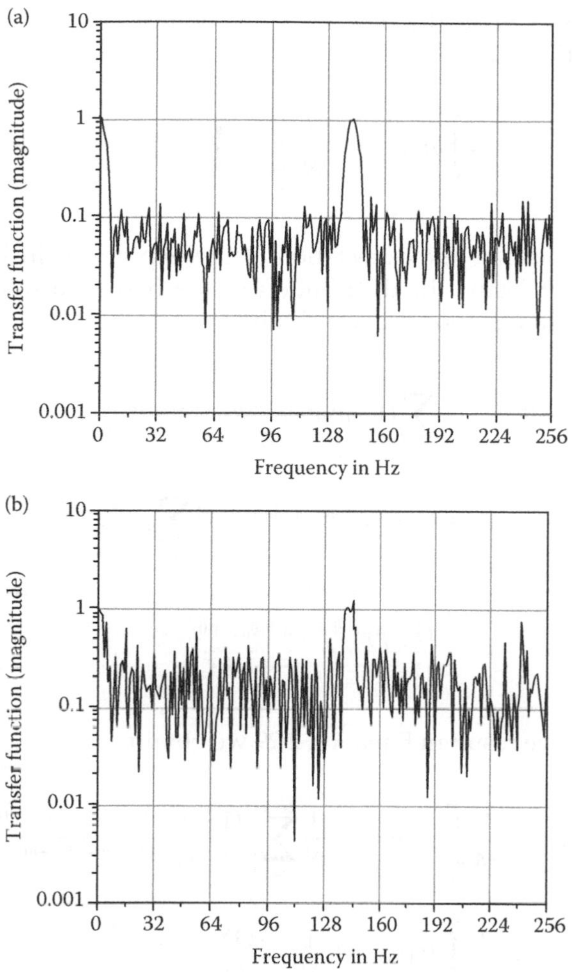

FIGURE 2.10
(a) Transfer function of a 16-sensor ULA and (b) transfer function of a seven-sensor minimum redundancy array of same length as ULA in (a). Sensor spacing = 15 m, wave speed = 1,500 m/sec and $\theta = 45°$.

by an amount equal to $\tau_n = \frac{\omega}{c} x_n \sin\theta$, where θ is the angle to which the array is steered

$$\begin{aligned} g(t) &= \frac{1}{N}\sum_{n=0}^{N-1} f_n(t-\tau_n) \\ &= \frac{1}{2\pi}\int_{-\infty}^{+\infty} dF(\omega) e^{j(\omega t)} \frac{1}{N}\sum_{n=0}^{N-1}\prod_{l=0}^{n} e^{-j(\omega/c)(\sin\theta_0 - \sin\theta)\Delta x_l}. \end{aligned} \tag{2.28}$$

We shall now evaluate the expected value of the delay and sum processor given by Equation 2.28

$$E\{g(t)\} = \frac{1}{2\pi}\int_{-\infty}^{+\infty} dF(\omega)e^{j(\omega t)}\frac{1}{N}\sum_{n=0}^{N-1}\prod_{l=0}^{n} E\{e^{-j(\omega/c)(\sin\theta_0-\sin\theta)\Delta x_l}\}. \tag{2.29}$$

We need to evaluate the expectation of the expression inside the braces in Equation 2.29. Using the distribution of the Bernoulli random variable (Equation 2.26) in Equation 2.29 we obtain

$$\begin{aligned} E\{e^{-j(\omega/c)(\sin\theta_0-\sin\theta)\Delta x_l}\} &= \sum_{k=1}^{\infty}(1-p)p^{k-1}e^{-j(\omega/c)(\sin\theta_0-\sin\theta)kd} \\ &= (1-p)e^{-j(\omega/c)d(\sin\theta_0-\sin\theta)}\sum_{k=1}^{\infty}p^{k-1}e^{-j(\omega/c)(\sin\theta_0-\sin\theta)(k-1)d} \\ &= \frac{(1-p)e^{-j(w/c)d(\sin\theta_0-\sin\theta)}}{1-pe^{-j(w/c)d(\sin\theta_0-\sin\theta)}}. \end{aligned} \tag{2.30}$$

Using the above result in Equation 2.29 we obtain

$$\begin{aligned} E\{g(t)\} &= \frac{1}{2\pi}\int_{-\infty}^{+\infty} dF(\omega)e^{j(\omega t)}\frac{1}{N}\sum_{n=0}^{N-1}\left[\frac{(1-p)e^{-j(\omega/c)d(\sin\theta_0-\sin\theta)}}{1-pe^{-j(\omega/c)d(\sin\theta_0-\sin\theta)}}\right]^n \\ &= \frac{1}{2\pi}\int_{-\infty}^{+\infty} dF(\omega)e^{j(\omega t)}\frac{1}{N}\frac{1-Q^N}{1-Q}, \end{aligned} \tag{2.31}$$

where Q stands for the quantity inside the brackets of Equation 2.31. Note that for $p=0$, $Q=e^{-j(\omega/c)d(\sin\theta_0-\sin\theta)}$ and Equation 2.31 reduces to a known expression for the response function of a ULA. The response of an array with missing sensors, given by

$$H(\omega\tau) = \frac{1}{N}\frac{1-Q^N}{1-Q},$$

where $\tau=(d/c)(\sin\theta_0-\sin\theta)$, is shown in Figure 2.11 for two different values of probability of malfunctioning. It is assumed that the total number of live sensors is the same in both cases, namely, 16. Notice that the magnitude response has fewer sidelobes, but the phase characteristics appear to be grossly different from those of a ULA, which is a linear function of $\omega\tau$

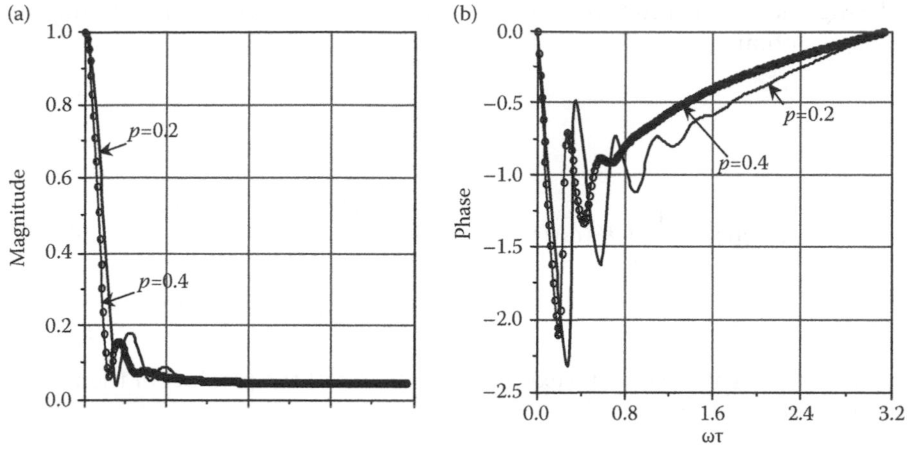

FIGURE 2.11
(a) Magnitude response and (b) phase response of an array with malfunctioning sensors. The total number of live sensors is assumed to be the same in both cases, namely, 16.

(see Equation 2.3). A nonlinear phase response results in a distortion of the received waveform particularly when it is broadband.

2.1.5.3 Random Array

A third type of nonuniform array is one where the sensors are spaced at random intervals, in particular, an exponential distribution for which a closed form solution can be derived. Let $x_n, n=0, 1,\ldots, N-1$ be the locations of the sensors; all of which, however, lie on a straight line. Let us assume an exponential distribution for the sensor spacing,

$$
\begin{aligned}
pdf(\Delta x_n) &= \beta e^{-\beta \Delta x} \quad \Delta x \geq 0 \\
&= 0 \qquad\quad \Delta x < 0,
\end{aligned}
\tag{2.32}
$$

where pdf stands for probability density function and β is a parameter in the exponential distribution. The output of the nth sensor may be written as in Equation 2.27. The delay and sum type of processing would result in the following array output (from Equation 2.29)

$$
E\{g(t)\}
$$

$$
= \frac{1}{2\pi}\int_{-\infty}^{+\infty} dF(\omega) e^{j(\omega t)} \frac{1}{N}\sum_{n=0}^{N-1}\prod_{l=0}^{n}\int_{0}^{\infty} pdf(\Delta x_l) e^{-j(\omega/c)(\sin\theta_0 - \sin\theta)\Delta x_l} d\Delta x_l. \tag{2.33a}
$$

Using the exponential distribution function (Equation 2.32) in Equation 2.33a, we obtain

$$E\{g(t)\}=\frac{1}{2\pi}\int_{-\infty}^{+\infty}dF(\omega)e^{j(\omega t)}\frac{1}{N}\frac{1-\left[\dfrac{\beta}{j\dfrac{\omega}{c}(\sin\varphi_0-\sin\varphi)+\beta}\right]^N}{1-\dfrac{\beta}{j\dfrac{\omega}{c}(\sin\varphi_0-\sin\varphi)+\beta}}. \tag{2.33b}$$

The array response may be expressed in terms of the product of the wavelength and parameter β,

$$H(\nu,\beta\lambda)=\frac{1}{N}\frac{1-\left[\dfrac{\beta\lambda}{j2\pi(\sin\theta_0-\sin\theta)+\beta\lambda}\right]^N}{1-\dfrac{\beta\lambda}{j2\pi(\sin\theta_0-\sin\theta)+\beta\lambda}}, \tag{2.33c}$$

where $v=2\pi(\sin\theta_0-\sin\theta)$. We have plotted the array response function for different values of $\beta\lambda$ in Figure 2.12. While the magnitude response is free from sidelobes, the phase response is highly nonlinear in the range $n=0.0$ to 1.6, where the magnitude response is significant. This behavior was also noticed in the case of an array with missing sensors.

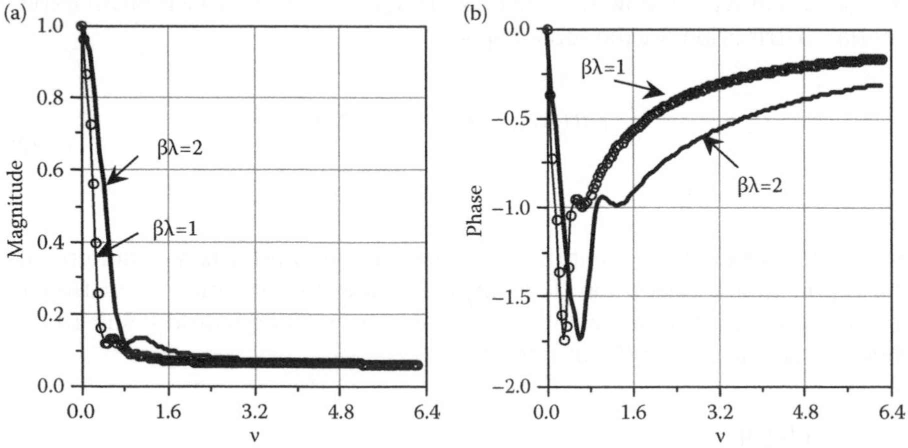

FIGURE 2.12
Frequency response of a random linear array. (a) Magnitude response and (b) phase response. Sixteen sensors spaced at random intervals having an exponential distribution.

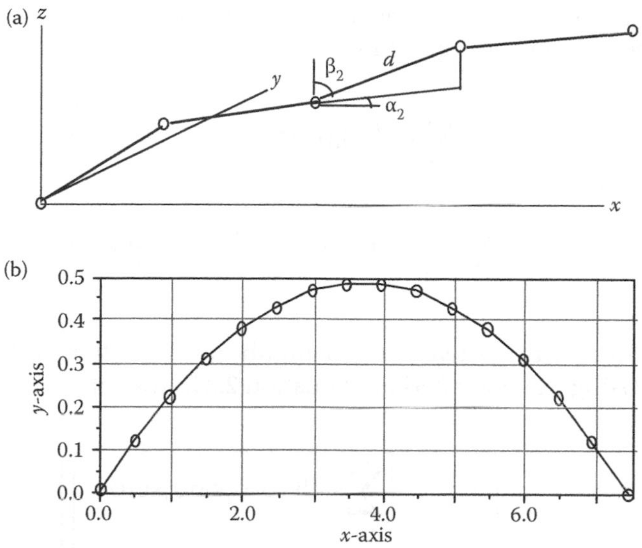

FIGURE 2.13
(a) Model of a flexible array where the adjacent sensors are held at fixed distance, but the azimuth and elevation of the line segment joining the sensors are random variables. (b) A bow-shaped 16-sensor array in x-y plane. The distance between the adjacent sensors is 0.5λ. All dimensions are in units of wavelength.

2.1.6 Flexible Array

We have so far considered a sensor array that is rigidly fixed to the ground or to a platform. We now consider an array where the sensors are held in position by means of a flexible rope that allows a sensor to move over a circular arc of fixed radius. The sensor spacing, however, remains unchanged. An important consequence of this freedom allowed to the array is to alter the shape of the array when it is being towed or it is under the influence of ocean currents. A commonly used array model is that M sensors are separated by straight-line segments of fixed length d [11] (see Figure 2.13a). Let (x_m, y_m, z_m) represent coordinates of the mth sensor with respect to $m-1$st sensor.

Since the distance between the sensors remains fixed, (x_m, y_m, z_m) must satisfy a relation $\sqrt{x_m^2 + y_m^2 + z_m^2} = d$, alternatively,

$$
\begin{aligned}
x_m &= d \sin\beta_m \sin\alpha_m \\
y_m &= d \sin\beta_m \cos\alpha_m \\
z_m &= d \cos\beta_m,
\end{aligned}
\tag{2.34a}
$$

where (α_m, β_m) are the azimuth and elevation of line segment joining the mth sensor and $m+1$st sensor (see Figure 2.13a). It is assumed that $x_0 = y_0 = z_0 = 0$.

Let a plane wavefront be incident on the array shown in Figure 2.13. The output of the mth sensor is given by

$$f_m(t) = s_0(t)\cos\left(\begin{array}{l}\omega_c t - \sum_{i=0}^{m} x_i \frac{\omega_c}{c}\sin\theta_0 \sin\varphi_0 \\ -\sum_{i=0}^{m} y_i \frac{\omega_c}{c}\sin\theta_0 \cos\varphi_0 - \sum_{i=0}^{m} z_i \frac{\omega_c}{c}\sin\theta_0\end{array}\right) + \eta_m(t), \quad (2.34b)$$

where (φ_0, θ_0) are, respectively, the azimuth and elevation of the incident wavefront. Using Equation 2.34a in Equation 2.34b we get

$$f_m(t) = s_0(t)\cos\left(\omega_c t - \omega_c \frac{d}{c}\left[\begin{array}{l}\sum_{i=0}^{m}\sin\beta_i \cos\alpha_i \sin\theta_0 \sin\varphi_0 \\ +\sum_{i=0}^{m}\sin\beta_i \sin\alpha_i \sin\theta_0 \cos\varphi_0 \\ +\sum_{i=0}^{m}\cos\beta_i \sin\theta_0\end{array}\right]\right) + \eta_m(t).$$

Transforming into a complex analytical signal (Equation 2.17d), the array output may be expressed in a matrix form

$$\mathbf{f}(t) = f_c(t) col\left\{\begin{array}{l}1, e^{-j\omega_c (d/c)[\gamma_1 \sin\theta_0 \sin\varphi_0 + \varepsilon_1 \sin\theta_0 \cos\varphi_0 + \xi_1 \sin\theta_0]}, \ldots, \\ e^{-j\omega_c (d/c)[\gamma_{M-1} \sin\theta_0 \sin\varphi_0 + \varepsilon_{M-1} \sin\theta_0 \cos\varphi_0 + \xi_{M-1} \sin\theta_0]}\end{array}\right\} + \eta(t), \quad (2.34c)$$

where $\gamma_m = \Sigma_{i=0}^{m} \sin\beta_i \cos\alpha_i$, $\varepsilon_m = \Sigma_{i=0}^{m} \sin\beta_i \sin\alpha_i$, and $\xi_m = \Sigma_{i=0}^{m} \cos\beta_i$. These parameters may be expressed in a recursive form

$$\begin{aligned}\gamma_m &= \gamma_{m-1} + \sin\beta_m \cos\alpha_m \\ \varepsilon_m &= \varepsilon_{m-1} + \sin\beta_m \sin\alpha_m \\ \xi_m &= \xi_{m-1} + \cos\beta_m.\end{aligned} \quad (2.34d)$$

When the array is perfectly linear, $\gamma_0 = \gamma_1 = ,\ldots, \gamma_{M-1} = \sin\beta_0 \cos\alpha_0$, $\varepsilon_0 = \varepsilon_1 = ,\ldots,$ $\varepsilon_{m-1} = \sin\beta_0 \sin\alpha_0$, and $\xi_0 = \sqrt{1-\gamma_0^2-\varepsilon_0^2}$.

We like to demonstrate the effect of array deformation on its response function. For this purpose we shall assume a 16-sensor ULA along the x-axis and that it is deformed into a bow-shaped curve in the x-y plane, as shown in

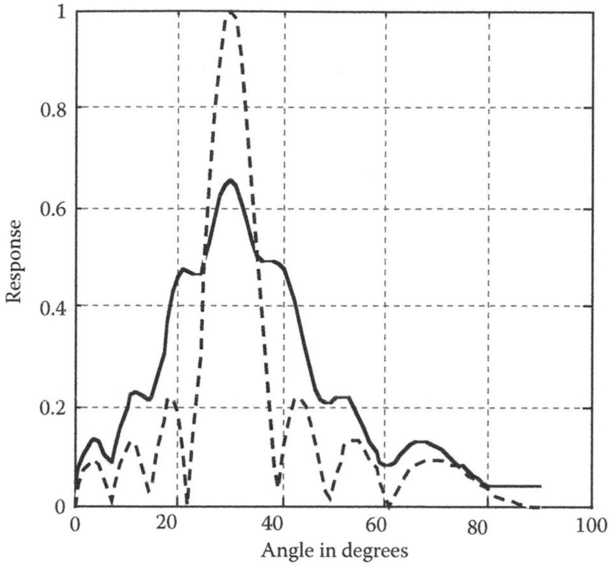

FIGURE 2.14
Response of a bow-shaped array (solid curve) shown in Figure 2.13b and the dashed curve is for undeformed array.

Figure 2.13b. The maximum displacement along the y-axis is one wavelength. The array response is shown in Figure 2.14. For comparison, we also show the response of an undeformed array. Notice the extremely broad main lobe that is fortunately located at a right position. If we further increase the deformation, for example, the bow height is increased to 2.3λ, the array response is found to be totally distorted. Even the main lobe is found to be wrongly placed.

2.2 Planar Array

A planar array has its sensors distributed over a plane. When a wavefront has two unknown parameters, azimuth and elevation, we need a planar array for estimation of a pair of parameters. Since a plane has two dimensions, there are many possible array geometries; some of these are illustrated in Figure 2.15. The natural extension of a ULA in two dimensions is a square or rectangular array where the sensors are placed on a square or rectangular grid. Other geometries are essentially sparse versions of the square or the rectangular array. We shall first study the rectangular array and then look into the sparse arrays, in particular, the circular array, which has found many applications.

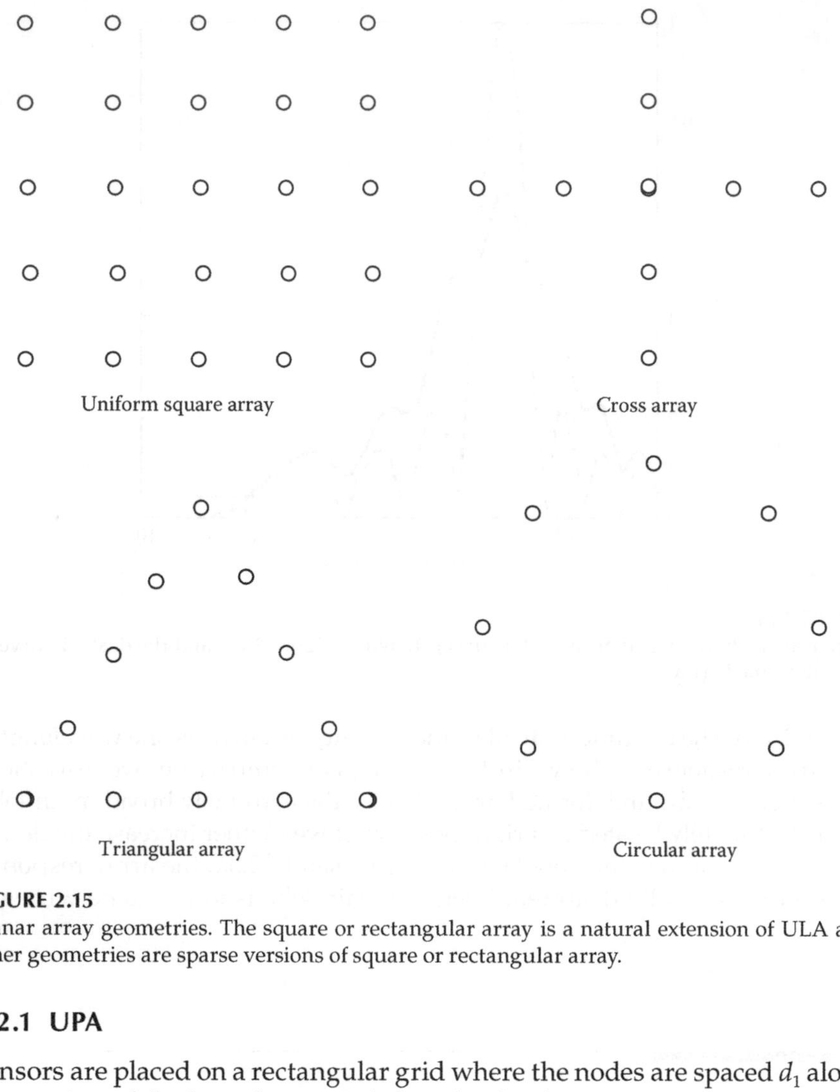

FIGURE 2.15
Planar array geometries. The square or rectangular array is a natural extension of ULA and other geometries are sparse versions of square or rectangular array.

2.2.1 UPA

Sensors are placed on a rectangular grid where the nodes are spaced d_1 along the x-axis and d_2 along the y-axis (when $d_1 = d_2 = d$ we get a square grid). Let a plane wavefront be incident at azimuth angle φ and elevation θ (see Figure 2.16). A plane wavefront, given by

$$f(t,x,y) = \frac{1}{2\pi}\int_{-\infty}^{+\infty} F(\omega) e^{j(\omega t - ux - vy)} d\omega, \tag{2.35}$$

where $u = (\omega/c)\sin\theta\cos\varphi$ and $v = (\omega/c)\sin\theta\sin\varphi$ is incident on a UPA. The output of the (m,n) th sensor is

$$f_{m_1 m_2}(t) = \frac{1}{2\pi}\int_{-\infty}^{+\infty} F(\omega) e^{j(\omega t - u m_1 d_1 - v m_2 d_2)} d\omega. \tag{2.36}$$

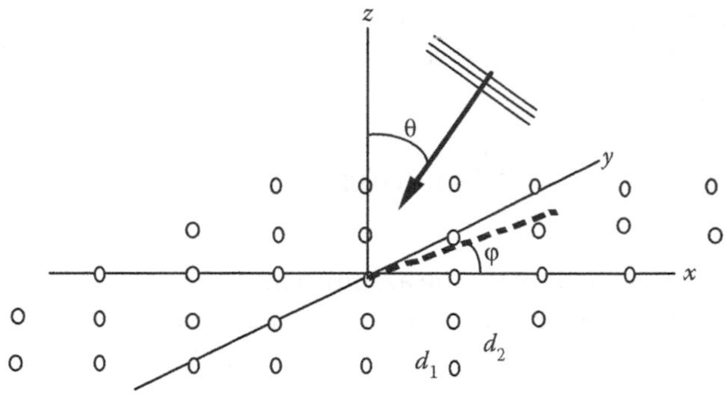

FIGURE 2.16
A plane wavefront incident on a UPA at azimuth φ and elevation θ.

All sensor outputs are summed in phase, yielding

$$g(t)=\frac{1}{M_1M_2}\sum_{m_1=1}^{M_1}\sum_{m_2=1}^{M_2}a_{m_1m_2}f_{m_1m_2}(t)$$

$$=\frac{1}{2\pi}\int_{-\infty}^{+\infty}F(\omega)e^{j\omega t}\frac{1}{M_1M_2}\sum_{m_1=1}^{M_1}\sum_{m_2=1}^{M_2}a_{m_1m_2}e^{-j(um_1d_1+vm_2d_2)}d\omega \tag{2.37}$$

$$=\frac{1}{2\pi}\int_{-\infty}^{+\infty}F(\omega)e^{j\omega t}H(ud_1,vd_2)d\omega,$$

where

$$H(ud_1,vd_2)=\frac{1}{M_1M_2}\sum_{m_1=1}^{M_1}\sum_{m_2=1}^{M_2}a_{m_1m_2}e^{-j(um_1d_1+vm_2d_2)}, \tag{2.38a}$$

which, for constant weighting coefficients, becomes

$$H(ud_1,vd_2)=\frac{\sin\left(\frac{M_1ud_1}{2}\right)}{M_1\sin\frac{ud_1}{2}}\frac{\sin\left(\frac{M_2vd_2}{2}\right)}{M_2\sin\frac{vd_2}{2}}e^{-j\frac{[(M_1-1)u+(M_2-1)v]}{2}}. \tag{2.38b}$$

The frequency response function of a UPA given by Equation 2.38b can be written as a product of two frequency response functions of two ULAs, one in the x direction and the other in the y direction. This is also true in the weighted sum case provided the weighting coefficients can be expressed as a product of two coefficient sets, that is, $a_{mn}=\alpha_m\beta_n$ where (α_{m1}, $m_1=0, 1, \ldots M_1-1$) and $(\beta_{m_2}, m_2=0,1,\ldots, M_2-1)$. This is the motivation behind the use of a cross array (Figure 2.15) in place of a UPA.

A UPA may be steered both in azimuth and elevation by means of appropriate delays introduced before summation. The delay to be introduced in the (m,n) th sensor is

$$\tau_{m_1,m_2} = \frac{m_1 d_1}{c}\sin\theta\cos\varphi + \frac{m_2 d_2}{c}\sin\theta\sin\varphi,$$

where φ and θ are, respectively, the azimuth and elevation to which the array is required to be steered. In place of Equation 2.37 we have

$$\begin{aligned} g(t) &= \sum_{m_1=1}^{M_1}\sum_{m_2=1}^{M_2} f_{m_1 m_2}(t+\tau_{m_1,m_2}) \\ &= \frac{1}{2\pi}\int_{-\infty}^{+\infty} F(\omega)e^{j\omega t}\sum_{m_1=1}^{M_1}\sum_{m_2=1}^{M_2} e^{-j[(u_0-u)m_1 d_1+(v_0-v)m_2 d_2)]}d\omega, \\ &= \frac{1}{2\pi}\int_{-\infty}^{+\infty} F(\omega)e^{j\omega t}H((u_0-u)d_1,(v_0-v)d_2)d\omega \end{aligned} \tag{2.39}$$

where

$$H((u_0-u)d_1,(v_0-v)d_2) = \frac{\sin\left(\frac{M_1(u_0-u)d_1}{2}\right)\sin\left(\frac{M_2(v_0-v)d_2}{2}\right)}{M_1\sin\frac{(u_0-u)d_1}{2}\; M_2\sin\frac{(v_0-v)d_2}{2}} e^{-j\frac{[(M_1-1)(u_0-u)+(M_2-1)(v_0-v)]}{2}},$$

where

$$u_0=(\omega/c)\sin\theta_0\cos\varphi_0,$$
$$v_0=(\omega/c)\sin\theta_0\sin\varphi_0,$$

and (φ_0, θ_0) are, respectively, azimuth and elevation angles.

Let the steering angles φ and θ be chosen such that $(d_1/\lambda)\sin\theta\cos\varphi=(k/M_1)$ and $(d_2/\lambda)\sin\theta\sin\varphi=(l/M_2)$, where k and l are integers, $0\le k\le M_1$ and $0\le l\le M_2$. Equation 2.39 may be expressed as a 2D spatial DFT of the wavefield incident on the array.

$$\begin{aligned} g(t) &= \frac{1}{2\pi}\int_{-\infty}^{+\infty} F(\omega)e^{j\omega t}\sum_{m_1=1}^{M_1}\sum_{m_2=1}^{M_2} e^{-j(u_0 m_1 d_1+v_0 m_2 d_2)}e^{j((2\pi k m_1/M_1)+(2\pi l m_2/M_2))}d\omega \\ &= \frac{1}{2\pi}\int_{-\infty}^{+\infty} F(\omega)e^{j\omega t}DFT\{e^{-j(u_0 m_1 d_1+v_0 m_2 d_2)}\}d\omega, \end{aligned} \tag{2.40a}$$

and

$$H((u_0-u)d_1,(v_0-v)d_2)=DFT\{e^{-j(u_0m_1d_1+v_0m_2d_2)}\}. \tag{2.40b}$$

This result is an extension of Equation 2.7 for a ULA to a UPA. The spatial frequencies u and v, which appear in Equation 2.40b, are linearly related to DFT the frequency numbers, k and l

$$u=\frac{2\pi k}{M_1d_1}, \quad v=\frac{2\pi l}{M_1d_1}.$$

Given the frequency numbers, we can get the azimuth and elevation angles. For example, assuming $d_1=d_2=(\lambda/2)$ and $M_1=M_2=M$, $\varphi=\sin^{-1}l/\sqrt{k^2+l^2}$ and $\theta=\sin^{-1}2\sqrt{k^2+l^2}/M$. Note that $0\le k,l\le M/2$ and $0\le\sqrt{k^2+l^2}\le M/2$. Thus, the acceptable domain for k and l is as shown in Figure 2.17. For those values of k and l lying outside this domain, we get nonpropagating waves (see Chapter 1, Section 1.2).

2.2.1.1 Random Array

An array with its sensors at a random location, modeled as an independent, identically distributed random variable, is known as a totally random array [12]. The root mean square (rms) error in the estimation of the azimuth angle

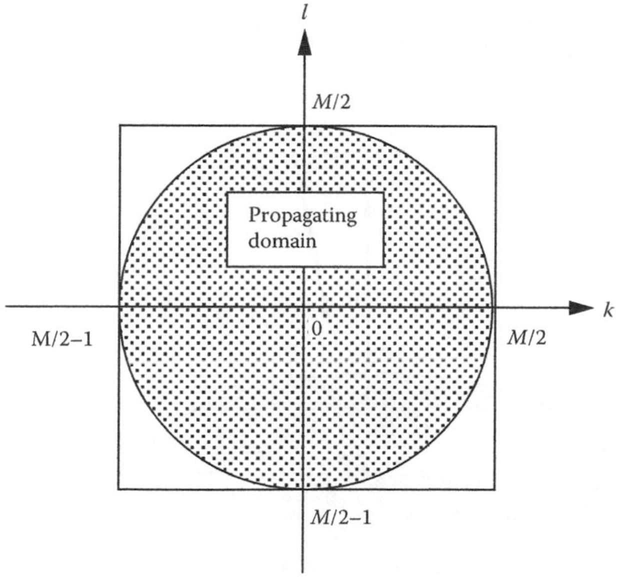

FIGURE 2.17
Acceptable domain for frequency numbers k and l. For those values of k and l lying outside this domain, we get nonpropagating waves.

(measured with respect to the x-axis) is given in Ref. [12] and reproduced here without derivation

$$\begin{aligned} rms\ error &= \sqrt{E\{(\hat{\varphi}_0 - \varphi_0)^2\}} \\ &\approx \frac{\lambda_0}{2\pi} \frac{1}{\sqrt{2snrM}} \frac{1}{\sigma}, \end{aligned}$$

where σ is the standard deviation of the random variable representing the location of the sensors. The approximate result above is valid for large M.

Interestingly, the rms error decreases if the random array is spread out over a large area. The practical interest in a random array arises in the context of sonobuoy arrays used in submarine detection.

2.2.2 UCA

The sensors may be placed on a plane in a polar grid. For a fixed radial distance we have a circle on which the sensors are placed, forming a circular array. Consider a circular array of radius a with M sensors, symmetrically placed on the circumference (see Figure 2.18). Let a plane wavefront be incident on the array at angles φ and θ. The output of the mth sensor is given by

$$\begin{aligned} f_m(t) &= \frac{1}{2\pi} \int_{-\infty}^{\infty} F(\omega) e^{j\omega t - j[(\omega a/c)(\cos\varphi \sin\theta \cos(2\pi m/M) + \sin\varphi \sin\theta \sin(2\pi m/M))]} d\omega \\ &= \frac{1}{2\pi} \int_{-\infty}^{\infty} F(\omega) e^{j\omega t - j[(\omega a/c)(\sin\theta \cos((2\pi m/M) - \varphi))]} d\omega. \end{aligned} \tag{2.41}$$

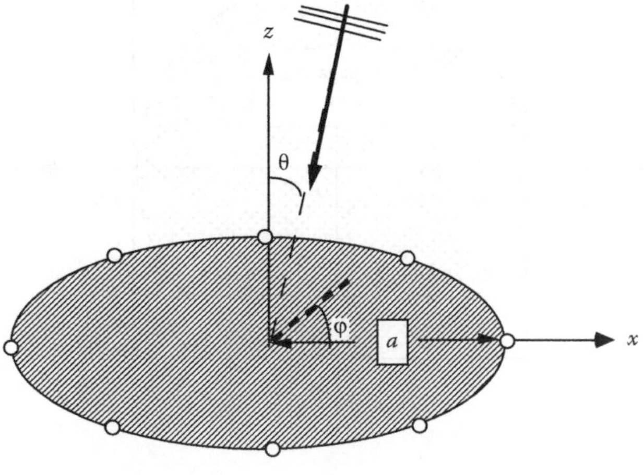

FIGURE 2.18
Sensors are uniformly spaced on the circumference of a circle of radius a. A plane wave is incident at an azimuth angle ϕ and an elevation angle θ.

Note that time is measured with respect to the time of arrival (TOA) of the wavefront at the center of the array. First, we evaluate the frequency response function. The sum of all outputs in the frequency domain is given by

$$
\begin{aligned}
g(t) &= \frac{1}{2\pi}\int_{-\infty}^{\infty} F(\omega)e^{j\omega t}\frac{1}{M}\sum_{m=0}^{M-1} e^{-j[(\omega a/c)(\sin\theta\cos((2\pi m/M)-\varphi))]}d\omega \\
&= \frac{1}{2\pi}\int_{-\infty}^{\infty} F(\omega)H(\omega,\varphi,\theta)e^{j\omega t}d\omega,
\end{aligned}
\tag{2.42}
$$

where the frequency response function $H(\omega, \varphi, \theta)$ is given by

$$
H(\omega,\varphi,\theta) = \frac{1}{M}\sum_{m=0}^{M-1} e^{-j[(\omega a/c)(\sin\theta\cos((2\pi m/M)-\varphi))]}. \tag{2.43a}
$$

For large M (e.g., $M > 48$ when $a = 6\lambda$ and $M > 32$ when $a = 4\lambda$) the summation in Equation 2.43a may be replaced by an integral and the result is

$$
H(\omega,\varphi,\theta) \approx J_0\left(\frac{\omega a}{c}\sin\theta\right). \tag{2.43b}
$$

We shall call such a UCA a fully populated array. The most interesting property of a circular array is that the frequency response function is independent of φ. The property arises from Equation 2.43b. Taking the distance to the first zero as the effective half width of the main lobe, the angular width will be equal to $\Delta\theta = \sin^{-1}(2.4048\,(c/\omega a))$. The height of the first (largest) sidelobe is 0.4028 at $\theta = \sin^{-1}(3.83\,(c/\omega a))$.

A circular array may be steered to any desired direction just like a ULA or a UPA. A delay τ_m is introduced at each sensor output before summation, where

$$
\tau_m = \left[\frac{a}{c}\left(\cos\varphi\sin\theta\cos\frac{2\pi m}{M} + \sin\varphi\sin\theta\sin\frac{2\pi m}{M}\right)\right],
$$

and φ and θ, respectively, are the desired azimuth and elevation angles. The delayed outputs of all sensors are summed, for the time being, without any weighting.

$$
\begin{aligned}
g(t) &= \frac{1}{M}\sum_{m=0}^{M-1} f_m(t+\tau_m) \\
&= \frac{1}{2\pi}\int_{-\infty}^{\infty} F(\omega)e^{j\omega t}\frac{1}{M}\sum_{m=0}^{M-1} e^{-j[(\omega a/c)(\sin\theta_0\cos((2\pi m/M)-\varphi_0)-\sin\theta\cos((2\pi m/M)-\varphi))]}d\omega,
\end{aligned}
\tag{2.44}
$$

where φ_0 and θ_0, respectively, are the unknown azimuth and elevation angles of the incident wavefront. Let

$$H\left(\frac{\omega a}{c},\theta_0,\varphi_0,\theta,\varphi\right)=\frac{1}{M}\sum_{m=0}^{M-1}e^{-j[(\omega a/c)(\sin\theta_0\cos((2\pi m/M)-\varphi_0)-\sin\theta\cos((2\pi m/M)-\varphi))]}.$$

The output power of the array, steered to any chosen direction φ and θ, is given by output power $=|F(\omega)H((\omega a/c),\theta_0,\varphi_0,\theta,\varphi)|^2$.

Earlier we noted that steering of an array is equivalent to spatial FT of the array output. This result holds in a slightly different form for a circular array. We will demonstrate how the spatial FT can be used for estimation of the azimuth and elevation angles [13]. Consider the spatial DFT of the circular array output.

$$\begin{aligned} g_k(t)&=\frac{1}{M}\sum_{m=0}^{M-1}f_m(t)e^{-j(2\pi m/M)k}\\ &=\frac{1}{2\pi}\int_{-\infty}^{\infty}F(\omega)e^{j\omega t}\frac{1}{M}\sum_{m=0}^{M-1}e^{-j[(\omega a/c)(\sin\theta_0\cos((2\pi m/M)-\varphi_0))]}e^{-j(2\pi mk/M)}d\omega\\ &=\frac{1}{2\pi}\int_{-\infty}^{\infty}F(\omega)H(\omega,\varphi_0,\theta_0)e^{j\omega t}d\omega. \end{aligned} \tag{2.45}$$

Taking the temporal FT of Equation 2.45, we obtain an important result,

$$G_k(\omega)\approx F(\omega)J_k\left(\frac{\omega a}{c}\sin\theta_0\right)e^{jk(\pi/2)}e^{-jk\varphi_0}, \tag{2.46}$$

which is valid for $k<k_{\max}\approx(\omega a/c)$ [15] and for sensor spacing approximately equal to $(\lambda/2)$ [14]. Consider the following quantity:

$$\frac{G_{k+1}(\omega)}{G_k(\omega)}=je^{-j\varphi_0}\frac{J_{k+1}\left(\frac{\omega a}{c}\sin\theta_0\right)}{J_k\left(\frac{\omega a}{c}\sin\theta_0\right)}. \tag{2.47}$$

Referring to the recurrence relation of Bessel functions [15],

$$J_{k+1}(x)=\frac{2k}{x}J_k(x)-J_{k-1}(x),$$

we can write

$$\frac{J_{k+1}(x)}{J_k(x)}=\frac{2k}{x}-\frac{J_{k-1}(x)}{J_k(x)},$$

which we use in Equation 2.47 and derive a basic result for the estimation of φ_0 and θ_0

$$-je^{j\varphi_0}\frac{G_{k+1}(\omega)}{G_k(\omega)}=\frac{2k}{\frac{\omega a}{c}\sin\theta_0}-je^{-j\varphi_0}\frac{G_{k-1}(\omega)}{G_k(\omega)}. \tag{2.48}$$

Equation 2.48 may be solved for φ_0 and θ_0. As an example, we consider a 16-sensor circular array of 3λ radius and a source in far-field emitting a band-limited random signal. The center frequency is 100 Hz and the bandwidth is 10 Hz. The azimuth and the elevation angle of the source are, respectively, 10° (0.1745 rad) and 45° (0.7854 rad). The sampling rate is 500 samples/sec. The estimates are averaged over all frequency bins lying within the bandwidth. The results are shown in Figure 2.19. Notice that the standard deviation of the estimates decreases considerably when a reference sensor is used at the center. The decrease is more pronounced at very low SNR, e.g., at 0 dB the decrease is by a factor of three or more. Analysis of errors has shown that the standard deviation is dominated by a few outliers that are caused by random noise in the array output. Unless these outliers are eliminated, the mean and the standard deviation of the estimate are severely affected. To overcome this problem, median in place of mean may be considered. It was observed through computer simulation that the median is a better estimate of the azimuth than the mean.

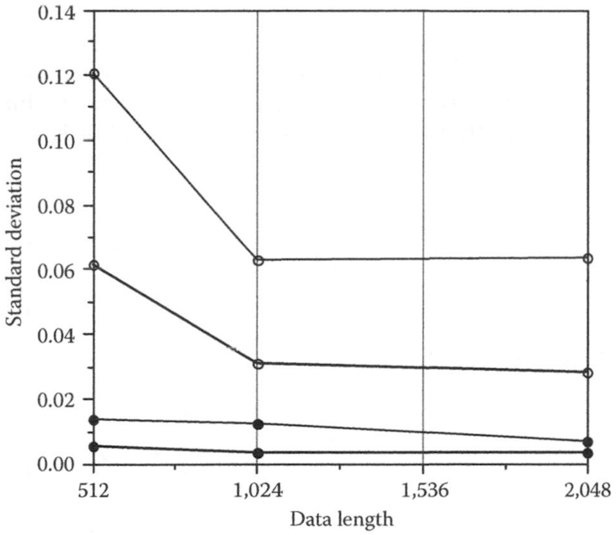

FIGURE 2.19
Standard deviation of azimuth and elevation estimates as a function of data length. Thick line: with a sensor at the center, thin line: without a sensor at the center, filled circle: azimuth, and empty circle: elevation. SNR = 10 dB.

When there is more than one source (say P sources), Equation 2.46 takes a form,

$$G_k(\omega)=\sum_{i=0}^{p-1} F_i(\omega)J_k\left(\frac{\omega a}{c}\sin\theta_i\right)e^{jk(\pi/2)}e^{-jk\varphi_i}+\eta_k(\omega). \tag{2.49a}$$

We shall rewrite Equation 2.49a in a matrix form. For this purpose, we define the following vectors and matrices:

$$\mathbf{G}_r(\omega)=col\{G_0(\omega), G_1(\omega),\ldots, G_{r-1}(\omega)\},$$

$$\mathbf{D}_i(\omega)=\text{diag}\left\{J_0\left(\frac{\omega a}{c}\sin\theta_i\right), J_1\left(\frac{\omega a}{c}\sin\theta_i\right), \cdots\ J_{r-1}\left(\frac{\omega a}{c}\sin\theta_i\right),\right\},$$

$$\mathbf{Z}_i=col\left\{1, e^{-j(\varphi_i-(\pi/2))}, e^{-j2(\varphi_i-(\pi/2))},\ldots, e^{-j(r-1)(\varphi_i-(\pi/2))}\right\},$$

$$\eta_r(\omega)=col\{\eta_0(\omega), \eta_1(\omega),\ldots, \eta_{r-1}(\omega)\},$$

where r is an integer ($p\le r\le M$).

$$\mathbf{G}_r(\omega)=\sum_{i=0}^{p-1}\mathbf{D}_i(\omega)\mathbf{Z}_iF_i(\omega)+\eta_r(\omega). \tag{2.49b}$$

Let us assume that all P sources emit stationary uncorrelated stochastic processes. Since the array output will also be a stationary stochastic process, we use spectral matrix that is given by $S_g=E\{\mathbf{G}_r(\omega)\mathbf{G}_r^H(\omega)\}$. From Equation 2.49b it follows that

$$S_g=\left[\mathbf{D}_0\mathbf{Z}_0,\mathbf{D}_1\mathbf{Z}_1,\ldots\mathbf{D}_{p-1}\mathbf{Z}_{p-1}\right]\begin{bmatrix} S_0 & & & \\ & S_1 & & \\ & & \ddots & \\ & & & S_{p-1}\end{bmatrix}\begin{bmatrix}(\mathbf{D}_0\mathbf{Z}_0)^H \\ (\mathbf{D}_1\mathbf{Z}_1)^H \\ \vdots \\ (\mathbf{D}_{p-1}\mathbf{Z}_{p-1})^H\end{bmatrix}+\sigma_\eta^2\mathbf{I}, \tag{2.50}$$

(rxr) (rxp) (pxp) (pxr)

Exploiting the above signal structure, a subspace approach has been developed in Refs. [14, 16] for the estimation of the azimuth and elevation.

2.2.2.1 Steering Vector

For circular array we define a steering vector as

$$\mathbf{a}(\varphi,\theta)$$
$$= col[e^{-j[(\omega a/c)(\sin\theta\cos(\varphi))]}, e^{-j[(\omega a/c)(\sin\theta\cos((2\pi/M)-\varphi))]},\ldots, e^{-j[(\omega a/c)(\sin\theta\cos((2\pi(M-1)/M)-\varphi))]}]. \tag{2.51}$$

Each term in the steering vector can be expanded in a series form [15]

$$e^{-j[(\omega a/c)(\sin\theta\cos((2\pi m/M)-\varphi))]} = \sum_{k=-\infty}^{\infty}(-j)^k J_k\left(\frac{\omega a}{c}\sin\theta\right)e^{jk((2\pi m/M)-\varphi)}.$$

Define a matrix

$$\{\mathbf{W}\}_{km} = e^{j(2\pi km/M)} \quad \begin{Bmatrix} m=0,1,\ldots,M-1 \\ k=0,\pm1,\ldots,\infty \end{Bmatrix},$$

and a vector

$$\{\mathbf{V}(\varphi,\theta)\}_k = J_k\left(\frac{\omega a}{c}\sin\theta\right)e^{-j((\pi/2)+\varphi)k}, \quad k=0,\pm1,\ldots,\infty.$$

In terms of matrix **W** and vector **V**, we can express Equation 2.51 as

$$\underset{M\times 1}{\mathbf{a}(\varphi,\theta)} = \underset{M\times\infty}{\mathbf{W}} \quad \underset{\infty\times 1}{\mathbf{V}(\varphi,\theta)}.$$

Observe that matrix, **W**, is independent of the azimuth and elevation angles, which are confined only to the vector $\mathbf{V}(\varphi,\theta)$. The size of $\mathbf{V}(\varphi,\theta)$ depends on the argument of the Bessel function, which may be approximated, for large order, as [15]

$$J_k\left(\frac{\omega a}{c}\sin\theta\right) \approx \frac{1}{\sqrt{2\pi k}}\left(\frac{e\frac{\omega a}{c}\sin\theta}{2k}\right)^k,$$

where $e=2.71828182$. Hence, the size of vector $\mathbf{V}(\varphi,\theta)$ will be of the order of $e(\omega a/c)\sin\theta$. For example, for $a=8\lambda$ and $\theta=(\pi/2)$ the size of the vector is about 140. The steering vector for a circular array possesses some interesting properties, namely, (a) $\mathbf{a}(\varphi,\theta) \neq \mathbf{a}(-\varphi,\theta)$, (b) $\mathbf{a}^*(\varphi,\theta)=\mathbf{a}(\pi-\varphi,\theta)$, and (c) $\mathbf{a}(\varphi,\theta)$ periodic in φ with period $2p$, and independent of sensor spacing.

Property (a) implies that a wavefront coming from the north can be distinguished from one coming from the south (north–south ambiguity). Property

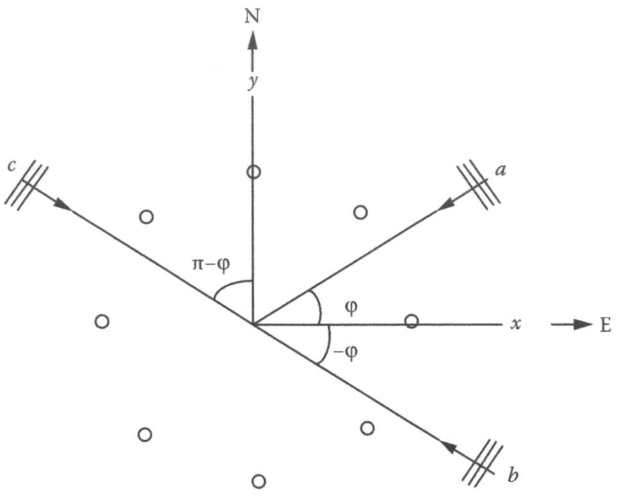

FIGURE 2.20
Circular array does not suffer from north-south ambiguity, that is, wavefronts *a* and *b* can be distinguished. There is no east-west ambiguity for complex signals, that is, wavefronts *a* and *c* can be distinguished.

(b) implies that a complex signal coming from the east can be distinguished from one coming from the west (east-west ambiguity; see Figure 2.20 for an illustration), and property (c) implies that for any sensor spacing there is no grating lobe in the range of $\pm\pi$. A circular array differs from a linear array in respect of properties (a) to (c).

2.2.2.2 Boundary Array

An array of sensors for localization of an object in the near-field region may take the form of a boundary array where the sensors are placed all around the object, as its approximate location is known before hand. A circular array enclosing a source is an example of a boundary array. With three coplanar sensors and accurate time delay measurements, it is possible to localize a point target in the plane of array. There is a vast literature on time delay estimation but, due to space limitation, the topic of time delay estimation will not be covered in this book. It is, on the other hand, possible for a boundary array to determine the time delays from the phase measurements. The basic idea is to consider a pair of sensors and estimate the phase difference at a fixed frequency ω. The source will fall on one of the phase trajectories drawn with a phase difference $\omega\Delta\tau=2\pi s+\phi$, where $\Delta\tau$ is the unknown time delay, s is an integer, and ϕ is the actual observed phase difference. To estimate the time delay $\Delta\tau$, we need to know the integer constant s in the range $\pm s_0$, where $s_0=\text{Int}[(\omega\Delta\tau)/(2\pi)]$ and Int[x] stands for the largest integer less than x. For a given ϕ, $\Delta\tau$ will assume a set of $2s_0+1$ values and for each value of s there corresponds a locus of points called the phase trajectory [17]. For every pair of adjacent sensors, we can draw a suite of trajectories, as shown in Figure 2.21a.

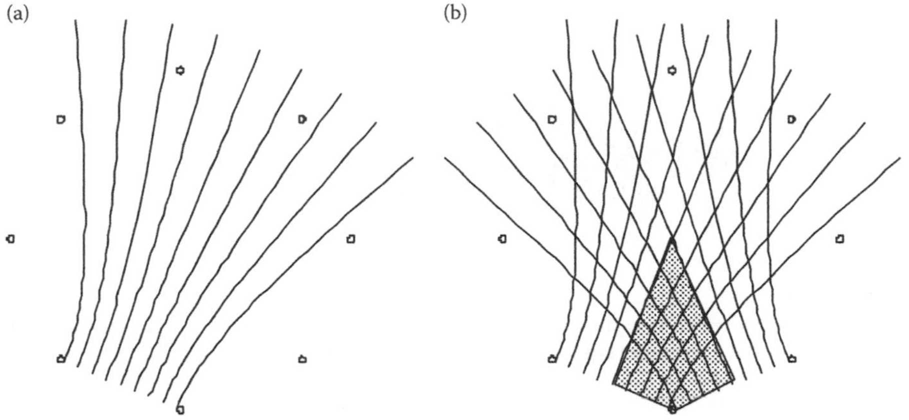

FIGURE 2.21
(a) For every pair of adjacent sensors we draw a suite of equiphase trajectories. (b) Intersection of two sets of equiphase trajectories. The unknown source must lie at one of the intersections. For every adjacent pair of sensors the search is carried out within the dotted quadrilateral.

The unknown source must lie on any one of the trajectories. Next, consider another sensor pair and draw another suite of trajectories. Any one of the points of intersection is a possible location of the unknown source (see Figure 2.21b). Since there are M pairs of sensors, there will be M suites of trajectories. The true position of the unknown source is then given by the intersection of all M trajectories, one from each suite. At the true source location, all phase estimates obtained from different pairs of sensors must match with the theoretically evaluated phases. Let $\hat{\phi}_1^k, k=0,1,\ldots,M-1$ be the estimated phases from M pairs of adjacent sensors and $\phi_1^k, k=0,1,\ldots,M-1$ be the theoretically computed phases. Define an error vector

$$\varepsilon = col\left\{(e^{j\phi_1^0} - e^{j\hat{\phi}_1^0}), (e^{j\phi_1^1} - e^{j\hat{\phi}_1^1}), \ldots, (e^{j\phi_1^{M-1}} - e^{j\hat{\phi}_1^{M-1}})\right\}, \tag{2.52}$$

and error power$=\varepsilon^H\varepsilon$. The error power will be zero at the true location of the source. This property may be utilized to spot the true location from all available intersections of any two suites of trajectories. Evaluate the error power at each and every intersection. That intersection which yields zero (or minimum) error power is the most likely location of the unknown source. Finally, to reconstruct the true phase we need to know the integer s, which may be obtained from the order of the trajectory for every pair of sensors passing through the source location. However, in any practical problem, this step may not be required as the interest is usually in localization and not in the true phase retrieval. The results of a numerical experiment are shown in Figure 2.22. The question of minimum size (number of sensors) of the array, required for unique localization, has not been answered. However, numerical experiments suggest a minimum array size of five sensors placed on the circumference of a large circle. It is not necessary for the array to be perfectly

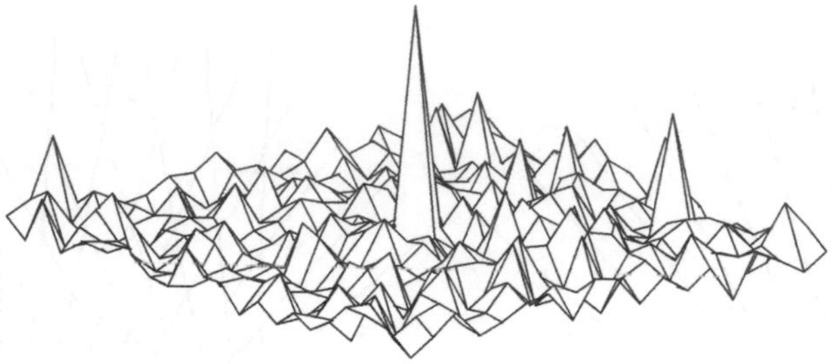

FIGURE 2.22
A plot of inverse error power. The source is at range 50 m and azimuth –60°. A circular array of radius 100 m and having eight equispaced sensors is assumed.

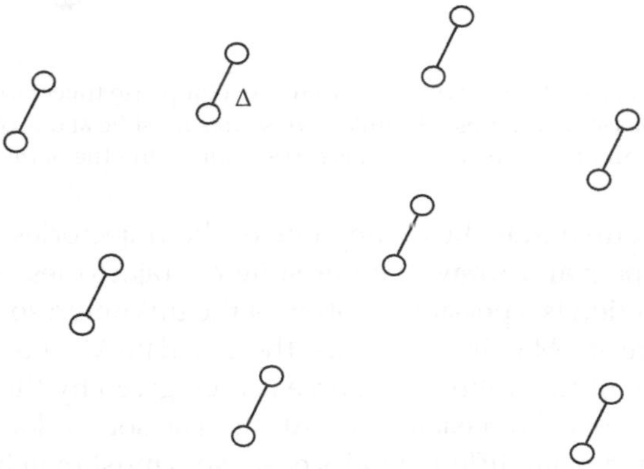

FIGURE 2.23
Randomly distributed planar dipole sensors. Each dipole consists of two identical sensors displaced by Δ (vector).

circular. Any closed curve will do provided the phase trajectories are drawn for each pair.

2.2.3 Distributed Dipoles

A planar array with sensors randomly distributed in a plane constitutes a random planar array. Here we shall consider a random planar array of dipole sensors. A dipole sensor consists of a pair of *identical* sensors displaced by Δ (a vector). An example of a dipole planar array is shown in Figure 2.23. Let $\mathbf{r}_i$, $i=0, 1,\ldots, M-1$ represent locations of M dipole sensors (the midpoint of the sensor pair). It may be emphasized that the spatial distribution of the dipoles can be quite arbitrary. Define the following data vector in a frequency domain:

$$\tilde{\mathbf{F}} = \begin{bmatrix} \mathbf{F}_1 \\ \mathbf{F}_2 \end{bmatrix}, \tag{2.53}$$

where

$$\mathbf{F}_1 = col\{F_0^+(\omega), F_1^+(\omega)...F_{M-1}^+(\omega)\}$$
$$\mathbf{F}_2 = col\{F_0^-(\omega), F_1^-(\omega)...F_{M-1}^-(\omega)\},$$

where superscript+refers to the upper sensor and superscript – refers to the lower sensor. Let P plane wavefronts from P sources be incident on the array

$$\mathbf{F}_1 = [\mathbf{a}_0, \mathbf{a}_1, \ldots, \mathbf{a}_{p-1}]\mathbf{F}(\omega) + d\eta_1(\omega), \tag{2.54}$$

where

$$\begin{aligned}\mathbf{a}_i &= col\left\{e^{-j(\Delta/2)\cdot\delta_i}, e^{-j(\omega/c)(\mathbf{r}_1+(\Delta/2))\cdot\delta_i}, \ldots, e^{-j(\omega/c)(\mathbf{r}_{M-1}+(\Delta/2))\cdot\delta_i}\right\} \\ &= col\left\{1, e^{-j(\omega/c)\mathbf{r}_1\cdot\delta_i}, \ldots, e^{-j(\omega/c)\mathbf{r}_{M-1}\cdot\delta_i}\right\}e^{-j(\Delta\cdot\delta_i/2)},\end{aligned}$$

where $\delta_i = col\{\sin\theta_i, \cos\theta_i\}$ and $\mathbf{F}(\omega) = [F_0(\omega), F_1(\omega)\ldots F_{p-1}(\omega)]^T$ is a vector whose components are FTs of the waveforms emitted by P sources. Equation 2.54 may be represented compactly as follows

$$\mathbf{F}_1 = \mathbf{A}\Gamma_+\mathbf{F}(\omega) + d\eta_1(\omega), \tag{2.55a}$$

where

$$\Gamma_+ = \mathrm{diag}\left\{e^{-j(\Delta\cdot\delta_0/2)}, e^{-j(\Delta\cdot\delta_1/2)}, \ldots, e^{-j(\Delta\cdot\delta_{p-1}/2)}\right\},$$

and

$$\mathbf{A} = \begin{bmatrix} 1 & 1 & \cdots & 1 \\ e^{-j(\omega/c)\mathbf{r}_1\cdot\delta_0} & e^{-j(\omega/c)\mathbf{r}_1\cdot\delta_1} & & e^{-j(\omega/c)\mathbf{r}_1\cdot\delta_{p-1}} \\ \cdot & & & \\ \cdot & & & \\ \cdot & & & \\ e^{-j(\omega/c)\mathbf{r}_{M-1}\cdot\delta_0} & e^{-j(\omega/c)\mathbf{r}_{M-1}\cdot\delta_1} & \cdots & e^{-j(\omega/c)\mathbf{r}_{M-1}\cdot\delta_{p-1}} \end{bmatrix}.$$

Similarly,

$$\mathbf{F}_2 = \mathbf{A}\Gamma_-\mathbf{F}(\omega) + d\eta_2(\omega), \tag{2.55b}$$

where

$$\Gamma_{-} = \mathrm{diag}\left\{e^{j(\Delta \cdot \delta_0/2)}, e^{j(\Delta \cdot \delta_1/2)}, \ldots, e^{j(\Delta \cdot \delta_{p-1}/2)}\right\}.$$

We shall now compute the spectral matrix of a dipole array output. For this, the incident signal will be treated as a stationary stochastic process. Then, in place of the ordinary FT we need to invoke the generalized FT of a stationary stochastic process (see Chapter 1, Section 1.4).

$$\begin{aligned}\frac{1}{2\pi}\mathbf{S}_{\tilde{f}}\,d\omega &= E\left\{\frac{1}{2\pi}d\tilde{\mathbf{F}}\frac{1}{2\pi}d\tilde{\mathbf{F}}^{H}\right\} = E\left\{\frac{1}{2\pi}\begin{bmatrix} d\mathbf{F}_1 \\ d\mathbf{F}_2 \end{bmatrix}\frac{1}{2\pi}\begin{bmatrix} d\mathbf{F}_1 \\ d\mathbf{F}_2 \end{bmatrix}^{H}\right\} \\ &= \frac{1}{2\pi}\begin{bmatrix} \mathbf{A}\mathbf{S}_f\mathbf{A}^H & \mathbf{A}\mathbf{S}_f\Gamma^H\mathbf{A}^H \\ \mathbf{A}\Gamma\mathbf{S}_f\mathbf{A}^H & \mathbf{A}\mathbf{S}_f\mathbf{A}^H \end{bmatrix} d\omega + \frac{1}{2\pi}\begin{bmatrix} \mathbf{I}\sigma_{\eta_1}^2 & 0 \\ 0 & \mathbf{I}\sigma_{\eta_2}^2 \end{bmatrix} d\omega,\end{aligned} \tag{2.56}$$

where

$$\Gamma = \Gamma_{-}\Gamma_{+}^{H} = \mathrm{diag}\left\{e^{j\omega\Delta\cdot\delta_0}, e^{j\omega\Delta\cdot\delta_1}, \ldots, e^{j\omega\Delta\cdot\delta_{p-1}}\right\},$$

$$\mathbf{S}_f = \mathrm{diag}\left\{S_{f_0}, S_{f_1}, \ldots, S_{f_{p-1}}\right\}.$$

Equation 2.56 may be expressed as

$$\mathbf{S}_{\tilde{f}} = \begin{bmatrix} \mathbf{A} \\ \mathbf{A}\Gamma \end{bmatrix}\mathbf{S}_f\begin{bmatrix} \mathbf{A} \\ \mathbf{A}\Gamma \end{bmatrix}^{H} + \begin{bmatrix} \mathbf{I}\sigma_{\eta_1}^2 & 0 \\ 0 & \mathbf{I}\sigma_{\eta_2}^2 \end{bmatrix}. \tag{2.57}$$

The DOA information is buried in Γ while the **A** matrix contains dipole location information, which is known. In Chapter 5, we shall show how Γ can be estimated using the subspace approach.

2.2.3.1 Electric Dipole Array

Now consider a ULA of electric dipoles. There are two possible arrangements: dipole axis along the *x*-axis or along the *y*-axis, as illustrated in Figure 2.24. Note that an electric dipole will measure the electric field component along its axis. The output will be proportional to the incident field when the dipole size is small ($< (\lambda/10)$). Let a plane electromagnetic (EM) wave with its wave vector in *y*-*z* plane be incident on both types of arrays (see Figure 2.24a and b) of electric dipoles. We shall simultaneously measure the *x* and *y* components of the electric field. The array outputs are given by [18].

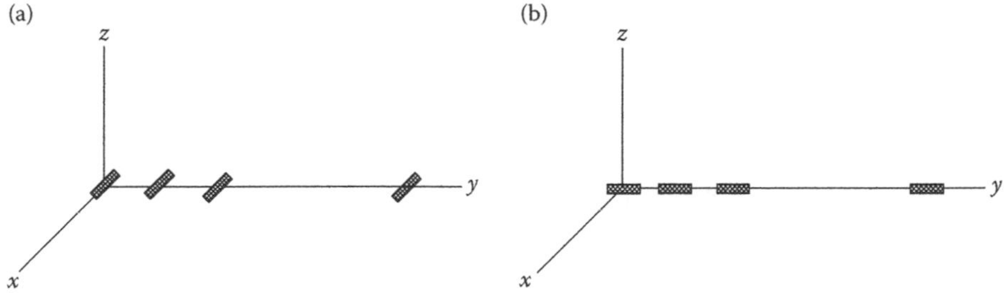

FIGURE 2.24
A ULA of electric dipoles (a) oriented in x direction and (b) in y direction.

(a) For dipoles oriented in the x direction

$$\begin{aligned}\mathbf{f}_x &= -\cos\gamma\left[1, e^{j(\omega d/c)\sin\theta}, e^{j2(\omega d/c)\sin\theta}, \ldots, e^{j(M-1)(\omega d/c)\sin\theta}\right]^T E e^{j(\omega t+\varphi)}.\\ &= -\cos\gamma\ \mathbf{a} E e^{j(\omega t+\varphi)}\end{aligned}$$

(b) For dipoles oriented in the y direction

$$\begin{aligned}\mathbf{f}_y &= \sin\gamma\cos\theta e^{j\kappa}\left[1, e^{j(\omega d/c)\sin\theta}, e^{j2(\omega d/c)\sin\theta}, \ldots, e^{j(M-1)(\omega d/c)\sin\theta}\right]^T E e^{j(\omega t+\varphi)},\\ &= \sin\gamma\cos\theta e^{j\kappa}\mathbf{a} E e^{j(\omega t+\varphi)}\end{aligned}$$

where θ is DOA and γ and κ are defined in terms of the polarization parameters (see Chapter 1, Section 1.1). When there are P wavefronts incident on a ULA of dipoles, the array output may be expressed as follows:

$$\mathbf{f}_x = \mathbf{A}\boldsymbol{\Gamma}_1\mathbf{S}_0 e^{j\omega t} + \eta_1(t), \tag{2.58}$$

where

$$\Gamma_1 = -\text{diag}\{\cos\gamma_0, \cos\gamma_1, \ldots, \cos\gamma_{P-1}\},$$

$$\mathbf{A} = \{\mathbf{a}_0, \mathbf{a}_1, \ldots, \mathbf{a}_{P-1}\},$$

$$\mathbf{S}_0 = col\{E_0 e^{j\varphi_0}, E_1 e^{j\varphi_1}, \ldots, E_{P-1} e^{j\varphi_{P-1}}\},$$

and $\eta_1(t) = col\{\eta_0'(t), \eta_1'(t), \ldots, \eta_{M-1}'(t)\}$ is the background noise vector. A similar expression for an array of dipoles oriented along the y-axis is given by,

$$\mathbf{f}_y = \mathbf{A}\boldsymbol{\Gamma}_2\mathbf{S}_0 e^{j\omega t} + \eta_2(t), \tag{2.59}$$

where

$$\Gamma_2 = \mathrm{diag}\{\sin\gamma_0 \cos\theta_0 e^{j\kappa_0}, \sin\gamma_1 \cos\theta_1 e^{j\kappa_1}, \ldots, \sin\gamma_{P-1} \cos\theta_{P-1} e^{j\kappa_{P-1}}\},$$

$$\eta_2(t) = col\{\eta_0''(t), \eta_1''(t), \ldots, \eta_{M-1}''(t)\}.$$

2.3 Distributed Sensor Array

Frequently it is not possible to build an array of sensors with regular geometry such as ULA or UPA. Sonobuoy array in open sea [12], sensor array used for environment sensing [19], and sensor network for hazardous contaminants localization [20] are some the examples where a randomly distributed sensor array is used. There are two types of distributed sensors; (a) the sensors covering a small area or volume of interest and they are all wired (physically or wireless) to a central processor (b) in the second type, the sensors cover a large area and they are not wired to a central processor. However, each sensor can communicate with its nearest neighbors. Such a sensor array is often referred to as an ad-hoc sensor network. In a regular array (e.g., ULA or UPA) the basic information used is the phase difference between successive sensors. In distributed sensor array the basic information used is TOA or time difference of arrival (TDOA). When the sensor array is wired to a central processor, there exist several complex methods for the estimation of TOA, including a maximum likelihood method [21]. In a sensor network, however, the TOA estimation is based on a relatively simple principle such as relative signal strength (RSS) with limited accuracy, or cross correlation between the transmitted signal (which is known) and the received signal, which yields better estimates but at the cost of higher computational complexity. The estimation of DOA requires an array of sensors, which is not practical because of limited available power. However, the DOA may be simply estimated, though not very accurately, from the ratio of RSS observed at two-directional antennas [22].

2.3.1 Wired Distributed Sensor Array

The sensors are randomly placed over a limited area or volume. The digital outputs are connected to a central processor by means of physical wire or wireless connection through telemetry. There are two possibilities, namely, a near-field case where a source is close to the array, even surrounded by sensors, and another far-field case where the source is far away from the array (see Figure 2.25). The near-field source is more likely to be strong compared to the far-field source, hence more easily localized.

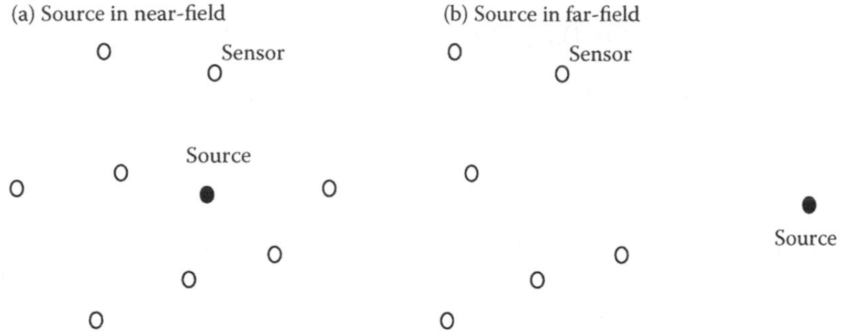

FIGURE 2.25
Distributed sensor array. (a) Near-field case. A source is surrounded by sensors. (b) Far-field case. The source is far away from the array, hence weaker.

Let M sensors be randomly located at (x_m, y_m), $m=0, 1,\dots, M-1$ in x-y plane. Arrange all digital outputs in a vector form,

$$\mathbf{f}(n) = col\{f_0(n), f_1(n),\dots, f_{M-1}(n)\}. \tag{2.60a}$$

Consider a source close to the sensors, a near-field case or away from the sensors, a far-field case. Our goal is, using the array outputs, to localize a source and to estimate source waveform (i.e., beamforming). Let $s(t)$ be the signal emitted by the source. The output of mth sensor is given by, assuming line-of-sight propagation,

$$f_m(t) = \alpha_m s(t-\tau_m) + \eta_m(t), \quad m = 0,1,\dots,M-1, \tag{2.60b}$$

where α_m and τ_m are attenuation and delay, respectively, of the signal and $\eta_m(t)$ is background noise. Let

$$w_{m,l}, \quad m=0, 1,\dots, M-1 \quad \text{and} \quad l=0, 1,\dots, L-1,$$

be weight coefficients. A weighted output array is defined as follows:

$$z(n) = \sum_{m=0}^{M-1}\sum_{l=0}^{L-1} w_{m,l} f_m(n-l), \tag{2.61}$$

which may be expressed in matrix form as

$$z(n) = \mathbf{w}^T\mathbf{f}, \tag{2.62}$$

where

$$\underset{ML\times 1}{\mathbf{w}} = \begin{Bmatrix} w_{0,0}, w_{0,1},\dots, w_{0,L-1}, w_{1,0}, w_{1,1},\dots, w_{1,L-1},\dots, \\ w_{M-1,0}, w_{M-1,1},\dots, w_{M-1,L-1} \end{Bmatrix}^T,$$

$$\mathbf{f} = \left\{ \begin{matrix} f_0(n), f_0(n-1),\ldots, f_0(n-L+1), f_1(n), f_1(n-1),\ldots, \\ f_1(n-L+1),\ldots, f_{M-1}(n), f_{M-1}(n-1),\ldots, f_{M-1}(n-L+1) \end{matrix} \right\}^T .$$
$$ML \times 1$$

Note that **w** and **f** are vectors obtained by *stacking* the weight coefficients and sensor outputs, respectively. The operation of arranging the columns of a matrix, for example, a space-time filter matrix, into a long vector is called stacking. This is often used in array signal processing. The operation of staking is given by

$$\mathbf{w} \Rightarrow \left\{ \mathbf{w}_0^T, \mathbf{w}_1^T, \ldots, \mathbf{w}_{M-1}^T \right\}^T ,$$

where $\mathbf{w}_m = \left\{ w_{m,0}, w_{m,1}, \ldots, w_{m,L-1} \right\}'$ is the mth column of a space-time filter matrix. We shall call the reverse operation de-stacking. Given a long filter vector of size $ML \times 1$, de-stacking involves creating a matrix of size $M \times L$.

The mean output power from Equation 2.62 is given by

$$E\{|z(n)|^2\} = \mathbf{w}^T E\{\mathbf{f}\mathbf{f}^T\}\mathbf{w} = \mathbf{w}^T \mathbf{c}_\mathbf{f} \mathbf{w}, \tag{2.63}$$

where $\mathbf{c}_\mathbf{f}$ is a ML×ML matrix, known as a multichannel covariance matrix of the array output. It may also be expressed as a block matrix of covariance and cross-covariance matrices. For example, for $M=3$ (three sensor array) $\mathbf{c}_\mathbf{f}$ is given by

$$\mathbf{c}_\mathbf{f} = \begin{bmatrix} \mathbf{c}_{f_0 f_0} & \mathbf{c}_{f_0 f_1} & \mathbf{c}_{f_0 f_2} \\ \mathbf{c}_{f_1 f_0} & \mathbf{c}_{f_1 f_1} & \mathbf{c}_{f_1 f_2} \\ \mathbf{c}_{f_2 f_0} & \mathbf{c}_{f_2 f_1} & \mathbf{c}_{f_2 f_2} \end{bmatrix}.$$

We like to maximize the mean output power subject to a condition that $\mathbf{w}^T\mathbf{w}=1$. The solution to the maximization problem above is straightforward; $\mathbf{w}_{opt}$ is an eigenvector of $\mathbf{c}_\mathbf{f}$ corresponding to its maximum eigenvalue [23]. By de-stacking $\mathbf{w}_{opt}$ we obtain the optimum weight matrix whose rows are the desired filters. The role of filtering is essentially to align all sensor outputs so that they are in phase. By combining all such aligned outputs, we form a beam pointing to the signal source or the strongest source when there is more than one signal source. Interestingly, it is not necessary to estimate the delays; it turns out to be a case of blind beamformation.

The different weight vectors seem to be derived by temporally shifting the weight vector of the reference sensor (which is marked as sensor #0). Thus, the relative TOA or TDOA may be obtained from the temporal shift of the weight vector of a sensor relative to the reference sensor. Further, the weight vector behaves like a bandpass filter with its pass band centered over the spectral peak of the transmitted signal. For narrowband signal, the TDOA is easily estimated from the phase delay measured at peak frequency relative to that of the weight vector at reference sensor. This approach will work if TDOAs satisfy an inequality: $|\omega_{peak} TDOA/\Delta t| \leq \pi$, where Δt is the sampling interval.

We shall demonstrate some of the nice properties of Yao's eigenvector-based weight vectors for TDOA estimation and blind beamformation through an example. We assume a four-sensor distributed array. The sensors are located at (0,0), (20,1), (15,20), and (0,15), being the x and y coordinates in meters. There are two uncorrelated sources, s_1 in the near-field range at (10,10) and s_2 source in the far-field range at (25,25), which serves the purpose of an interference source. The amplitude of the signal from the near-field source (s_1) is five time greater than that of the signal from the far-field source (s_2). The distributed array geometry and the position of the signal sources are shown in Figure 2.26.

Both sources emit narrow band uncorrelated signals with the same center frequency at 25 Hz and their bandwidth is equal to 10 Hz. The sampling interval was taken as Δt=0.005 sec. The wave speed was set to 1,500 m/sec (sound speed in sea). The filter length is L=64. From the eigenvector corresponding to the largest eigenvalue, we have computed filters to be applied to each sensor output.

All filters are of the same form except for a relative shift (see Figure 2.27), which enables us to compute the TDOA. The estimated and actual TDOAs are shown in Table 2.2. Estimated waveforms as filter output are shown in Figure 2.28a. We have switched off the interference source in the far-field. The waveform estimate, shown in Figure 2.28b, is very close to that in (a) (normalized mean square error (mse) is 0.009), which validates our claim that the filter focuses on the strong source.

2.3.2 Ad-hoc Sensor Network

A large sensor network, having a few hundred to a few thousand sensors and spread over a large area in some arbitrary form, is an ad-hoc sensor network. Sensors are not wired and there is no direct radio link to a central processor, but they are able to communicate with their immediate neighbors. To

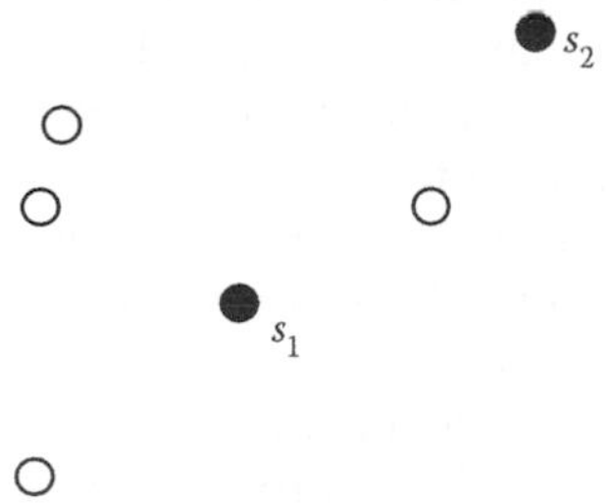

FIGURE 2.26
Four sensor distributed array (∘) and two sources (•).

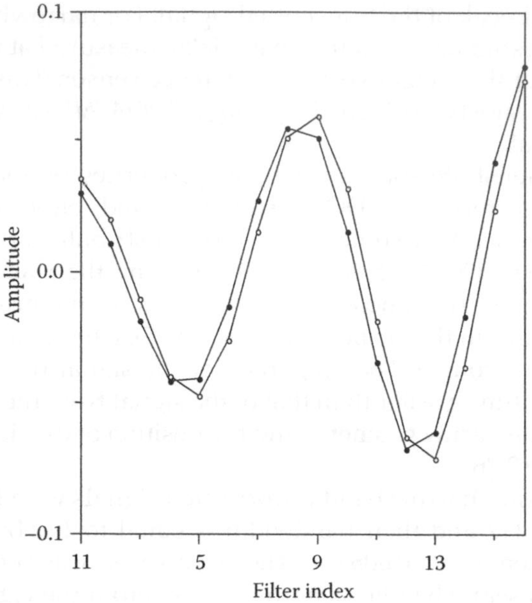

FIGURE 2.27
Filter coefficients for sensor one (thick line) and for sensor three (dashed line).

TABLE 2.2
TDOA estimates

	Sensor #1	Sensor #2	Sensor #3	Sensor #4
Estimated	0.0	−0.0862	−0.3741	−0.3932
Actual	0.0	−0.0918	−0.3949	−0.3949

maintain low power consumption and low cost, the sensors would necessarily have very limited on-board processing capability. An ad-hoc sensor network finds application in many human activities, such as meteorological data collection, tracking of animals in wildlife studies, surveillance work for military or antiterrorism, multihop communication, etc. A common requirement in all the above applications is that of being able to know the position of sensors, some of which may even be mobile. The most popular means of localization is the use of a global positioning system (GPS), but that requires additional hardware and power to operate the GPS hardware. Naturally, this is not a viable practical solution. Localization in ad-hoc sensor network is largely based on measurement of distances between pairs of nodes. The distance measurement is often done through TOA of radio signals transmitted from beacons located at known positions. This method requires precise clock synchronization. A simpler approach is to measure the received signal strength (RSS). Each sensor transmits a signal of fixed strength and all neighboring sensors measure the relative strength of the received signal.

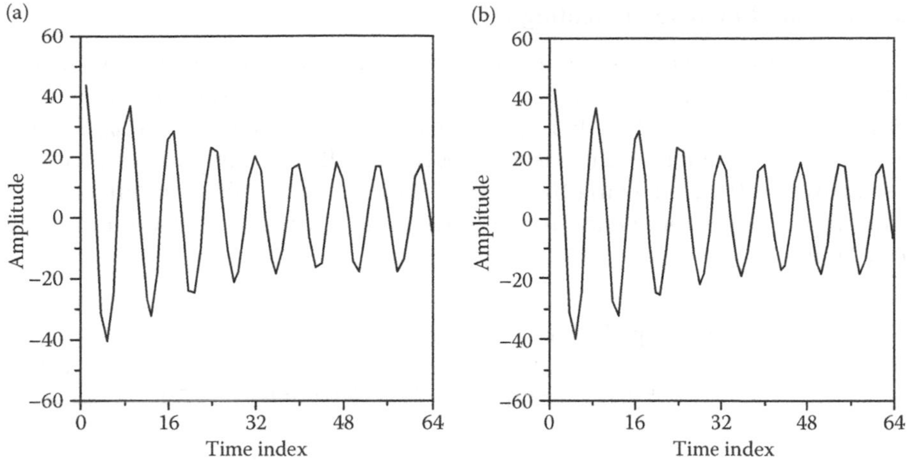

FIGURE 2.28
Estimated waveforms (a) with interference and (b) without interference. The estimated waveform, shown in (b), is very close to that in (a) (normalized mse is 0.009).

The RSS information can be translated into range provided the nature of the radio channel is known. Let $P_{i,j}$ be power in decibels (that is, $10\log_{10}$ (power in milliwatts)) received at sensor i transmitted by sensor j. It is assumed that $P_{i,j}$ is a Gaussian random variable with mean given by [24]

$$\bar{P}_{i,j} = P_0 - 10\alpha \log_{10}\left(\frac{d_{i,j}}{d_0}\right), \tag{2.64}$$

where P_0 is power measured in decibels at a distance d_0 from the transmitting sensor, $\bar{P}_{i,j}$ is mean power received at sensor i, and α is the path-loss exponent depending upon the channel, typically between two and four. In free space $\alpha=2$. Equation 2.64 leads to the following expression for range measurement

$$d_{i,j} = d_0 10^{(P_0 - \bar{P}_{i,j})/(10\alpha)}.$$

Maximum likelihood estimate of coordinates of ith sensor relative to jth sensor where a transmitter is located and its location is known, may be obtained by minimizing an expression [24]

$$\arg\min_{(x_i,y_i)} \sum_{i=1}^{N+m} \sum_{\substack{j\in H(i)\\ j\neq i}} \left(\ln \frac{d_{i,j}^2}{(x_i - x_j)^2 + (y_i - y_j)^2} \right)^2,$$

where $H(i)=\{j$: sensor j makes pairwise observation with sensor $j\}$ and m stands for the number of sensors with known locations, often called anchors or reference sensors.

2.3.3 Multidimensional Scaling (MDS)

An algorithm constructs a map of objects (in our case sensors) by using distances as dissimilarity between sensor nodes. Let M sensors be located at points $\mathbf{p}_i$, i=1, 2,..., M, where $\mathbf{p}_i$ is the coordinate vector, $\mathbf{p}_i=(x_i, y_i, z_i)$. The information available is the distance between all possible pairs of sensors, assuming for the time being they are all within communication range. Consider a pair of sensors i and j in 3D space and the distance between them is $D_{i,j}$. Then

$$\begin{aligned} D_{i,j}^2 &= \left\|\mathbf{p}_i - \mathbf{p}_j\right\|^2 = [(x_i - x_j)^2 + (y_i - y_j)^2 + (z_i - z_j)^2] \\ &= (x_i^2 + y_i^2 + z_i^2) + (x_j^2 + y_j^2 + z_j^2) - 2(x_i x_j + y_i y_j + z_i z_j). \end{aligned} \tag{2.65}$$

We make an important assumption that the center of coordinates is the geometrical center of sensor array, which implies that

$$\sum_i x_i = \sum_i y_i = \sum_i z_i = 0. \tag{2.66}$$

Consider the following relations

$$\begin{aligned} \bar{D}_i^{row} &= \sum_j D_{i,j}^2 = Mx_i^2 + My_i^2 + Mz_i^2 + \sum_j (x_j^2 + y_j^2 + z_j^2) \\ \bar{D}_j^{col} &= \sum_i D_{i,j}^2 = Mx_j^2 + My_j^2 + Mz_j^2 + \sum_i (x_i^2 + y_i^2 + z_i^2) \end{aligned} \tag{2.67}$$

$$\bar{D}_0 = \sum_i \sum_j D_{i,j}^2 = M\sum_i (x_i^2 + y_i^2 + z_i^2) + M\sum_j (x_j^2 + y_j^2 + z_j^2).$$

Note that the cross-terms in Equation 2.65 vanish on account of Equation 2.66. From Equation 2.67 it is easy to show that

$$\frac{\bar{D}_i^{row}}{M} + \frac{\bar{D}_j^{col}}{M} - \frac{\bar{D}_0}{M^2} = (x_i^2 + y_i^2 + z_i^2) + (x_j^2 + y_j^2 + z_j^2). \tag{2.68}$$

Using Equation 2.68 in Equation 2.65, we obtain an interesting relation, which forms the basis of the multidimensional scaling (MDS) algorithm,

$$D_{i,j}^2 - \frac{\bar{D}_i^{row}}{M} - \frac{\bar{D}_j^{col}}{M} + \frac{\bar{D}_0}{M^2} = -2(x_i x_j + y_i y_j + z_i z_j). \tag{2.69}$$

Define

$$B_{i,j} = -\frac{1}{2}[D_{i,j}^2 - \frac{\bar{D}_i^{row}}{M} - \frac{\bar{D}_j^{col}}{M} + \frac{\bar{D}_0}{M^2}] = (x_i x_j + y_i y_j + z_i z_j), \tag{2.70}$$

and a matrix **B**, where$\{\mathbf{B}\}_{i,j} = B_{i,j}$. Notice that **B** is a sum of outer products of three vectors,

$$\mathbf{x} = \{x_1, x_2, \ldots, x_M\}$$

$$\mathbf{y} = \{y_1, y_2, \ldots, y_M\}$$

$$\mathbf{z} = \{z_1, z_2, \ldots, z_M\},$$

that is,

$$\mathbf{B} = \mathbf{x}\mathbf{x}^T + \mathbf{y}\mathbf{y}^T + \mathbf{z}\mathbf{z}^T. \tag{2.71}$$

This is a useful result as the rank of **B**, in the noise-free case, will be equal to or less than three. The latter situation will arise when all sensor nodes are clustered or lie on a line or in a plane. Assuming that the above degenerate cases are ruled out, **B** will have three nonzero eigenvalues $\{\lambda_1, \lambda_2, \lambda_3\}$ and the corresponding eigenvectors $\{\mathbf{u}_1, \mathbf{u}_2, \mathbf{u}_3\}$ will be related to position coordinate vectors, **x**, **y**, and **z**. In fact, the relationship is given by

$$\mathbf{x} = \mathbf{u}_1\sqrt{\lambda_1}, \qquad \mathbf{y} = \mathbf{u}_2\sqrt{\lambda_2}, \qquad \mathbf{z} = \mathbf{u}_3\sqrt{\lambda_3}. \tag{2.72}$$

Note that **B** (Equation 2.70) is a symmetric matrix and hence its eigenvalues are real positive. But the eigenvectors are not unique. There is some normalized orthogonal matrix Ω (3×3) that will introduce arbitrary translation, rotation, and reflection of the estimated sensor location coordinates [25]. To illustrate the MDS algorithm we have considered six sensors, all located in a plane, with coordinates (0,0), (2,0), (3,1), (2,4), (0,4), (–1,1). The distance matrix as in Equation 2.65 and the **B** matrix as defined in Equation 2.70 were computed. The eigenvalues of the **B** matrix are: 17.33, 12.0, 0.0, 0.0, 0.0. The eigenvectors corresponding to nonzero eigenvalues were used to estimate the coordinates of the sensors. The results are shown in Figure 2.29, all estimated coordinates are scaled up by a factor of ten.

In the presence of noise or errors in the estimation of distance, the rank of the **B** matrix will be equal to N, that is, it is full rank, although the first three eigenvalues will be dominating and the corresponding eigenvectors will yield optimum estimates in the sense that

$$\sum_{i=1}^{N}\sum_{j=1}^{N}(D_{i,j}^2 - \hat{D}_{i,j}^2)^2 = \min,$$

where $\hat{D}_{i,j}^2$ is estimated distance [25].

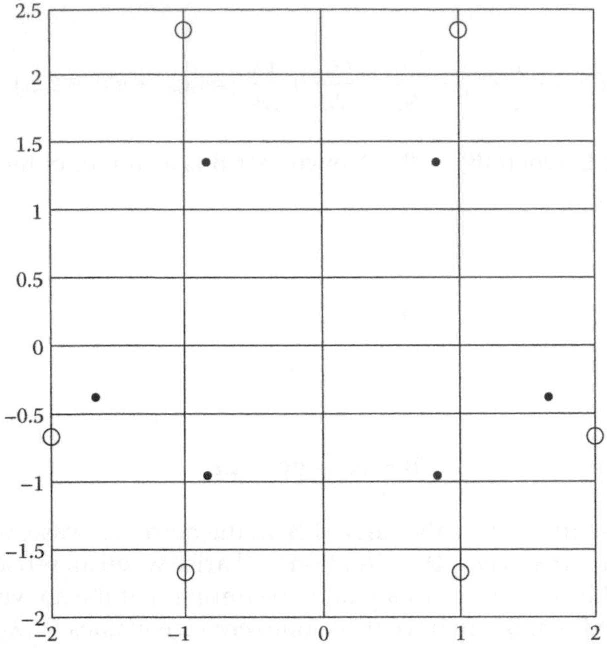

FIGURE 2.29
Assumed sensor locations are shown by empty circles and the estimated locations are shown by filled circles. We have scaled up the estimated coordinates by a factor of ten.

2.4 Broadband Sensor Array

Very often an array of sensors is required to receive broadband signals, which include both natural signals (e.g., seismic and sonar signals) or man-made signals (e.g., communication signals). Not only must the sensors be broadband, but also a special processing technique must be devised to exploit the broadband property of the signal. The bandwidth of the array response function depends upon the time taken by the wavefront to sweep across the array aperture, hence on the DOA for ULA but only on the diameter of UCA. A broadband signal may be considered as an aggregate of many narrowband signals covering the entire bandwidth. Since every narrowband signal is capable of determining the source parameters, we have considerable redundant information that may be used to fight against the noise and the model uncertainties. Alternatively, a broadband signal may also be represented by a collection of independent temporal samples (sampled at Nyquist rate) or snapshots, each carrying information about the source parameters. In this section, we shall introduce the concept of array bandwidth, and frequency and time snapshots.

2.4.1 Bandwidth of Array

The bandwidth of an array is the bandwidth of its transfer function, which for a ULA is given by Equation 2.3a

$$H(\omega\tau) = \frac{1}{M}\sum_{m=0}^{M-1} e^{j\omega(md/c)\sin\theta} = \frac{\sin((M/2)\omega\tau)}{M\sin(\omega\tau/2)} e^{j(M-1/2)\omega\tau}.$$

A plot of $H(\omega(d/c)\sin\theta)$ with θ held fixed but ω varied will be same as in Figure 2.2 except that the x-axis variation is now linear in ω, in place of the nonlinear variation in θ. If we define the bandwidth as one half of the distance between the first nulls, we obtain

$$\Delta\omega = \frac{2\pi}{M}\frac{c}{d\sin\theta} = \frac{2\pi}{(MD/c_x)} = \frac{2\pi}{\tau}, \tag{2.73}$$

where τ is the time required to sweep across the array aperture. Note that the bandwidth is infinite when the wavefront is incident on the broadside, that is, the array is steered in the direction of source; but it is equal to $2\pi c/Md$ when the wavefront is incident from the endfire. For a circular array, however, the wavefront has to sweep a constant aperture equal to the diameter of the circle, independent of the azimuth angle. This essential difference between a ULA and a UCA is illustrated in Figure 2.30. The bandwidth of a long ULA or a large diameter UCA is very small and hence much of the

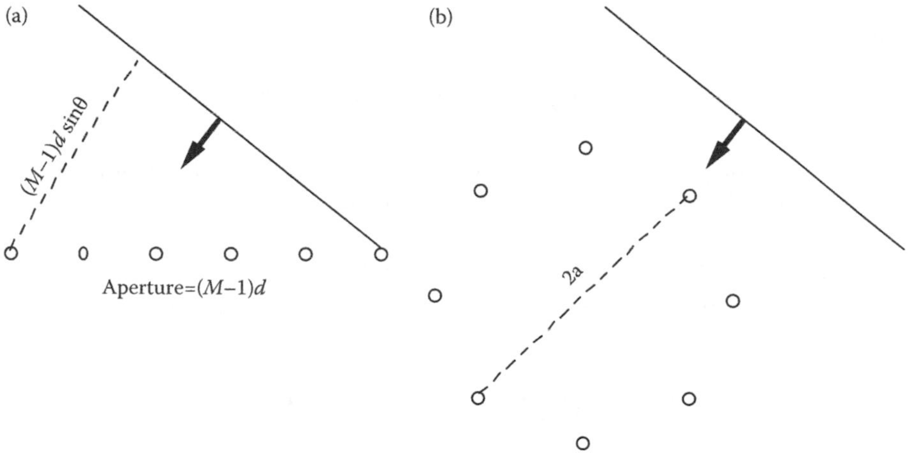

FIGURE 2.30
The essential difference between a ULA and UCA is that the effective aperture (dashed line) is azimuth-dependent for a ULA, but independent of the azimuth for a UCA. (a) Linear array, the effective aperture=$(M-1)d\sin\theta$. (b) Circular array, the effective aperture=diameter of the circle.

energy of a broadband source will be lost, unless the array is steered in the direction of the desired source. Further, as pointed out in Section 2.1, when there is more than one source radiating at different frequencies, there is a possibility for ambiguity. Let P narrowband sources with DOAs θ_i, $i=0,1,\ldots,P-1$ and center frequencies ω_i, $i=0, 1,\ldots, P-1$ be incident on a ULA. Further, we assume that the center frequency and DOA pair satisfy the following relation:

$$\omega_i \frac{d}{c}\sin\theta_i = \tau_0(\text{constant}), \quad i=0, 1,\ldots, P-1, \tag{2.74}$$

where τ_0 is the delay per sensor introduced by all sources. Now, through a process of sum and delay, the array is steered *simultaneously* to all sources, in other words, the array will "see" all P sources at the same time. When the sources are broadband with overlapping spectrum we can always find a set of frequencies that satisfies Equation 2.74. As a result, when the array is steered to one of the sources, the output may be contaminated with the power derived from other sources. Such an interference is unacceptable particularly when waveform estimation is the goal. Some of these problems can be overcome by introducing temporal processing over and above the delay and sum type of spatial processing.

A snapshot may be expressed in the frequency domain as

$$\mathbf{f}(t) = \frac{1}{2\pi}\int_{-\infty}^{\infty} \mathbf{a}(\omega,\theta_0)dF_0(\omega)e^{j\omega t}, \tag{2.75}$$

where

$$\mathbf{a}(\omega,\theta_0) = col\left\{1, e^{-j\omega(d\sin\theta_0/c)},\ldots, e^{-j\omega(M-1)(d\sin\theta_0/c)}\right\}.$$

Because of the properties of the steering vector listed in Sub-section 2.1.4.2, we can conclude the following about the broadband array output:

(a) $\mathbf{f}_{\theta_0}(t) = \mathbf{f}_{\pi-\theta_0}(t)$. Follows from property (a).

(b) $\mathbf{f}_{\theta_0}(t) \neq \mathbf{f}_{-\theta_0}(t)$. From property (b) and Equation 2.74 we can write

$$\mathbf{f}_{-\theta_0}(t) = \frac{1}{2\pi}\int_{-\infty}^{\infty} \mathbf{a}^*(\omega,\theta_0)dF_0(\omega)e^{j\omega t} \neq \mathbf{f}_{\theta_0}(t),$$

except in the unlikely event of $dF_0(\omega)$ being real. Thus, the east-west ambiguity shown for the narrowband complex signal does not apply to the broadband signals.

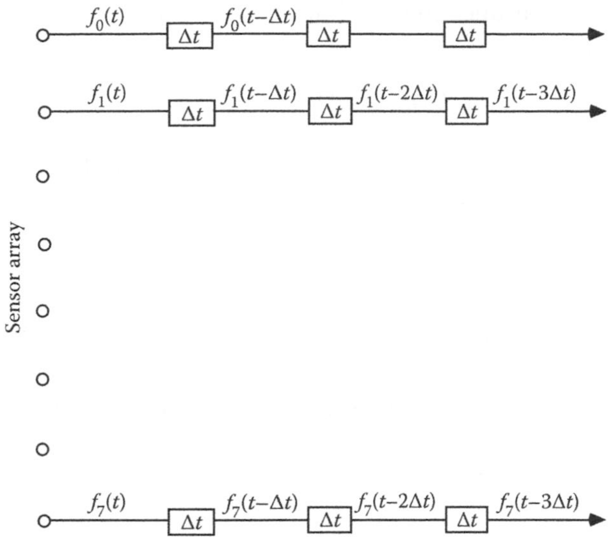

FIGURE 2.31
The output of an eight-sensor ULA is sampled at different time instants in the past. All outputs taken at the same time instant are grouped into a vector called a *temporal* snapshot. We have as many temporal snapshots as the number of samples or taps taken from each sensor output.

2.4.2 Broadband Signals

In the case of a broadband signal, since a snapshot may vary rapidly, it is necessary that many snapshots must be collected at different time instants in the past; for example, $\mathbf{f}(t)$, $\mathbf{f}(t-\Delta t)$, $\mathbf{f}(t-2\Delta t),\ldots,$ $\mathbf{f}(t-(N-1))\Delta t)$ are N past or delayed snapshots (see Figure 2.31). An alternate approach, in the case of a broadband signal, is to go over to the frequency domain (temporal frequency). The output of each sensor, consisting of N samples, is subjected to Fourier analysis (DFT). A collection of the Fourier coefficients, one from each sensor at a fixed frequency, constitutes a *frequency* snapshot. The array signal processing of broadband signals using the frequency snapshots closely follows the time domain approach for narrowband signals (after removing the carrier), widely used in radar signal processing. In place of a covariance matrix we use a spectral matrix, which indeed is a SCM. In this book, we shall emphasize the frequency domain approach, as the time domain approach for wideband signals turns out to be conceptually a bit more involved. We begin with time domain approach. First, let us introduce some new quantities required in the processing. All delayed snapshots are stacked one above the other to form one large vector of size $M \cdot N$,

$$\mathbf{f}_{MN}(t) = col\left\{\mathbf{f}^T(t),\, \mathbf{f}^T(t-\Delta t),\, \mathbf{f}^T(t-2\Delta t),\ldots,\, \mathbf{f}^T(t-(N-1)\Delta t))\right\}. \tag{2.76}$$

We define a covariance matrix, known as a spatio-temporal covariance matrix (STCM),

$$\mathbf{C}_{STCM} = E\{\mathbf{f}_{MN}(t)\mathbf{f}_{MN}^H(t)\}. \tag{2.77}$$

As an example, consider a two-sensor array ($M=2$) and two delayed snapshots ($N=2$). The STCM is given by

$$\mathbf{C}_{STCM} = \begin{bmatrix} C_f(0) & C_f(\tau_0) & C_f(\Delta t) & C_f(\Delta t+\tau_0) \\ C_f(-\tau_0) & C_f(0) & C_f(\Delta t-\tau_0) & C_f(\Delta t) \\ C_f(\Delta t) & C_f(\Delta t-\tau_0) & C_f(0) & C_f(\tau_0) \\ C_f(\Delta t+\tau_0) & C_f(\Delta t) & C_f(-\tau_0) & C_f(0) \end{bmatrix},$$

where $\tau 0=(d\sin\theta_0/c)$ is a propagation delay per sensor. We express the stacked vector by using Equation 2.76 in Equation 2.77,

$$\mathbf{f}_{stacked}(t) = \frac{1}{2\pi}\int_{-\infty}^{\infty} \boldsymbol{\zeta}(\omega)\otimes\mathbf{a}(\omega,\theta_0)dF_0(\omega)e^{j\omega t}, \tag{2.78}$$

where

$$\boldsymbol{\zeta}(\omega) = col\{1, e^{-j\omega\Delta t},\dots, e^{-j\omega(N-1)\Delta t}\},$$

and ⊗ stands for the Kronecker product. Define a direction vector $\mathbf{h}(\omega, \theta_0) = \zeta(\omega) \otimes \mathbf{a}(\omega, \theta_0)$ and rewrite Equation 2.65 as

$$\mathbf{f}_{stacked}(t) = \frac{1}{2\pi}\int_{-\infty}^{\infty} \mathbf{h}(\omega,\theta_0)dF_0(\omega)e^{j\omega t}, \tag{2.79}$$

which may be considered as an extended output of the array (stacked vector) due to a source at the azimuth angle θ_0. Using Equation 2.79, we can express STCM as

$$\begin{aligned}\mathbf{C}_{STCM} &= E\{\mathbf{f}_{stacked}(t)\mathbf{f}_{stacked}^H(t)\} \\ &= \frac{1}{2\pi}\int_{-\infty}^{\infty} \mathbf{h}(\omega,\theta_0)\mathbf{h}^H(\omega,\theta_0)S_{f_0}(\omega)d\omega.\end{aligned} \tag{2.80}$$

Let $d\mathbf{F}_{stacked}(\omega)$ be the generalized FT of the extended vector output, $\mathbf{f}_{stacked}(t)$,

$$d\mathbf{F}_{stacked}(\omega) = \mathbf{h}(\omega,\theta_0)dF_0(\omega). \tag{2.81}$$

Premultiply on both sides of Equation 2.81 with the steering vector in some desired direction and evaluate the expected value of the magnitude square. Dividing by $(M \cdot N)^2$ we get the output power,

$$\begin{aligned}\frac{1}{(M\cdot N)^2}\mathbf{h}^H(\omega,\theta)\mathbf{S}_{f_{stacked}}(\omega)\mathbf{h}(\omega,\theta) = \\ \left|\frac{\mathbf{h}^H(\omega,\theta)\mathbf{h}(\omega,\theta_0)}{M\cdot N}\right|^2 S_0(\omega) = \left|\frac{\mathbf{a}^H(\omega,\theta)\mathbf{a}(\omega,\theta_0}{M}\right|^2 S_0(\omega),\end{aligned} \tag{2.82}$$

where $\mathbf{S}_{f_{stacked}}(\omega)$ is the spectral matrix of an extended array signal. Interestingly, inspite of delayed snapshots used in deriving Equation 2.82, the output power remains the same as in Equation 2.19b. However, the STCM is different from the SCM. For example, reconsider the frequency-direction ambiguity due to the fact that if there is a set of DOAs and frequencies such that

$$\omega_i \frac{d}{c}\sin\theta_i = \tau_0(\text{constant}), \quad i = 0, 1, \ldots, P-1,$$

the steering vector $\mathbf{a}(\omega_i, \theta_i)$ remains unchanged. This, however, is not true for a ULA with a tapped delay line, as the steering vectors, $\mathbf{h}(\omega_i, \theta_i) = \zeta(\omega_i) \otimes \mathbf{a}(\omega_i, \theta_i)$, $i = 0, 1, \ldots, P-1$, will be different because of $\zeta(\omega_i)$. It may be noted that for a circular array there is no frequency-direction ambiguity of the type described above even when no tapped delay line is used [26].

2.5 Source and Sensor Arrays

In nonoptical imaging systems such as radar, sonar, seismic, and biomedical imaging systems, we often use an array of sources and an array of sensors. The main idea is to illuminate an element to be imaged from different directions and collect as much of the scattered radiation as possible for the purpose of imaging. To achieve this goal in a straightforward manner would require an impractically large array. It is known that a large aperture array can be synthesized from a small array of transmitters and receivers (transceivers). This leads to the concept of coarray used in radar imaging. In tomographic imaging, the element to be imaged has to be illuminated from all directions. We, therefore, need a boundary array of transceivers or a single source and a line of sensors going round the object as in some biomedical imaging systems. In seismic exploration, a source array is often used to provide the additional degree of freedom required to combat the high noise and the interference level, and also to estimate the wave speed in different rock layers.

2.5.1 Coarray

An imaging system is basically concerned with faithful mapping of a point target in the object space into a point in the image space. The point spread function (PSF) of the imaging system describes the mapping operation. Ideally, it is a delta function, but in practice a point may be mapped into a small area, the size of which will depend upon the aperture of the array used in the imaging system. The coarray is required to synthesize an arbitrary PSF using a finite array of sensors and sources. A source array (transmit array) is weighted with a complex function $t(x)$ and the sensor array (receive array) is also weighted with another complex weight function $r(y)$. The response function of a source array in the direction of the scatterer (in far-field) is given by

$$T(u,v)=\iint_{-\infty}^{+\infty} t(x)\delta(y)e^{j(ux+vy)}dxdy, \tag{2.83a}$$

and the response function of the sensor array in the direction of the scatterer is given by

$$R(u,v)=\iint_{-\infty}^{+\infty} r(y)\delta(x)e^{j(ux+vy)}dxdy, \tag{2.83b}$$

where $u=k\sin\theta\cos\varphi$, $v=k\sin\theta\sin\varphi$, and $k=(2\pi/\lambda)$ is the wavenumber. Let $\rho(u,v)$ be the reflection coefficient as a function of the azimuth angle φ and the elevation angle θ. The total response of the source/sensor array is given by

$$O(u,v)=T(u,v)\rho(u,v)R(u,v).$$

This is the image of a point scatterer in the frequency domain. If we assume that the scatterer is omnidirectional, that is, $\rho(u,v)=\rho_0$ (constant). The PSF of the array is given by the inverse FT of $T(u,v)\,R(u,v)$,

$$\begin{aligned}PSF&=\sum_x\sum_y t(x)\delta(y)r(q-y)\delta(p-x).\\&=t(p)r(q)\end{aligned}$$

Thus, we note that the combined PSF of a source array along the x-axis and a sensor array along the y-axis is equal to the product of two weighting functions; for example, when the source and sensor arrays are uniformly weighted, the PSF of the combined system is a rectangular function [27]. This is illustrated in Figure 2.32 for an L-shaped array. For an arbitrary distribution of sources and sensors the PSF is given by a convolution of the source and sensor weighting functions,

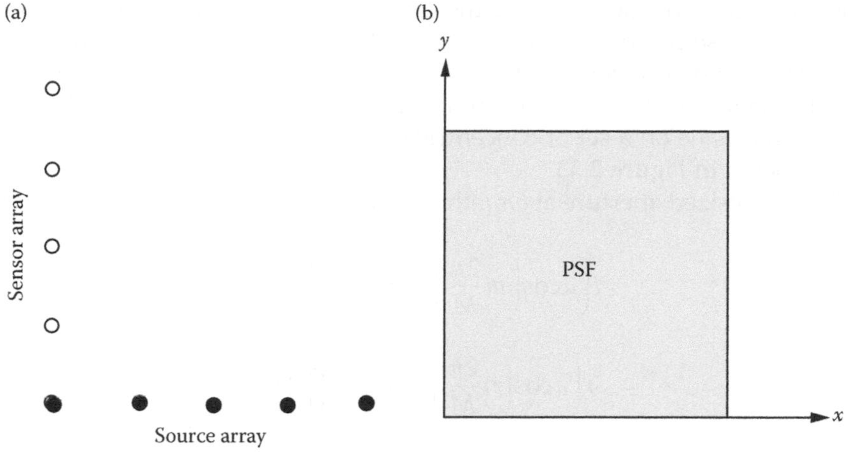

FIGURE 2.32
(a) Source and sensor array and (b) point spread function (PSF) on a grid, known as coarray.

$$\text{PSF} = \iint_{-\infty}^{+\infty} t(x,y)r(p-x,q-y)dxdy. \tag{2.84}$$

The PSF at point (p,q) is obtained by summing the product of two weight functions over two lines $p=x+x'$ and $q=y+y'$, where (x,y) refer to a point in the source array and (x',y') refer to a point in the sensor array. Such a coarray is also known as a sum coarray. Since we have a discrete set of point sources and sensors, the discrete version of Equation 2.84 is given by $\mathbf{R}^T\mathbf{T}$, where $\mathbf{T}=col\{t_0, t_1,\ldots, t_{M-1}\}$ and $\mathbf{R}=col\{r_0, r_1,\ldots, r_{M-1}\}$.

An arbitrary PSF may be synthesized through eigen-decomposition. Let $\mathbf{P}$ be the desired PSF given in the form of a matrix, which we shall assume to have a Hermitian symmetry. We can express its eigen-decomposition as $\mathbf{P} = \Sigma_{l=0}^{M-1}\lambda_l \mathbf{u}_l^H \mathbf{u}_l$, where λ_l is lth eigenvalue (real) and $\mathbf{u}_l$ is the corresponding eigenvector. We let $\mathbf{T}=\mathbf{R}=\mathbf{u}_l$, that is, both source and sensor arrays are weighted by the same eigenvector of the desired PSF matrix, and thus an image is formed. This step is repeated over all significant eigenvectors. All such images are linearly combined after weighting each with the corresponding eigenvalue [28].

Let us consider an example of a circular array of eight transceivers uniformly distributed on the circumference. A transceiver consists of a source and a sensor physically placed at the same place. The transceivers are located at

$$\begin{bmatrix} (a,0), \left(\frac{a}{\sqrt{2}}, \frac{a}{\sqrt{2}}\right), (0,a), \left(-\frac{a}{\sqrt{2}}, \frac{a}{\sqrt{2}}\right), (0,-a), \\ \left(-\frac{a}{\sqrt{2}}, -\frac{a}{\sqrt{2}}\right), (-a,0), \left(\frac{a}{\sqrt{2}}, -\frac{a}{\sqrt{2}}\right) \end{bmatrix},$$

where a stands for the radius of the circle. The sum coarray consists of 64 locations whose coordinates may be found by summing the first column and the first row entries from Table 2.3. The actual coordinates, thus obtained, are also shown in the table. There are in all $(N^2/2)+1$ distinct nodes. The coarray nodes lie on a set of concentric circles of radii (0, 0.76a, 1.141a, 1.85a, 2a), as shown in Figure 2.33.

The synthesized aperture at (m,n)th node of the circular coarray is given by

$$t\left(a\cos\left(m\frac{2\pi}{M}\right),a\sin\left(m\frac{2\pi}{M}\right)\right),$$

$$r\left(a\cos\left(n\frac{2\pi}{M}\right),a\sin\left(n\frac{2\pi}{M}\right)\right).$$

The PSF at (p,q), where $p=x_m+x_n$ and $q=y_m+y_n$, may be obtained from the discrete equivalent of Equation (2.83). Note that the coordinates of the mth point are (x_m, y_m) and those of the nth point are (x_n, y_n).

$$PSF(p,q)=t\left(a\cos\left(m\frac{2\pi}{M}\right),a\sin\left(m\frac{2\pi}{M}\right)\right)r\left(a\cos\left(m\frac{2\pi}{M}\right),a\sin\left(m\frac{2\pi}{M}\right)\right)$$

$$m=n$$

$$=\sum_{i=0}^{(M/2)-1}\left[\begin{array}{l} t\left(a\cos\left(i\frac{2\pi}{M}\right),a\sin\left(i\frac{2\pi}{M}\right)\right)r\left(a\cos\left(\left(i+\frac{M}{2}\right)\frac{2\pi}{M}\right),\right. \\ \left. a\sin\left(\left(i+\frac{M}{2}\right)\frac{2\pi}{M}\right)\right) \\ +r\left(a\cos\left(i\frac{2\pi}{M}\right),a\sin\left(i\frac{2\pi}{M}\right)\right)t\left(a\cos\left(\left(i+\frac{M}{2}\right)\frac{2\pi}{M}\right),\right. \\ \left. a\sin\left(\left(i+\frac{M}{2}\right)\frac{2\pi}{M}\right)\right) \end{array}\right] \tag{2.85}$$

$$|m-n|=\frac{M}{2}$$

$$=t\left(a\cos\left(m\frac{2\pi}{M}\right),a\sin\left(m\frac{2\pi}{M}\right)\right)r\left(a\cos\left(n\frac{2\pi}{M}\right),a\sin\left(n\frac{2\pi}{M}\right)\right)$$

$$+r\left(a\cos\left(m\frac{2\pi}{M}\right),a\sin\left(m\frac{2\pi}{M}\right)\right)t\left(a\cos\left(n\frac{2\pi}{M}\right),a\sin\left(n\frac{2\pi}{M}\right)\right).$$

$$m\neq n$$

$$|m-n|\neq\frac{M}{2}$$

TABLE 2.3

Sum coarray

Column entry → m	$(a, 0)$	$\left(\frac{a}{\sqrt{2}}, \frac{a}{\sqrt{2}}\right)$	$(0, a)$	$\left(-\frac{a}{\sqrt{2}}, \frac{a}{\sqrt{2}}\right)$	$(0, -a)$	$-\left(\frac{a}{\sqrt{2}}, \frac{a}{\sqrt{2}}\right)$	$(-a, 0)$	$\left(\frac{a}{\sqrt{2}}, -\frac{a}{\sqrt{2}}\right)$
$(a, 0)$	$(2a, 0)$	$a+\frac{a}{\sqrt{2}}, \frac{a}{\sqrt{2}}$	(a, a)	$a-\frac{a}{\sqrt{2}}, \frac{a}{\sqrt{2}}$	$(a, -a)$	$a-\frac{a}{\sqrt{2}}, -\frac{a}{\sqrt{2}}$	$(0, 0)$	$a+\frac{a}{\sqrt{2}}, -\frac{a}{\sqrt{2}}$
$\left(\frac{a}{\sqrt{2}}, \frac{a}{\sqrt{2}}\right)$	$a+\frac{a}{\sqrt{2}}, \frac{a}{\sqrt{2}}$	$\sqrt{2}a, \sqrt{2}a$	$\frac{a}{\sqrt{2}}, a+\frac{a}{\sqrt{2}}$	$(0, \sqrt{2}a)$	$\frac{a}{\sqrt{2}}, \frac{a}{\sqrt{2}}-a$	$(0, 0)$	$\frac{a}{\sqrt{2}}-a, \frac{a}{\sqrt{2}}$	$(\sqrt{2}a, 0)$
$(0, a)$	(a, a)	$\frac{a}{\sqrt{2}}, a+\frac{a}{\sqrt{2}}$	$(0, 2a)$	$-\frac{a}{\sqrt{2}}, a+\frac{a}{\sqrt{2}}$	$(0, 0)$	$-\frac{a}{\sqrt{2}}, a-\frac{a}{\sqrt{2}}$	$(-a, a)$	$\frac{a}{\sqrt{2}}, a-\frac{a}{\sqrt{2}}$
$\left(-\frac{a}{\sqrt{2}}, \frac{a}{\sqrt{2}}\right)$	$a-\frac{a}{\sqrt{2}}, \frac{a}{\sqrt{2}}$	$(0, \sqrt{2}a)$	$-\frac{a}{\sqrt{2}}, a+\frac{a}{\sqrt{2}}$	$-\sqrt{2}a, \sqrt{2}a$	$-\frac{a}{\sqrt{2}}, -a+\frac{a}{\sqrt{2}}$	$-\sqrt{2}a, 0$	$-a-\frac{a}{\sqrt{2}}, \frac{a}{\sqrt{2}}$	$(0, 0)$
$(0, -a)$	$(a, -a)$	$\frac{a}{\sqrt{2}}, \frac{a}{\sqrt{2}}-a$	$(0, 0)$	$-\frac{a}{\sqrt{2}}, -a+\frac{a}{\sqrt{2}}$	$(0.2a)$	$-\frac{a}{\sqrt{2}}, -a-\frac{a}{\sqrt{2}}$	$(-a, -a)$	$\frac{a}{\sqrt{2}}, -a-\frac{a}{\sqrt{2}}$
$-\left(\frac{a}{\sqrt{2}}, \frac{a}{\sqrt{2}}\right)$	$a-\frac{a}{\sqrt{2}}, -\frac{a}{\sqrt{2}}$	$(0, 0)$	$-\frac{a}{\sqrt{2}}, a-\frac{a}{\sqrt{2}}$	$-\sqrt{2}a, 0$	$-\frac{a}{\sqrt{2}}, -a-\frac{a}{\sqrt{2}}$	$-\sqrt{2}a, -\sqrt{2}a$	$-a-\frac{a}{\sqrt{2}}, -\frac{a}{\sqrt{2}}$	$(0, -\sqrt{2}a)$
$(-a, 0)$	$(0, 0)$	$\frac{a}{\sqrt{2}}-a, \frac{a}{\sqrt{2}}$	$(-a, a)$	$-a-\frac{a}{\sqrt{2}}, \frac{a}{\sqrt{2}}$	$(-a, -a)$	$-a-\frac{a}{\sqrt{2}}, -\frac{a}{\sqrt{2}}$	$(-2a, 0)$	$-a+\frac{a}{\sqrt{2}}, -\frac{a}{\sqrt{2}}$
$\left(\frac{a}{\sqrt{2}}, -\frac{a}{\sqrt{2}}\right)$	$a+\frac{a}{\sqrt{2}}, -\frac{a}{\sqrt{2}}$	$(\sqrt{2}a, 0)$	$\frac{a}{\sqrt{2}}, a-\frac{a}{\sqrt{2}}$	$(0, 0)$	$\frac{a}{\sqrt{2}}, -a-\frac{a}{\sqrt{2}}$	$0, -\sqrt{2}a$	$\frac{a}{\sqrt{2}}-a, -\frac{a}{\sqrt{2}}$	$\sqrt{2}a, -\sqrt{2}a$

↑ Row entry

Note: The sum coarray consists of 64 locations whose coordinates are obtained by summing the first column (left shaded column) and the respective entries from the first row (top shaded row). The coordinates are shown in clear cells.

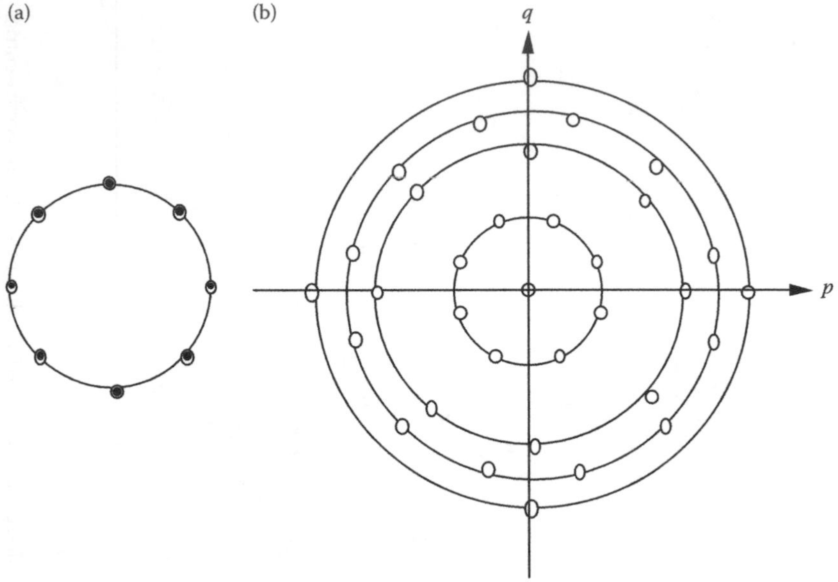

FIGURE 2.33
(a) A circular array (radius=1.5 cm) of eight transceivers. (b) Coarray: nodes are shown by empty circles. There are 33 nodes. The diameter of the outermost circle is 3.0 cm. The PSF is defined at the nodes. For uniform array, with constant source strength and sensor sensitivity, the PSF at the center is equal to $8t_0r_0$ and elsewhere it is equal to $2t_0r_0$.

As an illustration, consider uniform transmitter strength and receiver sensitivity; let $t=t_0$ and $r=r_0$. Clearly, since the array has a circular symmetry, PSF will also be radially symmetric. It is straightforward to show that

$$PSF(0,0)=Mt_0r_0 \quad \text{and} \quad PSF(\sqrt{p^2+q^2})=2t_0r_0. \tag{2.86}$$

The PSF has a tall peak at the center, four times the background level. In general

$$PSF(0,0)=Mt_0r_0 \quad \text{and} \quad PSF(\sqrt{p^2+q^2})=2t_0r_0.$$

Thus, for large M the PSF tends to a delta function, the ideal condition for a perfect imaging system. A reverse problem of synthesizing a circular array of transceivers given the desired PSF is proposed in Ref. [27].

2.5.2 Passive Imaging

In passive imaging, the object to be imaged is itself a source of energy, for example, a distant star or an earthquake deep inside the earth. It is of some interest to estimate the distribution of the energy as a function of azimuth and elevation or the spatial frequencies u and $v(u=(2\pi/\lambda)\sin q \cos j$ and

$v=(2\pi/\lambda)\sin q \sin j$). The imaging system consists of two receiving arrays each with its own weighting functions. Let $r_1(x,y)$ and $r_2(x,y)$ be the weighting functions. Let $P(u,v)$ be the source distribution. The outputs of the two arrays, in the frequency domain, can be written as

$$\begin{aligned} O_1(u,v) &= P(u,v)R_1(u,v) \\ O_2(u,v) &= P(u,v)R_2(u,v). \end{aligned} \tag{2.87}$$

We form a cross correlation of the two outputs. Using Equation 2.87 we obtain

$$\begin{aligned} I(u,v) &= O_1(u,v)O_2^*(u,v) \\ &= |P(u,v)|^2 R_1(u,v)R_2^*(u,v), \end{aligned} \tag{2.88a}$$

or

$$I(u,v) = |P(u,v)|^2 \int_p \int_q \left[\iint_{-\infty}^{+\infty} r_1(x,y) r_2(x+p,y+q)dxdy \right] e^{j(up+vq)} dpdq. \tag{2.88b}$$

The quantity inside the braces in Equation 2.88b represents the cross correlation of two weighting functions.

$$C_{r_1r_2}(p,q) = \iint_{-\infty}^{+\infty} r_1(x,y)r_2(x+p,y+q)dxdy. \tag{2.89}$$

Let $x'=x+p$ and $y'=y+q$, then the cross correlation may be looked upon as an integral of the product of weight functions at $r_1(x,y)$ and $r_2(x',y')$ for fixed p and q; alternatively, for a fixed difference, $p=x'-x$ and $q=y'-y$. Such a coarray is also known as a difference array. For example, for a L-shaped receiver array (sensors on both arms) the difference array is on a square grid in the fourth quadrant. Recall that the sum coarray is also on a square grid but in the first quadrant (see Figure 2.34).

2.5.3 Synthetic Aperture

A long aperture is the primary requirement for achieving high spatial resolution; however, there is a limit on the size of a physical array that one can deploy. This is particularly true in imaging systems where it is uneconomical to use a large physical array required to cover a large target area with high resolution. The main idea in synthetic aperture is that by moving a transceiver, preferably at a constant speed and in a straight line, an array of large aperture can be synthesized through subsequent processing. A simple illustration of this principle is shown in Figure 2.35. A transceiver is moving at a constant

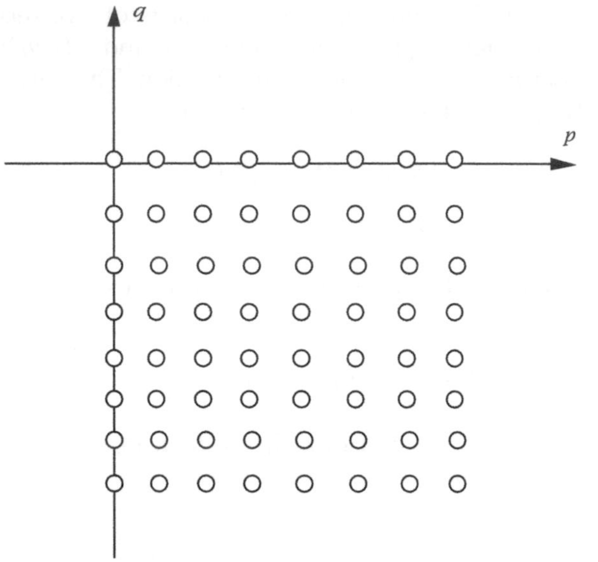

FIGURE 2.34
Difference coarray obtained from L-shaped receiver array shown in Figure 2.24a.

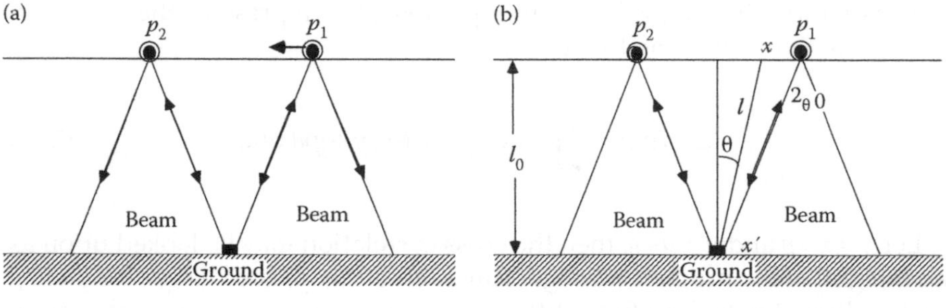

FIGURE 2.35
(a) A simple illustration of the principle of synthetic aperture. The energy scattered by a particle on the ground is received by a transceiver at p. (b) The two-way travel time from the scatterer at x' to the transceiver at x.

speed c_s in a straight line parallel to the ground. Both transmit and receiver beams are wide angle, as shown in the figure. At position p_1 the beam just begins to illuminate a scattering particle on the ground. The particle remains under illumination until the transceiver reaches position p_2. The energy scattered by the particle is received over an interval p_1p_2; hence the effective array aperture is equal to p_1p_2. This is the synthesized aperture. Let us now look into how to process the echoes received over p_1p_2. These echoes reach the receiver with different delays, depending upon the position of the transceiver. Secondly, since the transceiver is moving, there will be a Doppler shift proportional to the component of transceiver speed in the direction of the radius vector joining the transceiver and the scattering particle. From Figure 2.35b it can shown that $p_1p_2=2l_0 \tan\theta$. Let us assume that the transceiver is a directional sensor with effective aperture L, hence the beam width is 2θ where $\sin\theta=1/L$. The beamwidth is measured between the first two nulls of the array response function (see Sub-section 2.1.1). Therefore, the synthesized aperture may be expressed as

$$p_1p_2 = 2l_0\frac{\lambda}{\sqrt{L^2-\lambda^2}}, \tag{2.90}$$

where l_0 is the height of the moving transceiver above ground. The smallest size of an object that can be seen on the ground with the help of an array with aperture given by Equation 2.90 will be approximately equal to $\sqrt{L^2-\lambda^2}$. The underlying assumption is that $p_1p_2 \gg \lambda$, which would allow us to simplify the Rayleigh resolution criterion (see Chapter 4, Section 4.2) and then the result follows. The requirement that $p_1p_2 \gg \lambda$ is easily met by selecting a small antenna as a transmitter. The two-way (one-way in a passive system) travel time from the scatterer to the transceiver is given by

$$\frac{2l}{c} = \frac{2\sqrt{l_0^2+(x-x')^2}}{c} = \frac{2\sqrt{l_0^2+(c_st)^2}}{c}, \tag{2.91}$$

where we have expressed the horizontal distance as a product of time and transceiver speed, $(x-x')=c_st$. Let the transmitted waveform be a sinusoid (real), $\cos(\omega t+\varphi)$. The received waveform, which arrives after a delay of $(2l/c)$, is

$$f(t) = r_0\cos\left(\omega\left(t-\frac{2\sqrt{l_0^2+(c_st)^2}}{c}\right)+\varphi\right). \tag{2.92}$$

Since the phase of the received waveform is time dependent, the instantaneous frequency [1] will be different from the frequency of the transmitted waveform. This difference is the Doppler shift. The instantaneous frequency is given by

$$\omega(t) = \omega - \frac{2\omega c_s}{c}\frac{c_st}{\sqrt{l_0^2+(c_st)^2}}.$$

Hence, the Doppler shift is equal to

$$\Delta\omega = -\frac{2\omega c_s}{c}\frac{c_st}{\sqrt{l_0^2+(c_st)^2}} = -\frac{2\omega c_s}{c}\sin\theta. \tag{2.93}$$

In the case of passive synthetic aperture sonar, the Doppler shift can be used for estimation of the DOA of an unknown source [29].

At a fixed place and time, the signal received consists of a sum of the scattered wavefields from a patch on the ground that is coherently illuminated

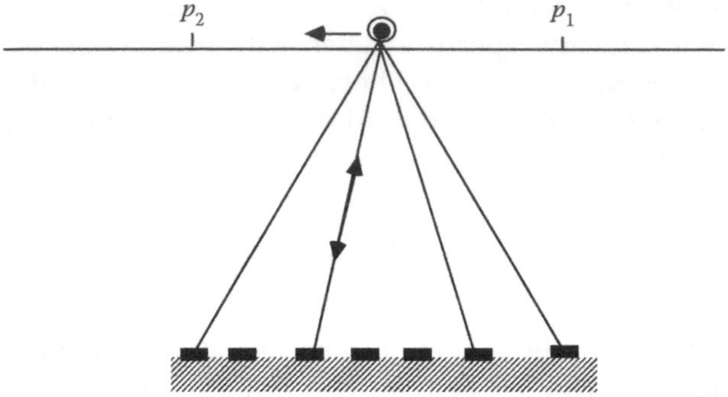

FIGURE 2.36
At a fixed place and time, the signal received consists of a sum of scattered wavefields from a patch of scatterers that is coherently illuminated. The width of the patch is equal to the synthesized aperture.

(see Figure 2.36). The size of the patch is equal to the synthesized aperture. The wavefield from different scatterers reach the sensor with delays as given by Equation 2.91. The receiver output is given by a convolutional integral [35],

$$f(x,t)=\int_{-\infty}^{\infty} r_0(x')w(x-x')\cos\left(\omega\left(t-\frac{2\sqrt{l_0^2+(x-x')^2}}{c}\right)+\varphi\right)dx', \tag{2.94}$$

where $w(x-x')=1$ for $|x-x'|\le(L/2)$ and $L=p_1p_2$, the length of the synthetic aperture. For $|x-x'| \ll l_0$ we can approximate Equation 2.94 as

$$f(x,t)=\int_{p_1}^{p_2} r_0(x')w(x-x')\cos\left(\omega\left(t-\tau_0-\frac{(x-x')^2}{cl_0}\right)+\varphi\right)dx', \tag{2.95a}$$

which simplifies to

$$\begin{aligned} f(x,t)=\cos(\omega(t-\tau_0))\int_{-\infty}^{\infty} r_0(x')w(x-x')\cos\left(\omega\left(\frac{(x-x')^2}{cl_0}\right)+\frac{\varphi}{2}\right)dx' \\ +\sin(\omega(t-\tau_0))\int_{-\infty}^{\infty} r_0(x')w(x-x')\sin\left(\omega\left(\frac{(x-x')^2}{cl_0}\right)+\frac{\varphi}{2}\right)dx'. \end{aligned} \tag{2.95b}$$

We can recover $r_0(x)$ from $f(x,t)$, using the FT method, that is,

$$\tilde{r}_0(x)=FT^{-1}\left\{\frac{FT\{\hat{f}(x)\}}{FT\{w(x)\cos(\omega\frac{x^2}{cl_0})\}}\right\}, \tag{2.96}$$

where

$$\hat{f}(x) = \int_{one\ period} f(x,t)\cos(\omega(t-\tau_0))dt,$$

and FT stands for the Fourier transform. For a large aperture, we get $\tilde{r}_0(x) \xrightarrow[L\to\infty]{} r_0(x)$, that is, an exact reconstruction. It is important to remember that the increased aperture and therefore the increased resolution is the result of the geometry of the data collection, as shown in Figure 2.35. Indeed, the transceiver need not move at all during pulse transmission and reception.

2.6 Multicomponent Sensor Array

While the nature of wave propagation is common to all types of waves, the physics of the waves can be quite different. For example, pressure waves or acoustic waves have just one vector, that is, pressure gradient, but the waves in solids (e.g., seismic waves) have two vectors, namely, pressure gradient and transverse stress vector orthogonal to the pressure gradient. The most complex wave phenomenon is in EM waves where we have coupled electric and magnetic vectors. The pressure waves and stress waves are characterized by three components of particle motion. The EM waves need to be characterized by six components (three electric and three magnetic). The components are inter-related depending upon azimuth, elevation, and nature of polarization. A multicomponent sensor array essentially exploits the relationship among the different components along with delays caused by wave propagation. In this section, we shall outline the relationship among the components and show how even a single multicomponent sensor can yield useful estimates of wave parameters.

2.6.1 Acoustic Vector Sensor

Acoustic vector sensors measure the amplitude and phase of acoustic pressure (scaler) and particle motion vector at a co-located point. Acoustic pressure gradient and particle velocity are related via Euler's equation

$$\nabla p(x,y,z,t) = -\rho_0 \frac{\partial \mathbf{v}}{\partial t}, \tag{2.97}$$

where $\mathbf{v}=[v_x, v_y, v_z]$ is the particle velocity vector and v_x, v_y, v_z are x, y, and z components, respectively. In polar coordinates the three components of particle velocity are related to the magnitude of the particle velocity as,

$$\begin{aligned} v_x &= |\mathbf{v}|\cos(\varphi)\sin(\theta) \\ v_y &= |\mathbf{v}|\sin(\varphi)\sin(\theta) \\ v_z &= |\mathbf{v}|\cos(\theta), \end{aligned} \tag{2.98}$$

where φ is azimuth (measured with respect to the x-axis) and θ is elevation (measured with respect to the z-axis) of a plane wave incident on a single vector sensor. The acoustic wave propagates in the direction of the particle velocity vector.

Consider a plane monochromatic pressure wave given by

$$p(t) = p_0 e^{-j(\omega/c)[\cos(\varphi)\sin(\theta)x+\sin(\varphi)\sin(\theta)y+\cos(\theta)z]+j\omega t},$$

where p_0 is the amplitude of the pressure wave. Using Equation 2.97 we can derive the particle velocity vector

$$\mathbf{v} = \frac{1}{\rho_0 c}[\cos(\varphi)\sin(\theta), \sin(\varphi)\sin(\theta), \cos(\theta)]p. \tag{2.99a}$$

From Equation 2.99a it is easy to show that

$$\frac{p_0}{|\mathbf{v}|} = \rho_0 c. \tag{2.99b}$$

$\rho_0 c$ is often known as the characteristic impedance of a medium. Finally, the output of a single acoustic vector sensor may be expressed as

$$p = p_0 e^{-j(\omega/c)[\cos(\varphi)\sin(\theta)x+\sin(\varphi)\sin(\theta)y+\cos(\theta)z]+j\omega t},$$

$$v_x = \frac{p_0}{\rho_0 c}\cos(\varphi)\sin(\theta)e^{-j(\omega/c)[\cos(\varphi)\sin(\theta)x+\sin(\varphi)\sin(\theta)y+\cos(\theta)z]+j\omega t},$$

$$v_y = \frac{p_0}{\rho_0 c}\sin(\varphi)\sin(\theta)e^{-j(\omega/c)[\cos(\varphi)\sin(\theta)x+\sin(\varphi)\sin(\theta)y+\cos(\theta)z]+j\omega t},$$

$$v_z = \frac{p_0}{\rho_0 c}\cos(\theta)e^{-j(\omega/c)[\cos(\varphi)\sin(\theta)x+\sin(\varphi)\sin(\theta)y+\cos(\theta)z]+j\omega t}, \tag{2.100a}$$

or in matrix form

$$\mathbf{f}(t) = \begin{bmatrix} p \\ v_x\rho_0 c \\ v_y\rho_0 c \\ v_z\rho_0 c \end{bmatrix} = \begin{bmatrix} 1 \\ \cos(\varphi)\sin(\theta) \\ \sin(\varphi)\sin(\theta) \\ \cos(\theta) \end{bmatrix} p(t). \tag{2.100b}$$

The output of an array of identical vector sensors may be easily obtained from Equation 2.100b by taking into account the propagation delay.

$$\left[\mathbf{f}_0^T\ \mathbf{f}_1^T \cdots \mathbf{f}_{M-1}^T\right]^T = \begin{bmatrix} 1 \\ e^{-j\omega\tau_1} \\ \vdots \\ e^{-j\omega\tau_{M-1}} \end{bmatrix} \otimes \begin{bmatrix} 1 \\ \cos(\varphi)\sin(\theta) \\ \sin(\varphi)\sin(\theta) \\ \cos(\theta) \end{bmatrix} p(t), \tag{2.100c}$$

where $\otimes$ stands for Kronecker product and $(\tau_1 \cdots \tau_{M-1})$ are propagation delays with respect to the zeroth sensor.

It may be observed that the amplitudes of the three particle velocity components normalized with respect to the amplitude of the pressure measurement depend upon the azimuth and elevation of incident wave. This property has been exploited for the estimation of azimuth and elevation. Using the four outputs of a single vector sensor it is possible to form a beam.

Let $w_1, w_2, w_3 w_4$ be weighting coefficients. The output power of the beam is given by

$$\begin{aligned} power &= \left|[w_1 p + w_2 v_x + w_3 v_y + w_4 v_z]\right|^2 \\ &= \left(\frac{p_0}{\rho_0 c}\right)^2 [w_1 \rho_0 c + w_2 \cos(\varphi)\sin(\theta) + w_3 \sin(\varphi)\sin(\theta) + w_4 \cos(\theta)]^2. \end{aligned} \tag{2.101}$$

It is shown in Ref. [30] that the beam output power is maximum in the look direction for the following choice of weighting coefficients,

$$\begin{aligned} w_1 &= \frac{1}{\rho_0 c}, \quad w_2 = \cos(\varphi_s)\sin(\theta_s), \\ w_3 &= \sin(\varphi_s)\sin(\theta_s) \quad w_4 = \cos(\theta_s), \end{aligned} \tag{2.102}$$

where φ_s and θ_s are steering angles. Substituting Equation 2.102 in Equation 2.101 we obtain the following sensor response (magnitude square)

$$power = \left(\frac{p_0}{\rho_0 c}\right)^2 \begin{bmatrix} 1 + \cos(\varphi_s)\sin(\theta_s)\cos(\varphi)\sin(\theta) \\ + \sin(\varphi_s)\sin(\theta_s)\sin(\varphi)\sin(\theta) + \cos(\theta_s)\cos(\theta) \end{bmatrix}^2, \tag{2.103}$$

which is plotted in Figure 2.37. The maximum of the bracketed quantity is four when the steering angles are in the look direction, that is, $\varphi_s = \varphi$ and $\theta_s = \theta$. Notice that there is no peak at $\varphi = -45°$ as in the case of ULA, on account of north–south ambiguity.

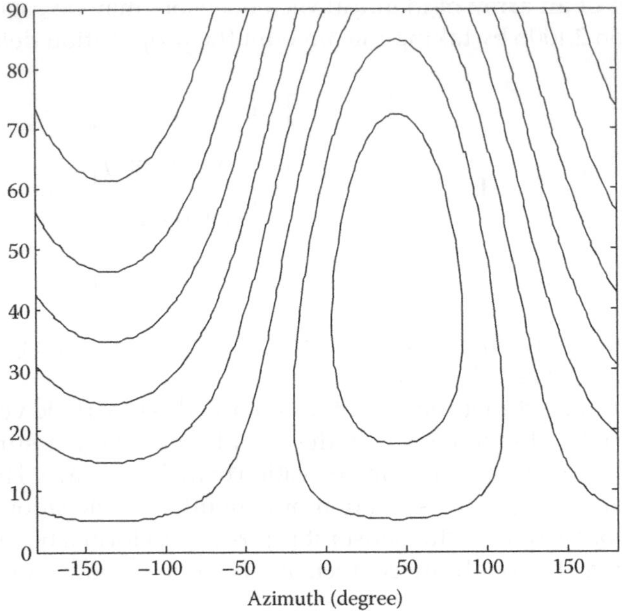

FIGURE 2.37
Beam output power of a four-component acoustic sensor. The look direction is at φ=45° and θ=45°. The maximum output power is in the look direction. Vertical axis represent elevation in degrees.

2.6.2 Seismic Vector Sensor

A geophone is a sensor used to measure a particle displacement vector. The displacement vector has two components, one in the direction of wave propagation and another perpendicular to the direction of wave propagation (see Equation 1.3a). The first component creates a so-called *p*-wave (pressure wave) and the second component creates an *s*-wave (shear wave). These two waves travel at different speeds and therefore arrive at different times, but they are not independent as one generates the other at interfaces of contrasting impedance. The three components measured by a geophone may either represent a pure *p*-wave or a pure *s*-wave or even a mixture of both in a heterogeneous medium. A pure *p*-wave is like an acoustic wave, which we have already considered in the previous section. We shall now concentrate on pure *s*-waves.

The shear wave displacement vector is a sum of two orthogonal components lying in a plane perpendicular to the direction of wave propagation, a transverse plane. Further, the components propagate with different phases, often time varying. The result is the tip of the vector traces a complex path known as a hodogram. This phenomenon is known as polarization. Let

u be the direction vector of plane wavefront incident on a three-component seismometer,

$$\mathbf{u} = \begin{bmatrix} \cos(\varphi)\sin(\theta) \\ \sin(\varphi)\sin(\theta) \\ \cos(\theta) \end{bmatrix}. \tag{2.104a}$$

We define a matrix whose columns are orthogonal and span the orthogonal space of **u**, that is, transverse plane. We choose

$$\mathbf{T} = \begin{bmatrix} \mathbf{v}_1 \ \mathbf{v}_2 \end{bmatrix} = \begin{bmatrix} -\sin(\varphi) & -\cos(\varphi)\cos(\theta) \\ \cos(\varphi) & -\sin(\varphi)\cos(\theta) \\ 0 & \sin(\theta) \end{bmatrix}. \tag{2.104b}$$

The transverse displacement vector $\mathbf{d}_s(t)$ will lie in the range space of **T**. Let $\xi(t)$ be some vector

$$\mathbf{d}_s(t) = \mathbf{T}\xi(t). \tag{2.105a}$$

There is a unique representation of $\xi(t)$ [31]

$$\xi(t) = \mathbf{Qw}x(t), \tag{2.105b}$$

where

$$\mathbf{Q} = \begin{bmatrix} \cos\theta_3 & \sin\theta_3 \\ -\sin\theta_3 & \cos\theta_3 \end{bmatrix} \qquad \mathbf{w} = \begin{bmatrix} \cos\theta_4 \\ j\sin\theta_4 \end{bmatrix}, \\ -\frac{\pi}{2} \le \theta_3 \le \frac{\pi}{2} \qquad -\frac{\pi}{4} \le \theta_4 \le \frac{\pi}{4} \tag{2.105c}$$

and $x(t)$ is the source signal, for example, let $x(t)=\exp(-j\omega_c t)$. Then, $\mathbf{w}\exp(-j\omega_c t)$ is a vector whose tip traces an elliptic hodogram (see Figure 2.38), described by $(x^2/\cos^2(\theta_4))+(y^2/\sin^2(\theta_4))$. We get linear polarization for $\theta_4=0$ and circular polarization for $\theta_4=(\pi/4)$. The polarization ellipse is traced anticlockwise when $\theta_4 > 0$ and clockwise when $\theta_4 < 0$. The role of the **Q** matrix is to rotate the axes of the elliptic hodogram by an angle equal to θ_3 in the transverse plane. Thus, **Qw** completely characterizes the polarization of the transverse displacement vector.

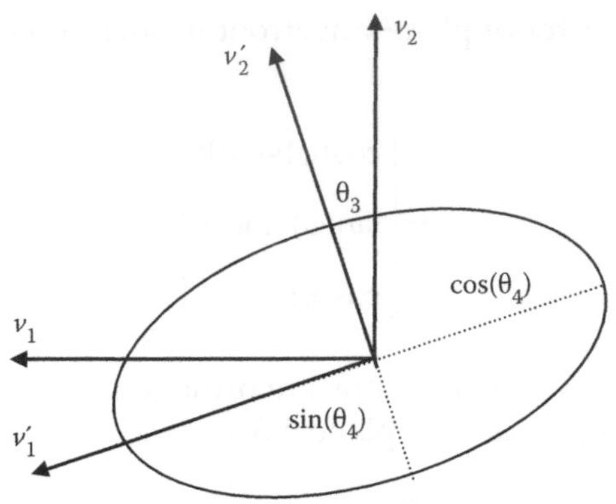

FIGURE 2.38
Elliptic polarization hodogram for transverse polarized waves. The direction of propagation is normal to the plane of paper.

Combining Equation 2.105a and Equation 2.105b the transverse displacement vector is given by

$$\mathbf{d}_s(t) = \mathbf{TQw}x(t). \tag{2.106}$$

2.6.3 Electromagnetic Vector Sensor

An EM field consists of electric and magnetic vectors, altogether six components. The governing equations are Maxwell's equations given by Equation 1.9. In free space and plane waves the Maxwell's equations take a simple form [31]. In the presence of plane wave, EM fields at two points in free space differ only in propagation delay. There is no spatial variation. For example, the electric fields at $\mathbf{r}_1$ and $\mathbf{r}_2$ are given by

$$\mathbf{E}(\mathbf{r}_1, t) = \mathbf{E}(\mathbf{r}_2, t - \tau),$$

where τ is the propagation delay. Because of this property for plane waves, we can express the divergence operator ∇ $(=(\partial/\partial x), (\partial/\partial y), (\partial/\partial z))$ as

$$\begin{aligned}\nabla &= -\frac{\partial t}{\partial x}\frac{\partial}{\partial t}, -\frac{\partial t}{\partial y}\frac{\partial}{\partial t}, -\frac{\partial t}{\partial z}\frac{\partial}{\partial t}\\ &= -\left(\frac{1}{c_x}, \frac{1}{c_y}, \frac{1}{c_z}\right)\frac{\partial}{\partial t}\\ &= -\frac{1}{c}[(\cos(\varphi)\sin(\theta), \sin(\varphi)\sin(\theta), \cos(\theta)]\frac{\partial}{\partial t}\\ &= -\frac{1}{c}\mathbf{u}\frac{\partial}{\partial t},\end{aligned} \tag{2.107}$$

where we have used the concept of apparent speed defined in Equation 1.21. Using Equation 2.107 in Equation 1.9, the Maxwell's equations in free space (conductivity = 0 and no free charge) assume the following form:

$$
\begin{array}{ll}
\mathbf{u} \times \frac{\partial}{\partial t}\mathbf{E} = \sqrt{\mu/\varepsilon}\,\frac{\partial}{\partial t}\mathbf{H} & \mathbf{u} \times \frac{\partial}{\partial t}\mathbf{H} = -\sqrt{\varepsilon/\mu}\,\frac{\partial}{\partial t}\mathbf{E} \\
\mathbf{u} \cdot \frac{\partial}{\partial t}\mathbf{E} = 0 & \mathbf{u} \cdot \frac{\partial}{\partial t}\mathbf{H} = 0, \\
\text{Faraday's law} & \text{Ampere's law}
\end{array}
\tag{2.108}
$$

where propagation direction vector **u** is defined in Equation 2.104a. The Faraday's law and Ampere's law are equivalent, as one can be derived from the other.

We shall express the measured electric and magnetic field components in terms of incident electric field, which we shall assume to be a harmonic plane wave,

$$\mathbf{E} = \mathbf{E}_0 \exp\left(j\left(\frac{\omega}{c}\mathbf{u}\cdot\mathbf{r} - \omega t\right)\right),$$

which we use in Equation 2.108 (Faraday's law) and obtain

$$\mathbf{H} = \sqrt{\varepsilon/\mu}\,\mathbf{u} \times \mathbf{E}, \tag{2.109a}$$

$$\mathbf{u} \cdot \mathbf{E} = 0. \tag{2.109b}$$

We can express vector cross-product in Equation 2.109a as a matrix operator

$$
\mathbf{u} \times \underline{\underline{\Delta}} \begin{bmatrix} 0 & -\cos(\theta) & \sin(\varphi)\sin(\theta) \\ \cos(\theta) & 0 & -\cos(\varphi)\sin(\theta) \\ -\sin(\varphi)\sin(\theta) & \cos(\varphi)\sin(\theta) & 0 \end{bmatrix} = \mathbf{R}
\tag{2.109c}
$$

Equation 2.109b implies that the electric must lie in a space ⊥ to **u**, that is, in the range space of matrix **T** defined in Equation 2.104b. Let $\xi(t)$ be some vector in the range space of **T**. Equation 2.109b implies that the electric field can be written as

$$
\mathbf{E}(t) = \begin{bmatrix} -\sin(\varphi) & -\cos(\varphi)\cos(\theta) \\ \cos(\varphi) & -\sin(\varphi)\cos(\theta) \\ 0 & \sin(\theta) \end{bmatrix} \xi(t), = \mathbf{T}\xi(t)
\tag{2.110a}
$$

using the above representation along with Equation 2.109c we obtain

$$\mathbf{H}(t) = \sqrt{\varepsilon/\mu}\mathbf{RT}\xi(t). \tag{2.110b}$$

As in Equation 2.105b, there is a unique representation for $\xi\ (t)=\mathbf{Q}\mathbf{w}x\ (t)$, which we use in Equation 2.110 and obtain

$$\mathbf{E}(t) = \mathbf{TQw}x(t), \tag{2.111a}$$

$$\mathbf{H}(t) = \sqrt{\varepsilon/\mu}\mathbf{RTQw}x(t). \tag{2.111b}$$

From Equation 2.111 it is evident that given the electric field we can obtain the magnetic field, but then we have to know the matrix **R**, which requires knowledge of DOA angles of the source. If there are multiple diversely polarized sources, above statement is no longer true. Thus, the electric and magnetic components provide independent information.

Finally, all six outputs of a vector sensor may be compactly represented as follows:

$$\begin{bmatrix} E_x \\ E_y \\ E_z \\ H_x \\ H_y \\ H_z \end{bmatrix} = \begin{bmatrix} -\sin(\varphi) & -\cos(\varphi)\cos(\theta) \\ \cos(\varphi) & -\sin(\varphi)\cos(\theta) \\ 0 & \sin(\theta) \\ -\gamma\cos(\varphi)\cos(\theta) & \gamma\sin(\varphi) \\ \gamma\sin(\varphi)\cos(\theta) & -\gamma\cos(\varphi) \\ \gamma\sin(\theta) & 0 \end{bmatrix} \mathbf{Qw}x(t), \tag{2.112}$$

where γ is a constant, which includes $\sqrt{\varepsilon/\mu}$ and any other sensor-dependent factors. We have computed the output beam power assuming the following parameters: $\gamma=1$, linear polarization $(\theta_4=(\pi/2))$, and linear polarization is along the x-axis, that is, $(\theta_3=0)$. The beam output power for a six-component EM sensor may be defined as in Equation 2.105 for an acoustic vector sensor. A plot of the beam power as a function of azimuth and elevation is shown in Figure 2.39. The peak output power lies in the look direction. There is no peak at $\varphi=-45°$ as in case of ULA, on account of north–south ambiguity (Chapter 1, Section 2.1). This is indeed a very useful property of vector sensor array. It has been demonstrated [32, 33] that an array of vector sensors performs better than a scalar sensor array as there is no north-south ambiguity and no grating lobes are present even when the spacing is greater than $\lambda_{\min}/2$.

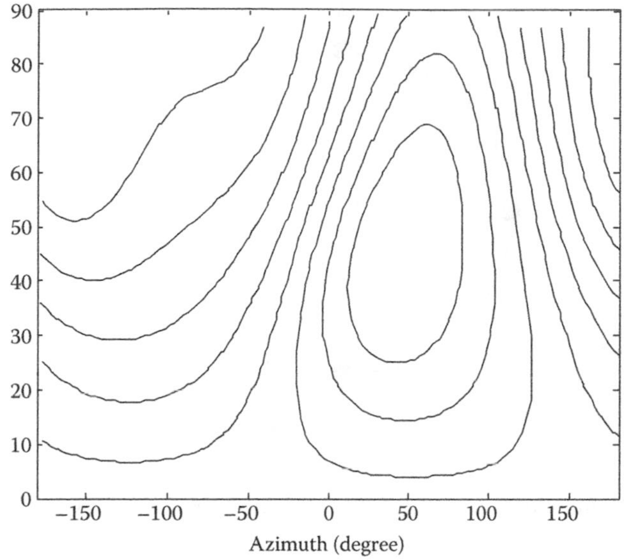

FIGURE 2.39
Beam output power of a six-component EM sensor. The look direction is at $\varphi=45°$ and $\theta=45°$. The maximum output power is in the look direction. Vertical axis represents elevation in degrees.

2.7 Exercises

(1) The angle of arrival of a broadband signal at ULA may be estimated from the spectrum of the sum of the output of all sensors (without delays) and the spectrum of the output of any one sensor. Find the bandwidth required to estimate the angle of arrival over a full range of 0–90°. What is the role of the sensor spacing?

(2) It is desired to measure the speed of wave propagation in a medium. A ULA with sensor spacing d meters is employed for this purpose. A broadband signal from a known direction is sweeping across the array. How do you go about estimating the speed of propagation?

(3) Consider a circular transmit and receive array, as shown in Figure 2.40. (a) Show the sum coarray along with weight coefficients, assuming that the physical array has unit coefficient. (b) Do the same for the difference coarray.

(4) Compare the outputs of crossed electric dipole arrays given by Equation 2.57 and Equation 2.58 with that of displaced identical sensors pairs (also known as dipoles) given by Equations 2.55a and b. Obtain an expression for spectral matrix, similar to the one given in Equation 2.57, of the output of the crossed dipole arrays, and show that Γ in Equation 2.57 is now equal to $\Gamma_1^{-1}\Gamma_2$.

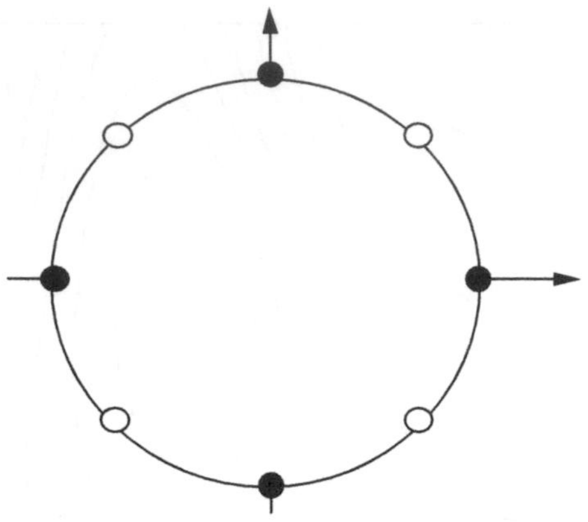

FIGURE 2.40
A circular sensor and source array. Compute sum and difference coarray.

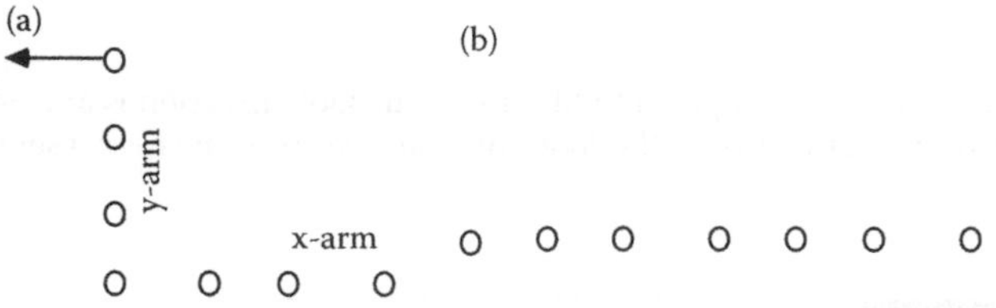

FIGURE 2.41
A right angle array is converted into a ULA by bending the y-arm through 90°.

(5) Consider a right angle array shown in Figure 2.41a, which may be converted into a ULA by bending the y-arm through 90°, as shown in Figure 2.41b. A plane wavefront is propagating in the plane of array. Derive the response of the right angle array and compare it with that of the converted ULA. Assume that the incident signal is narrowband.

(6) Consider an equilateral triangular array inscribed in a circle of radius$=64\lambda$. A point source is located at $r=32\lambda$ and $\varphi=45°$. Draw all phase trajectories passing through the point where the source is located.

(7) A ULA is placed on a 100-m deep sea-bed and a broadband acoustic source is placed 1,000 m away from the array and 50 m below the sea surface. Sketch the frequency wavenumber spectrum of the array signal. [Hint: Consider a direct and two reflected wavefronts.]

(8) A broadband wavefront is incident on a ULA. Show that there is no aliasing for any angle of incidence if the sensor spacing is$\leq(\lambda\text{min}/2)$.

(9) Consider an array of L component vector sensors ($L=6$ for EM sensors). P correlated plane wave signals are incident on an M vector sensor array. Compute a covariance matrix for each component and

average over all components. Show that the resulting smoothed covariance matrix will be P-rank matrix if $L > P$. This has been termed as polarization smoothing in Ref. [34].

References

1. P. S. Naidu, *Modern Spectrum Analysis of Time Series*, CRC Press Inc., Boca Rato, FL, 1996.
2. J. R. Williams, Fast beam-forming algorithm, *J. Acoust. Soc. Am.*, vol. 44, pp. 1454–1455, 1968.
3. S. H. Nawab, F. U. Dowla, and R. T. Lacoss, Direction determination for wide band signals, *IEEE Trans.*, ASSP-33, pp. 1114–1122, 1985.
4. M. Allam and A. Moghaddamjoo, Two dimensional DFT projection for wideband direction-of-arrival estimation, *IEEE Trans. Signal Processing*, vol. 43, pp. 1728–1732, 1995.
5. R. H. Tatham, Multidimensional filtering of seismic data, *IEEE Proc.*, vol. 72, pp. 1357–1369, 1984.
6. H. Brysk and D. W. McCowan, A slant-stack procedure for point-secure data, *Geophysics*, vol. 51, pp. 1370–1386, 1986.
7. C. W. Therrien, *Discrete Random Signals and Statistical Signal Processing*, Prentice Hall, Englewood Cliffs, NJ, 1992.
8. A. T. Moffet, Minimum redundancy linear arrays, *IEEE Trans. Antennas and Propagation*, vol. AP-16, pp. 172–175, 1968.
9. Y. Lee and S. U. Pillai, An algorithm for optimal placement of sensor elements, *ICASSP-88*, pp. 2674–2677, 1988.
10. N. M. J. Earwicker, Signal to noise ratio gain of sparse array processors, *J. Acoust. Soc. Am.*, vol. 88, pp. 1129–1134, 1980.
11. B. G. Quinn, R. F. Barrett, P. J. Kootsookos, and S. J. Searle, The estimation of the shape of an array using a hidden Markov model, *IEEE J. Ocean. Eng.*, vol. 18, pp. 557–564, 1993.
12. M. Hinich, Estimating signal and noise using a random array, *Acoust. Soc. Am.*, vol. 71, pp. 97–99, 1982.
13. M. P. Moody, Resolution of coherent sources incident on a circular antenna array, *Proc. IEEE*, vol. 68, pp. 276–277, 1980.
14. C. P. Mathews and M. D. Zoltowaski, Eigen structure technique for 2D angle estimation with uniform circular arrays, *IEEE Trans.*, SP-42, pp. 2395–2407, 1994.
15. M. Abramowitz and I. A. Stegun (Eds), *Handbook of Mathematical Functions*, Applied Mathematics Series 55, National Bureau of Standards, Gaithersburg, MD, 1964.
16. A. H. Tewfik and H. Wong, On the application of uniform linear array bearing estimation techniques to uniform circular arrays, *IEEE Trans.*, SP-40, pp. 1008–1011, 1992.
17. A. G. Peirsol, Time delay estimation using phase data, *IEEE Trans.*, ASSP-29, pp. 471–477, 1981.
18. J. Li and R. T. Compton Jr., Angle and polarization estimation in a coherent signal environment, *IEEE Trans. Aero. Elect. Sys.*, vol. 29, pp. 706–716, 1993.

19. M. Castillo-Effen, D. H. Quintela, R. Jordan, W. Westhoff, and W. Moreno, Wireless sensor networks for flash-flood alerting, *IEEE Conf. Proc.*, pp. 142–146, 2004.
20. X. Cui, T. Hardin, R. K. Ragade, and A. S. Elmaghraby, A swarm-based fuzzy logic control mobile sensor network for hazardous contaminants localization, *IEEE Conf. Proc.*, pp. 194–203, 2004.
21. C. Knapp and G. Carter, The generalized correlation method for estimation of time delay, *IEEE Trans. ASSP*, vol. 24, pp. 320–327, 1976.
22. N. Patwari, J. N. Ash, and S. Kyperountas, Locating the nodes, *IEEE Sig. Proc. Magazine*, pp. 54–69, 2005.
23. K. Yao, R. E. Hudson, C. W. Reed, D. Chan, and F. Lorenzelli, Blind beamforming on a randomly distributed sensor array system, *IEEE J. Select. Areas Commun.*, vol. 16, pp. 1555–1567, 1998.
24. N. Patwari, A. O. Hero III, M. Perkins, N. S. Correal and R. J. O'Dea, Relative location estimation in wireless sensor networks, *IEEE Trans. Signal Processing*, vol. 51, pp. 2137–2148, 2003.
25. S. T. Birchfield, Geometric microphone array calibration by multidimensional scaling, *IEEE ICASSP*, pp. V157–160, 2003.
26. C. U. Padmini and P. S. Naidu, Circular array and estimation of direction of arrival of a broadband source, *Signal Processing*, vol. 37, pp. 243–254, 1994.
27. R. J. Kozick, Linear imaging with source array on convex polygonal boundary, *IEEE Trans. Syst. Man. Cybern.*, vol. 21, pp. 1155–1166, 1991.
28. R. J. Kozick and S. A. Kassam, Coarray synthesis with circular and elliptical boundary arrays, *IEEE Trans. Image Process.*, vol. 1, pp. 391–404, 1992.
29. R. Williams and B. Harris, Passive acoustic synthetic aperture processing techniques, *IEEE J. Ocean. Eng.*, vol. 17, pp. 8–14, 1992.
30. B. A. Cray and A. N. Nuttal, Directivity factor for linear arrays of velocity sensors, *J. Acoust Soc. Am.*, vol. 110, pp. 324–331, 2001.
31. A. Nehorai and E. Paidi, Vector-sensor array processing for electromagnetic source localization, *IEEE Trans. Signal Processing*, vol. 42, pp. 376–398, 1994.
32. Y. T. Chen, G. H. Niezgoda, and S. P. Morton, Passive sonar detection and localization by matched velocity filtering, *IEEE J. Ocean. Eng.*, vol. 20, pp. 179–189, 1995.
33. C. E. Chen, F. Lorenzelli, R. E. Hudson, and K. Yao, Maximum likelihood DOA estimation of multiple wideband sources in the presence of nonuniform sensor noise, *EURASIP Journal on Advances in Signal Processing*, pp. 1–12, 2008.
34. D. Rahamim, J. Tabrikian, and R. Shvit, Source localization using vector sensor array in multipath environment, *IEEE Trans. Signal Processing*, vol. 52, pp. 3096–3103, 2004.
35. D. C. Munson Jr., An introduction to strip-mapping synthetic aperture radar, *IEEE Proc. ICASSP*, pp. 2245–2248, 1987.

3

Frequency Wavenumber Processing

In signal processing, extraction of a signal buried in noise has been a primary goal of lasting interest. A digital filter is often employed to modify the spectrum of a signal in some prescribed manner. Notably, the proposed filter is designed to possess unit response in the spectral region where the desired signal is residing and low response where the undesired signal and noise are residing. This strategy will work only when the spectrum of the desired signal does not overlap (or only partially overlap) with the spectrum of the undesired signal and noise. This is also true in wavefield processing. Spectrum shaping is necessary whenever the aim is to enhance certain types of wavefields and suppress other types of unwanted wavefields, often termed as noise or interference. Such a selective enhancement is possible on the basis of spectral differences in the frequency wavenumber (ω-k) domain. For example, it is possible, using a digital filter, to enhance the wavefield traveling at a speed different from that of the noise or the interference, as their spectra lie on different radial lines in the (ω-k) plane. Other situations where spectral shaping is required are (a) prediction of wavefields, (b) interpolation of wavefields, and (c) extrapolation of wavefield into space where the field could not be measured. In this chapter, we shall consider the design of pass filters, specially useful in wavefield processing such as a fan and a quadrant filter. When signal and noise are overlapping in the frequency wavenumber domain, simple pass filters are inadequate. Optimum filters such as Wiener filters are required. We cover this topic in some depth in view of its importance. Next, we introduce the concept of noise cancellation through prediction. In Chapter 6, we shall evaluate the effectiveness of some of the techniques described here.

3.1 Digital Filters in the ω-k Domain

In wavefield processing, digital filters are often used to remove the interference corrupting the signal of interest (SOI) which arrives from one or more known directions while the interference arrives from different directions, which are often unknown. The purpose of a digital filter is to accept the wave energy arriving from a given direction or an angular sector and to reject all wave energy (interference) lying outside the assumed passband,

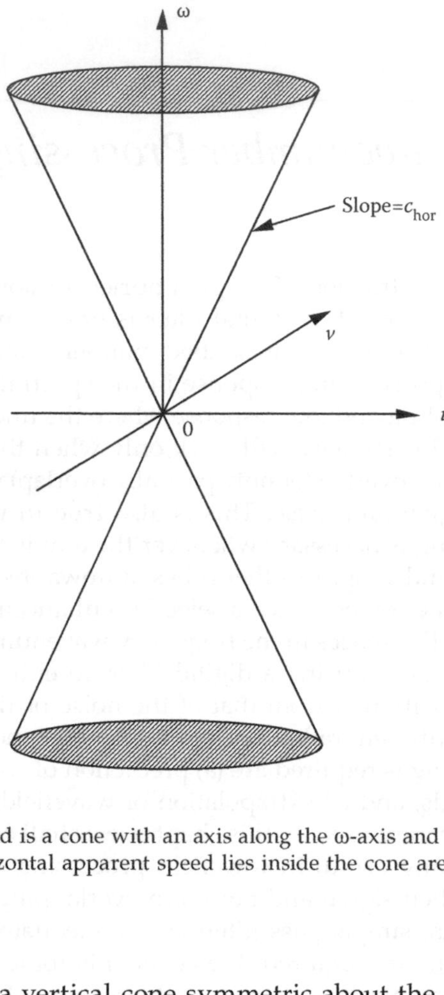

FIGURE 3.1
Fan filter. The passband is a cone with an axis along the ω-axis and the apex is at the center. All waves whose horizontal apparent speed lies inside the cone are allowed and others are attenuated.

which consists of a vertical cone symmetric about the temporal frequency axis. Such a filter is known as a fan filter in seismic exploration (Figure 3.1) or as a quadrant filter (Figure 3.6) when the axis is tilted. In this section we shall study in detail the fan and the quadrant filters. The issues connected with the design and implementation will be discussed. Finally, we shall also examine the role of sampling and the associated question of aliasing.

3.1.1 Two-Dimensional Filters

Two-dimensional filters are extensively used in image processing [1] and geophysical map processing [2]. However, because of the fundamental differences between the wavefield and the image signal, the 2D filters used in the respective applications are of different types. For example, in wavefield processing, the filter must satisfy the causality condition in time, but there is no such requirement in image processing. The causality requirement is waived whenever delay in the output can be tolerated or the data is prerecorded as in seismic exploration or nondestructive testing. Another important difference is that the spectrum of a propagating wavefield is necessarily confined to a conical domain in the frequency wavenumber space (see Chapter 1). Only locally generated disturbances (evanescent waves) and instrument-generated noise are not subject to the above restriction. Hence, in wavefield applications a typical lowpass filter has a shape of a hand-held fan. Such a filter is known as a fan filter, widely used in seismic data processing. The design and implementation of the fan filter is naturally of great interest in wavefield processing. As for other types of filters, such as lowpass filters with circular or elliptical passbands, a ring-type bandpass filter, Laplacian filter for edge detection, deblurring filter, etc., which are widely used in image processing, these are outside the scope of this book.

3.1.2 Fan Filters

The goal is to extract a broadband signal traveling with a horizontal apparent speed between c_{Hor} and ∞ (note $c_{Hor}=c/\sin\theta$). The desired fan filter will have the following transfer function:

$$H(s,\omega)=\begin{matrix} 1 & \begin{cases} |\omega|<\omega_{max} \\ s<\dfrac{|\omega|}{c_{Hor}} & \left(s=\sqrt{u^2+v^2}\right), \end{cases} \\ 0 & \text{Otherwise} \end{matrix} \tag{3.1}$$

which is illustrated in Figure 3.1. The filter will pass all propagating waves whose angle of elevation lies between 0° and $\pm\sin^{-1}(c/c_{Hor})$ (the elevation angle is measured with respect to the vertical axis). Fail and Grau [3] were first to introduce the fan filter in 1963. Independently, Embree [4] came up with a similar filter, which he called a "Pie Slice Filter." Since the filter has a circular symmetry in a spatial frequency plane, we need a series of concentric uniform circular arrays (UCA) (in effect, a planar array) to measure the direction of arrival (DOA) in the horizontal plane and horizontal apparent speed, when the wave speed is known. For the purpose of filter design it is enough if we consider any one radial line.

The filter may be rotated about the vertical axis to yield a circularly symmetric filter. For simplicity, we consider a single radial line of sensors coinciding with the x-axis and assume that all waves are propagating along the x-axis with different horizontal apparent speeds.

In order to compute the fan filter coefficients we take the inverse Fourier transform of the transfer function given in Equation 3.1,

$$h_{m,n} = \frac{1}{4\pi^2} \iint_{-\infty}^{+\infty} H(\omega,u) e^{j(\omega n\Delta t + um\Delta x)} d\omega du$$

$$= \frac{1}{4\pi^2} \int_{-\omega_{\max}}^{+\omega_{\max}} \int_{-(|\omega|/c_{\mathrm{Hor}})}^{|\omega|/c_{\mathrm{Hor}}} e^{j(\omega n\Delta t + um\Delta x)} d\omega du \tag{3.2}$$

$$= \frac{1}{2\pi^2} \int_{-\omega_{\max}}^{+\omega_{\max}} \frac{\sin\left(m\Delta x \frac{|\omega|}{c_{\mathrm{Hor}}}\right)}{m\Delta x} e^{j\omega n\Delta t} d\omega,$$

where Δt and Δx are temporal and spatial sampling intervals, respectively. Since $\sin(m\Delta x(|\omega|/c_{\mathrm{Hor}}))$ is a symmetric function, the last integral in Equation 3.2 can be simplified as

$$h_{m,n} = \frac{1}{2\pi^2 m\Delta x} \int_{0}^{+\omega_{\max}} \left[\sin\left(\frac{m\Delta x}{c_{\mathrm{Hor}}} - n\Delta t \right)\omega + \sin\left(\frac{m\Delta x}{c_{\mathrm{Hor}}} + n\Delta t \right)\omega \right] d\omega. \tag{3.3}$$

We are free to choose the spatial and temporal sampling intervals but within the constraints imposed by the sampling theorem. These are discussed in Section 3.2. Let the temporal sampling interval Δt be so chosen that $\Delta t = \Delta x / c_{\mathrm{Hor}}$. In practice Δx is held fixed; therefore, to alter c_{Hor} it is necessary to resample the whole signal with a different sampling interval, consistent with the above choice. Equation 3.3 is easily evaluated to yield

$$h_{m,n} = \frac{1}{2\pi^2 m\Delta x\Delta t} \left[\frac{1-\cos(m-n)\pi}{(m-n)} + \frac{1-\cos(m+n)\pi}{(m+n)} \right]. \tag{3.4}$$

Equation 3.4 takes a particularly simple form if we shift the origin to the halfway point between two sensors

$$h_{m,n} = \frac{1}{\pi^2 \Delta x \Delta t} \frac{1}{m^2 - n^2}, \tag{3.5}$$

where $m = \pm(1/2), \pm 1(1/2), \pm 2(1/2), \ldots$ (see Figure 3.2). Note that since $m \neq n$, $h_{m,n}$ will always be finite. The frequency response (magnitude) of a fan filter given by Equation 3.5 is shown in Figure 3.3.

3.1.3 Fast Algorithm

In order to implement the filter the following convolution sum will have to be evaluated

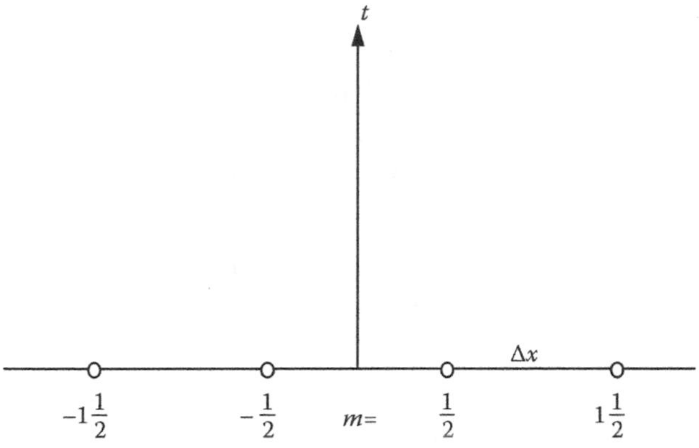

FIGURE 3.2
Centered array (ULA) with origin lying between two sensors. This is also the physical center of the array.

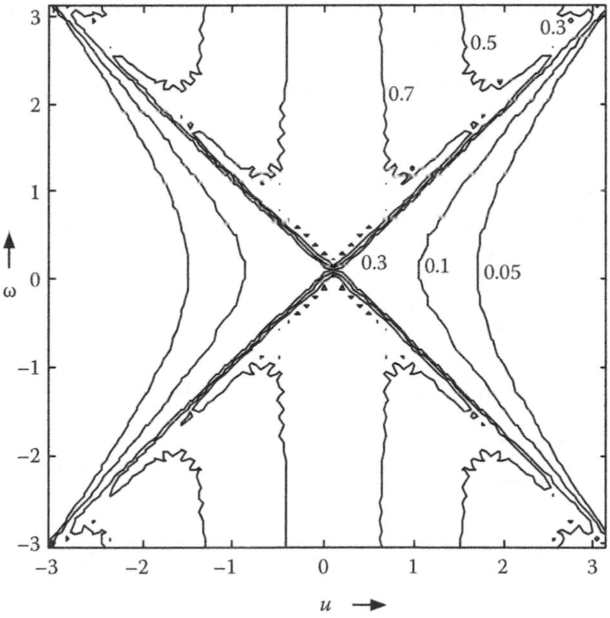

FIGURE 3.3
Frequency response (magnitude) of a fan filter (64 sensors and 64 time samples).

$$\hat{f}(n\Delta t)=\sum_{n'}\sum_{m} h_{m,n'} f(m\Delta x,(n-n')\Delta t). \tag{3.6}$$

The computational load for evaluating the convolutional sum can be greatly reduced by exploiting the filter structure [5]. We define a new index $\mu=(1+2m')$ where $m'=0, \pm1, \pm2, \ldots$. We note that the new index takes values, $\mu=\pm1, \pm3, \pm5, \ldots$, for different values of m'. Equation 3.5 may be reduced to

$$h_{m,n} = \frac{1}{\pi^2}\left(\frac{1}{\frac{\mu^2}{4} - n^2}\right),$$

which may be further split into two components

$$h_{m,n} = \frac{1}{\pi^2 \mu}\left(\frac{1}{\frac{\mu}{2} - n} + \frac{1}{\frac{\mu}{2} + n}\right)$$

$$= \frac{1}{\mu}\left[r_{\mu,n} + q_{\mu,n}\right].$$

The components $r_{\mu,n}$ and $q_{\mu,n}$ possess many symmetry properties as listed below:

- $r_{\mu,n}$ and $q_{\mu,n}$ are antisymmetric in time index (n) about a point $n=\mu/2$ and $n=-(\mu/2)$, respectively.
- $r_{\mu,n}$ and $q_{\mu,n}$ are of the same shape as $r_{1,n}$ except for the time shift. $r_{\mu,n}=r_{1,n-((\mu-1)/2)}$ and $q_{\mu,n}=-r_{1,n+((\mu+1)/2)}$.
- $q_{\mu,n}=-r_{\mu,n+\mu}$.

Using the last property we obtain $h_{m,n}=(1/\mu)\,[r_{\mu,n}-r_{\mu,n+\mu}]$ and using the second property we obtain

$$h_{m,n} = \frac{1}{\pi^2 \mu}\left(r_{1,n-((\mu-1)/2)} - r_{1,n+((\mu+1)/2)}\right). \tag{3.7}$$

Substituting Equation 3.7 in Equation 3.6 we obtain

$$\hat{f}(n\Delta t) = \sum_{n'}\sum_{\mu} \frac{1}{\pi^2 \mu}\left(r_{1,n'-((\mu-1)/2)} - r_{1,n'+((\mu+1)/2)}\right) f(\mu\Delta x,(n-n')\Delta t)$$

$$= \frac{1}{\pi^2}\sum_{n'} r_{1,n'} \sum_{\mu=-l}^{l} \frac{1}{\mu}\begin{bmatrix} f\left(\mu\Delta x, \left(n-n'-\frac{\mu-1}{2}\right)\Delta t\right) \\ -f\left(\mu\Delta x, \left(n-n'+\frac{\mu+1}{2}\right)\Delta t\right)\end{bmatrix}, \tag{3.8}$$

where $l=M-1$ and M (an even number) stands for the number of sensors. In Equation 3.8, there is only one convolution to be carried out in place of $M/2$ convolutions in Equation 3.6. The inner summation in Equation 3.8 stands for spatial filtering and the outer summation stands for temporal filtering. Let the output of the spatial filter be $f_1(n\Delta t)$, where

$$f_1(n\Delta t) = \sum_{\mu=-l}^{l} \frac{1}{\mu}\left[f\left(\mu\Delta x, \left(n-n'-\frac{\mu-1}{2}\right)\Delta t\right) - f\left(\mu\Delta x, \left(n-n'+\frac{\mu-1}{2}\right)\Delta t\right)\right].$$

Equation 3.8 takes the form

$$\hat{f}(n\Delta t) = \sum_{n'} r_{1,n'} f_1((n-n')\Delta t). \tag{3.9}$$

Taking z transform on both sides of Equation 3.9 we shall obtain

$$\hat{F}(z) = R_1(z)F_1(z),$$

where $\hat{F}(z)$, $F_1(z)$, $R_1(z)$ are z transforms of $\hat{f}(n)$, $f(n)$, $r_{1,n'}$ respectively. In particular,

$$\begin{aligned} R_1(z) &= \sum_{n=-N_1+1}^{N_1} \frac{1}{\left(\frac{1}{2}-n\right)} z^n \\ &= -z\sum_{n=0}^{N_1-1} \frac{1}{\left(\frac{1}{2}+n\right)} z^n + \sum_{n=0}^{N_1-1} \frac{1}{\left(\frac{1}{2}+n\right)} z^{-n} \\ &= -zR_{N_1}(z) + R_{N_1}(z^{-1}), \end{aligned} \tag{3.10}$$

where

$$R_{N_1}(z) = \sum_{n=0}^{N_1-1} \frac{1}{\left(\frac{1}{2}+n\right)} z^n,$$

and the number of time samples are assumed to be $2N_1$. Note that $R_{N_1}(z)$ acts on future time and $R_{N_1}(z^{-1})$ acts on past time. We assume that the data are prerecorded; hence, the question of causality is irrelevant. It is shown in Ref. [5] that as $N_1 \to \infty$, $R_{N_1}(z)$ may be approximated by a stable pole-zero filter, in particular,

$$\begin{aligned} &R_{N_1}(z) \to \frac{2(1-0.65465z)}{1-0.98612z+0.13091z^2} \\ &N_1 \to \infty. \end{aligned} \tag{3.11}$$

A comparison of the filter response $N_1 = 64$ and $N_1 \to \infty$ is shown in Figure 3.4.

A block diagram for fast realization of the fan is shown in Figure 3.5. The realization consists of two parts, the temporal processor (upper part) and the spatial processor (lower part). The temporal part has two pole-zero filters, one acting on past time and the other on future time. A pole-zero filter can be implemented recursively; for example, the left arm of the temporal processor may be implemented via a difference equation

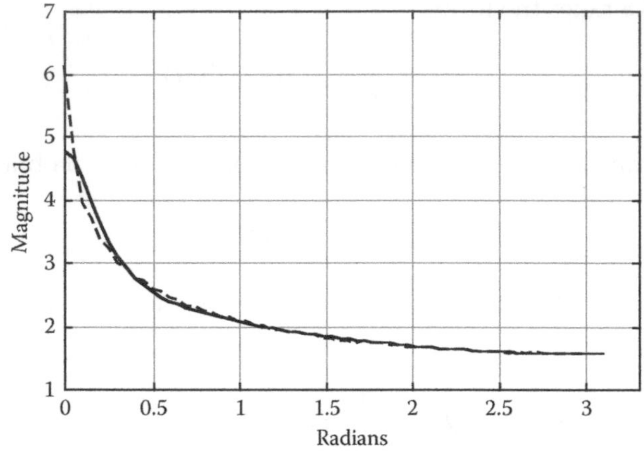

FIGURE 3.4
A comparison of the filter response $N_1=64$ (dashed line) and $N_1\rightarrow\infty$ (solid line).

$$\hat{f}_1(n\Delta t) - 0.98612\hat{f}_1((n+1)\Delta t) + 0.13091\hat{f}_1((n+2)\Delta t)$$
$$= 2f_1((n+1)\Delta t) - 1.3093 f_1((n+2)\Delta t),$$

and the right arm via

$$\hat{f}_r(n\Delta t) - 0.98612\hat{f}_r((n-1)\Delta t) + 0.13091\hat{f}_r((n-2)\Delta t)$$
$$= 2f_1(n\Delta t) - 1.3093 f_1(n-1)\Delta t).$$

The output of the temporal processor is finally given by $\hat{f}(n\Delta t) = \hat{f}_1(n\Delta t) + \hat{f}_r(n\Delta t)$.

3.1.4 Quadrant Filter

The fan filter is a half-plane filter; hence it is insensitive to the direction of propagation as long as the horizontal apparent speed lies within specified limits. A wave traveling with apparent speed c_{Hor} cannot be distinguished from a wave traveling with apparent speed of $-c_{Hor}$. For such a discriminatory property, the desired filter must possess a passband in only one quadrant, either the first or second quadrant. Note that for the filter to be real, the passband must be reflected diagonally into the opposite quadrant. An example of such a passband is shown in Figure 3.6. We will call such a filter a quadrant filter.

The passband is bounded with two radial lines with slopes a and b, respectively. Further, the passband is terminated by two horizontal lines $\omega = \pm\omega_{max}$, where ω_{max} is the cutoff temporal frequency. When $|b|>1$ and $|a|<1$ the passband will be terminated by two cutoff lines as shown in Figure 3.6. Draw a diagonal line and divide the passband into two triangles. Straightforward integration yields the filter impulse response function for $m\neq0$, $((m/a)(\Delta x/\Delta t)+n)\neq0$ and $((m/b)(\Delta x/\Delta t)+n)\neq0$ [6].

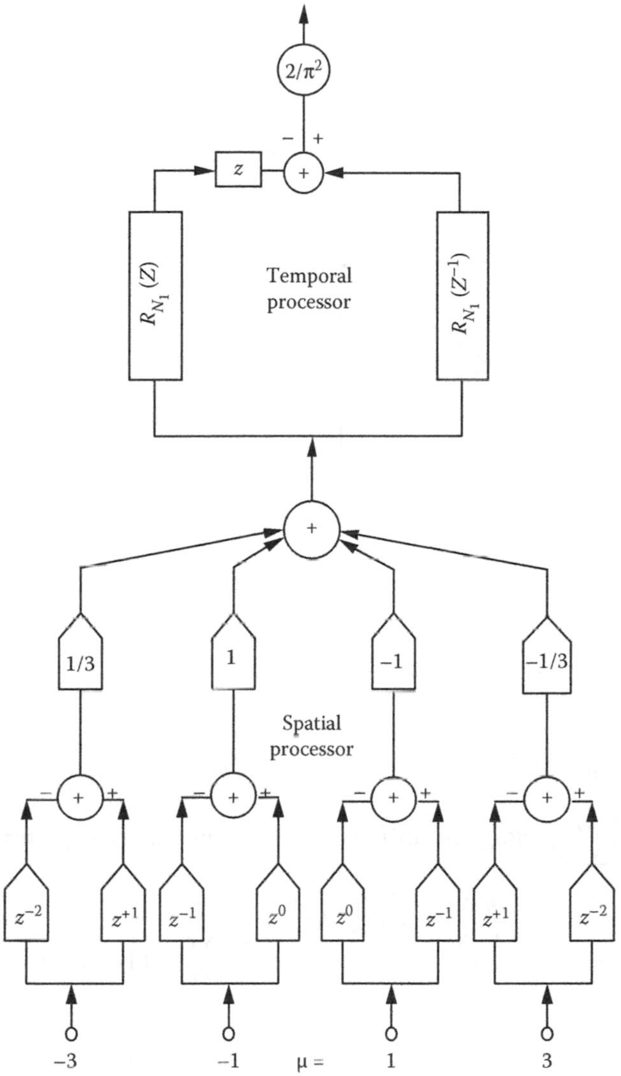

FIGURE 3.5
A block diagram of fan filter implementation. The filter structure has two parts, namely, spatial and temporal parts, which are independent of each other.

$$
\begin{aligned}
h_{m,n} &= \frac{1}{4\pi^2}\int_0^{\omega_{\max}}\left[\int_{\omega/b}^{\omega/a} 2\cos(m\Delta x u + n\Delta t\omega)du\right]d\omega \\
&= \frac{1}{2\pi^2 m\Delta x\Delta t}\left[\frac{\left(1-\cos\left(\frac{m}{a}\frac{\Delta x}{\Delta t}+n\right)\Delta t\omega_{\max}\right)}{\frac{m}{a}\frac{\Delta x}{\Delta t}+n} - \frac{\left(1-\cos\left(\frac{m}{b}\frac{\Delta x}{\Delta t}+n\right)\Delta t\omega_{\max}\right)}{\frac{m}{b}\frac{\Delta x}{\Delta t}+n}\right].
\end{aligned}
\tag{3.12a}
$$

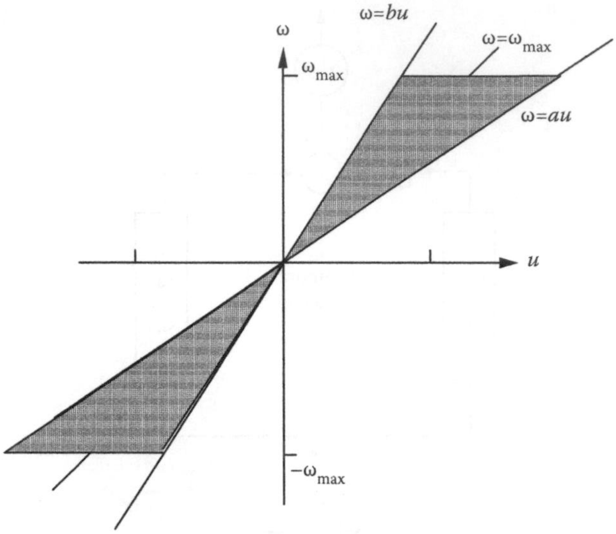

FIGURE 3.6
Quadrant filter in ω-u plane.

The special cases are given by

$$h_{0,0} = \frac{(b-a)\omega_{\max}^2}{4\pi^2 ab} \quad m=n=0$$

$$h_{m,0} = \frac{(b-a)}{2\pi^2 n^2 \Delta t^2 ab}\left[n\omega_{\max}\Delta t \sin(n\omega_{\max}\Delta t) + \cos(n\omega_{\max}\Delta t) - 1\right] \quad m=0, \quad n \neq 0,$$

$$h_{m,n} = \frac{-1}{2\pi^2}\frac{\left(1-\cos\left(\frac{m}{b}\frac{\Delta x}{\Delta t}+n\right)\omega_{\max}\Delta t\right)}{n\Delta x\Delta t\left(\frac{m}{b}\frac{\Delta x}{\Delta t}+n\right)} \quad \left(\frac{m}{a}\frac{\Delta x}{\Delta t}+n\right)=0, n\neq 0$$

$$\left(\frac{m}{b}\frac{\Delta x}{\Delta t}+n\right)\neq 0 \tag{3.12b}$$

$$h_{m,n} = \frac{1}{2\pi^2}\frac{\left(1-\cos\left(\frac{m}{a}\frac{\Delta x}{\Delta t}+n\right)\omega_{\max}\Delta t\right)}{n\Delta x\Delta t\left(\frac{m}{a}\frac{\Delta x}{\Delta t}+n\right)} \quad \left(\frac{m}{b}\frac{\Delta x}{\Delta t}+n\right)=0, n\neq 0$$

$$\left(\frac{m}{a}\frac{\Delta x}{\Delta t}+n\right)\neq 0.$$

We now set $\Delta x=1$ and $\Delta t=1$, hence $\Delta x/\Delta t=1$. Using these settings, Equation 3.12a reduces to Equation 3.6 in Ref. [6]. The highest temporal frequency is

$f_{hi}=(1/2)$ Hz (or $\omega_{hi}=\pi$ in angular frequency) and the corresponding lowest wavelength, $\lambda_{lowest}=2$. The sensor spacing will be 1. In real terms, consider a signal with $f_{hi}=1{,}500$ Hz propagating underwater where the wave speed is 1,500 m/sec. Since the wavelength is 1 m, the sensor spacing is $\Delta x=0.5$ m. The sampling interval is $\Delta t=(1/3)$ msec. Let us redefine units of time and distance. Let one unit of time be (1/3) msec and one unit of distance be 0.5 m. Here onwards we shall measure the time and the distance in terms of these newly defined units. The response function of a quadrant filter for the following parameters is shown in Figure 3.7. Parameters: 64 sensors, 64 time samples, $\omega_{max}=0.8\pi$, $a=1$, and $b=4$. In Equation 3.12a, if we let $a=-b=1$, $c_{Hor}=1$, $\Delta t=1$, and $\omega_{max}=\pi$, the resulting filter becomes a fan filter with its axis turned 90° and the filter coefficients are

$$h_{m,n}=\frac{1}{2m\pi^2}\left[\frac{(1-\cos(m+n)\pi)}{m+n}+\frac{(1-\cos(m-n)\pi)}{m-n}\right],$$

which indeed are the same as in Equation 3.4.

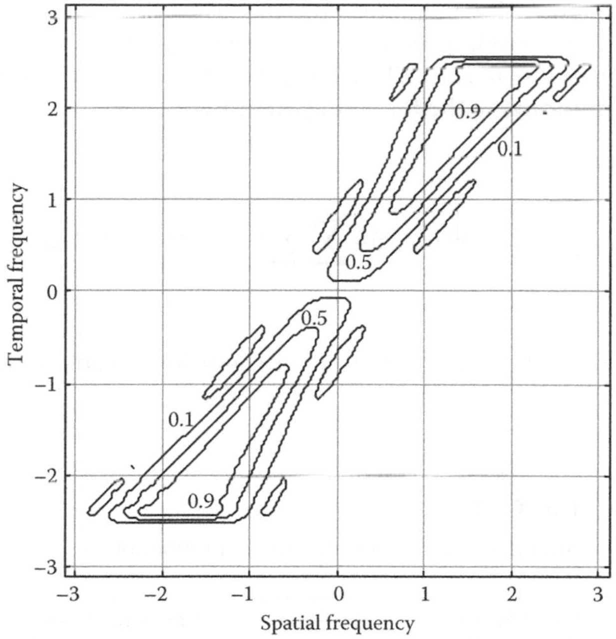

FIGURE 3.7
Frequency response of a quadrant filter ($a=1$, $b=2$). Sixteen sensors and 64 time samples are assumed. The maximum temporal frequency is 0.8π. The contour values are as shown in the figure.

3.1.5 Weighted Least Squares Filter

The filter size controls the root mean square (RMS) difference between the ideal infinite length filter and finite length filter. In general, the rms error decreases with increasing filter size but only slowly, as revealed in Table 3.1 where we considered a discrete fan filter whose ideal response is one in the region bounded by two rays with slopes 2 and 8 and zero outside. The maximum temporal frequency is set at 78% of the Nyquist frequency. The impulse response function of the quadrant filter is computed by taking the inverse discrete Fourier transform (DFT) of the ideal quadrant filter embedded in a large matrix (256×256) of zeros. The impulse response function is then truncated to 64 temporal samples and a variable number of spatial samples: 4, 8, 16, 32, and 64. Both rms error and maximum difference between the ideal filter (frequency) response and truncated filter (frequency) response, which is now interpolated to the same grid as the ideal impulse response, are shown in Table 3.1.

The response in the reject region, where the ideal response is zero, also decreases with increasing filter size. From a practical point of view, for a given filter size, the response in the pass region must be as close to unity as possible while the response in the reject region or at least in some identified parts of the reject region must be close to zero. This goal can be achieved by using a weighted least squares approach [7]. Let $W(\omega,u)$ be a weighting function which takes a large value in the reject region, where it is desired to have a response close to zero. Elsewhere, $W(\omega,u)=1$. The filter coefficients are obtained by minimizing the following quantity:

$$\frac{1}{4\pi^2}\int_{-(\pi/\Delta x)}^{\pi/\Delta x}\int_{-(\pi/\Delta t)}^{\pi/\Delta t} W(\omega,u)\left|H(\omega,u)-\sum_m\sum_n h_{mn}e^{-j(\omega n\Delta t-um\Delta x)}\right|^2 d\omega du=\min. \quad (3.13)$$

Minimization with respect to the filter coefficients leads to the following normal equations:

TABLE 3.1

rms error and maximum error vs number of sensors

No. of Sensors	rms Error	Maximum Error
4	0.1183	0.9959
8	0.0775	0.9958
16	0.0418	0.9912
32	0.0240	0.9743
64	0.0153	0.9303

$$\sum_{m'}\sum_{n'} h_{mn} r_{m-m',n-n'} = g_{mn}, \tag{3.14a}$$

where

$$r_{m,n} = \frac{1}{4\pi^2} \int_{-(\pi/\Delta x)}^{\pi/\Delta x} \int_{-(\pi/\Delta t)}^{\pi/\Delta t} W(\omega,u) e^{+j(\omega n\Delta t - um\Delta x)} d\omega du,$$

and

$$g_{m,n} = \frac{1}{4\pi^2} \int_{-(\pi/\Delta x)}^{\pi/\Delta x} \int_{-(\pi/\Delta t)}^{\pi/\Delta t} W(\omega,u) H(\omega,u) e^{+j(\omega n\Delta t - um\Delta x)} d\omega du.$$

Note that, when $W(\omega,u)=1$ over the entire plane, Equation 3.14a reduces to the expected result, namely,

$$h_{m,n} = \frac{1}{4\pi^2} \int_{-(\pi/\Delta x)}^{\pi/\Delta x} \int_{-(\pi/\Delta t)}^{\pi/\Delta t} H(\omega,u) e^{+j(\omega n\Delta t - um\Delta x)} d\omega du.$$

In order to understand the role of the weighting function, consider Equation 3.14a with finite limits on the summation signs

$$\sum_{m'=0}^{M-1}\sum_{n'=0}^{N-1} h_{mn} r_{m-m',n-n'} = g_{mn}, \tag{3.14b}$$

and take a finite Fourier transform on both sides of Equation 3.14b. We obtain

$$[H]_{\text{finite}}\,[W]_{\text{finite}} = [WH]_{\text{finite}} \tag{3.15}$$

where $[F]_{\text{finite}} = \sum_{m=0}^{M-1}\sum_{n=0}^{N-1} f_{mn} e^{-j(\omega m + un)}$ stands for the DFT for $\omega=(2\pi/M)k$ and $u=(2\pi/N)l$. Note that $[F]_{\text{finite}} \to F$ as $M \to \infty$ and $N \to \infty$. From Equation 3.15 we obtain

$$[H]_{\text{finite}} = \frac{[WH]_{\text{finite}}}{[W]_{\text{finite}}}. \tag{3.16}$$

Although WH is a bandpass filter (because H is a bandpass filter), $[WH]_{\text{finite}}$ will not be zero outside the passband. The presence of W in the denominator will however greatly reduce (if $W \gg 1$) the out of band magnitude of $[WH]_{\text{finite}}$. Hence, $[H]_{\text{finite}}$ on the left-hand side of Equation 3.16 will have low magnitude outside the passband. In Figure 3.8, we illustrate this

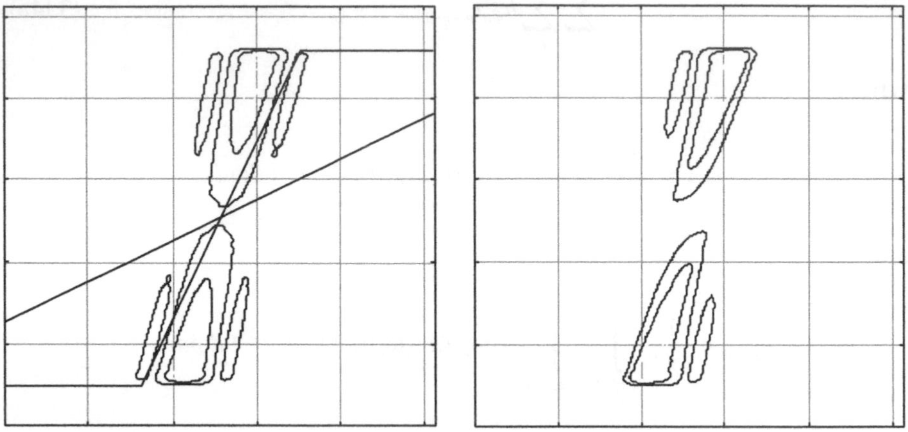

FIGURE 3.8
The role of a weighting function. (a) Unweighted case. (b) Weighted case (equal to six inside a fan-shaped region bounded by two radial lines). Notice how a sidelobe falling within the region of high weight function has been attenuated.

phenomenon. A quadrant filter with unit response in a region bounded by two radial lines with slopes 3 and 8 and with maximum temporal frequency 0.78 times the Nyquist frequency is considered. We selected a weighting function whose value is six in a fan-shaped region bounded by two radial lines with slopes 0.5 and 2 and with the maximum temporal frequency 0.78 times the Nyquist frequency and equal to one elsewhere. The impulse response functions of WH and W are now limited to 64×16 (64 temporal samples and 16 spatial samples). The finite impulse functions are next used to compute the weighted least squares filter as given in Equation 3.16. The plain (unweighted) finite filter response function along with the region where the weight function is six is shown in Figure 3.8a and the weighted least squares filter response function is shown in Figure 3.8b. Notice how a sidelobe falling within the region of the high weight function has been attenuated.

3.1.6 Aliasing Effect in Fan Filter and Quadrant Filter

In array signal processing there is a basic limitation imposed by discrete sensors. Consequently, the wavefield is sampled in the spatial dimension but not in the temporal dimension. It is possible to avoid temporal aliasing through a lowpass filtering and by sampling according to the sampling theorem, but it is a different story with the spatial aliasing, which is intimately related to the propagation speed. This phenomenon was briefly discussed in Chapter 1, Section 1.2. Here we shall examine the effect of spatial sampling on digital filters for wavefields, in particular, a quadrant filter whose pass region is defined by two radial lines with slopes c_{Lo} and c_{Hi}, as shown in Figure 3.9. We further assume that sensor outputs have been prefiltered to confine the spectra to $\pm\omega_{max}$.

The sampled version of the quadrant filter will be periodic both in spatial and temporal frequencies, but as the sensor outputs have been prefiltered, replication in the temporal frequency would not cause any aliasing. However, aliasing can take place due to the replication in spatial frequency. This is shown in Figure 3.9, where we have drawn three replications, that is, three rectangles including the principal rectangle. Notice the intrusion of a radial line with a slope c_{Lo} from the neighboring rectangles into the principal rectangle. Such an intrusion is the cause of aliasing. Clearly, the aliasing can be prevented if

$$c_{Lo} \geq \frac{\omega_{max}}{B_0} = \frac{\Delta x}{\Delta t}. \tag{3.17}$$

A similar requirement was shown (in Chapter 1, Section 1.2) to be necessary to avoid aliasing of a broadband signal traveling with a horizontal speed c_{Hor}.

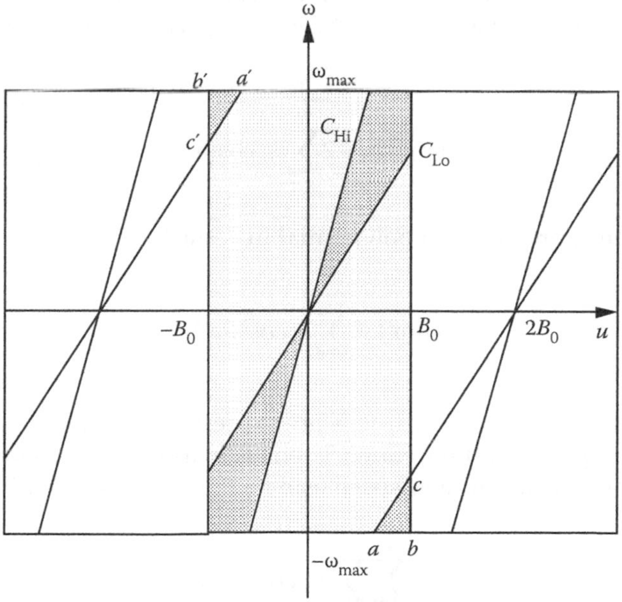

FIGURE 3.9
The aliasing is caused by intrusion of a radial line with slope c_{Lo} from the neighboring rectangles into the principal rectangle. Triangles *abc* and *a′b′c′* are the aliased pass regions. Aliasing can be prevented if an analog quadrant filter was sampled, according to Equation 3.17.

3.2 Mapping of 1D into 2D Filters

As in a one-dimensional filter, a sharp transition from passband to rejectband in a two-dimensional filter (in frequency wavenumber space) results in large ripples in passband and rejectband. To overcome this drawback, a sharp transition is replaced by a smooth transition, but this will degrade the quality of the filter. Thus, in the design of a pass filter there is a contradictory requirement, calling for an optimum design strategy of minimum transition width while maintaining the ripple height below a prescribed limit. This optimization problem has been largely solved in one dimension [8], but in two or higher dimensions the procedure of optimization becomes highly cumbersome and computationally intensive. It has been found that optimally designed 1D filters may be mapped into 2D filters having similar transition and ripple height properties as those of 1D filters [9]. The 2D filters of interest in the wavefield analysis are fan filter and quadrant filter. The fan filter may be obtained by transforming an optimally designed 1D filter.

3.2.1 McClellan's Transformation

Consider a real linear phase finite impulse response (FIR) filter $h_0, h_{\pm 1}, h_{\pm 2}, \ldots, h_{\pm N}$, where for linear phase we must have $h_k = h_{-k}$. The transfer function of the filter is given by

$$H(\omega') = h_0 + 2\sum_{k=1}^{N} h_k \cos(\omega' k), \tag{3.18}$$

which may be expressed as a polynomial in $\cos(\omega')$,

$$H(\omega') = \sum_{k=0}^{N} b_k \cos^k(\omega'), \tag{3.19}$$

where the coefficients b_k, $k=0, 1,\ldots, N$ are expressed in terms of the FIR filter coefficients. A point on the frequency axis is mapped onto a closed contour in the frequency wavenumber space using a linear transformation

$$\cos(\omega') = \sum_{p=0}^{P} \sum_{q=0}^{Q} t_{pq} \cos(p\omega)\cos(qu), \tag{3.20a}$$

or

$$\sin^2\left(\frac{\omega'}{2}\right) = \sum_{p=0}^{P} \sum_{q=0}^{Q} t'_{pq} \sin^{2p}\left(\frac{\omega}{2}\right)\sin^{2q}\left(\frac{u}{2}\right), \tag{3.20b}$$

where t_{pq} and t'_{pq} are yet to be determined coefficients. The shape of the contour depends upon the coefficients t_{pq} or t'_{pq}. For example, we obtain an approximate circular contour (see Figure 3.10) for

$$t_{00} = -\frac{1}{2},$$

$$t_{01} = t_{10} = t_{11} = \frac{1}{2}.$$

The mapping relation Equation 3.20a takes a form

$$\cos(\omega') = \frac{1}{2}\{-1 + \cos(u) + \cos(\omega) + \cos(u)\cos(\omega)\}. \tag{3.21}$$

Contours corresponding to a set of fixed values of $\cos(\omega')$ are shown in Figure 3.10. Circular approximation is better in the low frequency range. Using Equation 3.20a in Equation 3.19 we obtain a 2D filter transfer function,

$$H(\omega, u) = \sum_{k=0}^{N} b_k \left[\sum_{p=0}^{P} \sum_{q=0}^{Q} t_{pq} \cos(p\omega)\cos(qu) \right]^k. \tag{3.22}$$

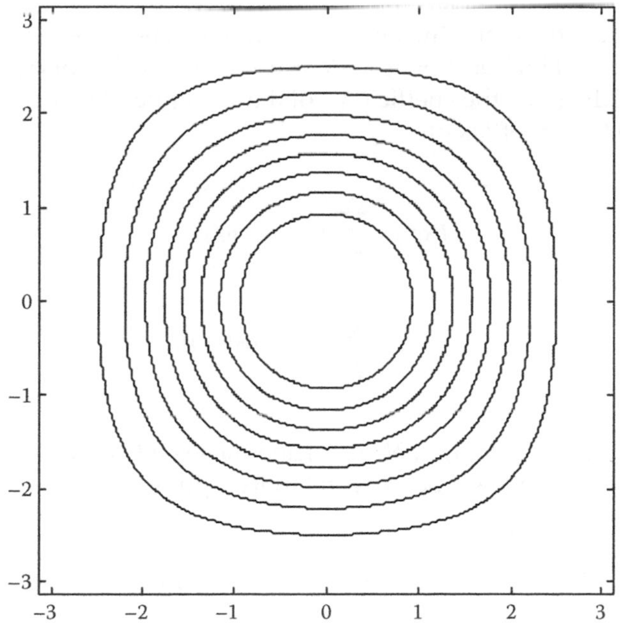

FIGURE 3.10
Mapping produced by Equation 3.21. The inner contours corresponding to low frequency are close to circular shape, but the outer contours corresponding to high frequency are only approximately circular.

The mapping function (Equation 3.20a) must satisfy the following conditions:

(a) When $\omega'=0$, $\omega=u=0$. This requires $\sum_{p=0}^{P}\sum_{q=0}^{Q} t_{pq}=1$. The condition ensures that a lowpass filter remains a lowpass filter even after the transformation, that is, $H(0,0)=H(0)$.

(b) When $\omega'=\pi$, $\omega=u=\pi$. This requires $\sum_{p=0}^{P}\sum_{q=0}^{Q} t_{pq}(-1)^{p+q}=-1$. The condition ensures that a highpass filter remains a highpass filter after transformation, that is, $H(\pi,\pi)=H(\pi)$.

(c) When $\omega=0$, $H(0,u)=\mathrm{H}(u)$. This requires $\sum_{p=0}^{P} t_{p1}=1$ and $\sum_{p=0}^{P} t_{pq}=0$ for $q\neq 1$.

(d) When $u=0$, $H(\omega,0)=H(\omega)$. This requires $\sum_{q=0}^{Q} t_{1q}=1$ and $\sum_{q=0}^{Q} t_{pq}=0$ for $p\neq 1$.

For circularly symmetric mapping, conditions (b) and (c) must hold good. For $P=Q=1$, a general solution for the transformation coefficients satisfying the above conditions is given by $t_{00}=-1/2$, $t_{01}=t_{10}=t_{11}=1/2$. By relaxing (b), the solution is given by

$$t_{00}=-a,\ t_{01}=t_{10}=a,\ t_{11}=1-a, \tag{3.23}$$

having one free constant which may be optimally chosen for the best fit.

The next important question is to determine the coefficients b_k, $k=0, 1,\ldots, N$ in Equation 3.19 given the coefficients of a digital filter. Let $h_0, h_1, h_2,\ldots, h_N$ be the given 1D filter coefficients.

$$\begin{aligned} H(\omega') &= \sum_{k=0}^{N} \tilde{h}_k \cos(k\omega') \\ &= \sum_{k=0}^{N} \tilde{h}_k T_k(\cos(\omega')), \end{aligned} \tag{3.24}$$

where $\tilde{h}_0=h_0$, $\tilde{h}_k=2h_k$, $k=1,\ldots, N$, and $T_n(\cos(\omega'))$ is a Chebyshev polynomial that may be expressed as a polynomial in its argument,

$$T_k(\cos(\omega')) = \sum_{m=0}^{k} c_m^k \cos^m(\omega'). \tag{3.25}$$

Using Equation 3.25 in Equation 3.24 and comparing with Equation 3.19 we obtain the following system of equations:

$$b_0 = \sum_{n=0}^{N} c_0^n \tilde{h}_n,\ b_1 = \sum_{n=1}^{N} c_1^n \tilde{h}_n, \ldots, b_k = \sum_{n=k}^{N} c_k^n \tilde{h}_n, \ldots, b_N = c_N^N \tilde{h}_N.$$

The coefficients c_k^n are listed in [10, p. 795]. Finally, the 2D filter transfer function for the case $P=Q=1$ can be written as

$$H(\omega,u) = \sum_{k=0}^{N} b_k \left[\{t_{00} + t_{01}\cos(u) + t_{10}\cos(\omega) + t_{11}\cos(u)\cos(\omega)\}\right]^k. \quad (3.26a)$$

Let $F(\omega,u)$ be the input and $G(\omega,u)$ be the output of a 2D filter,

$$\begin{aligned} G(\omega,u) &= \sum_{k=0}^{N} b_k \left[t_{00} + t_{01}\cos(u) + t_{10}\cos(\omega) + t_{11}\cos(u)\cos(\omega)\right]^k F(\omega,u) \\ &= \sum_{k=0}^{N} b_k \left[H_0(\omega,u)\right]^k F(\omega,u), \end{aligned} \quad (3.26b)$$

which may be written in a recursive fashion. Let

$$\begin{aligned} F_0(\omega,u) &= F(\omega,u) \\ F_1(\omega,u) &= \left[H_0(\omega,u)\right]F_0(\omega,u) \\ F_2(\omega,u) &= \left[H_0(\omega,u)\right]F_1(\omega,u) \\ &\ldots \\ &\ldots \\ &\ldots \\ F_N(\omega,u) &= \left[H_0(\omega,u)\right]F_{N-1}(\omega,u). \end{aligned} \quad (3.26c)$$

where $H_0(\omega,u)=t_{00}+t_{01}\cos(u)+t_{10}\cos(\omega)+t_{11}\cos(\omega)\cos(u)$. Note that $F_0(\omega,u)$, $F_1(\omega,u),\ldots, F_N(\omega,u)$ do not depend upon the filter coefficients, but only on mapping coefficients and the input. Using Equation 3.26c, the filter output may be written as

$$G(\omega,u) = \sum_{k=0}^{N} b_k F_k(\omega,u). \quad (3.27)$$

In filter implementation it is possible to generate $F_0(\omega,u)$, $F_1(\omega,u),\ldots, F_N(\omega,u)$ in a recursive manner and then combine them after weighting each with b_k. The filter structure is illustrated in Figure 3.11.

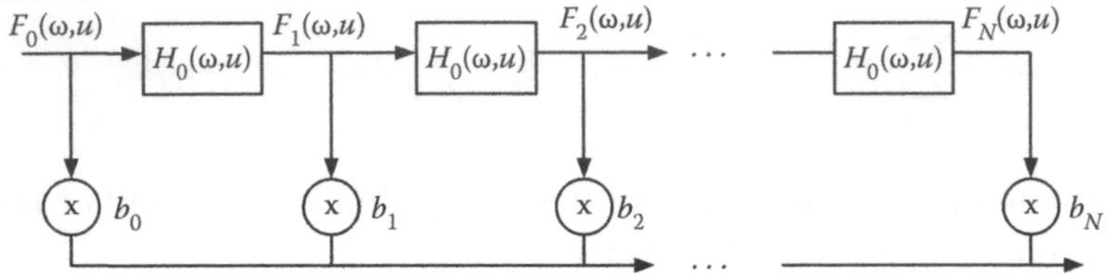

FIGURE 3.11
A recursive implementation of a 2D filter obtained by transforming a 1D filter. $H_0(\omega,u)$ solely depends upon the mapping coefficients while the coefficients and b_kS depend on 1D filter coefficients.

3.2.2 Fan Filter

To get fan-shaped contours the required mapping coefficients are

$$t_{00} = t_{11} = 0 \text{ and } t_{01} = -t_{10} = \frac{1}{2}, \tag{3.28a}$$

and the mapping function is given by [11]

$$\cos(\omega') = \frac{1}{2}[\cos(\omega) - \cos(u)]. \tag{3.28b}$$

The contours generated by Equation 3.28b are shown in Figure 3.12. The zero-valued diagonal contour corresponds to $\omega'=\pm(\pi/2)$. When $\omega'=0$, $\omega=0$, and $u=\pm\pi$; and when $\omega'=\pm\pi$, $\omega=\pm\pi$, and $u=0$. The positive contours correspond to $0\le|\omega'|\le(\pi/2)$ and the negative contours to $(\pi/2)<|\omega'|\le\pi$.

Ideally, a lowpass filter having a unit response in the range $\pm(\pi/2)$ and zero outside when mapped into two dimensions using the mapping function (Equation 3.28b) will result in a fan filter with a unit response in the top and bottom triangles and a zero response in the left and right triangles. The value of the 2D filter transfer function may be found by first computing the frequency corresponding to the index of each contour, that is, $\omega'=\cos^{-1}$ (contour index) and then evaluating the 1D filter response at the desired frequency.

It is interesting to note that the mapping coefficients given by Equation 3.28a do not satisfy any of the conditions listed on page 156. This is expected as the mapping function does not satisfy circular symmetry nor is it necessary to map a lowpass filter into a lowpass filter or a highpass filter into a highpass filter. A fan filter in 3D can also be obtained by transformation of a 1D filter. The transformation function is given by [12]

$$\cos(\omega') = \frac{1}{2}\cos(\omega) - \frac{1}{2}\cos(u) - \frac{1}{2}\cos(v) + \frac{1}{2}. \tag{3.29}$$

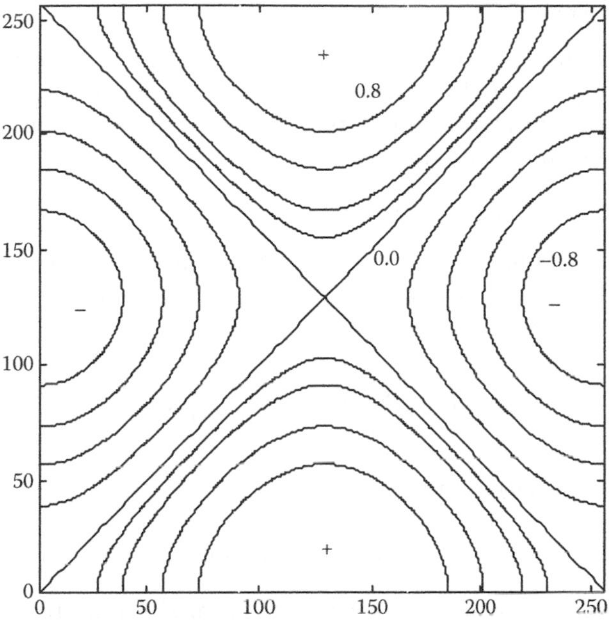

FIGURE 3.12
Contours generated by a mapping function given by Equation 3.28. The two diagonals make up the zero contour, which divides the display area into four triangles. The positive contours correspond to $0 \le |\omega'| \le \pi/2$ and the negative contours to $\pi/2 < |\omega'| \le \pi$.

3.3 Multichannel Wiener Filters

In Wiener filtering, the goal is to make the filter output as close as possible, in the least square sense, to a desired signal. The sensor output and the desired signal are assumed to be stationary stochastic signals, which are characterized through their covariance functions. The Wiener filters are named after Norbert Wiener who did pioneering work on the prediction of a trajectory of a moving object from its past observations [13]. A recursive algorithm for the solution of a discrete version of the Wiener–Hopf equation was developed by Levinson [14] and Durbin [15] in the context of time series model fitting. Multichannel extension was made by Wiggins and Robinson [16]. In this section, we briefly describe the Wiener filter as applied to array signals, where we like to extract a signal traveling in some *known* direction and to optimally suppress all other propagating waves and noise. Here, a straightforward solution of the Wiener–Hopf equation requires invertion of a large block Toeplitz covariance matrix, leading to a dramatic increase in the computational load over the single time series version. Therefore, it is worthwhile making some effort to understand the principles of the Levinson–Durbin recursive algorithm for the multichannel Wiener filter [16].

3.3.1 Planar Array

We consider a planar array of sensors, not necessarily uniformly distributed (see Figure 3.13). Let $f_p(t)$ be the output of the pth sensor located at (x_p, y_p). We shall model the output as a sum of two random processes, namely, a desired signal, $\xi_p(t)$, and unwanted noise, $\eta_p(t)$,

$$f_p(t)=\xi_p(t)+\eta_p(t), \quad p=0, 1,\ldots, P-1. \tag{3.30}$$

A multichannel filter $h_p(t)$, $p=0, 1,\ldots, P-1$ is sought such that the output, as given by Equation 3.31, is closest to the signal at one of the sensors, e.g., at $p=0$. The filter output, given by,

$$\hat{f}(t)=\sum_{p=0}^{P-1}\int_0^{\infty} h_p(\tau)f_p(t-\tau)d\tau, \tag{3.31}$$

must be closest to $\xi_0(t)$ in the sense

$$E\left\{\left|\hat{f}(t)-\xi_0(t)\right|^2\right\}=\min,$$

that is, the mean square error (mse) is minimum. This requires minimization of an expression for mse,

$$\begin{aligned}\text{mse} &= c_{\xi_0}(0)-2\sum_{p=0}^{P-1}\int_0^{\infty} h_p(\tau)c_{\xi_0 f_p}(\tau)d\tau \\ &+\sum_{p=0}^{P-1}\sum_{q=0}^{P-1}\int_0^{\infty}\int_0^{\infty} h_p(\tau)h_q(\tau')\left[c_{\xi_p\xi_q}(\tau-\tau')+c_{\eta_p\eta_q}(\tau-\tau')\right]d\tau d\tau'.\end{aligned} \tag{3.32}$$

FIGURE 3.13
A distribution of sensors in a planar array. All sensors are identical but for their position.

We shall minimize Equation 3.32 with respect to $h_p(\tau)$. So we need to differentiate Equation 3.32 with respect to $h_p(\tau)$ which occurs inside an integral. To see how such a differentiation may be carried out, let us express the integral as a limiting sum. For example,

$$\int_0^{\infty} h_p(\tau)c_{\xi_0\xi_p}(t-\tau)d\tau \to \Delta\tau \sum_{n=0}^{\infty} h_p(n\Delta\tau)c_{\xi_0\xi_p}(t-n\Delta\tau) \tag{3.33}$$

$$\Delta\tau \to 0.$$

Now, differentiate the sum on the right-hand side in Equation 3.33 with respect to $h_p(n\Delta\tau)$ for fixed n. The result is $\Delta\tau c_{\xi_0\xi_p}(t-n\Delta\tau)$. A similar approach is used to differentiate an expression involving a double integral. The derivative is first set to zero and then a limiting operation ($\Delta\tau \to 0$) is carried out. We finally obtain a set of equations known as normal equations:

$$\sum_{q=0}^{P-1} \int_0^{\infty} h_q(\tau')c_{f_pf_q}(\tau-\tau')d\tau' = c_{\xi_0 f_p}(\tau) \quad p=0,1,\ldots,P-1. \tag{3.34a}$$

The minimum mean square error (mmse) may be derived by using Equation 3.34a in Equation 3.32. We obtain

$$\text{mse}\big|_{\min} = c_{\xi_0}(0) - \sum_{p=0}^{P-1} \int_0^{\infty} h_p(\tau)c_{\xi_0 f_p}(\tau)d\tau. \tag{3.34b}$$

Let us express Equation 3.34 in discrete form. In order to do this we introduce the following vectors and matrix:

$$\mathbf{c}_{f_pf_q} = \begin{bmatrix} c_{f_pf_q}(0) & c_{f_pf_q}(\Delta\tau) & \cdots & c_{f_pf_q}((N-1)\Delta\tau) \\ c_{f_pf_q}(-\Delta\tau) & c_{f_pf_q}(0) & \cdots & c_{f_pf_q}((N-1)\Delta\tau) \\ \vdots & \vdots & & \vdots \\ \vdots & \vdots & & \vdots \\ \vdots & \vdots & & \vdots \\ c_{f_pf_q}((1-N)\Delta\tau) & c_{f_pf_q}((2-N)\Delta\tau) & \cdots & c_{f_pf_q}((2-N)\Delta\tau),\ldots,c_{f_pf_q}(0) \end{bmatrix},$$

$$\mathbf{h}_p = [h_p(0) \quad h_p(\Delta\tau) \quad h_p(2\Delta\tau) \quad \ldots \quad h_p((N-1)\Delta\tau)]^T$$

and

$$\mathbf{c}_{\xi_0 f_p} = \left[c_{\xi_0 f_p}(0) \quad c_{\xi_0 f_p}(\Delta\tau) \quad c_{\xi_0 f_p}(2\Delta\tau) \quad \ldots \quad c_{\xi_0 f_p}((N-1)\Delta\tau)\right]^T,$$

where $\Delta\tau$ is the sampling interval and $[\ ]^T$ stands for matrix or vector transpose. Equation 3.34 may now be expressed using the above vectors and matrices

$$\sum_{q=0}^{P-1} \mathbf{c}_{f_p f_q} \mathbf{h}_q = \mathbf{c}_{\xi_0 f_p}, \quad p = 0, 1, 2, \ldots, P-1, \tag{3.35a}$$

$$\text{mse}\big|_{\min} = c_{\xi_0}(0) - \sum_{p=0}^{P-1} \mathbf{c}_{\xi_0 f_p}^T \mathbf{h}_p. \tag{3.35b}$$

An alternate representation of Equation 3.34 is through block matrices defined as

$$\mathbf{h} = [\mathbf{h}(0) \quad \mathbf{h}(\Delta\tau) \quad \ldots \quad \mathbf{h}((N-1)\Delta\tau)]^T,$$

where

$$\mathbf{h}(n\Delta\tau) = [h_0(n\Delta\tau) \quad h_1(n\Delta\tau) \quad \ldots \quad h_{p-1}(n\Delta\tau)]^T$$

$$\mathbf{C}_0 = [\mathbf{c}_0(0) \quad \mathbf{c}_0(\Delta\tau) \quad \ldots \quad \mathbf{c}_0((N-1)\Delta\tau)]^T,$$

where

$$\mathbf{c}_0(n\Delta\tau) = \left[c_{\xi_0 f_0}(n\Delta\tau) \quad c_{\xi_0 f_1}(n\Delta\tau) \quad c_{\xi_0 f_2}(n\Delta\tau) \quad \ldots \quad c_{\xi_0 f_{P-1}}(n\Delta\tau)\right]^T,$$

$$\mathbf{C} = \begin{bmatrix} \mathbf{c}(0) & \mathbf{c}(\Delta\tau) & \cdots & \mathbf{c}((N-1)\Delta\tau) \\ \mathbf{c}(-\Delta\tau) & \mathbf{c}(0) & \cdots & \mathbf{c}((N-2)\Delta\tau) \\ \vdots & & & \vdots \\ \vdots & & & \vdots \\ \mathbf{c}((1-N)\Delta\tau) & & \cdots & \mathbf{c}(0) \end{bmatrix},$$

$$(PN \times PN)$$

where each element is a block matrix of the type shown below

$$\mathbf{c}(n\Delta\tau) = \begin{bmatrix} c_{f_0 f_0}(n\Delta\tau) & c_{f_0 f_1}(n\Delta\tau) & \cdots & c_{f_0 f_{P-1}}(n\Delta\tau) \\ c_{f_1 f_0}(n\Delta\tau) & c_{f_1 f_1}(n\Delta\tau) & \cdots & c_{f_1 f_{P-1}}(n\Delta\tau) \\ \vdots & \vdots & & \vdots \\ \vdots & \vdots & & \vdots \\ \vdots & \vdots & & \vdots \\ c_{f_{P-1} f_0}(n\Delta\tau) & c_{f_{P-1} f_0}(n\Delta\tau) & \cdots & c_{f_{P-1} f_{P-1}}(n\Delta\tau) \end{bmatrix},$$

$$(P \times P)$$

As an example, consider a three-sensor array with two time samples ($N = 2$). The above quantities, $\mathbf{C}$, $\mathbf{h}$, and $\mathbf{C}_0$, become

$$\mathbf{C}=\begin{bmatrix} c_{f_0f_0}(0) & c_{f_0f_1}(0) & c_{f_0f_2}(0) & c_{f_0f_0}(1) & c_{f_0f_1}(1) & c_{f_0f_2}(1) \\ c_{f_1f_0}(0) & c_{f_1f_1}(0) & c_{f_1f_2}(0) & c_{f_1f_0}(1) & c_{f_1f_1}(1) & c_{f_1f_2}(1) \\ c_{f_2f_0}(0) & c_{f_2f_1}(0) & c_{f_2f_2}(0) & c_{f_2f_0}(1) & c_{f_2f_1}(1) & c_{f_2f_2}(1) \\ c_{f_0f_0}(-1) & c_{f_0f_1}(-1) & c_{f_0f_2}(-1) & c_{f_0f_0}(0) & c_{f_0f_1}(0) & c_{f_0f_2}(0) \\ c_{f_1f_0}(-1) & c_{f_1f_1}(-1) & c_{f_1f_2}(-1) & c_{f_1f_0}(0) & c_{f_1f_1}(0) & c_{f_1f_2}(0) \\ c_{f_2f_0}(-1) & c_{f_2f_1}(-1) & c_{f_2f_2}(-1) & c_{f_2f_0}(0) & c_{f_2f_1}(0) & c_{f_2f_2}(0) \end{bmatrix},$$

$$\mathbf{h}=[h_0(0), h_1(0), h_2(0), h_0(1), h_1(1), h_2(1)]^T,$$

and

$$\mathbf{C}_0=\left[c_{\xi_0f_0}(0), c_{\xi_0f_1}(0), c_{\xi_0f_2}(0), c_{\xi_0f_0}(1), c_{\xi_0f_1}(1), c_{\xi_0f_2}(1)\right]^T.$$

Equation 3.35 may be expressed in a compact form as

$$\mathbf{Ch}=\mathbf{C}_0, \tag{3.36a}$$

$$\text{mse}\big|_{\min}=c_{\xi_0}(0)-\mathbf{C}_0^T\mathbf{h}. \tag{3.36b}$$

3.3.2 Frequency Domain

Sometimes it is advantageous to express the normal equations (Equation 3.34) in the frequency domain. Taking the Fourier transform on both sides of Equation 3.34, we obtain the following result:

$$\sum_{q=1}^{P-1} H_q(\omega)S_{f_pf_q}(\omega)=S_{\xi_0f_p}(\omega), \quad p=0, 1,\dots, P-1, \tag{3.37a}$$

where $H_q(\omega)$ is the transfer function of the qth filter given by

$$H_q(\omega)=\int_0^{\infty} h_q(\tau)e^{j\omega\tau}d\tau,$$

$S_{f_pf_q}(\omega)$ is the cross-spectrum between $f_q(t)$ and $f_p(t)$ and similarly $S_{\xi_0f_p}(\omega)$ is the cross-spectrum between $\xi_0(t)$ and $f_p(t)$. Similarly, mmse in the frequency domain may be obtained from Equation 3.34b

$$\text{mse}\big|_{\min}=\int_{-\infty}^{\infty}\left\{S_{\xi_0}(\omega)-\sum_{p=0}^{P-1}H_p(\omega)S_{\xi_0f_p}(-\omega)\right\}d\omega. \tag{3.37b}$$

To write Equation 3.37 in a matrix form, define the following vectors and matrix,

$$\mathbf{S}_f(\omega)=\begin{bmatrix} S_{0,0}(\omega) & S_{0,1}(\omega) & \cdots & S_{0,P-1}(\omega) \\ S_{1,0}(\omega) & S_{1,1}(\omega) & \cdots & S_{1,P-1}(\omega) \\ \vdots & \vdots & & \vdots \\ \vdots & \vdots & & \vdots \\ \vdots & \vdots & & \vdots \\ S_{P-1,0}(\omega) & S_{P-1,1}(\omega) & \cdots & S_{P-1,P-1}(\omega) \end{bmatrix},$$

$$\mathbf{H}(\omega)=[H_0(\omega) \quad H_1(\omega) \quad \ldots \quad H_{P-1}(\omega)]^T,$$

and

$$\mathbf{S}_0(\omega)=[S_{00}(\omega) \quad S_{01}(\omega) \quad \ldots \quad S_{0P-1}(\omega)]^T.$$

The normal equations in the frequency domain (Equation 3.37) may be expressed in a compact form

$$\mathbf{S}_f(\omega)\,\mathbf{H}(\omega)=\mathbf{S}_0(\omega), \tag{3.38a}$$

for all ω in the range ±∞. Formally, the solution of Equation 3.38 may be expressed as

$$\mathbf{H}(\omega)=\mathbf{S}_f^{-1}(\omega)\mathbf{S}_0(\omega). \tag{3.38b}$$

Now consider a plane wave sweeping across an array of sensors. Let the background noise be spatially white. The spectral matrix for this model is given by

$$\mathbf{S}_f(\omega)=\mathbf{A}(\omega)\,S_\eta(\omega), \tag{3.39}$$

where

$$\mathbf{A}(\omega)=\begin{bmatrix} 1+T(\omega) & T(\omega)e^{j(u_0x_{1,0}+v_0y_{1,0})} & \cdots & T(\omega)e^{j(u_0x_{P-1,0}+v_0y_{P-1,0})} \\ T(\omega)e^{-j(u_0x_{1,0}+v_0y_{1,0})} & 1+T(\omega) & \cdots & T(\omega)e^{j(u_0x_{P-1,1}+v_0y_{P-1,1})} \\ \vdots & \vdots & & \vdots \\ \vdots & \vdots & & \vdots \\ \vdots & \vdots & & \vdots \\ T(\omega)e^{-j(u_0x_{P-1,0}+v_0y_{P-1,0})} & T(\omega)e^{-j(u_0x_{P-1,1}+v_0y_{P-1,1})} & \cdots & 1+T(\omega) \end{bmatrix},$$

$T(\omega)=(S_0(\omega)/S_\eta(\omega))$, $x_{pq}=x_p-x_q$, $y_{pq}=y_p-y_q$, $S_0(\omega)$ is the signal spectrum, and $S_\eta(\omega)$ is the noise spectrum common to all sensors. Similarly, the vector on the right-hand side of Equation 3.38 may be expressed as

$$\mathbf{S}_0(\omega) = \left[1 \quad e^{-j(u_0 x_{1,0}+v_0 y_{1,0})} \quad e^{-j(u_0 x_{2,0}+v_0 y_{2,0})} \quad \dots \quad e^{-j(u_0 x_{P-1,0}+v_0 y_{P-1,0})}\right] S_0(\omega)$$
$$= \mathbf{B}(\omega) S_0(\omega). \tag{3.40}$$

Using Equations 3.39 and 3.40 in Equation 3.38, we obtain special normal equations

$$\mathbf{A}(\omega)\mathbf{H}(\omega) = \mathbf{B}(\omega)\frac{S_0(\omega)}{S_\eta(\omega)} \tag{3.41a}$$
$$= \mathbf{B}(\omega)T(\omega),$$

or

$$\mathbf{H}(\omega) = \mathbf{A}^{-1}(\omega)\mathbf{B}(\omega)T(\omega). \tag{3.41b}$$

We note that $\mathbf{A}(\omega)$ has a useful structure in that it can be expressed as

$$\mathbf{A}(\omega) = \mathbf{I} + T(\omega)\mathbf{a}(\omega)\mathbf{a}^H(\omega), \tag{3.42a}$$

where

$$\mathbf{a}(\omega) = \left[e^{j(u_0 x_0+v_0 y_0)} \quad e^{j(u_0 x_1+v_0 y_1)} \quad e^{j(u_0 x_2+v_0 y_2)} \quad \dots \quad e^{j(u_0 x_{P-1}+v_0 y_{P-1})}\right]^T. \tag{3.42b}$$

3.3.3 Constrained Minimization

Let the filtered signal output of an array of sensors be given by

$$\hat{f}(t) = \sum_{p=0}^{P-1} \int_0^\infty h_p(\tau)\xi_p(t-\tau)d\tau. \tag{3.43}$$

Assume a signal model where a plane wave is sweeping across the array maintaining its waveform unchanged. The outputs of any two sensors differ only in propagation delays. The signal at the pth sensor may be given by $\xi_p(t)=\xi_0(t-\tau_p)$, where τ_p is the propagation delay at the pth sensor ($\tau_0=0$ when delays are measured with respect to the 0th sensor). Since the geometry of the array, speed of propagation, and DOA are known or can be estimated independently, the propagation delays are presumed to be known. The output of each sensor is advanced to make it in phase with the output of the reference sensor. Equation 3.43 reduces to

$$\hat{f}(t)=\sum_{p=0}^{P-1}\int_0^\infty h_p(\tau)\xi_0(t-\tau)d\tau$$
$$=\int_0^\infty\left[\sum_{p=0}^{P-1}h_p(\tau)\right]\xi_0(t-\tau)d\tau. \tag{3.44}$$

The filters are to be chosen to satisfy the condition that $\hat{f}(t)=\xi_0(t)$. From Equation 3.44 it is clear that this constraint can be satisfied if

$$\sum_{p=0}^{P-1}h_p(t)=\delta(t), \tag{3.45a}$$

or in the frequency domain

$$\sum_{p=0}^{P-1}H_p(\omega)=1. \tag{3.45b}$$

Thus, for distortion free extraction of waveforms, the filters must satisfy Equation 3.45 [17]. While an individual filter transfer function is allowed to be of any form, the sum must be a constant.

Another type of constraint arises when it is required to minimize the background noise power. The noise in the output of an array processor is given by

$$\hat{\eta}(t)=\sum_{p=0}^{P-1}\int_0^\infty h_p(\tau)\eta_p(t-\tau)d\tau,$$

and the noise power is given by

$$\sigma_{\hat{\eta}}^2=\sum_p\sum_q\int_0^\infty\int_0^\infty h_p(\tau)h_q(\tau')E\{\eta_p(t-\tau)\eta_q(t-\tau')\}d\tau d\tau'$$
$$=\sum_p\sum_q\int_0^\infty\int_0^\infty h_p(\tau)h_q(\tau')c_{pq}(\tau'-\tau)d\tau d\tau' \tag{3.46}$$
$$=\sum_p\sum_q\frac{1}{2\pi}\int_0^\infty H_p(\omega)H_q^*(\omega)S_{\eta_{pq}}(\omega)d\omega.$$

For spatially and temporally uncorrelated noise Equation 3.46 reduces to

$$\sigma_{\hat{\eta}}^2 = \sum_{p=0}^{P-1} \frac{1}{2\pi} \int_0^{\infty} \left|H_p(\omega)\right|^2 d\omega\, \sigma_{\eta}^2.$$

The noise power in the array output shall be minimum whenever the filter transfer functions satisfy the condition

$$\sum_{p=0}^{P-1} \frac{1}{2\pi} \int_0^{\infty} \left|H_p(\omega)\right|^2 d\omega = \min. \tag{3.47}$$

The trivial solution, namely, $H_p(\omega)=0$ for all p, is not acceptable as it will set the output signal power also to zero. The constraint Equation 3.45 or Equation 3.47 is usually imposed along with other constraints. Sometimes the output noise power is set to a given fraction noise reduction factor (NRF) of the input noise power.

$$\sum_{p=0}^{P-1} \frac{1}{2\pi} \int_0^{\infty} \left|H_p(\omega)\right|^2 d\omega = \text{NRF} < 1. \tag{3.48}$$

3.4 Wiener Filters for Uniform Linear Array (ULA) and (UCA)

We shall now turn to some specific array geometries; in particular, we consider ULA and UCA.

3.4.1 ULA

As in the last section (Section 3.3), the signal model assumed here is a plane wavefront sweeping across a linear array of sensors spaced at an interval d on the x-axis. The noise is assumed to be spatially white. In Equation 3.41a the matrix $\mathbf{A}(\omega)$ and the column $\mathbf{B}(\omega)$ take the following form:

$$\mathbf{A}_{\text{ULA}}(\omega) = \begin{bmatrix} 1+T(\omega) & T(\omega)e^{ju_0 d} & \cdots & T(\omega)e^{j(u_0(P-1)d} \\ T(\omega)e^{-ju_0 d} & 1+T(\omega) & \cdots & T(\omega)e^{j(u_0(P-2)d} \\ \vdots & \vdots & & \vdots \\ \vdots & \vdots & & \vdots \\ \vdots & \vdots & & \vdots \\ T(\omega)e^{-ju_0(P-1)d} & T(\omega)e^{-ju_0(P-2)d} & \cdots & 1+T(\omega) \end{bmatrix},$$

and

$$\mathbf{B}_{\text{ULA}} = \text{col}\left[1 \quad e^{-ju_0 d} \quad e^{-ju_0 2d} \quad \ldots \quad e^{-ju_0 (P-1)d}\right].$$

$\mathbf{A}_{\text{ULA}}(\omega)$ can be expressed as $(\mathbf{I}+T(\omega)\mathbf{a}(\omega)\mathbf{a}^H(\omega))$ (see Equation 3.42), where the vector $\mathbf{a}(\omega)$ for a ULA is given by

$$\mathbf{a}(\omega) = \text{col}\left[1 \quad e^{-ju_0 d} \quad e^{-ju_0 2d} \quad \ldots \quad e^{-ju_0 (P-1)d}\right].$$

Remember that in a ULA the right most sensor is conventionally taken as the reference sensor. Using the Woodbury's identity [18], we obtain its inverse in a closed form,

$$\mathbf{A}_{\text{ULA}}^{-1}(\omega) = \frac{1}{1+PT(\omega)} \times \begin{bmatrix} 1+(P-1)T(\omega) & -T(\omega)e^{ju_0 d} & \cdots & -T(\omega)e^{ju_0 (P-1)d} \\ -T(\omega)e^{-ju_0 d} & 1+(P-1)T(\omega) & \cdots & -T(\omega)e^{ju_0 (P-2)d} \\ \vdots & \vdots & & \vdots \\ \vdots & \vdots & & \vdots \\ \vdots & \vdots & & \vdots \\ -T(\omega)e^{-ju_0 (P-1)d} & -T(\omega)e^{-ju_0 (P-2)d} & \cdots & 1+(P-1)T(\omega) \end{bmatrix}. \tag{3.49}$$

Using Equation 3.49 in Equation 3.41b, we obtain transfer functions of Wiener filters

$$\begin{aligned} H_0(\omega) &= \frac{T(\omega)}{1+PT(\omega)} \\ H_1(\omega) &= \frac{T(\omega)}{1+PT(\omega)} e^{-ju_0 d} \\ H_2(\omega) &= \frac{T(\omega)}{1+PT(\omega)} e^{-j2u_0 d} \\ &\vdots \\ H_{P-1}(\omega) &= \frac{T(\omega)}{1+PT(\omega)} e^{-ju_0 (P-1)d}. \end{aligned} \tag{3.50}$$

The frequency wavenumber response of the Wiener filter for $T=4$ and 16-sensor array (ULA) is shown in Figure 3.14. Notice that whenever $PT(\omega) \gg 1$, that is, either $T(\omega)=(S_0(\omega)/S_\eta(\omega)) \gg 1$, or $P \gg 1$, or both, or the spectra of the signal and noise are nonoverlapping, the Wiener filter reduces to a simple delay filter

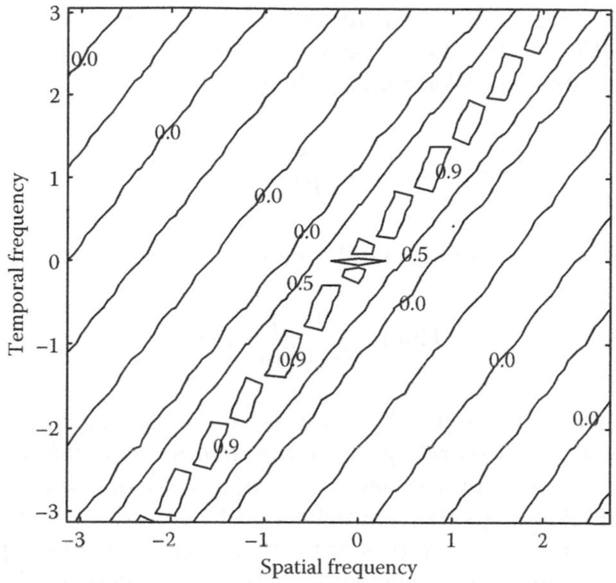

FIGURE 3.14
Frequency wavenumber response of the Wiener filter for ULA. It is assumed that the direction of arrival (DOA) is known. In the example, it is equal to 30°.

$$H_k(\omega) = \frac{1}{P} e^{-jku_0 d}. \tag{3.51}$$

3.4.2 Uniform Circular Array

The sensors are uniformly spaced on the circumference of a circular array of radius R units (see Chapter 2 for more on circular arrays). For circular array the matrix $\mathbf{A}(\omega)$ takes the form,

$$\mathbf{A}_{\mathrm{UCA}}(\omega) = \mathbf{I} + T(\omega)\mathbf{a}(\omega)\mathbf{a}^H(\omega), \tag{3.52a}$$

where

$$\mathbf{a}(\omega) = \mathrm{col}\left[e^{-ju_0R} \quad e^{-j[u_0R\cos(2\pi/P)+v_0R\sin(2\pi/P)]} \quad \ldots \quad e^{-j[u_0R\cos(P-1)(2\pi/P)+v_0R\sin(P-1)(2\pi/P)]}\right], \tag{3.52b}$$

and

$$\mathbf{B}_{\mathrm{UCA}}(\omega) = \mathbf{a}(\omega). \tag{3.53}$$

The reference point is at the center of the circle; however, there is no sensor physically present there. The Wiener filter is designed to predict the waveform as seen by a hypothetical sensor kept at the center of the circle. To solve for the Wiener filter (Equation 3.41b) we need to invert the $\mathbf{A}(\omega)$ matrix. We shall once again use Woodbury's identity and obtain

$$\mathbf{A}_{\text{UCA}}^{-1}(\omega) = \mathbf{I} - \frac{T(\omega)}{1+PT(\omega)}\left[\mathbf{a}(\omega)\mathbf{a}^H(\omega)\right]. \tag{3.54}$$

Using Equations 3.53 and 3.54 in Equation 3.41b we get

$$\mathbf{H}(\omega) = \frac{T(\omega)}{1+PT(\omega)}\mathbf{a}(\omega). \tag{3.55}$$

The Wiener filter for a circular array is similar to that for a linear array except for the difference arising out of the definition of $\mathbf{a}(\omega)$. The frequency wavenumber response of a 16-sensor circular array of radius 10λ is shown in Figure 3.15. Although the mainlobe width is narrow, the sidelobe level is quite high. This is clearly brought out in a cross-sectional plot passing through the maximum (see Figure 3.16). When we increase the number of sensors to 64 the sidelobe level is brought down considerably, but the main lobe width remains practically unchanged. It may be emphasized that the sensors need not be spaced at ≤0.5λ as in a ULA [19]. What is gained by

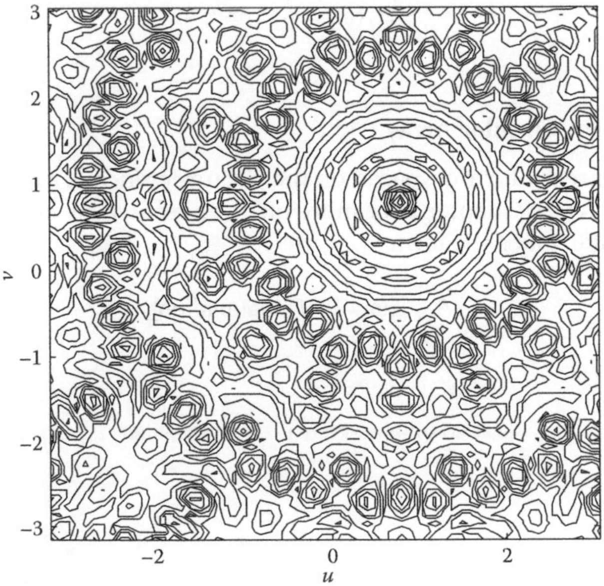

FIGURE 3.15
Frequency wavenumber response of Wiener filter for UCA. The DOAs are known (azimuth = elevation = 45° and $\omega = \pi/2$). Sixteen sensors and constant SNR = 4 are assumed.

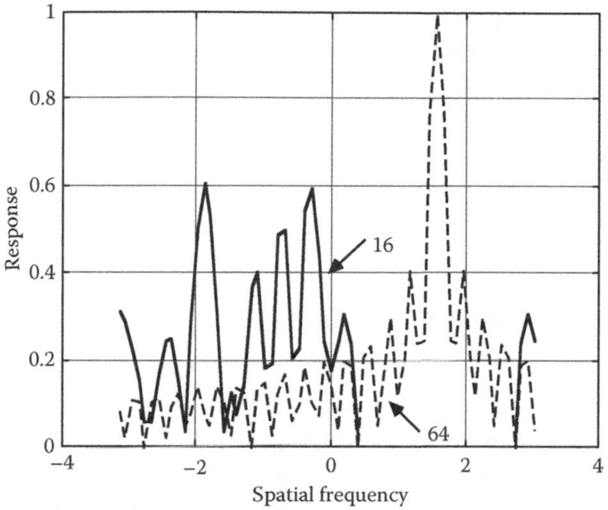

FIGURE 3.16
A cross section of the Wiener filter response taken through the peak. Number of sensors is 16 and 64. The sidelobe level relative to the peak has been reduced when the number of sensors is increased from 16 to 64. Radius of circular aperture is 10 units.

increasing the number of sensors (keeping the radius fixed) is the reduction of the sidelobe level. In contrast, in case of a ULA, by increasing the number of sensors the array aperture is increased, which in turn sharpens the mainlobe but does not reduce the sidelobe level.

3.4.3 Robustification

The Wiener filters given by Equations 3.51 and 3.55 require a knowledge of $\mathbf{a}(\omega)$ for which we need to know the apparent speed of the wavefront sweeping across the array. Prior to waveform estimation, it is a common practice to estimate the DOA of a wavefront. (This topic is covered in some detail in Chapter 5.) The DOA estimation is not without error. Hence, it would be nice if the Wiener filters were made robust so that the degradation in its performance was minimum. We shall confine to a linear array (ULA). An error in DOA estimation will introduce an error in the wavenumber. For a ULA, the erroneous wavenumber may be expressed as $(u_0+\varepsilon)$, where ε is a uniformly distributed random variable. Naturally, there will be an error in each element of $\mathbf{A}_{ULA}(\omega)$ and $\mathbf{B}_{ULA}(\omega)$. We shall use the stochastically averaged $\mathbf{A}_{ULA}(\omega)$ and $\mathbf{B}_{ULA}(\omega)$ matrices in Equation 3.41b. This approach was suggested in Ref. [20] in the context of optimal velocity filters in seismic exploration. Observe that in each element of the matrices there is an extra term which does not permit the matrix to be written as an outer product of two vectors as in Equation 3.52a.

$$\bar{\mathbf{A}}_{\mathrm{ULA}}(\omega) = \begin{bmatrix} 1+T(\omega) & T(\omega)\mathrm{sinc}(\varepsilon_0 d)e^{ju_0 d} & \cdots & T(\omega)\mathrm{sinc}(\varepsilon_0 (P-1)d)e^{ju_0 (P-1)d} \\ T(\omega)\mathrm{sinc}(\varepsilon_0 d)e^{-ju_0 d} & 1+T(\omega) & \cdots & T(\omega)\mathrm{sinc}(\varepsilon_0 (P-2)d)e^{ju_0 (P-2)d} \\ \vdots & \vdots & & \vdots \\ \vdots & \vdots & & \vdots \\ T(\omega)\mathrm{sinc}(\varepsilon_0 (P-1)d)e^{-ju_0 (P-1)d} & T(\omega)\mathrm{sinc}(\varepsilon_0 (P-2)d)e^{-ju_0 (P-2)d} & \cdots & 1+T(\omega) \end{bmatrix}, \tag{3.56a}$$

and

$$\bar{\mathbf{B}}_{\mathrm{ULA}} = \mathrm{col}\left[1 \quad \mathrm{sinc}(\varepsilon_0 d)e^{-ju_0 d} \quad \mathrm{sinc}(2\varepsilon_0 d)e^{-ju_0 2d} \quad \ldots \quad \mathrm{sinc}((P-1)\varepsilon_0 d)e^{-ju_0 (P-1)d}\right]. \tag{3.56b}$$

The effectiveness of the proposed approach is demonstrated through the frequency wavenumber response of the Wiener filters before and after robustification. We compute the frequency wavenumber response given that the DOA is known to an accuracy of ±2.0°. The plot is shown in Figure 3.17a for a ULA with 16 sensors and constant signal to noise ratio (SNR=4). Notice the splitting of the peak, particularly in the higher temporal frequency range. Next, we compute the frequency wavenumber response of a UCA designed to tolerate an error of 2°. The frequency wavenumber response is shown in Figure 3.17b. The main lobe shape remains practically unchanged, but there is an increase in the sidelobe level. Cross-sectional plots passing through the maximum of the response function of the UCA with and without DOA error are shown in Figure 3.18. While the shape of the main lobe remains practically unchanged, the sidelobe level seems to have slightly increased. This is the price one has to pay for the lack of exact knowledge of the DOA.

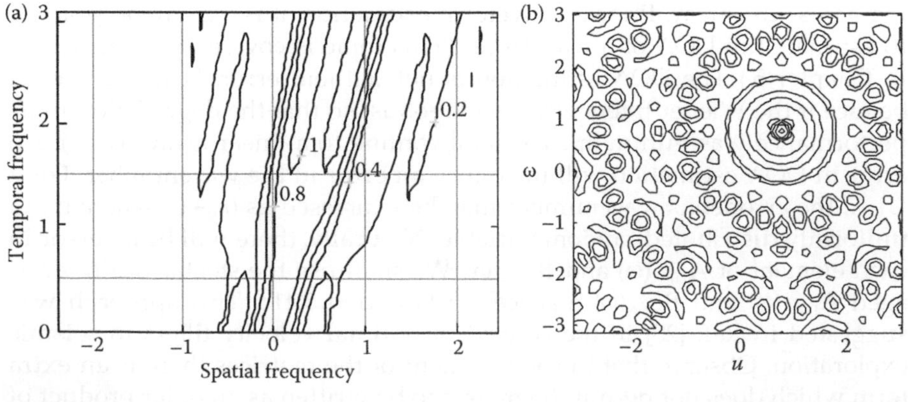

FIGURE 3.17
Frequency wavenumber response of Wiener filter when the error in DOA estimate is ±2.0°. (a) ULA and (b) UCA. Sixteen sensors and 64 time samples are assumed.

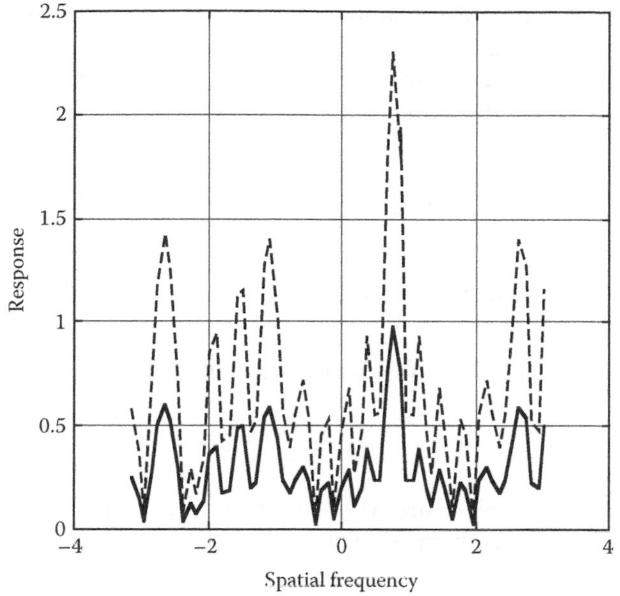

FIGURE 3.18
A cross section through the maximum (v_0=0.7854) for UCA. Solid line shows response when there is no error and the dashed line shows one with DOA error (±2.0°). Sixteen sensors and 64 time samples are assumed.

3.4.4 Levinson–Durbin Algorithm

In Equation 3.36, in order to solve for **h**, we have to invert a large covariance matrix (e.g., with the array size, M=24, and filter length, N=16, the size of the covariance matrix will be 384×384). The computational load for inverting such a large matrix will be very high. We shall outline a recursive method applicable to a ULA. For any other geometry of an array, the covariance matrix becomes a block symmetric matrix, but it become Toeplitz only for a ULA. This important property enables us to devise a recursive algorithm, known as Levinson-Durbin algorithm, which requires inversion of a matrix of size $M\times M$ in place of a matrix of size $MN\times MN$. Briefly, the algorithm is as follows [16]: Let $\mathbf{h}_N$ ($\mathbf{h}_N=\text{col}[\mathbf{h}(0)\ \mathbf{h}(\Delta\tau)\ldots\mathbf{h}((N-1)\ \Delta\tau)]$) be the solution of the Nth order normal equations (Equation 3.36). Let us now increase the size of the covariance matrix by padding one row and one column of covariance matrices

$$\begin{bmatrix} \mathbf{C}_N & \cdots & \mathbf{c}(N) \\ \vdots & & \vdots \\ \mathbf{c}(N) & \cdots & \mathbf{c}(0) \end{bmatrix} \begin{bmatrix} \mathbf{h}_N \\ \\ \mathbf{0} \end{bmatrix} = \begin{bmatrix} \mathbf{C}_0 \\ \\ \gamma_N \end{bmatrix}, \tag{3.57}$$

where

$$\underset{(M\times 1)}{\gamma_N} = \underset{(M\times MN)}{[\mathbf{c}(N) \quad \ldots \quad \mathbf{c}(1)]}\underset{(MN\times 1).}{\mathbf{h}_N}$$

Note that the square matrix on the left-hand side of Equation 3.57 is a block covariance matrix of size $M(N+1)\times M(N+1)$; therefore, Equation 3.57 is similar to Equation 3.36 but of order $N+1$. Indeed, if we subtract Equation 3.57 from Equation 3.36 of the same order, we shall obtain

$$\mathbf{C}_{N+1}\left[\mathbf{h}_{N+1}-\begin{bmatrix}\mathbf{h}_N\\ \mathbf{0}\end{bmatrix}\right]=\begin{bmatrix}\mathbf{0}\\ \mathbf{c}_0(N)-\gamma_N\end{bmatrix}. \tag{3.58}$$

Define a set of auxiliary coefficient matrices

$$\mathbf{b}_N=\text{col}\{\mathbf{b}_{NN},\mathbf{b}_{NN-1},\ldots,\mathbf{b}_{N1},\mathbf{I}\}$$

$$(N+1)M\times M,$$

where $\mathbf{b}_{NN-i}$, $i=0,\ldots,N-1$ are $M\times M$ matrices yet to be defined, but they satisfy the following recurrence relation:

$$\mathbf{h}_{N+1}(i)=\mathbf{b}_{NN-i}\mathbf{h}_{N+1}(N)+\mathbf{h}_N(i),\quad i=0,1,\ldots,N-1, \tag{3.59}$$

$$\mathbf{C}_{N+1}\begin{bmatrix}\mathbf{b}_{NN}\mathbf{h}_{N+1}(N)\\ \mathbf{b}_{NN-1}\mathbf{h}_{N+1}(N)\\ \vdots\\ \mathbf{h}_{N+1}(N)\end{bmatrix}=\begin{bmatrix}\mathbf{0}\\ \beta_N\mathbf{h}_{N+1}(N)\end{bmatrix}, \tag{3.60a}$$

where $\beta_N\,\mathbf{h}_{N+1}(N)=\mathbf{c}_0(N)-\gamma_N$. Eliminating $\mathbf{h}_{N+1}(N)$ from both sides of Equation 3.60a we obtain

$$\mathbf{C}_{N+1}\mathbf{b}_N=\begin{bmatrix}\mathbf{0}\\ \beta_N\end{bmatrix}. \tag{3.60b}$$

We shall once again increase the order of $\mathbf{C}_{N+1}$ in Equation 3.60b by padding one more row and one more column. We get

$$\mathbf{C}_{N+2}\begin{bmatrix}\mathbf{0}\\ \mathbf{b}_N\end{bmatrix}=\begin{bmatrix}\beta'_N\\ \beta'_N\\ \beta_N\end{bmatrix}, \tag{3.60c}$$

where

$$\beta'_N=[\mathbf{c}(1)\quad \mathbf{c}(2)\quad \ldots\quad \mathbf{c}(N+1)]\mathbf{b}_N$$

$$(M\times M(N+1))\;\; M(N+1)\times M.$$

we shall now introduce another set of auxiliary coefficients, $\mathbf{a}_N$=col{$\mathbf{I}$, $\mathbf{a}_{N1}$, $\mathbf{a}_{N2}, \ldots, \mathbf{a}_{NN}$} as a solution of the following system of equations:

$$\mathbf{C}_{N+1}\mathbf{a}_N = \begin{bmatrix} \alpha_N \\ \mathbf{0} \end{bmatrix}, \tag{3.61a}$$

where

$$\alpha_N = [\mathbf{c}(0)\mathbf{I} + \mathbf{c}(1)\mathbf{a}_{N1} + \cdots + \mathbf{c}(N)\mathbf{a}_{NN}]$$
$$M \times M.$$

Let us now increase the size of $\mathbf{C}_{N+1}$ in Equation 3.61a by padding one more row and one more column of covariance matrices. We obtain

$$\mathbf{C}_{N+2}\begin{bmatrix} \mathbf{a}_N \\ \mathbf{0} \end{bmatrix} = \begin{bmatrix} \alpha_N \\ \mathbf{0} \\ \alpha'_N \end{bmatrix}, \tag{3.61b}$$

where

$$\alpha'_N = [\mathbf{c}(N+1)\mathbf{I} + \mathbf{c}(N)\mathbf{a}_{N1} + \cdots + \mathbf{c}(1)\mathbf{a}_{NN}]$$
$$M \times M.$$

We linearly combine Equations 3.60c and 3.61b such that the resulting equation is the (N+2)th order equivalent of Equation 3.60b. Let the linear combination be given by

$$\mathbf{C}_{N+2}\left\{\begin{bmatrix} \mathbf{0} \\ \mathbf{b}_N \end{bmatrix} + \begin{bmatrix} \mathbf{a}_N \\ \mathbf{0} \end{bmatrix}\delta_N\right\} = \left\{\begin{bmatrix} \beta'_N \\ \mathbf{0} \\ \beta_N \end{bmatrix} + \begin{bmatrix} \alpha_N \\ \mathbf{0} \\ \alpha'_N \end{bmatrix}\delta_N\right\}, \tag{3.62a}$$

where δ_N is a $M \times M$ matrix of constants which we shall select in such a manner that the right-hand side of Equation 3.62a resembles the right-hand side of Equation 3.60b. This may be achieved by requiring

$$\beta'_N + \alpha_N\delta_N = \mathbf{0},$$

or

$$\delta_N = -\alpha_N^{-1}\beta'_N. \tag{3.62b}$$

The resulting equation is equivalent to Equation 3.60 but of order N+2. Then, we have

$$\mathbf{b}_{N+1} = \begin{bmatrix} \mathbf{0} \\ \mathbf{b}_N \end{bmatrix} + \begin{bmatrix} \mathbf{a}_N \\ \mathbf{0} \end{bmatrix} \delta_N, \tag{3.63a}$$

$$\beta_{N+1} = \beta_N + \alpha'_N \delta_N. \tag{3.63b}$$

Further, we take a linear combination of Equations 3.60c and 3.61b such that the resulting equation resembles Equation 3.61a but it is of order $N+2$. This may be achieved by requiring

$$\Delta_N \beta_N + \alpha'_N = 0,$$

or

$$\Delta_N = -\beta_N^{-1} \alpha'_N, \tag{3.64a}$$

where Δ_N is also a $M \times M$ matrix of constants for linear combination. Now we have

$$\mathbf{a}_{N+1} = \begin{bmatrix} \mathbf{a}_N \\ \mathbf{0} \end{bmatrix} + \begin{bmatrix} \mathbf{0} \\ \mathbf{b}_N \end{bmatrix} \Delta_N, \tag{3.64b}$$

$$\alpha_{N+1} = \alpha_N + \beta'_N \Delta_N. \tag{3.64c}$$

Equations 3.63a and 3.64b form a set of recursive relations to compute $\mathbf{a}_{N+1}$ and $\mathbf{b}_{N+1}$ given $\mathbf{a}_N$ and $\mathbf{b}_N$. Similarly, Equations 3.63b and 3.64c form another set of recursive relations to compute α_{N+1} and β_{N+1} given α_N and β_N. The initial conditions are:

(1) $\mathbf{b}_{0,0} = \mathbf{a}_{0,0} = \mathbf{I}$

(2) $\alpha_0 = \beta_0 = \mathbf{c}(0)$

(3) $\mathbf{h}_0 = 0$

Finally, to compute the filter coefficients we need $\mathbf{h}_{N+1}(N)$ which we get from

$$\mathbf{h}_{N+1}(N) = [\beta_N]^{-1} (\mathbf{c}_0(N) - \gamma_N).$$

It must be emphasized that, since the Levinson–Durbin algorithm exploits the Toeplitz property of the covariance matrix, the array signal must be stationary and sufficiently large data must be available for the estimation of statistically averaged covariance matrix. Such a filter will naturally be optimum to an ensemble of time series having the same second order structure. In practice, this is often difficult to realize; consequently, the Toeplitz character is lost and we cannot use the Levinson–Durbin algorithm altogether. In

Chapter 6, we shall describe a deterministic least squares approach that uses the actual measured signal in place of covariance function.

3.5 Predictive Noise Cancellation

Noise suppression is based on the principle of differential spectral properties of signal and noise. A filter is often used to maximally attenuate the noise power but at the same time to minimally attenuate the signal power. There is an alternate approach to noise attenuation, known as noise cancellation. This involves prediction of the noise that is actually corrupting the signal. For prediction, we need a sample of noise that is not contaminated with the signal, but which is correlated with the noise present in the actual observed signal plus noise sample. In principle, it is possible to devise a Wiener filter for the prediction of the unknown noise from the sample of noise that is correlated with the unknown. A signal free noise sample can be obtained as follows:

(1) Place an extra sensor within the correlation distance of the noise but away from the signal source. This is possible when the signal source is in the near-field region but the noise sources are in the far-field region, for example, when a speaker is close to a microphone and the noise sources are far away from the microphone.

(2) Both signal and noise are in the far-field region but reach the microphone array from different directions. An array of sensors may be simultaneously used to receive the signal and the noise coming from different directions. However, since the array response is finite in a direction other than the direction to which it is tuned, some amount of noise will leak into the array output. The noise, which has leaked into the array output, will be strongly correlated with the incident noise and hence it may be predicted using the noise output when the array is steered in the direction of the noise.

In the above approach the noise is canceled electronically. It is also possible to achieve the noise cancellation acoustically. This would, however, require use of many coherently generated noise sources whose combined effect is to produce a noise wavefront with a phase equal but opposite to that of the noise present in the observed waveform.

3.5.1 Signal Source in Near Field

Consider a speaker close to a microphone (M_1) and another microphone (M_2) away from the speaker but well within the correlation distance of the noise (see Figure 3.19). It is assumed that all noise sources are in the far-field region.

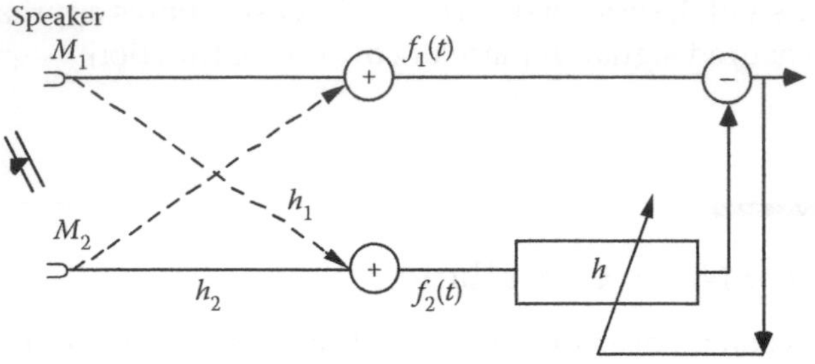

FIGURE 3.19
Source is in near field and noise is in far field. Microphone M_2 (reference microphone) receives very little of the signal.

Let $f_1(t)$ be the output of M_1, $f_2(t)$ be the output of M_2, $\xi_0(t)$ be the signal emitted by the speaker, and $\eta_1(t)$ be the noise in M_1. The signal and noise in M_2 are related to those in M_1 through impulse response functions, $h_1(t)$ and $h_2(t)$.

$$f_1(t) = \xi_0(t) + \eta_1(t)$$

$$f_2(t) = \int_0^{\infty} \xi_0(t-t')h_1(t')dt' + \int_0^{\infty} \eta_1(t-t')h_2(t')dt'. \tag{3.65}$$

The output of M_2 is passed through a prediction filter, $h_{pred}(t)$, which is found by minimizing a quantity,

$$E\left\{\left| f_1(t) - \int_0^{\infty} f_2(t'-t)h_{pred}(t')dt' \right|^2\right\} = \min,$$

or in the frequency domain by minimizing the following

$$\frac{1}{2\pi}\int_{-\infty}^{\infty}\left[S_1(\omega) + S_2(\omega)\left|H_{pred}(\omega)\right|^2 - S_{12}(\omega)H^*_{pred}(\omega) - S^*_{12}(\omega)H_{pred}(\omega)\right]d\omega.$$

We obtain

$$H_{pred}(\omega) = \frac{S_{12}(\omega)}{S_2(\omega)} = \frac{S_0(\omega)H_1^*(\omega) + S_{\eta_1}(\omega)H_2^*(\omega)}{S_0(\omega)\left|H_1(\omega)\right|^2 + S_{\eta_1}(\omega)\left|H_2(\omega)\right|^2}, \tag{3.66a}$$

and

$$\text{Error}\big|_{min} = \frac{1}{2\pi}\int_{-\infty}^{\infty}\left[S_1(\omega) - S_2(\omega)\left|H_{pred}(\omega)\right|^2\right]d\omega. \tag{3.66b}$$

3.5.1.1 Some Special Cases

(1) $h_1(t)=0$: The signal from the speaker does not reach M_2. $h_2(t)=\delta(t-\tau_0)$: The noise reaching the microphone M_2 is simply a delayed version of the noise reaching microphone M_1. For this case, the prediction filter is simply a delay filter, $H_{\text{pred}}(\omega)=e^{j\omega\tau_0}$ and the minimum error is equal to the signal power,

$$\text{Error}\big|_{\min}=\frac{1}{2\pi}\int_{-\infty}^{\infty} S_0(\omega)d\omega.$$

In this special case complete noise cancellation takes place.

(2) $h_1(t)=h_2(t)\neq 0$: The output of the reference microphone is a filtered version of the output of M_1. For this case, the prediction filter is given by

$$H_{\text{pred}}(\omega)=\frac{1}{H_1(\omega)},$$

and the minimum error is equal to zero. Apparently, both signal and noise are canceled and the output power is zero. For noise cancellation to take place we must have $h_2(t)>h_1(t)$.

Define gain G as a ratio of SNR at the output to SNR at the input. The output power is given by

$$\begin{aligned}\text{Output power} &=\frac{1}{2\pi}\int_{-\infty}^{\infty}[S_1(\omega)-S_2(\omega)|H_{\text{pred}}(\omega)|^2]d\omega\\
&=\frac{1}{2\pi}\int_{-\infty}^{\infty}\left\{S_0(\omega)+S_{\eta_1}(\omega)-(S_0(\omega)|H_1(\omega)|^2+S_{\eta_1}(\omega)|H_2(\omega)|^2)|H_{\text{pred}}(\omega)|^2\right\}d\omega\\
&=\frac{1}{2\pi}\int_{-\infty}^{\infty}\left\{S_0(\omega)(1-|H_1(\omega)|^2|H_{\text{pred}}(\omega)|^2)\right\}d\omega \quad \text{signal power}\\
&+\frac{1}{2\pi}\int_{-\infty}^{\infty}\left\{S_{\eta_1}(\omega)(1-|H_2(\omega)|^2|H_{\text{pred}}(\omega)|^2)\right\}d\omega \quad \text{noise power.}\end{aligned}$$

The SNR at the output as a function of frequency is given by

$$\text{SNR}_{\text{output}}=\frac{\left\{S_0(\omega)(1-|H_1(\omega)|^2|H_{\text{pred}}(\omega)|^2)\right\}}{\left\{S_{\eta_1}(\omega)(1-|H_2(\omega)|^2|H_{\text{pred}}(\omega)|^2)\right\}},$$

from which the gain as defined here turns out to be

$$G=\frac{[\mathrm{SNR}]_{\mathrm{output}}}{[\mathrm{SNR}]_{\mathrm{input}}}=\frac{1-|H_1(\omega)|^2|H_{\mathrm{pred}}(\omega)|^2}{1-|H_2(\omega)|^2|H_{\mathrm{pred}}(\omega)|^2}. \tag{3.67}$$

For $G>1$ we must have $|H_1(\omega)|^2<|H_2(\omega)|^2$.

3.5.2 Source in Far Field

Both signal and noise sources are in the far-field region, but the DOAs of their wavefronts are different. Let τ_0 and τ_1 be the incremental delays produced by the signal wavefront and noise wavefront, respectively. The array can be steered to receive the signal or noise at the same time (see Figure 3.20) [21]. Let $f_1(t)$ be the output of an array when it is steered to the signal wavefront and $f_2(t)$ be the output of an array when steered to the noise wavefront. Since the array response function has finite sidelobes, some amount of wave energy will leak through the sidelobes. Hence, we model the array output as

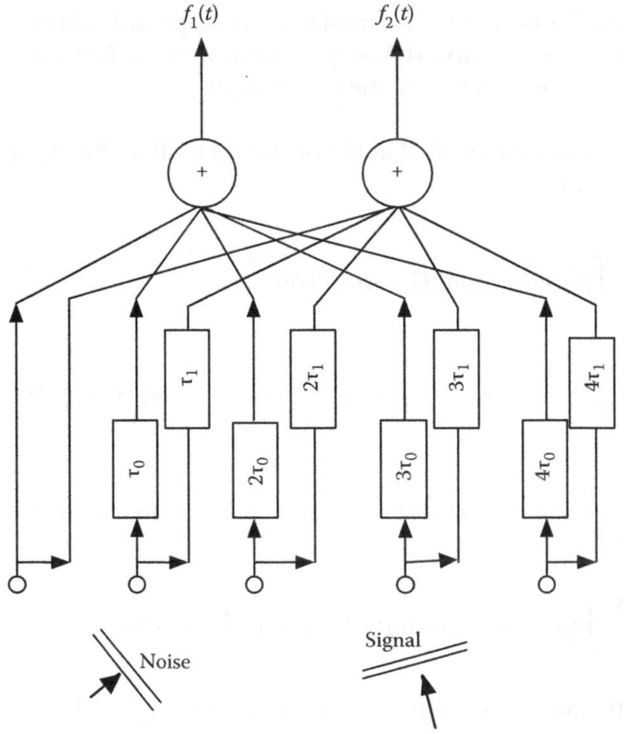

FIGURE 3.20
An array of sensors can be steered simultaneously in the direction of the signal and in the direction of noise. When the array is steered in the direction of signal the output $f_1(t)$ is mostly signal and when it is steered in the direction of noise the output $f_2(t)$ is mostly noise.

$$f_1(t) = \xi_0(t) + \frac{1}{2\pi}\int_{-\infty}^{\infty} N_1(\omega)H(\omega(\tau_0 - \tau_1))e^{j\omega t}\,d\omega, \tag{3.68a}$$

and

$$f_2(t) = \eta_1(t) + \frac{1}{2\pi}\int_{-\infty}^{\infty} \Xi_0(\omega)H(\omega(\tau_1 - \tau_0))e^{j\omega t}\,d\omega, \tag{3.68b}$$

where $\Xi_0(\omega)$ is the Fourier transform of the signal and $N_1(\omega)$ is that of the noise. By comparing Equation 3.65 with Equation 3.68 it is possible to write

$$H_1(\omega) = H(\omega(\tau_1 - \tau_0)), \tag{3.69a}$$

and

$$H_2(\omega) = \frac{1}{H(\omega(\tau_0 - \tau_1))}. \tag{3.69b}$$

Using Equation 3.69 in Equation 3.66a, we obtain a filter to predict the noise in $f_1(t)$,

$$\begin{aligned} H_{\text{pred}}(\omega) &= \frac{S_{\xi_0}(\omega)H^*(\omega(\tau_1 - \tau_0)) + \dfrac{S_{\eta_1}(\omega)}{H^*(\omega(\tau_0 - \tau_1))}}{S_{\xi_0}(\omega)\left|H(\omega(\tau_1 - \tau_0))\right|^2 + \dfrac{S_{\eta_1}(\omega)}{\left|H(\omega(\tau_0 - \tau_1))\right|^2}} \\ &= \frac{1 + \text{SNR}_{\text{input}}}{1 + \text{SNR}_{\text{input}}\left|H(\omega(\tau_1 - \tau_0))\right|^2} H(\omega(\tau_0 - \tau_1)). \end{aligned}$$

Let us now consider a few special cases:

(1) When $\text{SNR}_{\text{input}} \gg 1$ and $\text{SNR}_{\text{input}}\,|H(\omega(\tau_1 - \tau_0))|^2 \gg 1$

$$H_{\text{pred}}(\omega) \approx \frac{1}{H(\omega(\tau_1 - \tau_0))}.$$

If this filter is used on $f_2(t)$ (see Equation 3.68b) for predicting the noise in $f_1(t)$, the signal component will be restored, causing the cancellation of the signal.

(2) $\text{SNR}_{\text{input}}\,|H(\omega(\tau_1 - \tau_0))|^2 \ll 1$

$$H_{\text{pred}}(\omega) \approx (1 + \text{SNR}_{\text{input}})H(\omega(\tau_0 - \tau_1)).$$

If this filter is used on $f_2(t)$ (see Equation 3.68b) for predicting the noise component in $f_1(t)$, the noise component will be largely canceled without canceling the signal. As an illustration, we consider two pure sinusoidal signals (of same frequency) arriving with different DOAs (0° and 5.7°) at a ULA of 16 sensors spaced at $\lambda/2$ spacing. The second sinusoid arrives 50 time units later with an amplitude of 0.8. Figure 3.21a shows a sum of the two tones as received by the first sensor. The array is steered in the direction of the first sinusoid and at the same time in the direction of the second sinusoid. The array outputs are described by Equation 3.68, which is now considerably simplified for pure sinusoidal inputs.

$$\begin{aligned} f_1(t) &= s_1(t) + s_2(t)\, H(\omega_0\,(\tau_0 - \tau_1)) \\ f_2(t) &= s_2(t) + s_1(t)\, H(\omega_0\,(\tau_0 - \tau_1)) \end{aligned} \tag{3.70a}$$

where $s_1(t)$ and $s_2(t)$ are the first and the second sinusoid, respectively, and ω_0 is the frequency of the sinusoids. Solving Equation 3.70 we obtain, for $|H(\omega_0\,(\tau_0 - \tau_1))|^2 < 1$,

$$\begin{aligned} s_1(t) &= \frac{f_1(t) - f_2(t)H(\omega_0(\tau_0 - \tau_1))}{1 - H^2(\omega_0(\tau_0 - \tau_1))} \\ s_2(t) &= \frac{f_2(t) - f_1(t)H(\omega_0(\tau_0 - \tau_1))}{1 - H^2(\omega_0(\tau_0 - \tau_1))}. \end{aligned} \tag{3.70b}$$

The results are shown in Figure 3.21b and c.

We explore another possibility, where, instead of steering the array in the direction of interference, we steer a null in the direction of a signal with finite response in the direction of interference [22]. Equation 3.68b may be written as

$$f_2(t) = \frac{1}{2\pi}\int_{-\infty}^{\infty} \mathbf{H}_{\mathrm{null}}^H(\omega)\mathbf{N}_1(\omega)e^{j\omega t}d\omega, \tag{3.71a}$$

where

$$\mathbf{H}_{\mathrm{null}}(\omega) = \left[\mathbf{I} - \frac{\mathbf{a}_0(\omega)\mathbf{a}_0^H(\omega)}{M}\right]\mathbf{a}_1(\omega),$$

is a filter with a null in the direction of the signal. Noting that $\mathbf{N}_1(\omega) = \mathbf{a}_1(\omega)\mathrm{N}_1(\omega)$, Equation 3.71a may be written as

$$f_2(t) = \frac{1}{2\pi}\int_{-\infty}^{\infty} M\left[1 - \frac{|\mathbf{a}_1^H(\omega)\mathbf{a}_0(\omega)|^2}{M^2}\right]\mathrm{N}_1(\omega)e^{j\omega t}d\omega. \tag{3.71b}$$

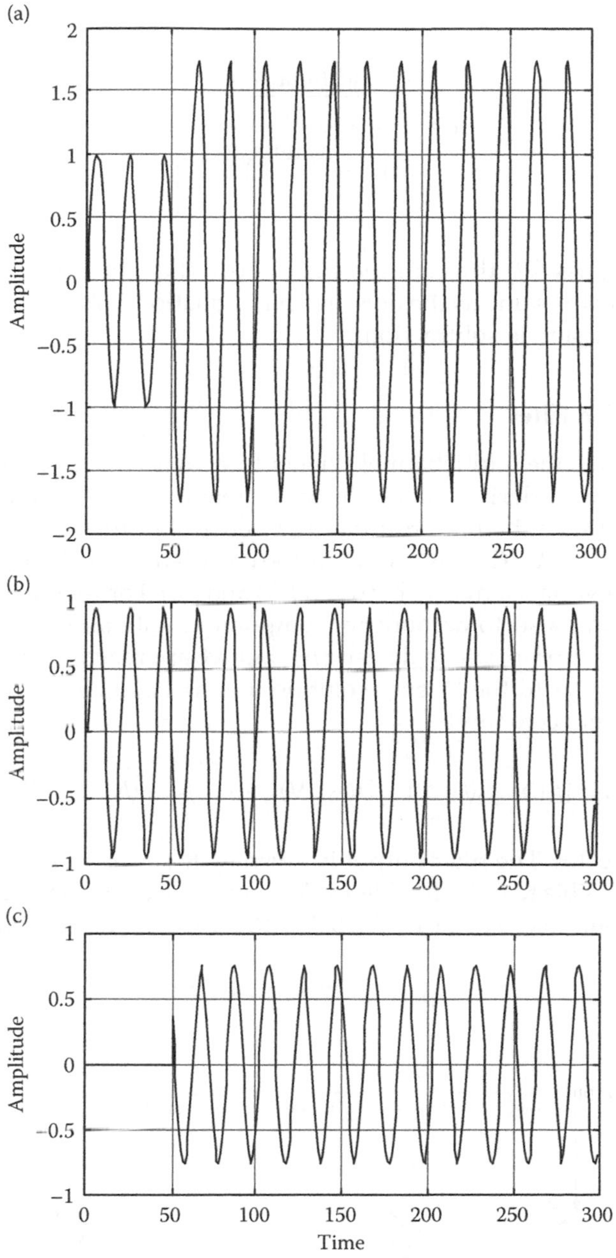

FIGURE 3.21
(a) Sum of two sinusoids, (b) first sinusoid after subtraction, and (c) second sinusoid after subtraction.

Define a filter

$$w = \frac{\dfrac{\mathbf{a}_0^H(\omega)\mathbf{a}_1(\omega)}{M}}{M\left[1 - \dfrac{\left|\mathbf{a}_1^H(\omega)\mathbf{a}_0(\omega)\right|^2}{M^2}\right]}. \tag{3.71c}$$

If we now pass $f_2(t)$ through the filter above (Equation 3.71c), the output will be exactly the same as the noise term in Equation 3.68a; therefore, it may be removed by simple subtraction.

3.5.3 Adaptive Filter

We have derived the multichannel Wiener filter in Section 3.3. A single channel version may be derived along the same lines (see Ref. [22] for derivation). Here we shall state the final result. The Wiener filter that predicts $f_1(t)$ from $f_2(t)$ is given by $\mathbf{h} = \mathbf{C}_{f_2}^{-1}\mathbf{C}_{f_1f_2}$, where $\mathbf{C}_{f_2}$ is the covariance matrix of $f_2(t)$ and $\mathbf{C}_{f_2f_1}$ is the cross-covariance matrix between $f_2(t)$ and $f_1(t)$. For real time estimation of the filter and also to account for temporal variations in the covariance functions, it is appropriate to devise an adaptive approach, which in the limiting case reduces to the Wiener solution. Let $\mathbf{h}=[h_0, h_1, h_2, \ldots, h_{N-1}]^T$ be the prediction filter vector and

$$\mathbf{f}_2=[f_2(t) \quad f_2(t-\Delta t) \quad f_2(t-2\Delta t) \quad \cdots \quad f_2(t-(N-1)\Delta t)]^T,$$

be the data vector. The filter output is given by $\mathbf{h}^T\mathbf{f}_2$, which is required to be as close as possible to $f_1(t)$. This filter is known as a transversal filter acting on the delayed outputs (see Figure 3.22). For this we need to minimize the mse

$$E\{|\varepsilon(t)|^2\}=E\{|f_1(t)-\mathbf{h}^T\mathbf{f}_2|^2\},$$

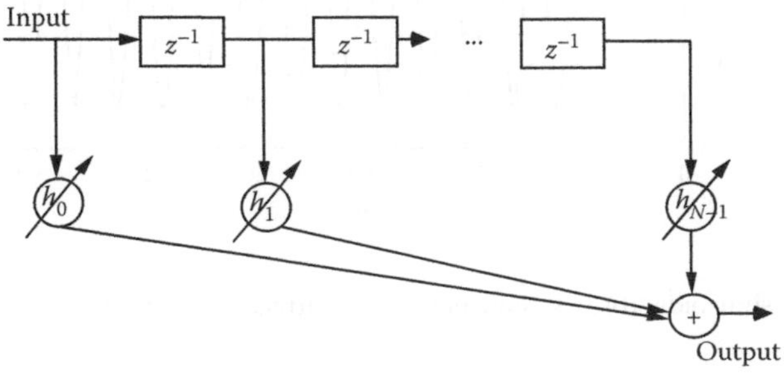

FIGURE 3.22
Structure of transversal filter. The filter coefficients are made adaptable to the changing input.

with respect to the prediction filter coefficients. To minimize the error power, we need to compute a gradient of $E\{|\varepsilon(t)|^2\}$ with respect to $h_0, h_1, h_2, \ldots, h_{N-1}$ and go down the path of the steepest descent until a minimum (possibly a local minimum) is encountered. The difficulty, however, is in estimating $E\{|\varepsilon(t)|^2\}$, which requires averaging over a finite interval (ideally infinite) of time. Instead, in the least mean squares (LMS) algorithm, it is proposed to use $|\varepsilon(t)|^2$ in place of $E\{|\varepsilon(t)|^2\}$. The gradient of $|\varepsilon(t)|^2$ is now easily computed,

$$\begin{aligned}\nabla|\varepsilon(t)|^2 &= 2\varepsilon(t)\left\{\frac{\partial\varepsilon(t)}{\partial h_0}, \frac{\partial\varepsilon(t)}{\partial h_1}, \ldots, \frac{\partial\varepsilon(t)}{\partial h_{N-1}}\right\} \\ &= -2\varepsilon(t)\mathbf{f}_2(t).\end{aligned} \tag{3.72}$$

In the steepest descent search method, the current filter vector is adjusted by an amount proportional to negative of the gradient of the error function, $\nabla|\varepsilon(t)|^2$ [23]. The idea of adaptation is illustrated in Figure 3.23. The current filter vector is updated by an amount proportional to the product of prediction error and current input,

$$\mathbf{h}_{i+1} = \mathbf{h}_i + 2\mu\varepsilon(t)\mathbf{f}_2(t), \tag{3.73}$$

where μ is a gain constant that regulates the speed of adaptation.

It is interesting to note that the filter vector converges to the Wiener filter, that is, as $i \to \infty\, \mathbf{h}_i \to \mathbf{C}_{f_2}^{-1}\mathbf{C}_{f_1 f_2}$. To show this, consider the expected value of $\mathbf{h}_{i+1}$

$$\begin{aligned}E\{\mathbf{h}_{i+1}\} &= E\{\mathbf{h}_i\} + E\{2\mu\varepsilon(t)\mathbf{f}_2(t)\} \\ &= E\{\mathbf{h}_i\} + 2\mu E\{\mathbf{f}_2(t)(f_1(t) - \mathbf{f}_2^T(t)\mathbf{h}_i)\} \\ &= E\{\mathbf{h}_i\} + 2\mu\mathbf{C}_{\mathbf{f}_2 f_1} - 2\mu\mathbf{C}_{\mathbf{f}_2}E\{\mathbf{h}_i\},\end{aligned} \tag{3.74}$$

where we have assumed that $\mathbf{h}_i$ and $\mathbf{f}_2(t)$ are independent. Let $\tilde{\mathbf{h}}$ represent filter coefficients obtained by solving the normal equation (Equation 3.36), that is, $\tilde{\mathbf{h}} = \mathbf{C}_{\mathbf{f}_2}^{-1}\mathbf{C}_{\mathbf{f}_2 f_1}$. Equation 3.74 reduces to

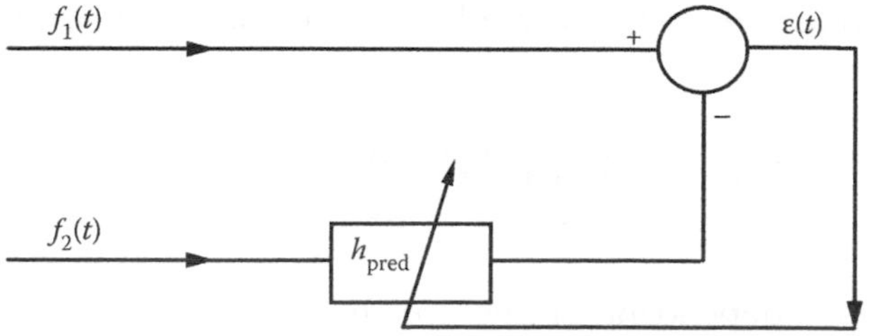

FIGURE 3.23
Idea of an adaptive prediction filter. The prediction error modifies the filter in order to reduce the prediction error.

$$E\{\mathbf{h}_{i+1}-\tilde{\mathbf{h}}\}=E\{\mathbf{h}_i-\tilde{\mathbf{h}}\}+2\mu\tilde{\mathbf{h}}\mathbf{C}_{\mathbf{f}_2}-2\mu\mathbf{C}_{\mathbf{f}_2}E\{\mathbf{h}_i\}, \tag{3.75a}$$

$$E\{\Delta\mathbf{h}_{i+1}\}=E\{\Delta\mathbf{h}_i\}-2\mu\mathbf{C}_{\mathbf{f}_2}E\{\Delta\mathbf{h}_i\}, \tag{3.75b}$$

$$E\{\Delta\mathbf{h}_{i+1}\}=(\mathbf{I}-2\mu\mathbf{C}_{\mathbf{f}_2})E\{\Delta\mathbf{h}_i\}. \tag{3.75c}$$

Let us use the eigen-decomposition of the covariance matrix, $\mathbf{C}_{\mathbf{f}_2}=\mathbf{V}\boldsymbol{\Lambda}\mathbf{V}^H$, in Equation 3.75c and obtain

$$E\{\mathbf{V}^H\Delta\mathbf{h}_{i+1}\}=(\mathbf{I}-2\mu\boldsymbol{\Lambda})\,E\{\mathbf{V}^H\,\Delta\mathbf{h}_i\}. \tag{3.76}$$

The solution of the difference equation (Equation 3.76) is given by $E\{\mathbf{V}^H\Delta\mathbf{h}_i\}=(\mathbf{I}-2\mu\boldsymbol{\Lambda})^i\,\chi_0$ where χ_0 is the initial condition. As $i\to\infty$ $E\{\mathbf{V}^H\Delta\mathbf{h}_i\}\to 0$ provided $(\mathbf{I}-2\mu\boldsymbol{\Lambda})^i\to 0$ for $i\to\infty$. This is possible if $|(1-2\mu\lambda_l)|<1$ for all l. This can be easily achieved if we were to select μ such that $0<\mu<(1/\lambda_{max})$, where λ_{max} stands for the maximum eigenvalue of the covariance matrix. Note that $\lambda_{max}\le tr\{\mathbf{C}_{\mathbf{f}_2}\}$ is the sum of the eigenvalues of the covariance matrix. Hence, $E\{\Delta\mathbf{h}_i\}\to 0$ as $i\to\infty$. From this result it follows that $E\{\mathbf{h}_i\}\to\tilde{\mathbf{h}}$ as $i\to\infty$.

3.6 Exercises

(1) A tapered fan filter is defined as [3],

$$H(u,\omega)=\frac{1}{|\omega|}\text{rect}\left(\frac{u}{\omega}\right)\bullet\text{rect}\left(\frac{u}{2\omega}\right),$$

where • stands for the convolution in u. Sketch the filter frequency response function. Compute the impulse response function.

(2) In the weighted least squares filter design, the mmse, which is given by

$$\varepsilon^2_{\min}=\frac{1}{4\pi^2}\int_{-\pi}^{+\pi}\!\!\int\left|\frac{WH-[WH]_{\text{finite}}}{W}\right|^2 du\,dv,$$

becomes independent of the weight function as the filter size increases, ideally at infinity.

(3) The output of a ULA with its sensors spaced 1 m apart is sampled at the rate of 5 kHz (Nyquist rate). A quadrant filter is desired with

upper and lower cutoff speeds 7 and 3 km/sec, respectively. Sketch the pass regions including the aliased part, if any.

(4) The following are the low pass 1D filter coefficients:

$h(0)=0.52$,

$h(1)=0.3133176$,

$h(2)=-0.01808986$,

$h(3)=-0.09138802$,

$h(4)=0.01223454$,

$h(5)=0.04000004$,

$h(6)=-0.001945309$,

$h(7)=-0.014112893$,

Compute the coefficients b_k, $k=0,1,\ldots,N$ appearing in Equation 3.26b on page 157 and then, using these coefficients, evaluate a circular and a fan-shaped 2D filter.

(5) A UCA is split into two UCAs. The first UCA has all even sensors and the second UCA has all odd sensors. Show that the response of the first UCA may be obtained by rotating the response of the second UCA through an angle equal to angular separation between the sensors. Using this property, give a heuristic explanation on the behavior of the sidelobe as a function of the number of sensors (see p. 169).

References

1. H. C. Andrews and B. R. Hunt, *Digital Image Restoration*, Prentice-Hall, Englewood Cliffs, NJ, 1977.
2. P. S. Naidu and M. P. Mathew, *Geophysical Potential Field Analysis – A Digital Signal Processing Approach*, vol. 5, *Advances in Geophysical Exploration*, Elsevier Science, Amsterdam, 1998.
3. J. P. Fail and G. Grau, Les filtere en eventail, *Geophys. Prosp.*, vol. 11, pp. 131–163, 1963.
4. P. Embree, J. P. Burg, and M. M. Backus, Wide-band velocity filtering – The pie slice process, *Geophys.*, vol. 28, pp. 948–974, 1963.
5. S. Treitel, J. L. Shanks and C. W. Frasier, Some aspects of fan filtering, *Geophys.*, vol. 32, pp. 789–800, 1967.
6. S. C. Pei and S.-B. Jaw, Two dimensional general fan-type FIR digital filter design, *Signal Process.* (Eurosip), vol. 37, pp. 265–274, 1994.
7. R. A. Wiggins, ω-k filter design, *Geophys. Prosp.*, vol. 14, pp. 427–440, 1966.
8. J. G. Proakis and D. G. Manolakis, *Digital Signal Processing* (Chapter 8), Second Edition, Prentice-Hall of India Private Ltd., New Delhi, 1995.
9. J. H. McClellan, The design of two dimensional digital filter by transformation, *Proceedings of 7th Annual Princeton Conference on Information Sciences and Systems*, pp. 247–251, 1973.

10. M. Abramowitz and I. A. Stegun (Eds), *Handbook of Mathematical Functions*, Applied Mathematics Series 55, National Bureau of Standards, Gaithersburg, MD, 1964.
11. J. H. McClellan and T. W. Parks, Equiripple approximation of fan filter, *Geophysics*, vol. 37, pp. 573–583, 1972.
12. D. E. Dudgeon, Fundamentals of digital array processing, *Proc. IEEE*, vol. 65, pp. 898–905, 1975.
13. N. Wiener, Extrapolation, *Interpolation and Smoothing of Stationary Time Series*, The MIT Press, Cambridge, MA, 1949.
14. N. Levinson, The Wiener RMS (root mean square) error criterion in filter design and prediction, *J. Math. Phys.*, vol. 25, pp. 261–278, 1947 (see also Appendix B in [13]).
15. J. Durbin, The fitting of time series models, *Rev. Int. Statist. Inst.*, vol. 28, pp. 233–243, 1960.
16. R. A. Wiggins and E. A. Robinson, Recursive solution to the multichannel filtering problem, *J. Geophys. Res.*, vol. 70, pp. 1885–1891, 1965.
17. P. E. Green, E. J. Kelly Jr., and M. J. Levin, A comparison of seismic array processing methods, *Geophys. J. Roy. Soc.*, vol. 11, pp. 67–84, 1966.
18. S. M. Kay, *Modern Spectrum Analysis*, Prentice-Hall, Englewood Cliffs, NJ, p. 24, 1989.
19. C. Usha Padmini and Prabhakar S. Naidu, Circular array and estimation of direction of arrival of a broadband source, *Signal Process.*, vol. 37, pp. 243–254, 1994.
20. R. C. Sengbush and M. R. Foster, Design and application of optimal velocity filters in seismic exploration, *IEEE Trans.*, G21, pp. 648–654, 1972.
21. A. Cantoni and L. C. Godara, Performance of a postbeamformer interference canceller in the presence of broadband directional signals, *J. Acoust. Soc. Am.*, vol. 76, pp. 128–138, 1984.
22. L. C. Godara, A robust adaptive array processor, *IEEE Trans.*, vol. CAS-34, pp. 721–730, 1987.
23. B. Widrow and J. D. Stearms, *Adaptive Signal Processing*, Prentice-Hall, Englewood Cliffs, NJ, p. 47, 1985.

4

Source Localization: Frequency Wavenumber Spectrum

In this chapter, we consider the most important problem in sensor array signal processing, which is estimation of the coordinates of a source emitting a signal (passive localization) or a point target illuminated by an external signal (active localization). A point in three-dimensional space is defined by three parameters, namely, range (r), azimuth (φ), and elevation (θ). The range is often measured by means of return time of travel in active systems and by means of time delays measured at a number of sensors in passive systems. The azimuth and elevation angles are obtained from the measurements of direction of arrival (DOA) by an array of sensors. A horizontal array of sensors is required for azimuth measurement and a vertical array for elevation measurement. The basic quantity used for estimation of location parameters is the frequency wavenumber (ω, k) spectrum (see Chapter 2). A source is assumed to be present where there is a concentration of power. We shall describe three different methods; namely, beamformation, Capon spectrum, and maximum entropy (ME) spectrum. The last two methods fall under the nonlinear category while the first method belongs to the linear category. The important difference between the linear and nonlinear methods lies in their response to an input, which consists of a sum of two or more uncorrelated signals. The output of a linear method will be a sum of the spectra of input signals, but the output of a nonlinear method may contain an additional cross-term. In spite of this drawback, the nonlinear methods have become quite popular [1]. In the last section, we show how to exploit the special features of the wavenumber Doppler frequency spectrum of ground clutter and slow-moving targets using a space-time filter for detection of ground targets.

4.1 Frequency Wavenumber Spectrum

A wavefield produced by sources in the far-field region may be expressed as a sum of plane waves with random phase (see plane wave decomposition in Chapter 1, Section 2). The quantity of interest is power (or energy when transient waves are involved) as a function of azimuth and elevation. This is the frequency wavenumber (ω, k) spectrum introduced in Chapter 2. As the array is of limited size, the task of estimating the frequency wavenumber spectrum becomes too ambitious. What can be estimated with reasonable certainty is the spectral matrix by treating the array output as a multichannel time series. The spectral matrix is indeed related to the frequency wavenumber spectrum. Fortunately, for the purpose of source localization it is enough if we can accurately estimate the spectral matrix.

4.1.1 Spectral Representation of the Wavefield

As in Chapter 2, we shall model the wavefield as a stochastic process; hence the basic tool will be the spectral representation of the wavefield,

$$f(x,y,z,t) = \frac{1}{(2\pi)^3} \iiint_{-\pi}^{+\pi} dF(\omega,u,v) e^{j(\omega t - ux - vy - \sqrt{k^2 - s^2} z)},$$

where $k=\omega/c$ and $s=\sqrt{u^2+v^2}$. Using the stochastic properties of the spectral representation of a wavefield, the frequency wavenumber (ω, k) spectrum may be given by

$$\begin{aligned} &\frac{1}{(2\pi)^3} S_f(\omega,u,v) d\omega du dv \\ &= E\left\{ \frac{1}{(2\pi)^3} dF(\omega,u,v) \frac{1}{(2\pi)^3} dF^*(\omega,u,v) \right\}. \end{aligned} \tag{4.1}$$

Note that $S_f(\omega, u, v) \geq 0$ and it satisfies symmetry relations

$$S_f(\omega,+u,-v) = S_f(\omega,-u,+v)$$

$$S_f(\omega,u,v) = S_f(\omega,-u,-v).$$

Further, the (ω, k) spectrum, for a propagating wavefield, must satisfy a condition, $S_f(w, u, v)=0$ for $\sqrt{u^2+v^2} > k$. The spectrum is also defined as the Fourier transform of the spatio-temporal covariance function,

$$S_f(\omega,u,v) = \frac{1}{(2\pi)^3} \iiint_{-\infty}^{\infty} c_f(\bar{x},\bar{y},\tau) e^{j(\omega\tau - u\bar{x} - v\bar{y})} d\bar{x} d\bar{y} d\tau, \tag{4.2}$$

where

$$c_f(\bar{x},\bar{y},\tau) = E\{f(x,y,t) f^*(x+\bar{x}, y+\bar{y}, t+\tau)\},$$

is the spatio-temporal covariance function of the wavefield on the $z=0$ surface. Consider an example of a wideband plane wavefront. Let c_x and c_y be the apparent wave speed in the x and y directions, respectively. The spectrum of the wavefront is given by

$$S_f(\omega, u, v) = S_0(\omega)\,\delta(\omega - c_x u)\,\delta(\omega - c_y v), \tag{4.3}$$

where $S_0(\omega)$ is the spectrum of the temporal waveform. The spectrum of a plane wavefront given in Equation 4.3 is a line in the (ω, k) space passing through the origin with direction cosines $\alpha = \sin\theta\cos\varphi$, $\beta = \sin\theta\sin\varphi$, and $\gamma = \cos\theta$ (see Figure 4.1). Note that $c_x = c/\alpha$ and $c_y = c/\beta$. A stochastic wavefield may be modeled as a sum of uncorrelated plane waves; therefore, its spectrum will consist of a set of radial line segments.

4.1.2 Aliasing

In the spatial domain a wavefield is always measured at a set of discrete points (using point detectors) while in the temporal domain the wavefield is measured as a continuous function of time or over dense sample points, as required. Sampling in the spatial domain is dictated by the cost of deploying a large array of sensors, but in the temporal domain the sampling rate may be as high as required, only at a marginally higher cost. Hence, the phenomenon of aliasing in the spatial domain becomes important. Consider a uniform linear array (ULA) of infinite length with interelement spacing equal to d. The highest spatial frequency beyond which there is a reflection of power resulting in aliasing is π/d. This phenomenon is illustrated in Figure 3.9 with reference to a digital filter. Let us assume that the signal has been prefiltered to limit the temporal spectrum to a band, $\pm\omega_{max}$. The effect of spatial sampling

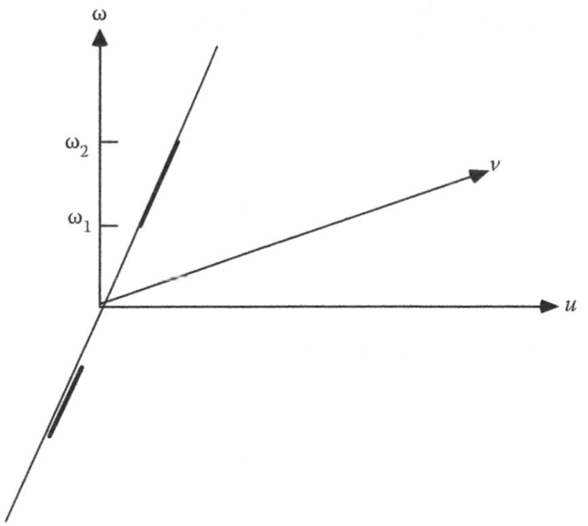

FIGURE 4.1
Frequency-wavenumber spectrum of a broadband plane wavefront. It lies on a line passing through the center and has direction cosines α, β, and γ. The temporal frequency bandwidth is from ω_1 to ω_2.

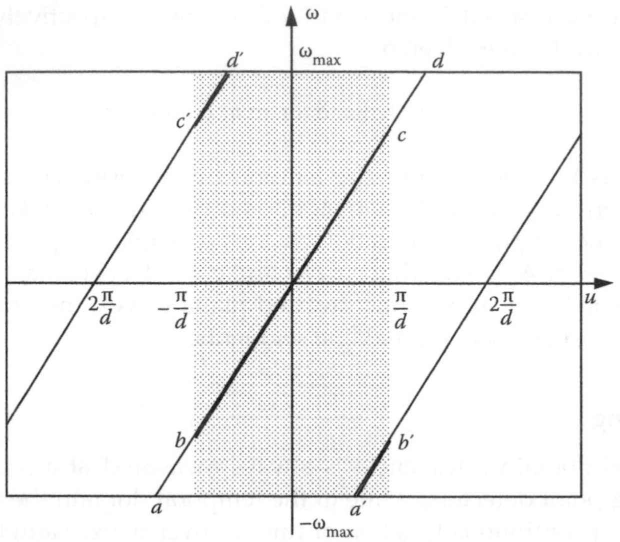

FIGURE 4.2
Aliasing error due to spatial sampling of a broadband plane wave.

on the spectrum of a broadband plane wave is illustrated in Figure 4.2. The (ω, *k*) spectrum lies on a line *abcd*. The segments *ab* and *cd* lie outside the principal domain, but reappear as *c′d′* and *a′b′* as shown. Since, in practice we have a finite array, the line spectrum will be broadened. To show this, consider the Fourier transform of the output of a finite ULA

$$\begin{aligned} F(\omega,k) &= \sum_{m=0}^{M-1}\int_{-\infty}^{+\infty} f(t,md)e^{-j\omega t}dt\, e^{-j(2\pi/M)mk} \\ &= \sum_{m=0}^{M-1} e^{jm((\omega/c_x)d-(2\pi/M)k)}\int_{-\infty}^{+\infty} f_0(t)e^{-j\omega t}dt \\ &= F_0(\omega)H\left(\frac{\omega}{c_x}d-\frac{2\pi}{M}k\right), \end{aligned} \tag{4.4}$$

where $H((\omega, c_x)\, d - (2\pi/M)\, k)$ is the response function of a ULA of length *M*. Note that

$$H\left(\frac{\omega}{c_x}d-\frac{2\pi}{M}k\right) \rightarrow \delta\left(\frac{\omega}{c_x}d-u\right),$$

$$\text{as } M \rightarrow \infty$$

where $(2\pi/M)\, k \rightarrow u$ as $M, k \rightarrow \infty$. To avoid the aliasing error in the spectrum, the temporal sampling interval and sensor spacing must satisfy the relation shown in Equation 1.34b, which, when expressed in terms of ω_{max} and d, reduces to

$$\omega_{max}\frac{d}{\pi} \le \frac{c}{\sin\theta}. \tag{4.5a}$$

For fixed ω_{max}, d, and c, in order to avoid aliasing error, the angle of incidence will have to satisfy the following inequality,

$$\theta \le \sin^{-1}\left(\frac{\pi}{\omega_{max}}\frac{c}{d}\right). \tag{4.5b}$$

From Equation 4.5b it may be seen that for $d=\lambda/2$ and $\Delta t=\pi/\omega_{max}$ there is no aliasing for any angle of incidence. Aliasing error will occur whenever the above requirements are not satisfied. As an example, consider a stochastic plane wave incident on a ULA with 16 sensors at an angle of 45°. The sensors are 15 m apart. The bandwidth of the waveform is ±100 Hz and it is sampled with a sampling interval of 0.005 sec. The aliasing error is present in the top left and bottom right corners (see Figure 4.3).

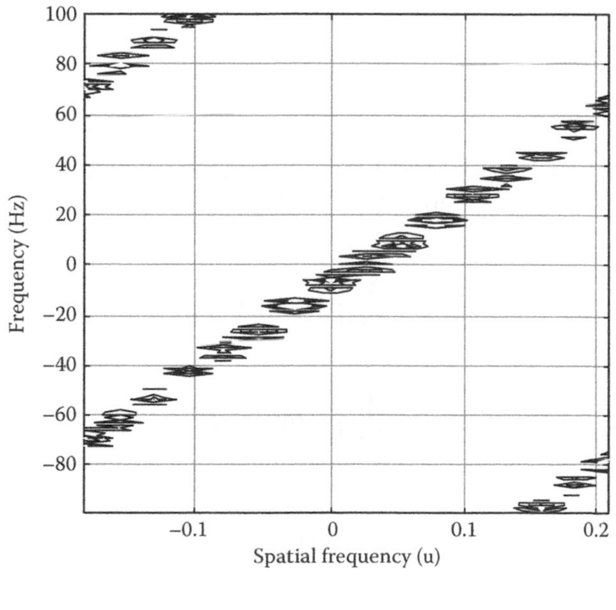

FIGURE 4.3
Aliasing error due to spatial undersampling. A stochastic plane wave is incident on a ULA at 45°. The ULA consists of 16 sensors spaced 15 m apart. The wave speed is 1,500 m/sec. The bandwidth of the waveform is±50 Hz and it is sampled with a sampling interval of 0.005 sec.

4.1.3 Spectral Matrix

The output of an array of sensors may be treated as a collection of time series or vector time series. A spectral matrix whose elements are the spectra and cross-spectra of a pair of outputs provides a complete characterization, particularly when the outputs are Gaussian. We would like to relate a spectral matrix to the (ω, k) spectrum. Note that the output of the mth sensor, $f_m(t)$, has the following spectral representation:

$$f_m(t) = f(t, x = md) = \frac{1}{4\pi^2} \iint_{-\infty}^{\infty} dF(\omega, u) e^{j(\omega t - umd)}. \tag{4.6}$$

Using Equation 4.6, the cross-covariance function between two sensor outputs is given by

$$\begin{aligned} c_{mn}(\tau) &= E\{f_m(t) f_n(t+\tau)\} \\ &= \frac{1}{4\pi^2} \iint_{-\infty}^{\infty} s_f(\omega, u) e^{j\omega\tau} e^{-j(m-n)ud} d\omega du. \end{aligned} \tag{4.7}$$

Further, the spectral representation of a cross-covariance function [2] is

$$c_{mn}(\tau) = \frac{1}{2\pi} \int_{-\infty}^{\infty} s_{mn}(\omega) e^{j\omega\tau} d\omega. \tag{4.8}$$

Comparing Equations 4.7 and 4.8, we obtain the following relation between elements of the spectral matrix and (ω, k) spectrum:

$$\begin{aligned} S_{mn}(\omega) &= \frac{1}{2\pi} \int_{-\infty}^{\infty} S_f(\omega, u) e^{jd(m-n)u} du \\ S_{mm}(\omega) &= \frac{1}{2\pi} \int_{-\infty}^{\infty} S_f(\omega, u) du. \end{aligned} \tag{4.9a}$$

The reverse relation, that is, (ω, k) spectrum in terms of the elements of spectral matrix, is

$$S_f(\omega, u) = \sum_{m=-\infty}^{\infty} \sum_{n=-\infty}^{\infty} S_{mn}(\omega) e^{-jd(m-n)u}. \tag{4.9b}$$

The spectral matrix has Hermitian symmetry. Additionally, for a ULA, it has Toeplitz symmetry. Consider an example of a stochastic plane wave incident on a ULA. The output of the mth sensor is given by

$$f_m(t) = \frac{1}{2\pi}\int_{-\infty}^{\infty} dF(\omega)e^{j\omega(t-m(d/c)\sin\theta)},$$

and the cross-covariance function between two outputs is given by

$$c_{mn}(\tau) = \frac{1}{2\pi}\int_{-\infty}^{\infty} S_f(\omega)e^{j\omega(\tau-(m-n)(d/c)\sin\theta)}d\omega. \tag{4.10}$$

From Equation 4.10, the cross-spectrum between the outputs of the mth and nth sensors is given by

$$S_{mn}(\omega) = S_f(\omega)\, e^{-j\omega(m-n)(d/c)\sin\theta}, \tag{4.11a}$$

which may also be expressed as

$$S_{mn}(\omega) = \frac{1}{2\pi}\int_{-\infty}^{\infty} S_f(\omega)\delta(\omega-u)e^{-jd(m-n)u}du, \tag{4.11b}$$

where $u=(\omega/c)\sin\theta$. Comparing Equation 4.11b with Equation 4.9, we obtain $S_f(\omega, u) = S_f(\omega)\,\delta(\omega-u)$.

The spectral matrix for this model has a very useful representation, that is, as an outer product of two vectors

$$\mathbf{S}_f(\omega) = S_f(\omega)\mathbf{a}(\omega)\mathbf{a}^H(\omega), \tag{4.12a}$$

where

$$\mathbf{a}(\omega) = \text{col}\{1, e^{-j\omega(d/c)\sin\theta}, \ldots, e^{-j(M-1)\omega(d/c)\sin\theta}\},$$

is the direction vector of the incident plane wave. Equation (4.12a) may be easily generalized for P uncorrelated sources,

$$\mathbf{S}_f(\omega) = \sum_{i=0}^{P-1} S_{f_i}(\omega)\mathbf{a}_i(\omega)\mathbf{a}_i^H(\omega),$$

which we shall express in matrix form. Define the following matrices:

$$\mathbf{S}_0(\omega) = \text{diag}\{S_{f_0}(\omega), S_{f_1}(\omega), \ldots, S_{f_{P-1}}(\omega)\},$$

$$\mathbf{A}(\omega) = [\mathbf{a}_0(\omega), \mathbf{a}_1(\omega), \ldots, \mathbf{a}_{P-1}(\omega)].$$

The spectral matrix for a case of P uncorrelated waves and uncorrelated white background noise is given by

$$\mathbf{S}_f(\omega) = \mathbf{A}(\omega)\mathbf{S}_0(\omega)\mathbf{A}^H(\omega) + \sigma_\eta^2\mathbf{I}. \tag{4.12b}$$

The columns of matrix A (ω) possess an interesting structure (for a ULA only), namely, each column can be expressed as powers of a constant, col $\{1, \mu^1, \mu^2, \ldots, \mu^{M-1}\}$, where μ is a constant. Let μ_n and μ_m be the constants corresponding to mth and nth columns. If $\mu_m \neq \mu_n$ for all m and n, $m \neq n$ the matrix; $\mathbf{A}(\omega)$ will have full column rank, that is, equal to P. Such a matrix is also known as a Vandermonde matrix [3]. Even when the plane waves are correlated, as in multipath propagation, the representation of a spectral matrix given by Equation 4.12b still holds good, with the only difference that $\mathbf{S}_0(\omega)$ is no longer a diagonal matrix. The nondiagonal terms will represent cross-spectra between the outputs of sources. An important consequence of the nondiagonal character of $\mathbf{S}_0(\omega)$ is the loss of the Toeplitz character (see Equation 4.9) of the spectral matrix. Interestingly, the Toeplitz character of the spectral matrix is lost whenever the sensors of the ULA are disturbed. In a more general situation where we have a large number of random plane waves incident on a ULA, the spectral matrix is a spatial covariance matrix at a fixed frequency.

4.1.3.1 Propagation Matrix

The matrix $\mathbf{A}(\omega)$ is partitioned as follows:

$$\mathbf{A}(\omega) = \begin{bmatrix} \mathbf{A}_1\,(P \times P) \\ \mathbf{A}_2\,(M - P \times P) \end{bmatrix}.$$

Since $\mathbf{A}(\omega)$ has full column rank, there is a unique linear operator known as the propagation matrix, $\Gamma(P+M-P)$, such that $\Gamma^H\mathbf{A}_1 = \mathbf{A}_2$, which may also be written as

$$\mathbf{A}^H \begin{bmatrix} \Gamma \\ -\mathbf{I} \end{bmatrix} = \mathbf{A}^H\mathbf{Q} = 0. \tag{4.13a}$$

It follows that $\mathbf{Q}$ spans the null space of $\mathbf{A}$. Now let us use the partitioned $\mathbf{A}(\omega)$ matrix in Equation 4.12b

$$\mathbf{S}_f(\omega) = \begin{bmatrix} \mathbf{G}_1 & \mathbf{H}_1 \\ \mathbf{H}_1 & \mathbf{H}_2 \end{bmatrix} = \overset{\qquad\quad P \qquad\qquad\quad M-P}{\begin{bmatrix} \mathbf{A}_1\mathbf{S}_0\mathbf{A}_1^H & \mathbf{A}_1\mathbf{S}_0\mathbf{A}_2^H \\ \mathbf{A}_2\mathbf{S}_0\mathbf{A}_1^H & \mathbf{A}_2\mathbf{S}_0\mathbf{A}_2^H \end{bmatrix}}\begin{matrix} P \\ M-P \end{matrix} + \sigma_\eta^2\mathbf{I}.$$

Note that the spectral matrix has also been partitioned so that

$$\begin{aligned} \mathbf{G}_1 &= \mathbf{A}_1\mathbf{S}_0\mathbf{A}_1^H + \sigma_\eta^2\mathbf{I}_P, \quad \mathbf{G}_2 = \mathbf{A}_2\mathbf{S}_0\mathbf{A}_1^H \\ \mathbf{H}_1 &= \mathbf{A}_1\mathbf{S}_0\mathbf{A}_2^H, \quad \mathbf{H}_2 = \mathbf{A}_2\mathbf{S}_0\mathbf{A}_2^H + \sigma_\eta^2\mathbf{I}_{M-P}, \end{aligned} \tag{4.13b}$$

where I_p stands for a unit matrix of size $P \times P$. It may be shown from Equation 4.13b that

$$\mathbf{G}_2 = \Gamma^H\mathbf{A}_1\mathbf{S}_0\mathbf{A}_1^H = \Gamma^H(\mathbf{G}_1 - \sigma_\eta^2\mathbf{I}_P),$$

and hence,

$$\Gamma^H = \mathbf{G}_2(\mathbf{G}_1 - \sigma_\eta^2\mathbf{I}_P)^{-1}. \tag{4.13c}$$

Thus, the propagation matrix may be derived from the partitioned spectral matrix. The background noise variance is assumed to be known (see Section 4.6, Exercise 4). **Q** may be used to find the DOA in place of eigenvectors corresponding to noise eigenvalues as in the MUSIC algorithm (described in Chapter 5) [4, 5].

4.1.4 Eigenstructure

The spectral matrix possesses interesting eigenstructure. $\mathbf{S}_f(\omega)$ is a Hermitian symmetric Toeplitz (only for a ULA) matrix; hence its eigenvalues are real. Further, they are positive as shown below: Let $\mathbf{e}(\omega)$ be some arbitrary vector and consider a quadratic form $\mathbf{e}^H(\omega)\ \mathbf{S}_f(\omega)\ \mathbf{e}(\omega)$. It follows from Equation 4.13a for any $\mathbf{e}(\omega)$,

$$\begin{aligned} \mathbf{e}^H(\omega)\mathbf{S}_f(\omega)\mathbf{e}(\omega) &= \sum_{i=0}^{P-1} S_{f_i}(\omega)\mathbf{e}^H(\omega)\mathbf{a}_i(\omega)\mathbf{a}_i^H(\omega)\mathbf{e}(\omega) \\ &= \sum_{i=0}^{P-1} S_{f_i}(\omega)\left|\mathbf{e}^H(\omega)\mathbf{a}_i(\omega)\right|^2 \geq 0. \end{aligned}$$

Therefore, a spectral matrix is always positive definite or positive semi-definite (when $P<M$ and there is no noise), that is, all its eigenvalues are positive or zero. Let $\mathbf{v}_m$, $m=0, 1,\ldots, M-1$ be the eigenvectors of $\mathbf{S}_f(\omega)$ and the corresponding eigenvalues be λ_m, $m=0, 1,\ldots, M-1$. From Equation (4.13b) it follows that

$$\underbrace{\mathbf{v}_m^H\mathbf{S}_f(\omega)\mathbf{v}_m}_{\lambda_m} = \underbrace{\mathbf{v}_m^H\mathbf{A}(\omega)\mathbf{S}_0(\omega)\mathbf{A}^H(\omega)\mathbf{v}_m}_{\alpha_m} + \sigma_\eta^2\underbrace{\mathbf{v}_m^H\mathbf{I}\mathbf{v}_m}_{\sigma_\eta^2},$$

hence,

$$\lambda_m = \alpha_m + \sigma_\eta^2, \tag{4.14a}$$

where α_m is an eigenvalue of the noise-free spectral matrix. Note that $\mathbf{A}(\omega)\mathbf{S}_0(\omega)\mathbf{A}^H(\omega)$ is a rank P matrix, as all P columns of the Vandermonde matrix A (ω) are independent, and $\mathbf{S}_0(\omega)$ is a $P\times P$ diagonal matrix. Hence, the remaining $M-P$ eigenvalues must be equal to zero, that is, $\alpha_P, \alpha_{P+1},\ldots, \alpha_{M-1}=0$. It therefore follows that

$$\mathbf{v}_m^H\mathbf{A}(\omega)\mathbf{S}_0(\omega)\mathbf{A}^H(\omega)\mathbf{v}_m = 0 \quad m = P, P+1,\ldots,M-1. \tag{4.14b}$$

Since $\mathbf{S}_0(\omega)$ is positive definite, by assumption, for Equation 4.14b to be valid we must have

$$\mathbf{v}_m^H\mathbf{A} = 0, \quad \text{for } m = P, P+1,\ldots,M-1, \tag{4.14c}$$

that is, $\mathbf{v}_m$, $m=P, P+1,\ldots, M-1$ is perpendicular to direction vectors of all incident wavefronts. Equations 4.14a and c are of great importance, as will be evident in the sequel (Chapter 5). Some of these properties of a spectral matrix were first noted by Nakhamkin et al. [25] in connection with the separation of seismic wavefronts.

4.1.5 Frequency Wavenumber Spectrum

When a large number of random plane waves are incident on an array, the (ω, k) spectrum may be computed from the spectral matrix using Equation 4.9b. Writing Equation 4.9b in matrix form for a finite array we obtain

$$S(\omega, \theta) = \mathrm{a}^H \mathrm{S}_f(\omega)\ \mathrm{a}. \tag{4.15a}$$

Since the above spectrum is analogous to the Blackman–Tuckey (BT) spectrum in time series literature [2], we shall call it a BT frequency wavenumber spectrum.

The (ω, k) spectrum turns into an angular spectrum when integrated over the bandwidth of the incident signal for a fixed angle of propagation

$$S(\theta) = \frac{1}{2\omega_{\max}} \int_{-\omega_{\max}}^{\omega_{\max}} S_f\left(\omega, \frac{\omega}{c}\sin\theta\right) d\omega. \tag{4.15b}$$

In the (ω, k) domain, the integration is carried out over a radial line sloping at angle θ (see Figure 4.4). A peak in the angular spectrum is an indication of wave energy arriving from a direction where the peak is found.

The angular spectrum defined in Equation 4.15b is also the beam power integrated over the frequency bandwidth as a function of the look angle.

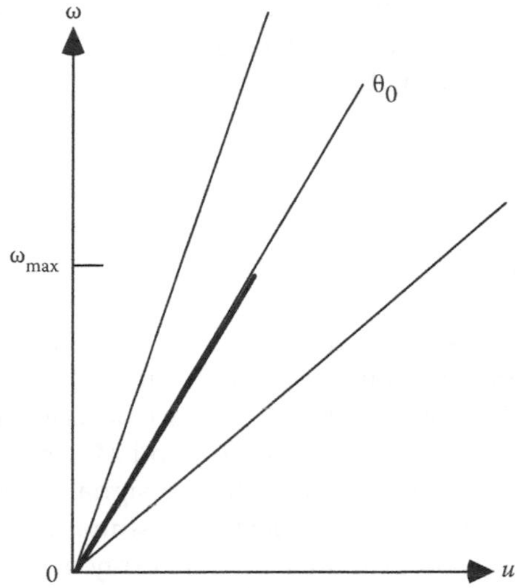

FIGURE 4.4
In the frequency wavenumber plane the spectrum is averaged over a series of radial lines. The spectrum of a plane wave that is incident at angle θ_0 is shown by the bold line.

4.1.6 Parametric Spectrum

The frequency wavenumber spectrum defined in Equation 4.15a assumes a plane wave model, which is more appropriate in open space (see Chapter 1 for wavefield representation in open and bounded space). In bounded space the wavefronts are far from planar. Such nonplanar wavefronts may be represented in terms of source location parameters measured with reference to the bounded space geometry. For example, in shallow water the source location parameters are range, depth (measured from the surface), and azimuth. In place of frequency wavenumber spectrum, where the parameters of interest are frequency and wavenumbers that depend on azimuth and elevation, we introduce a similarly defined quantity called parametric spectrum

$$S(\omega, \theta) = \mathbf{a}^H(\theta)\mathbf{S}_f(\omega)\mathbf{a}(\theta), \tag{4.16}$$

where θ now stands for generalized location parameters and $\mathbf{a}(\theta)$ is the wavefield that the array would sense if the source were to be located at θ. $\mathbf{S}(\omega, \theta)$ is computed over a range of values of θ spanning the entire parameter space. The actual source position is indicated by the position of the maximum in $\mathbf{S}(\omega, \theta)$. Evidently, $\mathbf{a}(\theta)$must be computed for the assumed geometry and the boundary conditions there on. Since the central idea is to match the computed field with the observed field, the processing is also known as matched field processing. It was first introduced by Bucker [6] in 1976 and since then much research effort has been devoted toward its development as a tool in underwater detection. An account of this effort is summarized in a monograph [7]. The chief obstacle in the use of matched field processing lies in the requirement of an exact knowledge of the propagation environment for the purpose of computing the wavefield.

4.2 Beamformation

In this section, we shall deal with the estimation of the (ω, k) spectrum, starting with the beamformation both in time and frequency domains including fast Fourier transform (FFT) based method. Next, we describe nonlinear methods: Capon's maximum likelihood (ML) method and ME method. The nonlinear methods provide higher resolution when the signal-to-noise ratio (SNR) is reasonably high. The importance of spectrum estimation arises on account of the fact that a signal wavefield approaching an array of sensors in a particular direction will produce a strong peak in the frequency wavenumber spectrum. Given the peak position, we can estimate the center frequency of radiation and the direction of approach, that is, the DOA (azimuth and elevation).

4.2.1 Narrowband Beamformation

A beam in a desired direction is formed by introducing delays before summation. The required delay per sensor in a ULA is equal to $\tau=(d/c)\sin\theta$ and in a uniform circular array (UCA) the delay for the mth sensor is equal to $\tau_m=(a/c)\sin\theta\cos((2\pi m/M)-\varphi)$. In analog beamformation, introduction of continuously varying delays is achieved through analog delay lines, but in digital beamformation the delays can be achieved only as integral steps of sampling time units. Consider a ULA with sensor spacing d equal to $\lambda/2$ and time sampling interval, Δt, equal to d/c. As noted in the previous section, there will be no aliasing, spatial or temporal for this choice of parameter. However, we can form just one beam, namely, for $\theta=0$ (excluding endfire beam). Clearly, to form more beams we need more samples between two Nyquist samples. Assume that we have sampled at q times the Nyquist rate, that is, we have q equispaced samples between two Nyquist samples, which will enable us to form beams at angles, θ_i, $i=0, 1,\ldots, q-1$, where $\theta_i=\sin^{-1}(i/q)$. For example, let $q=8$, the beam angles are 0°, 7.18°, 14.48°, 22.04°, 30.0°, 38.68°, 48.59°, 61.04°. Evidently, only a fixed number of beams can be formed for a given oversampling rate. It is not possible to form a beam in any arbitrary direction. The situation with a UCA is far more difficult as for no DOA can a uniformly sampled sensor output be used for beamformation. For example, consider a UCA of 16 sensors and $a/c=8$ time samples the delays to be introduced in the sensor outputs, in units of the temporal sampling interval, for $\theta=90°$ and $\phi=0$ are 8.00, 7.39, 5.66, 3.06, 0.0, −3.06, −5.65, −7.39, −8.00, −7.39, −5.66, −3.06, 0.0, 3.06, 5.66, 7.39 (rounded to second decimal place). All these delays are with respect to a hypothetical sensor at the center of the circle. Notice that the delays are not in integral steps of the sampling interval. This leaves us with the only alternative of nonuniform sampling through interpolation of uniformly sampled sensor output. To minimize the computational load, a simple linear interpolation has been suggested [8].

4.2.1.1 Narrowband Signals

For narrowband signals, the delays applied to the sensor outputs before summation may be expressed in terms of phase rotation. A narrowband signal output of the mth sensor may be represented as

$$f_m(t) \approx \frac{1}{2\pi}\int_{\omega_0-(\Delta\omega/2)}^{\omega_0+(\Delta\omega/2)} F_{nb}(\omega)e^{j\omega(t-m(d/c)\sin\theta_0)}d\omega$$

$$= \frac{1}{2\pi}\int_{-(\Delta\omega/2)}^{\Delta\omega/2} F(\omega_0+\delta\omega)e^{j[\omega_0(t-m(d/c)\sin\theta_0)+\delta\omega(t-m(d/c)\sin\theta_0)]}d\delta\omega \tag{4.17a}$$

$$= e^{-j\omega_0 m(d/c)\sin\theta_0)} f_{nb}(t),$$

where subscript nb stands for narrowband. The approximation is valid when the bandwidth satisfies the condition, $\Delta w\ (m\ (d/c)\ \sin\theta_0) \ll 2\pi$ for all m, which implies that the time taken for a wave to sweep across the array must be much less than the inverse of the bandwidth, expressed in Hertz. In vector notation, Equation 4.17a may be expressed as

$$\mathbf{f}(t) = \mathbf{a}_0 f_{nb}(t), \tag{4.17b}$$

where $\mathbf{a}_0 = \left[1, e^{-j\omega_0(d/c)\sin\theta_0}, \ldots, e^{-j\omega_0(M-1)(d/c)\sin\theta_0}\right]$ is the direction vector on the incident wavefront. The delays applied to sensor outputs may be expressed in terms of a vector dot product. Define a vector, $\mathbf{a} = \left[1, e^{-j\omega_0(d/c)\sin\theta}, \ldots, e^{-j\omega_0(M-1)(d/c)\sin\theta}\right]$, known as the steering vector, which will rotate the phase of each sensor output by an amount equal to $\omega_0 m\ (d/c)\ \sin\theta$ for the mth sensor. Thus, a narrowband beam is formed in the direction θ as

$$\mathbf{a}^H\mathbf{f}\ (t) = \mathbf{a}^H\mathbf{a}_0 f_{nb}\ (t),$$

or in terms of beam power, that is, the (ω, k) spectrum is given by

$$S(\omega_0,\theta) = E\left\{\left|\mathbf{a}^H\mathbf{a}_0 f_{\mathrm{nb}}(t)\right|^2\right\} = \left|\mathbf{a}^H\mathbf{a}_0\right|^2 \sigma_f^2. \tag{4.17c}$$

4.2.1.2 Window

The sensor outputs are often weighted before summation, the purpose being to reduce the sidelobes of the response function just as in spectrum estimation where a window is used to reduce the sidelobes and thereby reduce the power leakage. As this topic is extensively covered under spectrum estimation (e.g., see Refs. [2, 9]), we shall not pursue it any further. Instead, we would like to explain the use of a weight vector to reduce the background noise variance or to increase the SNR. Let us select a weight vector, w, such that the signal amplitude is preserved but the noise power is minimized.

$$\mathbf{w}^H\mathbf{a}_0=1 \text{ and } \mathbf{w}^H\mathbf{c}_\eta\mathbf{w}=\min, \tag{4.18a}$$

where $\mathbf{c}_\eta$ is the noise covariance function. The solution to the constrained minimization problem in Equation 4.18a results in

$$\mathbf{w}=\frac{\mathbf{c}_\eta^{-1}\mathbf{a}_0}{\mathbf{a}_0^H\mathbf{c}_\eta^{-1}\mathbf{a}_0}. \tag{4.18b}$$

It may be observed that for spatially white noise $\mathbf{c}_\eta=\sigma_\eta^2\mathbf{I}$, therefore, $\mathbf{w}=\mathbf{a}_0/M$. In other words, the weights are simply phase shifts or delays as in beamformation. The variance of the noise in the output is equal to $\sigma_{\hat{\eta}}^2=\sigma_\eta^2/M$.

A weight vector may be chosen to maximize the SNR. The output power of an array with a weight vector **w**, when there is no noise, is given by $\mathbf{w}^H\mathbf{c}_s\mathbf{w}$ and when there is no signal, the output power is $\mathbf{w}^H\mathbf{c}_\eta\mathbf{w}$. We select that weight vector which will maximize the output power ratio

$$\frac{\mathbf{w}^H\mathbf{c}_s\mathbf{w}}{\mathbf{w}^H\mathbf{c}_\eta\mathbf{w}}=\max \text{ (with respect to } \mathbf{w}).$$

The solution is simply the generalized eigenvector corresponding to the largest eigenvalue of the pencil matrix $[\mathbf{c}_s, \mathbf{c}_\eta]$ [10]. There is a large body of knowledge on how to obtain an optimum weight vector that meets different types of constraints. A brief review of the relevant literature is given in Ref. [10].

4.2.1.3 Rayleigh Resolution

When two wavefronts are simultaneously incident on an array, naturally we would like to measure their DOA. For this to be possible, the spectrum given by Equation 4.17c must show two distinct peaks. Let $f_{nb_1}(t)$ and $f_{nb_2}(t)$ be two uncorrelated signals incident at angles θ_1 and θ_2, with the center frequencies being the same for both signals. The beam power is given by

$$s(\omega_0,\theta)=\left|\mathbf{a}^H\mathbf{a}_1\right|^2\sigma_{f_1}^2+\left|\mathbf{a}^H\mathbf{a}_2\right|^2\sigma_{f_2}^2.$$

In order that each signal gives rise to a distinct peak, $\left|\mathbf{a}^H\mathbf{a}_1\right|^2\sigma_{f_1}^2$ when plotted as a function of θ should not overlap with $\left|\mathbf{a}^H\mathbf{a}_2\right|^2\sigma_{f_2}^2$. A condition for nonoverlap is necessarily arbitrary as the array response to an incident wavefront is strictly not limited to a fixed angular range. The Rayleigh resolution criterion states that two wavefronts are resolved when the peak of the array response due to the first wavefront falls on the first zero of the array response due to the second wavefront. The first zero is located at an angle, $\sin^{-1}(\lambda/Md)$, away from the DOA (broadside). Thus, two wavefronts are resolved, according to the Rayleigh resolution criterion, when their DOA differ by $\sin^{-1}(\lambda/Md)$. An example of resolution is shown in Figure 4.5. For a UCA we can derive a simple expression when it is fully populated. In this case, its response function is a Bessal function of 0th order (see Equation 2.43b). The first zero of the Bessal function of 0th order is at 2.4048. Two wavefronts are said to be resolved, according to the Rayleigh resolution criterion, when the angular separation is greater than $\sin^{-1}(1.2024\lambda/\pi a)$. Let us compare the resolution properties of a ULA and a UCA having *equal* aperture, for example, a 16-sensor ULA with 7.5λ aperture and the corresponding UCA with a radius equal to 3.75λ but fully populated with more than 32 sensors. The relative performance is shown in Table 4.1. The performance of the UCA is marginally better than that of the ULA. Beamformation in the frequency domain requires the 2D Fourier transform. For a fixed temporal frequency, the magnitude of the spatial Fourier transform coefficients is related to the power of a wave coming from a direction that may be computed from the spatial frequency number, $\theta = \sin^{-1}(k/M)$ (see Chapter 2, Section 2.1), where k is the spatial frequency number. Here too, only a finite number of fixed beams are formed. This number is equal to the number of sensors. However, the discrete Fourier

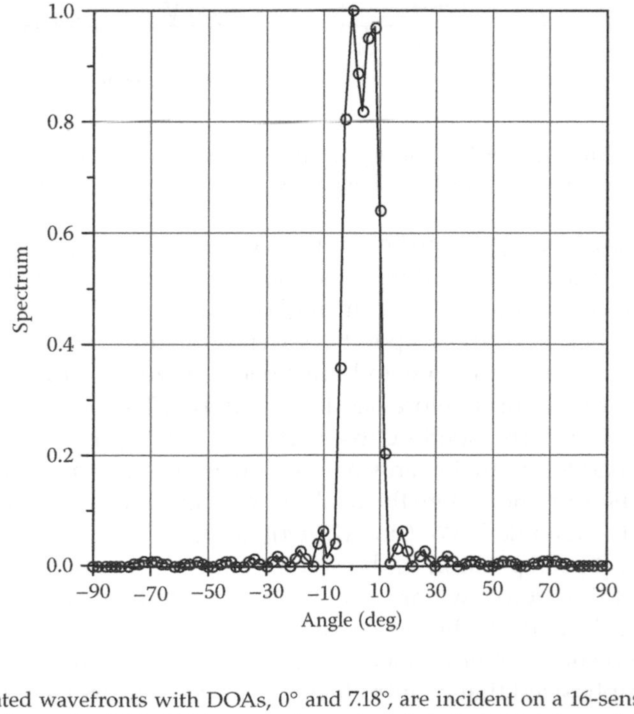

FIGURE 4.5
Two uncorrelated wavefronts with DOAs, 0° and 7.18°, are incident on a 16-sensor ULA. The waves are clearly resolved. The DOAs were chosen to satisfy the Rayleigh resolution criterion.

TABLE 4.1

The Rayleigh resolution angle as a function of the number of sensors (ULA with $\lambda/2$ Sensor Spacing)

No. of Sensors	Rayleigh Resolution Angle in Degrees (ULA)	Rayleigh Resolution Angle in Degrees (UCA)
4	30	30.61
8	14.48	12.67
16	7.18	5.84
32	3.58	2.83
64	1.79	1.39

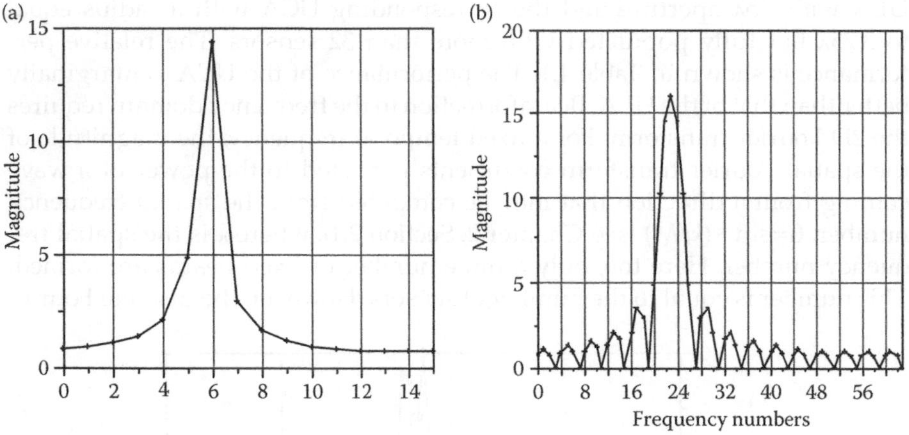

FIGURE 4.6
The role of padding zeros or introducing dummy sensors is to correctly position the peak (a) before padding zeros and (b) after padding zeros. The beam width remains unchanged.

transform allows interpolation between fixed beams through a simple means of padding zeros or placing dummy sensors giving no output. Consider an example of a wavefront incident at an angle of 21.06° on a 16-sensor array. The output is subjected to the temporal Fourier transform. The spatial Fourier transform is performed before and after padding zeros. In Figure 4.6a, the spatial Fourier transform before padding is shown. The peak appears at frequency number 6 corresponding to an angle of 22.02° ($\sin^{-1}(6/16)$). Next, the sequence is padded with 48 zeros before Fourier transformation. The result is shown in Figure 4.6b, where the peak appears at frequency number 23 corresponding to an angle 21.06° ($\sin^{-1}(23/64)$), which is the correct figure. Note that the correct peak position lies between frequency numbers 5 and 6 (closer to 6). By padding zeros, we are able to interpolate between the frequency numbers 5 and 6 and are thus able to capture the peak at its correct position. Further, the peak is better defined, but the peak width remains unchanged. It may be emphasized that by introducing dummy sensors (zeros) we cannot achieve higher resolution.

4.2.1.4 Sources of Error

In practical beamformation we encounter several sources of phase errors, such as those caused by sensor position errors, variable propagation conditions, sensor and associated electronics phase errors, quantization error in the phase shifter, etc. The array response is highly prone to such phase errors. Nominally, the array response may be expressed as

$$H(\omega) = \mathbf{a}^H\left(\omega\frac{d}{c}\sin\theta\right)\mathbf{a}\left(\omega\frac{d}{c}\sin\theta_0\right),$$

where θ_0 is the DOA of the incident wave and θ is the steering angle. We shall model two types of phase errors, namely, those caused by position errors, and phase errors caused by all other sources lumped into one. The corrupted direction vector has the following form:

$$\tilde{\mathbf{a}} = \text{col}\left\{e^{-j\phi_0},\ e^{-j(\omega((d+\Delta d_1)/c)\sin\theta_0+\phi_1)},\dots,\ e^{-j(\omega(M-1)((d+\Delta d_{M-1})/c)\sin\theta_0+\phi_{M-1})}\right\}, \quad (4.19)$$

where Δd_1 is the position error of the ith sensor and ϕ_1 is the phase error. We have assumed that the first sensor is a reference sensor and hence there is no position error. We shall illustrate the effect of position and phase errors on the array response function. We assume that the ULA has 16 sensors that are equispaced but with some position error. Let $d = (\lambda/2)$ and Δd be a uniformly distributed random variable in the range $\pm(\lambda/16)$. The resulting response is shown in Figure 4.7. The array response due to phase errors, caused by other factors, is shown in Figure 4.8. The phase errors seem to cause less harm

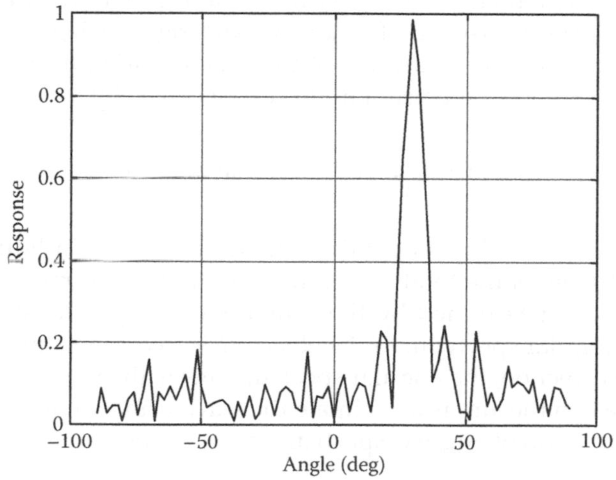

FIGURE 4.7
Response of a ULA with position errors that are uniformly distributed in the range $\pm\lambda/4$ (solid curve). Compare this with the response of the ULA without any position errors (dashed curve).

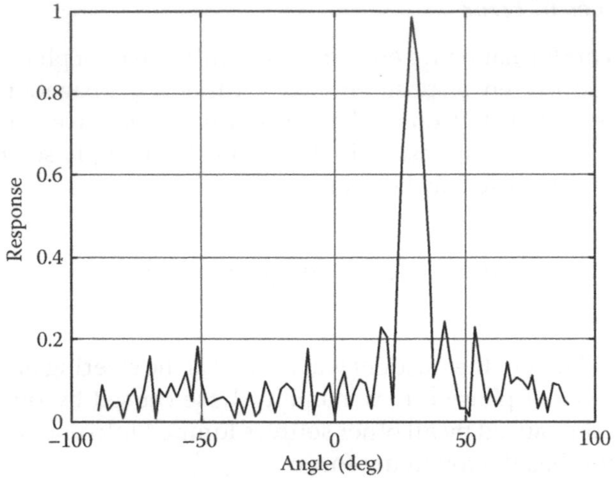

FIGURE 4.8
Response of a ULA with phase errors that are uniformly distributed in the range $\pm\lambda/4$.

compared to the position errors. The sensor position and phase errors largely affect the sidelobe structure of the response function while the main lobe position and the width remain unchanged.

4.2.2 Broadband Beamformation

Beamformation with a broadband signal can also be written in a form similar to that for a narrowband signal. We must first Fourier transform (temporal) the broadband signal output from each sensor and treat each Fourier coefficient as a Fourier transform of a narrowband signal whose bandwidth is approximately equal to the inverse of the time duration of the signal. The frequency wavenumber spectrum in this case is given by

$$S(\omega,\theta) = E\left\{\left|\mathbf{a}^H\mathbf{a}_0 F(\omega)\right|^2\right\} = \left|\mathbf{a}^H\mathbf{a}_0\right|^2 S_f(\omega). \tag{4.20}$$

By integrating over the temporal frequency as shown in Equation 4.15b, we get an estimate of the total power arriving from the direction θ (over an angular interval determined by the array). We had previously called this quantity an angular spectrum. If the objective is to estimate power received from a given direction, the angular spectrum meets the requirement. On the other hand, when the aim is to estimate the waveform arriving from a given direction estimation of angular spectrum is not enough.

4.2.2.1 Delayed Snapshots

In Chapter 2, we introduced the concept of delayed snapshots to represent a broadband wavefield. We shall make use of that representation in devising a

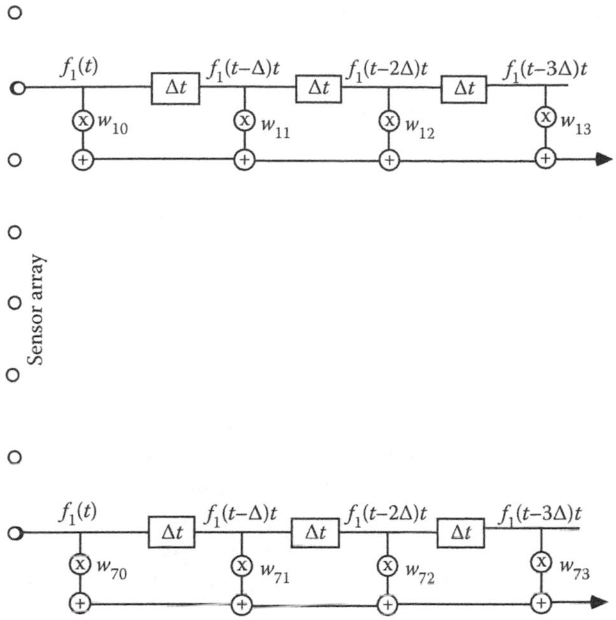

FIGURE 4.9
The structure of a 2D filter for broadband beamformation. $M=8$ and $N=4$.

2D spatio-temporal filter for beamformation. A desired frequency wavenumber response is shown in Figure 3.6, where we have a passband lying between two radial lines with prescribed slopes and a horizontal line representing the maximum temporal frequency. Let w_{mn}, $m=0, 1, \ldots, M-1$ and $n=0, 1, \ldots, N-1$, where M stands for the number of sensors and N for the number of delayed snapshots, be the required finite 2D filter whose response is as close to the desired frequency wavenumber response as possible. A block diagram showing the filter structure is given in Figure 4.9.

We shall express the frequency wavenumber response of the filter in a matrix form. For this purpose we define the following quantities:

$$\mathbf{w}=\text{col}\{w_{0,0}, w_{1,0}, \ldots, w_{M-1,0}, w_{0,1}, w_{1,1}, \ldots, w_{M-1,1}, \ldots, w_{0,N-1}, w_{1,N-1}, \ldots, w_{M-1,N-1}\}, \tag{4.21a}$$

$$\mathbf{A}=\text{col}\,\{1, e^{jud}, \ldots, e^{ju\,(M-1)\,d}; e^{j\omega\Delta t}, e^{j\,(ud+\omega\Delta t)}, \ldots, e^{j\,(ud\,(M-1)+\omega\Delta t)}, \ldots; e^{j\omega\,(N-1)\Delta t}, e^{j\,(ud+\omega\,(N-1)\Delta t}, \ldots, e^{j\omega\,(N-1)\Delta t}; e^{j\,(ud\,(M-1)+\omega\,(N-1)\Delta t)}\}. \tag{4.21b}$$

It is easy to show that the response function can be expressed as an inner product of two vectors defined in Equation 4.21

$$\begin{aligned} H(u,\omega) &= \sum_{m=0}^{M-1}\sum_{n=0}^{N-1} w_{mn} e^{-j(umd+\omega n\Delta t)} \\ &= \mathbf{A}^H \mathbf{w}. \end{aligned} \tag{4.22}$$

The power output of the filter (Equation 4.22) is given by

$$\text{Output power} = \mathbf{w}^H \left[\frac{1}{4\pi^2} \int_{\omega_1}^{\omega_2} \int_{\omega/a}^{\omega/b} \mathbf{A}\mathbf{A}^H du d\omega \right] \mathbf{w}, \tag{4.23a}$$

which we like to maximize with respect to $\mathbf{w}$ in relation to the total energy,

$$\text{Total power} = \mathbf{w}^H \left[\frac{1}{4\pi^2} \int_{\omega_{min}}^{\omega_{max}} \int_{-\pi/d}^{\pi/d} \mathbf{A}\mathbf{A}^H du d\omega \right] \mathbf{w}, \tag{4.23b}$$

where a and b are slopes of the radial lines defining the passband (see Figure 3.6), ω_2 and ω_1 are, respectively, the upper and the lower cutoff frequency for the beam, and ω_{max} and ω_{min} refer to the maximum and minimum frequency present in the signal, respectively. The problem may be expressed as a problem in maximization of a ratio, $\mathbf{w}^H \Gamma_1 \mathbf{w}/\mathbf{w}^H \Gamma_2 \mathbf{w} = \lambda$, which is achieved by solving the following generalized eigen-decomposition problem:

$$\Gamma_1 \mathbf{w} = \lambda \Gamma_2 \mathbf{w}, \tag{4.24}$$

where $\Gamma_1 = (1/4\pi^2)\int_{\omega_1}^{\omega_2}\int_{\omega/a}^{\omega/b} \mathbf{A}\mathbf{A}^H du d\omega$ and $\Gamma_2 = (1/4\pi^2)\int_{\omega_{min}}^{\omega_{max}}\int_{-\pi/d}^{\pi/d} \mathbf{A}\mathbf{A}^H du d\omega$. The solution is given by the eigenvector corresponding to the largest eigenvalue and the maximum relative power is equal to the largest eigenvalue.

To evaluate Γ_1 and Γ_2, we must first simplify the elements of $\mathbf{A}\mathbf{A}^H$,

$$\begin{aligned} \left[\mathbf{A}\mathbf{A}^H\right]_{pq} &= e^{j(p_1 ud + p_2 \omega \Delta t)} e^{-j(q_1 ud + q_2 \omega \Delta t)} \\ &= e^{j((p_1 - q_1)ud + (p_2 - q_2)\omega \Delta t)}, \end{aligned}$$

where $p_1 = p - \text{Int}\,[p/M]M$, $p_2 = \text{Int}\,[p/M]$, $p = 0, 1, \ldots M-1$ and Int $[x]$ stands for the largest integer less than or equal to x; q_1 and q_2 are similarly defined. The elements of Γ_1 may be evaluated as in Equation 3.12a and those of Γ_2 are given as

$$[\Gamma_2]_{p,q} = \frac{1}{d}\operatorname{sin}c((p_1 - q_1)\pi) \begin{bmatrix} \dfrac{\omega_{max}}{\pi}\operatorname{sin}c((p_2 - q_2)\omega_{max}\Delta t) \\ -\dfrac{\omega_{min}}{\pi}\operatorname{sin}c((p_2 - q_2)\omega_{min}\Delta t) \end{bmatrix}. \tag{4.25}$$

The principle of maximizing the power in the passband was first suggested in Ref. [11] in the context of optimum window for spectrum estimation of time series. Later, this principle with additional constraints on the magnitude and derivative has been applied in beamformation [12, 13]. In Figure 4.10, we show

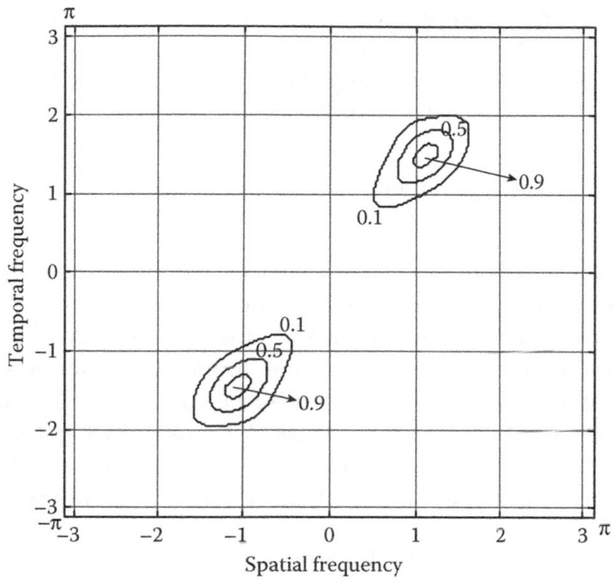

FIGURE 4.10
Response of filter for broadband beamformation. The slopes of the radial lines are $a=1$ (45°) and $b=2$ (63.4°); $\omega_1=0$ and $\omega_2=0.6\pi$. The maximum energy in the passband is 94% ($\lambda_{max}=0.94$). Sixteen sensors and sixteen delayed samples. $\Delta x=1$ and $\Delta t=1$.

a numerical example of spatio-temporal filter for broadband beamformation. The desired response is unity in the region bounded by two radial lines and the upper and the lower frequency cutoff lines. The actual response of the maximum energy filter is contoured in the same figure. It is observed that the maximum sidelobe level is about 4 dB less than that in the simple quadrant filter shown in Figure 3.7.

4.3 Capon's ω-k Spectrum

We consider a stochastic plane wave incident on a ULA. Let us represent the array output in a matrix form,

$$\mathbf{f}(t)=\frac{1}{2\pi}\int_{-\infty}^{\infty} d\mathbf{F}(\omega)e^{j\omega t}, \tag{4.26a}$$

where

$$d\mathbf{F}(\omega)=dF_0(\omega)\mathbf{a}\left(\omega\frac{d}{c}\sin\theta\right)+d\boldsymbol{\eta}(\omega), \tag{4.26b}$$

where $dF_0(\omega)$ is the generalized Fourier transform of the stochastic waveform, $\mathbf{a}(\omega\,(d/c)\sin\theta)$ is the direction vector, and $d\eta(\omega)$ is the background noise. From Equation 4.26 and using the properties of the generalized Fourier transform we can derive an expression for the spectral matrix

$$\mathbf{S}_f(\omega) = S_0(\omega)\mathbf{a}\left(\omega\frac{d}{c}\sin\theta\right)\mathbf{a}^H\left(\omega\frac{d}{c}\sin\theta\right) + \mathbf{S}_\eta(\omega),$$

where both $\mathbf{S}_f(\omega)$ and $\mathbf{S}_\eta(\omega)$ are $M\times M$ matrices, but $S_0(\omega)$ is a scalar. We would like to find a weight vector $\mathbf{w}$, acting on the array output such that it minimizes the power output of the array and is transparent to all waves propagating through a narrow cone with a cone angle $\Delta\theta$ and its axis pointing in the direction of θ_0. Thus, the beamwidth is made intentionally wider to allow for possible variation in the DOA. This model is useful when the direction of the incoming wavefront is likely to be different from the assumed or given direction. Translated into a mathematical statement we obtain

$$\mathbf{w}^H\mathbf{S}_f(\omega)\mathbf{w} = \min \qquad (4.27a)$$

$$\mathbf{w}^H\Gamma\mathbf{w} = 1,$$

where

$$\Gamma = \frac{1}{\Delta\theta}\int_{\theta_0-(\Delta\theta/2)}^{\theta_0+(\Delta\theta/2)} \mathbf{a}\left(\omega\frac{d}{c}\sin\theta\right)\mathbf{a}^H\left(\omega\frac{d}{c}\sin\theta\right)d\theta, \qquad (4.27b)$$

and $\mathbf{S}_f(\omega)$ is the array signal spectrum under the assumption that the source bearing lies in the range $\theta_0 \pm \Delta\theta/2$.

The constrained minimization problem specified in Equation 4.27 is solved by the Lagrange method,

$$\mathbf{w}^H\mathbf{S}_f(\omega)\,\mathbf{w} + \lambda\,(1 - \mathbf{w}^H\Gamma\mathbf{w}) = \min. \qquad (4.28)$$

Differentiating Equation 4.28 with respect to w and setting the derivative to zero we obtain

$$\mathbf{S}_f\mathbf{w} = \lambda\Gamma\mathbf{w}, \qquad (4.29a)$$

or

$$\lambda^{-1}\mathbf{w} = \mathbf{S}_f^{-1}\Gamma\mathbf{w}. \qquad (4.29b)$$

From Equation 4.29b it is clear that $\mathbf{w}$ is an eigenvector of $\mathbf{S}_f^{-1}\Gamma$ and the corresponding eigenvalue is λ^{-1}. Note that from Equation 4.29, $\mathbf{w}^H\mathbf{S}_f\mathbf{w} = \lambda$, that is, equal to the output power of the array weighted by vector $\mathbf{w}$. In order that the array output power is minimum, we must select the largest eigenvalue of $\mathbf{S}_f^{-1}\Gamma$ and the corresponding eigenvector as the weight vector, $\mathbf{w}$.

4.3.1 Special Case

Let $\Delta\theta=0$, that is, the beamwidth is zero. Therefore,

$$\Gamma=\mathbf{a}\left(\omega\frac{d}{c}\sin\theta_0\right)\mathbf{a}^H\left(\omega\frac{d}{c}\sin\theta\right).$$

Equation 4.29b now becomes

$$\lambda^{-1}\mathbf{w}=\mathbf{S}_f^{-1}\mathbf{a}\left(\omega\frac{d}{c}\sin\theta_0\right)\mathbf{a}^H\left(\omega\frac{d}{c}\sin\theta_0\right)\mathbf{w}. \tag{4.30}$$

By premultiplying both sides by $\mathbf{a}^H\omega(d/c)$ $(\sin\theta_0)$ we find that

$$\lambda^{-1}=\mathbf{a}^H\left(\omega\frac{d}{c}\sin\theta_0\right)\mathbf{S}_f^{-1}\mathbf{a}\left(\omega\frac{d}{c}\sin\theta_0\right). \tag{4.31a}$$

It turns out that

$$\mathbf{w}=\frac{\mathbf{S}_f^{-1}\mathbf{a}\left(\omega\frac{d}{c}\sin\theta_0\right)}{\mathbf{a}^H\left(\omega\frac{d}{c}\sin\theta_0\right)\mathbf{S}_f^{-1}\mathbf{a}\left(\omega\frac{d}{c}\sin\theta_0\right)}, \tag{4.31b}$$

satisfies Equation 4.30. We can express the array output power, which we shall call Capon spectrum,

$$s_{\text{Cap}}(\omega,\theta_0)=\frac{1}{\mathbf{a}^H\left(\omega\frac{d}{c}\sin\theta_0\right)\mathbf{S}_f^{-1}\mathbf{a}\left(\omega\frac{d}{c}\sin\theta_0\right)}. \tag{4.32}$$

Capon [14], who first suggested the above measure of spectrum, however, called it maximum likelihood spectrum. It is also known as the minimum variance distortionless response (MVDR) beamformer or a linearly constrained minimum variance (LCMV) beamformer [15]. Since θ is related to the spatial frequency, $u=\omega\ (d/c)\sin\theta$, $s_{\text{Cap}}(\omega,\theta)$ is indeed a (ω, k) spectrum as a function of θ or u.

4.3.2 Resolution

The Capon spectrum has a better resolution compared to the BT ω-k spectrum. We shall demonstrate this by considering two uncorrelated wavefronts in the presence of white noise. The spectral matrix is given by

$$\mathbf{S}_f(\omega)=s_0\mathbf{a}_0(\omega,\theta_0)\mathbf{a}_0^H(\omega,\theta_0)+s_1\mathbf{a}_1(\omega,\theta_1)\mathbf{a}_1^H(\omega,\theta_1)+\sigma_\eta^2\mathbf{I}, \tag{4.33}$$

where θ_0 and θ_1 are directions of arrival, s_0 and s_1 are powers of two plane wavefronts, and σ_η^2 is noise variance. The inverse of the spectral matrix in Equation 4.33 may be computed following the procedure described in Ref. [16].

$$\mathbf{S}_f^{-1}(\omega) = \mathbf{V}^{-1} - s_1 \frac{\mathbf{V}^{-1}\mathbf{a}_1(\omega,\theta_1)\mathbf{a}_1^H(\omega,\theta_1)\mathbf{V}^{-H}}{1 + s_1\mathbf{a}_1^H(\omega,\theta_1)\mathbf{V}^{-1}\mathbf{a}_1(\omega,\theta_1)}, \tag{4.34a}$$

where

$$\mathbf{V}^{-1} = \frac{1}{\sigma_\eta^2}\left[\mathbf{I} - \frac{\frac{s_0}{\sigma_\eta^2}\mathbf{a}_0(\omega,\theta_0)\mathbf{a}_0{}^H(\omega,\theta_0)}{1 + \frac{s_0}{\sigma_\eta^2}M}\right]. \tag{4.34b}$$

Using Equation 4.34a in Equation 4.32, we obtain the Capon spectrum for the two-source model,

$$s_{\text{Cap}}(\omega,\theta) = \frac{\sigma_\eta^2}{M - \frac{\frac{s_0}{\sigma_\eta^2}\left|\mathbf{a}^H\mathbf{a}_0\right|^2}{1+\frac{s_0}{\sigma_\eta^2}M} - \frac{\frac{s_1}{\sigma_\eta^2}\left|\mathbf{a}^H\mathbf{a}_1 - \frac{\frac{s_0}{\sigma_\eta^2}\mathbf{a}^H\mathbf{a}_0\mathbf{a}_0^H\mathbf{a}_1}{1+\frac{s_0}{\sigma_\eta^2}M}\right|^2}{1+\frac{s_1}{\sigma_\eta^2}\left(M - \frac{\frac{s_0}{\sigma_\eta^2}\left|\mathbf{a}_0^H\mathbf{a}_1\right|^2}{1+\frac{s_0}{\sigma_\eta^2}M}\right)}}, \tag{4.35}$$

where, for the sake of compactness, we have dropped the arguments of the vectors $\mathbf{a}$, $\mathbf{a}_0$, and $\mathbf{a}_1$. When the steering vector points to one of the sources, for example, when $\mathbf{a}=\mathbf{a}_0$

$$s_{\text{Cap}}(\omega,\theta_0) = \frac{\sigma_\eta^2}{\frac{M}{1+\frac{s_0}{\sigma_\eta^2}M} - \frac{\frac{s_1}{\sigma_\eta^2}\left|\mathbf{a}_0^H\mathbf{a}_1 - \frac{\frac{s_0}{\sigma_\eta^2}M\mathbf{a}_0^H\mathbf{a}_1}{1+\frac{s_0}{\sigma_\eta^2}M}\right|^2}{1+\frac{s_1}{\sigma_\eta^2}\left(M - \frac{\frac{s_0}{\sigma_\eta^2}\left|\mathbf{a}_0^H\mathbf{a}_1\right|^2}{1+\frac{s_0}{\sigma_\eta^2}M}\right)}} \approx s_0 + \frac{\sigma_\eta^2}{M}, \tag{4.36a}$$

and when $\mathbf{a}=\mathbf{a}_1$

$$s_{\text{Cap}}(\omega,\theta_1)=\sigma_\eta^2\frac{1+\dfrac{s_1}{\sigma_\eta^2}\left(M-\dfrac{\dfrac{s_0}{\sigma_\eta^2}\left|\mathbf{a}_0^H\mathbf{a}_1\right|^2}{1+\dfrac{s_0}{\sigma_\eta^2}M}\right)}{M-\dfrac{\dfrac{s_0}{\sigma_\eta^2}\left|\mathbf{a}_1^H\mathbf{a}_0\right|^2}{1+\dfrac{s_0}{\sigma_\eta^2}M}}\approx s_1+\frac{\sigma_\eta^2}{M}. \tag{4.36b}$$

The approximation shown in Equation 4.36 is valid for $\left|\mathbf{a}_1^H\mathbf{a}_0\right|\ll M$. From the above it is clear that when the wavefronts are well resolved, the peak amplitude approximately equals the power of the source. The noise power is reduced by a factor equal to the number of sensors.

We will examine the resolution power of the Capon spectrum. Consider again two equal power wavefronts incident at angles θ_0 and θ_1. The peaks corresponding to two wavefronts are resolved when a valley is formed in between them. Let $s_{\text{Cap}}(\omega,\tilde{\theta})$ be the spectrum at $\tilde{\theta}$ midway between θ_0 and θ_1. Define the ratio ρ as

$$\rho=\frac{s_{\text{Cap}}(\omega,\theta_0)}{s_{\text{Cap}}(\omega,\tilde{\theta})}$$

$$=\frac{\left\{1+\dfrac{s_1M}{\sigma_\eta^2}\left(1-\alpha\left|\dfrac{\mathbf{a}_0^H\mathbf{a}_1}{M}\right|^2-(1+\alpha)\left|\dfrac{\tilde{\mathbf{a}}^H\mathbf{a}_1}{M}\right|^2+2\alpha\,\text{Re}\left[\dfrac{\tilde{\mathbf{a}}^H\mathbf{a}_1\mathbf{a}_1^H\mathbf{a}_0\mathbf{a}_0^H\tilde{\mathbf{a}}}{M^3}\right]\right)-\alpha\left|\dfrac{\tilde{\mathbf{a}}^H\mathbf{a}_1}{M}\right|^2\right\}}{1-\alpha\left|\dfrac{\mathbf{a}_0^H\mathbf{a}_1}{M}\right|^2}. \tag{4.37a}$$

Assuming $(s_1M/\sigma_\eta^2)\gg 1$, Equation 4.37a simplifies to

$$\approx\frac{\left\{1+\dfrac{s_1M}{\sigma_\eta^2}\left(1-\left|\dfrac{\mathbf{a}_0^H\mathbf{a}_1}{M}\right|^2-2\left|\dfrac{\tilde{\mathbf{a}}^H\mathbf{a}_1}{M}\right|^2+2\,\text{Re}\left[\dfrac{\tilde{\mathbf{a}}^H\mathbf{a}_1\mathbf{a}_1^H\mathbf{a}_0\mathbf{a}_0^H\tilde{\mathbf{a}}}{M^3}\right]\right)\right\}}{1-\left|\dfrac{\mathbf{a}_0^H\mathbf{a}_1}{M}\right|^2}, \tag{4.37b}$$

where $\alpha=s_1M/(1+s_1M)\approx 1$ for $(s_1M/\sigma_\eta^2)\gg 1$. A valley is formed if $\rho>1$. Let $\mathbf{a}_0=\mathbf{a}_1=\tilde{\mathbf{a}}$, that is, when two wavefronts merge into a single wavefront we

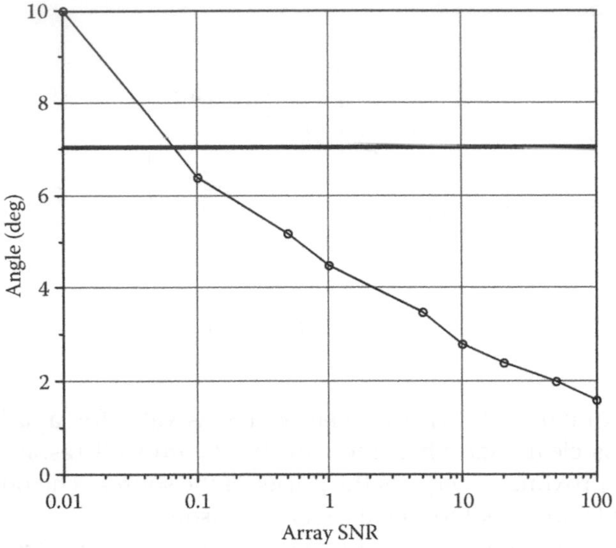

FIGURE 4.11
Resolution properties of the Capon spectrum as a function of array SNR. A 16-sensor ULA is assumed. The angles of incidence are 30° and 30°+angular separation, as shown on the y-axis. Simple beamformation (BT ω-k spectrum) yields a resolution of 7.18°, shown in the figure by a thick line for comparison. Resolution gain by the Capon spectrum is possible only for high array SNR.

notice that $\rho=1$ for all SNR, which means that these two wavefronts can never be resolved. Next, let $|\mathbf{a}_0^H\mathbf{a}_1/M|^2=|\tilde{\mathbf{a}}^H\mathbf{a}_1/M|^2\approx 0$, which means that the wavefronts are well separated. Then, $\rho=1+(s_1M/\sigma_\eta^2)>1$ except when $s_1=0$. The wavefronts can then always be resolved. All the above conclusions follow from commonsense. We now consider two wavefronts with DOAs $\pm(\Delta\theta/2)$, respectively and compute ρ for different s_1M/σ_η^2 and $\Delta\theta$. A plot of $\Delta\theta$ for which ρ is just >1 as a function of array SNR is shown in Figure 4.11.

4.3.3 Robust Beamformation

The sensitivity of beamformation to errors in the sensor position and other phase errors has been demonstrated in Figures 4.7 and 4.8. These drawbacks may be reduced through an appropriate choice of weighting coefficients. In this section, we shall show how such coefficients can be obtained [17] following a constraint used in deriving Capon's filter and the associated spectrum. Let $\mathbf{w}=\text{col}\,\{w_0, w_1, \ldots w_{M-1}\}$ be a coefficient vector. The array response may be expressed as

$$H(\omega)=\mathbf{w}^H\mathbf{a}\left(\omega\frac{d}{c}\sin\theta\right). \tag{4.38}$$

We shall model two types of phase errors, namely, those caused by position errors and those caused by all other sources of errors lumped into a single phase error. The steering vector is given by Equation 4.19. We have assumed that the first sensor is a reference sensor and hence there is no position error. Let H_0 be the desired response, e.g., equal to 1, and $\tilde{H}$ be the corrupted response,

$$\tilde{H} = \mathbf{w}^H \tilde{\mathbf{a}}. \tag{4.39}$$

The weighting coefficients are selected in order to minimize the mean square difference between H_0 and $\tilde{H}$, defined as

$$\text{mse} = \int \cdots \int \Omega(\zeta_0, \zeta_1, \ldots, \zeta_{2M-2}) \left| H_0 - \mathbf{w}^H \tilde{\mathbf{a}} \right|^2 d\zeta_0 d\zeta_1 \ldots d\zeta_{2M-2}, \tag{4.40a}$$

where $\Omega(\zeta_0, \zeta_1, \ldots, \zeta_{2M-2})$ is the probability density function of the random variables appearing in Equation 4.19. We can rewrite Equation 4.40a in a compact form

$$\text{mse} = \mathbf{w}^H \mathbf{Q} \mathbf{w} - (H_0 \mathbf{P}^H \mathbf{w} + H_0^H \mathbf{w}^H \mathbf{P}) + \left| H_0 \right|^2, \tag{4.40b}$$

where

$$\mathbf{P} = \int \ldots \int \Omega(\zeta_0, \zeta_1, \ldots, \zeta_{2M-2}) \tilde{\mathbf{a}} d\zeta_0 d\zeta_1 \ldots d\zeta_{2M-2},$$

$$\mathbf{Q} = \int \ldots \int \Omega(\zeta_0, \zeta_1, \ldots, \zeta_{2M-2}) \tilde{\mathbf{a}} \tilde{\mathbf{a}}^H d\zeta_0 d\zeta_1 \ldots d\zeta_{2M-2}.$$

The mean square error (mse) is minimum for $\mathbf{w}_0$, which is a solution of

$$\mathbf{Q}\mathbf{w}_0 = \mathbf{P}. \tag{4.41}$$

We shall rewrite Equation 4.40b in terms of $\mathbf{w}_0$

$$\text{mse} = (\mathbf{w}_0 - \mathbf{w})^H \mathbf{Q} (\mathbf{w}_0 - \mathbf{w}) + \left| H_0 \right|^2 - \mathbf{w}_0^H \mathbf{Q} \mathbf{w}_0. \tag{4.42}$$

An increased robustness in beamformation is sought by requiring that the output power of the beamformer be minimum [17],

$$\text{output power} = \mathbf{w}^H \mathbf{S}_f \mathbf{w} = \min, \tag{4.43a}$$

subject to a quadratic constraint on the weight vector, namely,

$$(\mathbf{w}_0 - \mathbf{w})^H \mathbf{Q} (\mathbf{w}_0 - \mathbf{w}) \le \varepsilon^2, \tag{4.43b}$$

where $\varepsilon^2 = \text{mse} - \left| H_0 \right|^2 + \mathbf{w}_0^H \mathbf{Q} \mathbf{w}_0$ is a prescribed number that represents an excess error over the minimum that can be achieved by satisfying

Equation 4.41. Note that $\mathbf{S}_f$ is a spectral matrix of the array output. Using the standard primal-dual method we can solve the constrained optimization problem [18]. The solution is given by

$$\mathbf{w}=\mathbf{w}_0-(\mathbf{S}_f+\lambda\mathbf{Q})^{-1}\mathbf{S}_f\mathbf{w}_0, \tag{4.44}$$

where λ is a Lagrange multiplier and it is given as a root of the following rational function:

$$\mathbf{w}_0^H\mathbf{S}_f(\mathbf{S}_f+\lambda\mathbf{Q})^{-1}\mathbf{Q}(\mathbf{S}_f+\lambda\mathbf{Q})^{-1}\mathbf{S}_f\mathbf{w}_0=\varepsilon^2. \tag{4.45}$$

A simple approach to solve for λ is to plot λ vs. ε^2 for the given $\mathbf{S}_f$, $\mathbf{P}$, and $\mathbf{Q}$, and pick a value of λ for a prescribed ε^2. It is demonstrated in Ref. [17] that for a circular array consisting of two concentric rings with random inter-ring spacing, the array gain remains practically unaltered if the actual spacing is well within the bounds used in the design of the weight coefficients. However, the sidelobe characteristics of the response function of the weight coefficients are not known.

4.3.4 High-Resolution Capon Spectrum

The resolution capability of Capon's frequency wavenumber spectrum may be improved by noise cancellation by subtracting an estimated white noise power from the diagonal elements of the spectral matrix [19] and stretching the eigenvalue spread of the spectral matrix [20]. A predetermined quantity is subtracted from the diagonal elements of the spectral matrix, thus increasing the ratio between the maximum and minimum eigenvalues. This process is called stretching the eigenvalue spread. The spectral matrix must, however, remain positive definite. Consider a model of P plane waves and white background noise. The spectral matrix, given in Equation 4.12b, is subjected to stretching of the eigenvalue spread by subtracting a fixed number σ_0^2 from the diagonal elements,

$$\begin{aligned}\tilde{\mathbf{S}}_f(\omega)&=\mathbf{S}_f(\omega)-\sigma_0^2\mathbf{I}\\&=\mathbf{v}\boldsymbol{\Gamma}_0\mathbf{v}^H,\end{aligned} \tag{4.46}$$

where

$$\begin{aligned}\Gamma_0&=\operatorname{diag}\{\alpha_m+\sigma_\eta^2-\sigma_0^2,\ m=1,2,\ldots,M-1\}\\&=\operatorname{diag}\{\gamma_m,\ m=1,2,\ldots,M-1\}.\end{aligned}$$

We now introduce an improved Capon spectrum by using the spectral matrix whose eigenvalue spread has been stretched in Equation 4.32 to obtain

$$\tilde{s}_{\text{Cap}}(\omega,\theta)=\frac{1}{\mathbf{a}^H\left(\omega\frac{d}{c}\sin\theta\right)\tilde{\mathbf{S}}_f^{-1}\mathbf{a}\left(\omega\frac{d}{c}\sin\theta\right)}$$
$$=\frac{1}{\mathbf{a}^H\mathbf{v}\Gamma_0^{-1}\mathbf{v}^H\mathbf{a}}=\frac{1}{\sum_{m=0}^{M-1}\frac{1}{\gamma_m}\left|\mathbf{a}^H\mathbf{v}_m\right|^2}. \tag{4.47}$$

By selecting σ_0^2 as close to σ_η^2 as possible, we can make $\gamma_m \approx 0$, $m=P, P+1,\ldots, M-1$ and $\gamma_m \approx \alpha_m$, $m=0, 1,\ldots, P-1$. Next we note that from Equation 4.14c, $\mathbf{a}^H\mathbf{v}_m=0$ for $m=P, P+1,\ldots, M-1$ and when $\mathbf{a}$ is equal to one of the direction vectors of the incident wavefronts. Conversely, when $\mathbf{a}$ does not belong to that set of direction vectors, $\mathbf{a}^H\mathbf{v}_m \neq 0$. Let us rewrite Equation 4.47 as

$$\tilde{s}_{\text{Cap}}(\omega,\theta)=\frac{1}{\sum_{m=0}^{P-1}\frac{1}{\gamma_m}\left|\mathbf{a}^H\mathbf{v}_m\right|^2+\sum_{m=P}^{M-1}\frac{1}{\gamma_m}\left|\mathbf{a}^H\mathbf{v}_m\right|^2}. \tag{4.48}$$

In the denominator of Equation 4.48, the second term dominates whenever $\mathbf{a}$ does not belong to the set of direction vectors of the incident wavefronts and vice versa. Hence,

$$\tilde{s}_{\text{Cap}}(\omega,\theta)=s_{\text{Cap}}(\omega,\theta), \quad \theta=\theta_i, \quad i=0,1,\ldots,P-1,$$

and

$$\tilde{s}_{\text{Cap}}(\omega,\theta)\approx 0, \quad \theta\neq\theta_i.$$

As an example, consider a single source, that is, $P=1$, then $|\mathbf{a}^H\mathbf{v}_0|^2=M$ and $\alpha_0=Mp_0$. On account of Equation 4.14c, $\mathbf{a}^H\mathbf{v}_m=0$ (but $\gamma_m\neq 0$) for $m=1, 2,\ldots, M-1$. Therefore, $\tilde{s}_{\text{Cap}}(\omega,\theta_0)$ is equal to p_0 (power of the incident wavefront). For all other values of θ $\tilde{s}_{\text{Cap}}(\omega,\theta)\approx 0$. To demonstrate the resolution power of the high-resolution Capon spectrum, we consider a 16-sensor ULA and two wavefronts incident at angles 30° and 35° and 0 dB SNR. The resolution of the Capon spectrum, as shown in Figure 4.12, has dramatically improved when the eigenvalue spread is increased from 30 to 1, 000. The above result is for an error-free spectral matrix, that is, with infinite data. To study the effect of finite data, a numerical experiment was carried out [20] on an eight-sensor ULA with two wavefronts incident at angles 45° and 53° in the presence of white background noise (–5 dB). The spectral matrix was computed by averaging over ten independent segments. The Capon spectrum is barely able to resolve the peaks, but the high-resolution Capon spectrum shows two clear peaks (see Figure 2 in Ref. [20]). The minimum resolvable angular separation between two uncorrelated wavefronts as a function of SNR is shown in Figure 4.13.

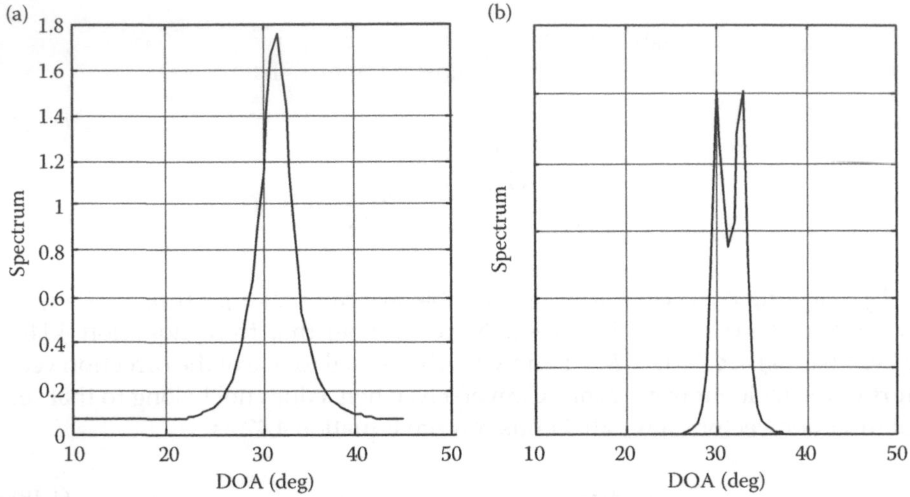

FIGURE 4.12
(a) Capon spectrum. (b) High-resolution Capon spectrum. Eigenvalue spread is 1, 000. Sixteen-sensor ULA with sensor spacing of $\lambda/2$ and two uncorrelated wavefronts incident at 30° and 33° were assumed. The high-resolution Capon spectrum yields correct amplitude and DOA information.

The spectral matrix was computed using 10 data segments. The eigenvalue spread was set at 2,000 (note that the Rayleigh resolution is equal to 16.6°). Further, a study of the bias, the mse, and the probability of resolution was also carried out in the above experiment. The results are summarized in Table 4.2. The total bias is a sum of the bias (magnitude) in both peaks. Similarly, the total standard deviation is a sum of the standard deviations of both peaks. The probability of resolution was computed by noting the number of times the peaks were clearly resolved in a hundred trials. Finally, by increasing the eigenvalue spread the spectral peaks become sharper, but soon instability sets in while inverting the modified spectral matrix.

4.4 ME ω-k Spectrum

The ME spectrum is another example of the nonlinear spectrum estimation method, originally developed for time series. The basic idea is to find a frequency wavenumber spectrum that is consistent with the observed spectral matrix of the array signal, but is maximally noncommittal on the wavefield that has not been observed simply because the array happened to be finite. The requirement of being maximally noncommittal is translated into maximization of entropy [2]. As in time series, there is an implied prediction of the wavefield outside the array aperture. Many of the properties of the ME spectrum of time series naturally hold good in the wavefield analysis. In this section, we shall elaborate on some of these aspects of the ME spectrum.

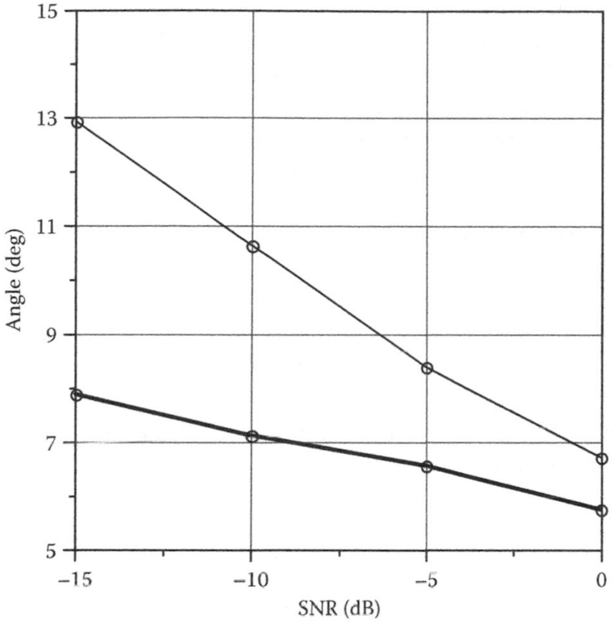

FIGURE 4.13
Resolution capability of high-resolution Capon spectrum as a function of SNR. Thick line: high-resolution Capon spectrum and thin line: Capon spectrum.

TABLE 4.2

Comparison of performance of capon spectrum and its high-resolution version

Method	Total Bias	Total Std. Dev.	Probability of Resolution
Capon spectrum	1.42°	0.86°	58.3%
High resolution Capon spectrum	0.69°	0.80°	92.2%

Notes: Parameters: eight-sensor ULA, two wavefronts with angular separation=6° incident on broadside, SNR=0 dB, number of segments used=50, number of trials=500, and eigenvalue spread=2,000.

4.4.1 ME

We shall start with an assumption that the spectral matrix of the array signal is available. An element of the spectral matrix and the frequency wavenumber spectrum are related through Equation 4.9a, reproduced here for convenience

$$S_{mn}(\omega)=\frac{1}{2\pi}\int_{-\infty}^{\infty}S_f(\omega,u)e^{-jd(m-n)u}du.$$

The entropy gain is given by

$$\Delta H=\frac{1}{4\pi}\int_{-\infty}^{\infty}\log S_f(\omega,u)du. \tag{4.49}$$

It is proposed to find $S_f(\omega, u)$ that satisfies Equation 4.9a and maximizes the entropy gain (Equation 4.49). We shall keep the temporal frequency ω fixed throughout. The above statement of estimation using the principle of ME is similar to that in the estimation of time series [2]. The approach to maximization is essentially the same as in the time series case. We shall only briefly outline the approach, leaving out all the details that may be found in Ref. [2]. Maximization of entropy is achieved when the spectrum can be expressed in a parametric form,

$$S_f(\omega,u)=\frac{1}{\sum_{p=-p_0}^{p=p_0}\lambda_p e^{jup}}, \tag{4.50}$$

where λ_p, $p=0, \pm1, \pm2, \pm p_0$ are Lagrange coefficients. Let us further express the denominator in Equation 4.50 as

$$\begin{aligned} S_f(\omega,u)&=\frac{1}{H(u)H^H(u)} \\ &=\frac{1}{\sum_{m=0}^{p_0}h_m e^{jum}\sum_{m=0}^{p_0}h_m e^{-jum}}. \end{aligned} \tag{4.51}$$

The phase of $H(u)$ is yet to be specified. We will choose $H(u)$ as a minimum phase function whose poles and zeros are to the left of an imaginary axis in the complex frequency plane. The coefficients h_m, $m=0, 1,\ldots, p_0$ may be obtained by solving

$$\mathbf{S}_f\mathbf{H}=\frac{1}{h_0}\delta, \tag{4.52a}$$

where S_f is a spectral matrix, $H=\text{col}\{h_0, h_1, h_2,\ldots, h_{p0}\}$ and $\delta=\text{col}\{1, 0, 0,\ldots, 0\}$. The solution of Equation 4.52a is given by

$$\mathbf{H}=\frac{1}{h_0}\mathbf{S}_f^{-1}\delta. \tag{4.52b}$$

Notice that $\mathbf{S}_f^{-1}\delta$ refers to the first column of $\mathbf{S}_f^{-1}$ and h_0 is the first element of that column. Using Equation 4.52b in Equation 4.51, we can express the ME spectrum as

$$S_{\text{ME}}(\omega,u)=\frac{h_0^2}{\left|\mathbf{a}^H\mathbf{S}_f^{-1}\delta\right|^2}, \tag{4.52c}$$

where $\mathbf{a}=\text{col}\{1, e^{ju}, e^{j2u},\ldots, e^{jp_0u}\}$ is the steering vector.

We shall look at an alternate approach that will lead to an equation identical to Equation 4.52b. At a fixed temporal frequency, the output of an array (ULA) may be expressed as a sum of complex sinusoids,

$$f_m(t) = \frac{1}{2\pi} \int_{-\infty}^{\infty} \sum_{p=0}^{P} dG_p(\omega) e^{j\omega(t - m(d/c)\sin\theta_p)}. \tag{4.53a}$$

Let

$$dF_m(\omega) = \sum_{p=0}^{P} dG_p(\omega) e^{-jm(\omega d/c)\sin\theta_p}, \quad m = 0, 1, \ldots, M-1, \tag{4.53b}$$

where we have assumed that P plane wavefronts are incident on a ULA. The sources are assumed to radiate stationary stochastic but uncorrelated signals. Clearly, Equation 4.53b is a sum of P random spatial complex sinusoids. A sum of P random sinusoids (real) are known as a deterministic random process and it can be predicted without error from $2P$ past samples [2]. In the case of P complex sinusoids, we will require P past samples for prediction. Error-free prediction is not possible when there is background noise. The prediction equation is given by

$$dF_m(\omega) + \sum_{p-1}^{P} h_p dF_{m-p}(\omega) = \eta_m, \tag{4.54a}$$

where η_m is the prediction error. We express Equation 4.54a in a matrix form,

$$\mathbf{H}^H d\mathrm{F}_m(\omega) = \eta_m, \tag{4.54b}$$

where $d\mathrm{F}_m(\omega) = \mathrm{col}\{dF_m, d\mathrm{F}_{m-1}, \ldots, d\mathrm{F}_{m-p0}\}$ and $\mathrm{h}_0 = 1$. The prediction error is given by

$$\mathbf{H}^H \mathbf{S}_f(\omega) \mathbf{H} = \sigma_\eta^2, \tag{4.55}$$

which we wish to minimize, subject to the constraint that $h_0 = 1$ or $\mathbf{H}^H \delta = 1$. This leads to the following equation:

$$\mathbf{S}_f(\omega)\mathbf{H} = \sigma_\eta^2 \delta,$$

or

$$\mathbf{H} = \sigma_\eta^2 \mathbf{S}_f^{-1}(\omega)\delta, \tag{4.56}$$

which is identical to Equation 4.52b, except for a scale factor. Using Equation 4.56 in Equation 4.55, the minimum prediction error is equal to

$\sigma_\eta^2\big|_{\min} = 1/\delta^H \mathbf{S}_f^{-1}(\omega)\delta$. The prediction filter vector corresponding to minimum error is given by

$$\mathbf{H} = \frac{\mathbf{S}_f^{-1}(\omega)\delta}{\delta^H \mathbf{S}_f^{-1}(\omega)\delta}. \tag{4.57}$$

The ME spectrum defined in Equation 4.51 may be expressed as follows:

$$S_{\mathrm{ME}}(\omega, u) = \frac{\left(\delta^H \mathbf{S}_f^{-1}(\omega)\delta\right)^2}{\left|\mathbf{a}^H \mathbf{S}_f^{-1}(\omega)\delta\right|^2}. \tag{4.58}$$

Note that the numerator of Equation 4.58 is equal to the first element of the first column of $\mathbf{S}_f^{-1}$. The ME spectrum given by Equation 4.52c and the spectrum obtained by minimizing the prediction error (Equation 4.58) are identical. Thus, the ME principle leads to a simple interpretation in the form of linear prediction.

4.4.2 Resolution

We shall now examine some of the properties of the ME spectrum and compare them with those of the Capon spectrum. As earlier, we shall consider two wavefronts incident on a ULA. The spectral matrix is given by Equation 4.33. Using the inversion formula (Equation 4.34) in Equation 4.52c, we obtain an expression for the ME spectrum for two wavefronts in the presence of white noise.

$$s_{\mathrm{ME}}(\omega, \theta) = \frac{h_0^2 \sigma_\eta^4}{\mathrm{den}}, \tag{4.59}$$

where

$$\mathrm{den} = \left| 1 - \frac{\frac{s_0}{\sigma_\eta^2}\mathbf{a}^H\mathbf{a}_0}{1+\frac{s_0}{\sigma_\eta^2}M} - \frac{\frac{s_1}{\sigma_\eta^2}\left[\mathbf{a}^H\mathbf{a}_1 - \frac{\frac{s_0}{\sigma_\eta^2}\mathbf{a}^H\mathbf{a}_0\mathbf{a}_0^H\mathbf{a}_1}{1+\frac{s_0}{\sigma_\eta^2}M}\right]\left[1-\frac{\frac{s_0}{\sigma_\eta^2}\mathbf{a}_1{}^H\mathbf{a}_0}{1+\frac{s_0}{\sigma_\eta^2}M}\right]}{1+\frac{s_1}{\sigma_\eta^2}\left(M - \frac{\frac{s_0}{\sigma_\eta^2}\left|\mathbf{a}_0^H\mathbf{a}_1\right|^2}{1+\frac{s_0}{\sigma_\eta^2}M}\right)} \right|^2,$$

where all symbols are as in Equation 4.35. An example of ME of two uncorrelated wavefronts incident at 30° and 35° on a 16-sensor ULA is shown in Figure 4.14a and the corresponding Capon spectrum is shown in Figure 4.14b. In both cases we have assumed white background noise and SNR=1.0.

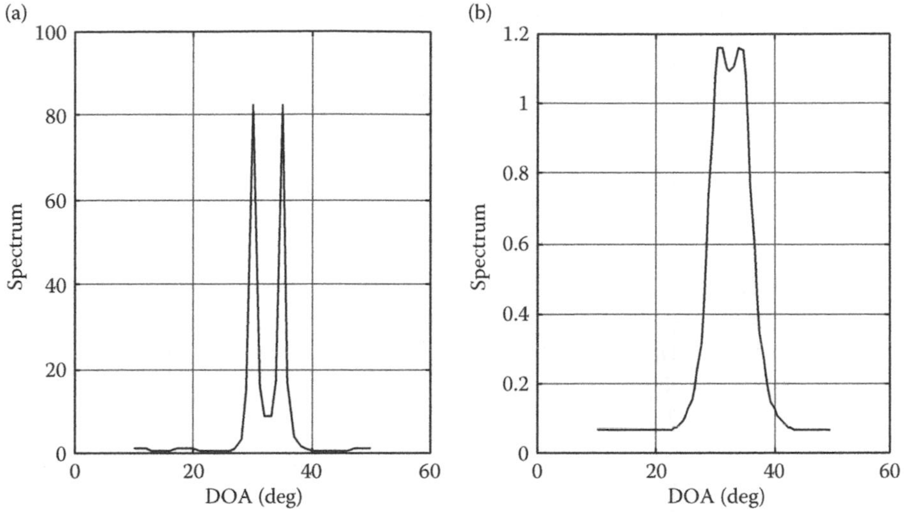

FIGURE 4.14
(a) Maximum entropy spectrum and (b) Capon spectrum. Two unit amplitude plane wavefronts are incident at angles 30° and 35° on a 16-sensor ULA. The amplitude of the peaks of Capon spectrum is close to the actual amplitude, but the amplitude of ME spectrum is much higher. According to Equation 4.53, valid for large separation, the computed amplitude is equal to 200.

Let us now evaluate Equation 4.59 at $\theta=\theta_0$ and $\theta=\theta_1$ under the assumption that $\left|\mathbf{a}_0^H\mathbf{a}_1\right|^2 \approx 0$. We obtain

$$s_{\mathrm{ME}}(\omega,\theta_0) = h_0^2(\sigma_\eta^2 + s_0 M)^2, \tag{4.60a}$$

$$s_{\mathrm{ME}}(\omega,\theta_1) = h_0^2(\sigma_\eta^2 + s_1 M)^2. \tag{4.60b}$$

The height of the spectral peak grows with the array SNR, increasing to infinity as $sM \to \infty$. This is demonstrated in Figure 4.15 for the two wavefront models, whose spectrum is plotted in Figure 4.14.

From Figure 4.14a and b, it may be conjectured that the ME spectrum has a better resolution property than that of the Capon spectrum. The depth of the valley for the ME spectrum is much larger than the one for the Capon spectrum. We have carried out a series computation to find out the minimum SNR required to resolve two equal amplitude wavefronts separated by a specified angular distance. The criterion for resolution was the formation of a nascent valley between the spectral peaks. While this is not a quantitative criterion, it serves the purpose of comparison. The results are shown in Figure 4.16, which may be compared with Figure 4.11 for the Capon spectrum. Clearly, the ME spectrum has a better resolution capability, but its peak amplitude does not bear any simple relation to the actual spectrum value. In the time series context, it was shown in Ref. [21] that the Capon spectrum and ME spectrum of different orders are related,

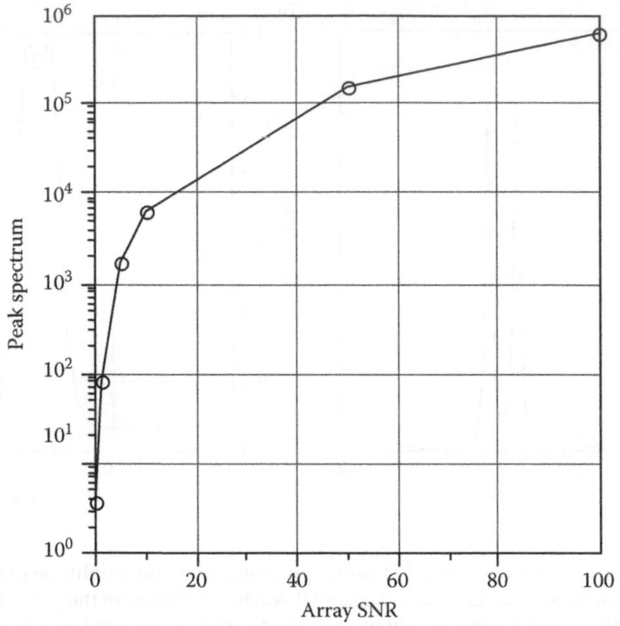

FIGURE 4.15
Peak spectrum as a function of array SNR, that is, MSNR. Two equal amplitude uncorrelated wavefronts incident at 30° and 35° on a 16-sensor ULA.

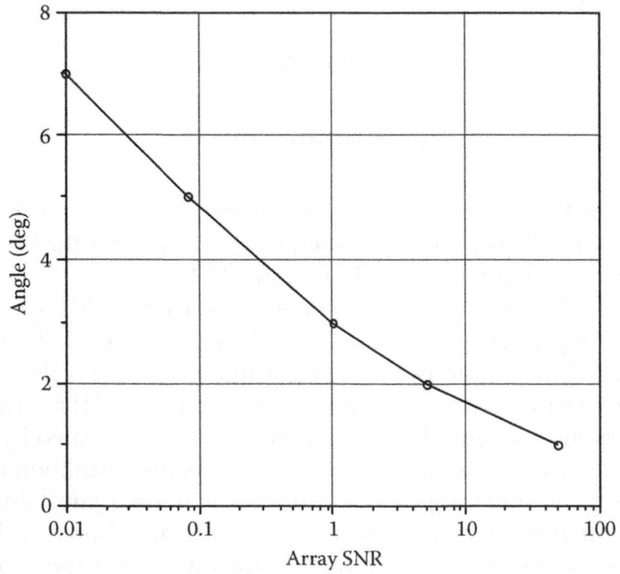

FIGURE 4.16
Resolution property of ME spectrum. Two equal amplitude uncorrelated wavefronts are incident on a 16-sensor ULA. The directions of arrival are 30° and 30° + the angular separation, as shown on the y-axis.

$$\frac{1}{S_{\text{Cap}}(\omega)} = \frac{1}{M}\sum_{m=0}^{M-1}\frac{1}{S_{\text{ME}}^{m}(\omega)}, \tag{4.61}$$

where M stands for the size of the covariance matrix. In array processing, M stands for array size.

4.4.3 Finite Data Effects

So far we have tacitly assumed that the spectral matrix is known and the incident wavefield and noise confirm with the assumed model; for example, the wavefronts are planar and uncorrelated and the noise is white. In practice, however, the spectral matrix needs to be computed from the available data. Since the spectral matrix is a statistically defined quantity involving the operation of expectation, there is bound to be some error in its estimation when only finite length data is available. The effect of the errors in the spectral matrix on wavenumber spectrum has been a subject of investigation by many researchers [22, 23]. Here, we shall briefly outline the important results. The mean and variance of the BT frequency wavenumber spectrum (linear), the Capon frequency wavenumber spectrum (nonlinear), are tabulated in Table 4.3. Unfortunately, we do not have simple expressions for mean and variance of ME spectrum estimates; only experimental results are available [22]. The ME spectrum is more variable and hence it needs much larger data to stabilize. It is reported in Ref. [22] that to get a stable ME spectrum, the number of snapshots must be far greater than M and M^2SNR. The effect of finite data will result in (a) a loss in resolution, that is, closely spaced wavefronts cannot be resolved; (b) a shift in the position of peaks, that is, an erroneous estimate of the DOA.

4.4.4 Iterative Inversion of Spectral Matrix

The most important computational step in Capon and ME methods is the computation of an inverse of the spectral matrix (covariance matrix for narrowband signal). The spectral matrix is computed as an average of the outer product of the frequency snapshots. The output of an array is divided into equal duration

TABLE 4.3

Mean and variance of frequency wavenumber spectrum estimated from finite data

Method	Mean	Variance
BT spectrum	$S(\omega, \theta)$	$\frac{S^2(\omega,\theta)}{N}$
Capon spectrum	$\frac{N-M+1}{N}S_{\text{Cap}}(\omega,\theta)$	$\frac{N-M+1}{N^2}S_{\text{Cap}}^2(\omega,\theta)$

Note: N stands for number of independent snapshots and M for array size.

and overlapping segments and each one is Fourier transformed. A frequency snapshot of an array, analogous to the time snapshot, is an output of an array at a fixed (temporal) frequency. Let $\mathbf{F}_i(\omega)$ be the ith frequency snapshot obtained by Fourier transforming the ith segment of the array output. The spectral matrix is estimated as

$$\hat{\mathbf{S}}_f(\omega) = \frac{1}{N}\sum_{1}^{N} \mathbf{F}_i(\omega)\mathbf{F}_i^H(\omega).$$

When a new time snapshot arrives, a new segment is formed with the newly received snapshot and the past snapshots to form a required length segment and then a new frequency snapshot is formed. The spectral matrix is updated by incorporating the outer product of the newly formed frequency snapshot,

$$\begin{aligned}\hat{\mathbf{S}}_f^{N+1}(\omega) &= \frac{1}{N+1}\left[\sum_{1}^{N} \mathbf{F}_i(\omega)\mathbf{F}_i^H(\omega) + \mathbf{F}_{N+1}(\omega)\mathbf{F}_{N+1}^H(\omega)\right] \\ &= \frac{N}{N+1}\hat{\mathbf{S}}_f^N(\omega) + \frac{1}{N+1}\mathbf{F}_{N+1}(\omega)\mathbf{F}_{N+1}^H(\omega).\end{aligned} \tag{4.62a}$$

The recursion may be commenced with an initial value $\hat{\mathbf{S}}_f^1(\omega) = \mathbf{F}_1(\omega)\mathbf{F}_1^H(\omega)$. We can obtain a recursive expression for the (ω, k) spectrum by using Equation 4.62a in Equation 4.16

$$S^{N+1}(\omega,\theta) = \left[\frac{N}{N+1}S^N(\omega,\theta) + \frac{1}{N+1}\left|\mathbf{a}^H\left(\omega\frac{d}{c}\sin\theta\right)\mathbf{F}_{N+1}(\omega)\right|^2\right]. \tag{4.62b}$$

Using the matrix inversion formula given in Equation 4.34, we can get a recursive relation between the inverse of spectral matrix $\hat{\mathbf{S}}_f^N(\omega)$ and $\hat{\mathbf{S}}_f^{N+1}(\omega)$

$$\left[\hat{\mathbf{S}}_f^{N+1}(\omega)\right]^{-1} = \frac{N+1}{N}\left\{\left[\hat{\mathbf{S}}_f^N(\omega)\right]^{-1} - \frac{\frac{1}{N}\left[\hat{\mathbf{S}}_f^N(\omega)\right]^{-1}\mathbf{F}_{N+1}(\omega)\mathbf{F}_{N+1}^H(\omega)\left[\hat{\mathbf{S}}_f^N(\omega)\right]^{-H}}{1+\frac{1}{N}\mathbf{F}_{N+1}^H(\omega)\left[\hat{\mathbf{S}}_f^N(\omega)\right]^{-1}\mathbf{F}_{N+1}(\omega)}\right\}. \tag{4.62c}$$

The above recursion can be commenced only after $\hat{\mathbf{S}}_f^N(\omega)$ becomes full rank. This will require a minimum of M frequency snapshots. There is yet another recursive approach to spectral matrix inversion. It is based on diagonalization of the spectral matrix through a transformation,

$$\Gamma = \mathbf{Q}\hat{\mathbf{S}}_f^N(\omega)\mathbf{Q}^H, \tag{4.63a}$$

or equivalently,

$$\hat{\mathbf{S}}_f^N(\omega) = \mathbf{Q}\Gamma\mathbf{Q}^H, \tag{4.63b}$$

where $\Gamma = \text{diag}\{\gamma_1, \gamma_1, \ldots, \gamma_M\}$ and $\mathbf{Q}$ is the upper triangular matrix of prediction coefficients that are computed through a recursive algorithm [24]. From Equation 4.63b, we can express the inverse of the spectral matrix as

$$\left[\hat{\mathbf{S}}_f^N(\omega)\right]^{-1} = \mathbf{Q}\Gamma^{-1}\mathbf{Q}^H. \tag{4.64}$$

4.5 Doppler-Azimuth Processing

In airborne radar, the signals from slow-moving ground targets and the clutter from the stationary ground scatterers are found to be distributed differently in Doppler frequency and azimuth plane. This fact has been exploited to detect weak signals from slow-moving targets on the ground in the presence of strong clutter from the ground. The tool used is space-time adaptive processing (STAP), which is extensively covered in a recent monograph by Klemm [28]. Our aim here is to dwell upon the basic algorithm used in STAP. We consider a ULA mounted on an aircraft moving at a constant speed parallel to the ground. The array is oriented in the direction of aircraft motion, hence its broadside is perpendicular to the direction of aircraft motion (see Figure 4.17). This is often known as sidelooking configuration, which perhaps is the most easy to understand, among other possibilities.

4.5.1 Sidelooking Array

Consider a ULA of M sensors, which are spaced at $\lambda_c/2$ where λ_c is carrier wavelength. Let a transmitter transmit a train of N coherent pulses of the form

$$\begin{aligned} s(t) &= A_0 E(t) e^{-j2\pi f_c (t+nT)} \\ &= A_0 E(t-nT) e^{-j2\pi f_c t}, \quad n = 0, \ldots, N-1. \end{aligned}$$

where A_0 is the amplitude, $E(t)$ is the signal envelope, f_c is the carrier frequency, and T is the pulse interval. The ground-scattered signal received at mth sensor at time nT is given by

$$s_{mn}(\phi) = A(\phi)\exp\left[\begin{array}{l} -j\dfrac{2\pi}{\lambda_c}(((x_m + 2vnT)\cos(\phi) + y_m \sin(\phi))\sin(\theta) \\ \\ -z_m \cos(\theta)) \end{array}\right]$$

$$m = 0, \cdots, M-1 \quad n = 0, \cdots, N-1 \tag{4.65}$$

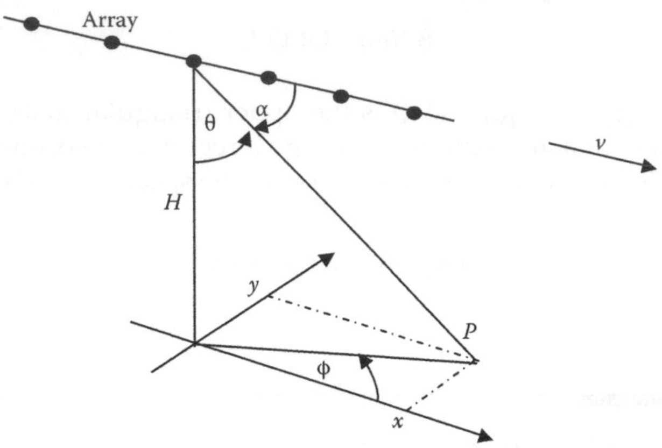

FIGURE 4.17
Sidelooking configuration of an airborne antenna array moving at a constant speed v. P is a fixed scatterer or a slow-moving ground target. ϕ and θ are the azimuth and elevation of P. The direction of arrival (DOA) is α.

where (x_m, y_m, z_m) are coordinates of mth sensor. The Doppler frequency shift is

$$\begin{aligned} f_{\text{D}} &= \frac{2v}{c} f_c \cos(\phi)\sin(\theta) \\ &= \frac{2v}{\lambda_c}\cos(\phi)\sin(\theta) = \frac{2v}{\lambda_c}\cos(\alpha). \end{aligned} \tag{4.66}$$

Note that the Doppler frequency shift is a function of ϕ and θ and lies in the range $\pm 2v/\lambda_c$. Doppler bandwidth B is defined as $4v/\lambda_c$. Significance of Equation 4.65 is that the Doppler frequency and the DOA are related. As a result of which the received scatter power will be confined to a diagonal in the Doppler-frequency and DOA plane.

A signal reflected by a moving target is given by

$$\begin{aligned} s_{mn}^{t}(\phi_t) &= A\exp\left[-j\frac{2\pi}{\lambda_c}((2v_{\text{rad}}nT + (x_n\cos(\phi_t) + y_n\sin(\phi_t))\sin(\theta) - z_n\cos(\theta))\right] \\ & m = 0,\ldots,M-1 \quad n = 0,\ldots,N-1, \end{aligned} \tag{4.67}$$

where v_{rad} is the radial target velocity relative to the array and ϕ_t is the target direction. The essential difference between Equation 4.67 and Equation 4.65 is that the Doppler frequency shift is now independent of ϕ and θ. Consequently, a slow-moving ground target is represented by a point in the Doppler frequency and DOA plane. Finally, a wideband jammer is simply represented by a line parallel to the Doppler frequency axis. A schematic of the power distribution due to all above sources is shown in Figure 4.18.

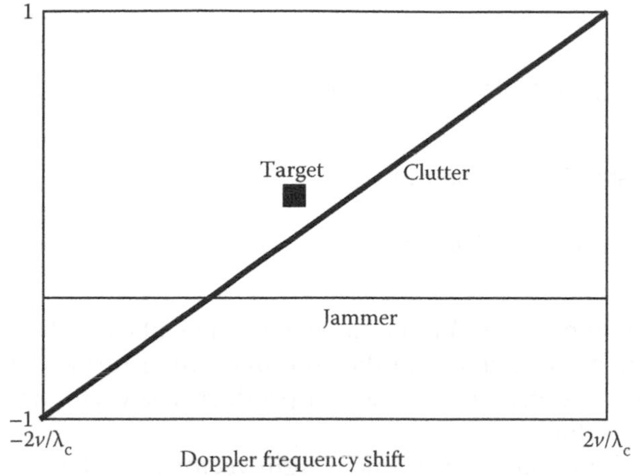

FIGURE 4.18
A schematic distribution of target, clutter plus jammer power spectrum in the Doppler-frequency and DOA plane. The y-axis represents the cosine of DOA.

Evidently, the target is fully masked by the clutter spectrum and partially by the jammer spectrum. Then temporal processing will not help to extract the target in the presence of the clutter or jammer. Nor will spatial processing (beamforming) help to extract the target in the presence of the clutter, which is present over the entire range of DOAs. The solution lies in the joint use of space and time processing. An example of space-time processing is a fan filter, which is described in Sections 3.1 and 3.2. But the distribution of signal and noise is inappropriate for the application of a fan filter. We need to devise a different type of space-time filtering for extraction of a target in the presence of ground clutter. Since the clutter spectrum is a narrow ridge, the desired filter has to have a narrow clutter notch so that even a slow target falls into the passband.

There exists extensive literature on this topic known as STAP [29–32]. In the next section, we shall briefly describe space-time processing, but without the adaptive aspects required for practical implementation [33].

4.5.2 Space-Time Filtering

Consider a data model wherein the signal is a reflection from a target and the background interference consists of the clutter, jammer interference, and measurement noise. We arrange the array output as a long vector by stacking all N snapshots into a MN column vector,

$$\mathbf{f} = A\mathbf{s} + \eta, \tag{4.68}$$

where

$$\mathbf{s} = \mathbf{a} \otimes \mathbf{b},$$

where

$$\mathbf{a}=\left[1, e^{-j(2\pi/\lambda_c)v_{\text{radial}}T}, e^{-j(2\pi/\lambda_c)v_{\text{radial}}2T},\ldots, e^{-j(2\pi/\lambda_c)v_{\text{radial}}(N-1)T}\right],$$

is a temporal steering vector and

$$\mathbf{b}=\left[1, e^{-j(2\pi/\lambda_c)d\cos\phi_t\sin\theta}, e^{-j(2\pi/\lambda_c)2d\cos\phi_t\sin\theta},\ldots, e^{-j(2\pi/\lambda_c)(M-1)d\cos\phi_t\sin\theta}\right],$$

is a spatial steering vector. A is a reflection coefficient (complex) of the target. The array is along the x-axis and the antennas are uniformly spaced d units apart. Note that $\otimes$ represents Kronecker product. The vector $\mathbf{a}\otimes\mathbf{b}$ is a space-time steering vector.

We now wish to find a space-time weight vector w so that the dot product with the data vector f is as close to the unknown amplitude A for the assumed steering vector $\mathbf{s}$,

$$\mathbf{w}^H\mathbf{f}=A\mathbf{w}^H\mathbf{s}+\mathbf{w}^H\eta.$$

For this to be possible we require that

$$\begin{aligned}|\mathbf{w}^H\mathbf{s}|^2&=1\\ \mathbf{w}^H\eta\eta^H\mathbf{w}&=\min.\end{aligned}\tag{4.69}$$

Constrained minimization in Equation 4.69 is easily solved using the method of Lagrange. We give the final result only

$$\mathbf{w}=\frac{\mathbf{C}_\eta^{-1}\mathbf{s}}{\mathbf{s}^H\mathbf{C}_\eta^{-1}\mathbf{s}},\tag{4.70a}$$

and

$$\text{Output power}=A^2+\frac{1}{\mathbf{s}^H\mathbf{C}_\eta^{-1}\mathbf{s}},\tag{4.70b}$$

where $\mathbf{C}_\eta=E\{\eta\eta^H\}$ is the covariance matrix of the background noise, which includes the clutter, jammer interference, and measurement noise. $\mathbf{C}_\eta$ is not known in general. It may, however, be estimated from the array output in the absence of any target. Given finite data, the estimate of covariance matrix will be approximate. Inversion of such a matrix will be strongly influenced by finite data errors, particularly in small eigenvalues. The usual strategy is to compute pseudoinverse using only significant eigenvalues. To understand the role of the significant eigenvalues of $\mathbf{C}_\eta$, we express Equation 4.70a using eigenvector decomposition of $\mathbf{C}_\eta$.

$$\begin{aligned}\mathbf{w} &= \gamma \sum_{k=1}^{MN} \mathbf{u}_k \lambda_k^{-1} \mathbf{u}_k^H \mathbf{s} \\ &= \frac{\gamma}{\lambda_0}\left[\mathbf{s} - \sum_{k=1}^{MN} \mathbf{u}_k (\mathbf{u}_k^H \mathbf{s}) + \sum_{k=1}^{MN} (\mathbf{u}_k^H \mathbf{s}) \frac{\lambda_0}{\lambda_k} \mathbf{u}_k\right] \\ &= \frac{\gamma}{\lambda_0}\left[\mathbf{s} - \sum_{k=1}^{MN} \left(1 - \frac{\lambda_0}{\lambda_k}\right)(\mathbf{u}_k^H \mathbf{s}) \mathbf{u}_k\right],\end{aligned} \tag{4.71}$$

where $\gamma = 1/(\mathbf{s}^H \mathbf{C}_\eta^{-1} \mathbf{s})$ and λ_0 is the smallest eigenvalue of $\mathbf{C}_\eta$. The filter output using Equation 4.71 is given by

$$\mathbf{w}^H \mathbf{f} = \frac{\gamma}{\lambda_0} \mathbf{s}^H \left[\mathbf{I} - \sum_{k=1}^{MN} \left(1 - \frac{\lambda_0}{\lambda_k}\right)(\mathbf{u}_k \mathbf{u}_k^H)\right] \mathbf{f}. \tag{4.72a}$$

We now assume that the covariance matrix $\mathbf{C}_\eta$ has a rank r and $\lambda_k \gg \lambda_0$ for $1 \le k \le r$. Then Equation 4.72a may be expressed as

$$\mathbf{w}^H \mathbf{f} \approx \frac{\gamma}{\lambda_0} \mathbf{s}^H \left[\mathbf{I} - \sum_{k=1}^{r} (\mathbf{u}_k \mathbf{u}_k^H)\right] \mathbf{f}. \tag{4.72b}$$

The representation above has an interesting interpretation. The second term inside the square brackets in Equation 4.72b is an orthogonal projection of the data vector onto the subspace spanned by the dominant eigenvectors of $\mathbf{C}_\eta$. This is the best prediction in the least squares sense of the noise component in the data vector. The predicted noise component is then subtracted from the data vector.

Note that $\mathbf{C}_\eta$ is a large matrix of size $MN \times MN$. Its inversion could be very expensive and slow. Therefore, for practical implementation, adaptive techniques have been developed leading to the well-known STAP [28].

4.6 Exercises

(1) The spatial undersampling has resulted in aliasing of the frequency wavenumber spectrum, as illustrated in Figures 4.2 and 4.3. Now consider temporal undersampling. Sketch the (ω, k) spectrum of a stochastic plane wave that has been undersampled temporally.

(2) Apply the principle of Rayleigh resolution to wideband signals. In Chapter 2, we have shown how the DOA of an incident wideband signal can be estimated from the position of the spectral peaks of the transfer functions. Show that for resolution the wavefronts must be separated by an angle greater than $\Delta\theta$, where

$$\Delta\theta = \frac{\tan\theta}{M}.$$

(3) Show that the power output of a beamformer given by Equation 4.16 is always greater than the Capon spectrum. Use Schwarze inequality (see Ref. [23, p.19]).

(4) The noise variance is often required to be known but in practice this is not likely to be known. It has to be estimated from the observed data. One possible approach is to use the partitioning of the spectral matrix as on page 197. Show that

$$\sigma_\eta^2 = \frac{tr(\mathbf{H}_2\pi)}{tr(\pi)},$$

where $\pi = \mathbf{I}_{M-P} - \mathbf{G}_2\mathbf{G}_2^\#$, and $\mathbf{G}_2^\#$ is a pseudoinverse of G_2.

(5) A plane wavefront is incident on a ULA from a known direction θ_0 in the presence of colored background noise. Consider the following noise reduction strategy. The array is steered to the known direction along with a weight vector. The steering vector is $\mathbf{w}_1 = \Gamma\mathbf{w}_0$, where $\mathbf{w}_0$ is an unknown weight vector and

$$\Gamma = \text{diag}\left\{1, e^{-j\omega(d/c)\sin\theta_0}, \ldots, e^{-j(M-1)\omega(d/c)\sin\theta_0}\right\}.$$

It is proposed to find $\mathbf{w}_0$, which will minimize the noise variance and at the same time preserve the signal power. Show that $\mathbf{w}_0$ is equal to the eigenvector corresponding to the least eigenvalue of $\Gamma^H\mathbf{S}_\eta\Gamma$, where $\mathbf{S}_\eta$ is the spectral matrix of the noise.

(6) It was shown on page 201 that the weight vector that maximizes the SNR is the eigenvector corresponding to the largest eigenvalue of a pencil matrix $[\mathbf{c}_s, \mathbf{c}_\eta]$. Let a single wavefront be incident on the array. The covariance matrix, when there is no noise, is $\mathbf{c}_s = s_0\mathbf{a}_0\mathbf{a}_0^H$. Show that $\mathbf{w} = \alpha\mathbf{c}_\eta^{-1}\mathbf{a}_0$, where α is a normalizing constant and the maximum SNR is equal to $s_0\mathbf{a}_0^H\mathbf{c}_\eta^{-1}\mathbf{a}_0$.

(7) Let $\mathbf{B}$ be a $M\times Q$ ($Q<M$) matrix satisfying a property $\mathbf{B}^H\mathbf{B}=\mathbf{I}$. We define a reduced array output $\mathbf{G}_i(\omega)=\mathbf{B}^H\mathbf{F}_i(\omega)$, where $\mathbf{F}_i(\omega)$ is a frequency snapshot (see Equation 4.62a). Show that the eigenstructure

of the spectral matrix of the reduced output is identical to that of the spectral matrix of the normal array output, in particular $\mathbf{B}^H\mathbf{a} \perp \mathbf{v}_\eta$. This property forms the basis for the beamspace subspace method, where **B** acts as a spatial filter to restrict the incident energy to a preselected angular sector [26, 27].

(8) Let the columns of **B** be the eigenvectors corresponding to the significant eigenvalues of **Q** in Equation 1.70. Assume that noise sources are distributed over an angular sector $\pm\Delta$ ($\theta_0=0$). Show that the reduced noise (as defined in Exercise 7) becomes white.

References

1. S. Haykin (Ed.), *Nonlinear Methods of Spectrum Analysis*, Springer-Verlag, Berlin, 1979.
2. P. S. Naidu, *Modern Spectrum Analysis of Time Series*, CRC Press, Boca Raton, FL, 1996.
3. G. H. Golub and C. F. Van Loan, *Matrix Computations*, The Johns Hopkins University Press, Baltimore, MD, 1983.
4. S. Marcos, A. Marsal, and M. Benidie, The propagation method for source bearing estimation, *Signal Process.*, vol. 42, pp. 121–138, 1995.
5. J. Munier and G. Y. Deliste, Spatial analysis using new properties of the cross spectral matrix, *IEEE Trans. Signal Process.*, vol. 37, pp. 746–749, 1991.
6. H. P. Bucker, Use of calculated sound fields and matched field detection to locate sound sources in shallow water, *J. Acoust. Soc. Am.*, vol. 59, pp. 368–373, 1976.
7. A. Tolstoy, *Matched Field Processing*, World Scientific, Singapore, 1995.
8. I. D. Rathjen, G. Boedecker, and M. Siegel, Omnidirectional beam forming for linear antennas by means of interpolated signals, *IEEE J. Ocean. Eng.*, vol. OE-10, pp. 360–368, 1985.
9. F. J. Harris, On the use of windows for harmonic analysis with the discrete Fourier transform, *Proc. IEEE*, vol. 66, pp. 51–83, 1978.
10. R. Monzingo and T. Miller, *Introduction to Adaptive Arrays*, Wiley and Sons, New York, 1980.
11. A. Eberhard, An optimal discrete window for calculation of power spectra, *IEEE Trans.*, AU-21, pp. 37–43, 1973.
12. K. Yao, Maximum energy windows with constrained spectral values, *Signal Process.*, vol. 11 pp. 157–168, 1986.
13. D. Korompis, K. Yao, and F. Lorenzelli, Broadband maximum energy array with user imposed spatial and frequency constraints, *IEEE ICASSP*, pp. IV-529–532, 1994.
14. J. Capon, High resolution frequency-wavenumber spectrum analysis, *Proc. IEEE*, vol. 57, pp. 1408–1418, 1969.
15. B. D. Van Veen and K. M. Buckley, Beamforming: A versatile approach to spatial filtering, *IEEE ASSP Mag.*, April, pp. 4–24, 1988.
16. H. Cox, Resolving power and sensitivity to mismatch of optimum array processors, *J. Acoust. Soc. Am.*, vol. 54, pp. 771–785, 1973.

17. M. H. Er and A. Cantoni, A unified approach to the design of robust narrowband antenna processor, *IEEE Trans. Antenn. Propag.*, vol. 38, pp. 17–23, 1990.
18. D. G. Luenberger, *Optimization by Vector Space Methods*, Wiley, New York, 1969.
19. J. Munier and G. R. Deliste, Spatial analysis in passive listening using adaptive techniques, *Proc. IEEE*, vol. 75, pp. 1458–1471, 1987.
20. P. S. Naidu and V. V. Krishna, Improved maximum likelihood (IML) spectrum estimation: Performance analysis, *J. IETE*, vol. 34, pp. 383–390, 1988.
21. J. P. Burg, The relationship between maximum entropy spectra and maximum likelihood spectra, *Geophysics*, vol. 37, pp. 375–376, 1972.
22. S. R. De Graaf and D. H. Johnson, Capability of array processing algorithms to estimate source bearings, *IEEE Trans.*, ASSP-33, pp. 1368–1379, 1985.
23. S. U. Pillai, *Array Signal Processing*, Springer-Verlag, New York, 1989.
24. T. S. Durrani and K. C. Sharman, Eigenfilter approaches to adaptive array processing, *IEE Proc.*, vol. 130, Pts. F & H, pp. 22–28, 1983.
25. S. A. Nakhamkin, L. G. Tyurikov, and A. V. Malik, Construction models and decomposition algorithm of seismic fields from the characteristic numbers and vectors of spectral matrices, Izvestia, *Earth Phys.*, No. 12, pp. 785–791, 1975.
26. G. Bienvenu and L. Kopp, Decreasing high resolution method sensitivity by conventional beamformer preprocessing, *IEEE ICASSP*-84, pp. 33.2.1–33.2.4, 1984.
27. P. Forster and G. Vezzosi, Application of spheroidal sequences to array processing, *IEEE ICASSP*-87, pp. 2268–2271, 1987.
28. R. Klemm, *Application of Space Time Adaptive Processing*, IEE, London, 2002.
29. R. Klemm, Introduction to space-time adaptive processing, *IEE, Electronics and Communication Journal*, pp. 5–12, 1999.
30. J. Ward, Space-time adaptive processing with sparse antenna arrays, *IEEE Conf. Proc.*, pp. 1537–1541, 1998.
31. J. Ward, Space-time adaptive processing for airborne radar, *IEEE Conf. Proc.*, pp. 2809–2812, 1995.
32. W. Melvin, A STAP overview, *IEEE A&E Systems Mag.*, vol. 19, pp. 19–35, 2004
33. T. K. Sarkar, Space-time adaptive processing using circular arrays, *IEEE Antenn. Propag. Mag.*, vol. 43, pp. 139–143, 2001.

5

Source Localization: Subspace Methods

The location parameters are estimated directly without having to search for peaks as in the frequency-wavenumber spectrum (the approach described in Chapter 4). In open space, the direction of arrival (DOA), that is, the azimuth or elevation or both, is estimated using the subspace properties of the spatial covariance matrix or spectral matrix. Multiple signal classification (MUSIC) is a well-known algorithm where we define a positive quantity that becomes infinity whenever the assumed parameter(s) is equal to the true parameter. We shall call this quantity a spectrum even though it does not possess the units of power as in the true spectrum. The MUSIC algorithm in its original form does involve scanning and searching, often very fine scanning lest we may miss the peak. Later extensions of the MUSIC, like root MUSIC, ESPRIT, etc. have overcome this limitation of the original MUSIC algorithm. When a source is located in a bounded space, such as a duct, the wavefront reaching an array of sensors is necessarily nonplanar due to multipath propagation in the bounded space. In this case, all three position parameters can be measured by means of an array of sensors. But the complexity of the problem of localization is such that a good prior knowledge of the channel becomes mandatory for successful localization. In active systems, since one has control over the source, it is possible to design waveforms that possess the property that is best suited for localization; for example, a binary phase shift key (BPSK) signal with its narrow autocorrelation function is best suited for time delay estimation. Source tracking of a moving source is another important extension of the source localization problem. With a linear array, one can only estimate the azimuth but not the elevation, which is required for localization in 3D space. In the last section, we take the problem of elevation estimation using a circular array and an L-shaped array.

5.1 Subspace Methods (Narrowband)

Earlier (Chapter 4, Section 4.1) we showed an interesting property of eigenvalues and eigenvectors of a spectral matrix of a wavefield that consists of uncorrelated plane waves in the presence of white noise. Specifically, Equation 4.14a and b form the basis for the signal subspace method for DOA estimation. For convenience, the two equations are reproduced here

$$\lambda_m = \alpha_m + \sigma_\eta^2$$

$$\mathbf{v}_m^H \mathbf{A} = \mathbf{0}, \quad \text{for} \quad m = P, P+1, \ldots, M-1.$$

Equation 4.14b implies that the space spanned by the columns of **A**, that is, the direction vectors of incident wavefronts, is orthogonal to the space spanned by the eigenvectors, $\mathbf{v}_m$, $m=P, P+1,\ldots,M-1$, often known as noise subspace, $\mathbf{v}_\eta$. The space spanned by the columns of **A** is known as signal subspace, $\mathbf{v}_s$. Consider the space spanned by a steering vector, $\mathbf{a}(\omega,\theta)=\text{col}\{1, e^{-j\omega(d/c)\sin\theta}, e^{-j2\omega(d/c)\sin\theta}, \ldots, e^{-j(M-1)\omega(d/c)\sin\theta}\}$, as the steering angle is varied over a range $\pm(\pi/2)$. The intersection of the array manifold with the signal subspace yields the direction vectors of the signals.

5.1.1 MUSIC

On account of Equation 4.14b, a steering vector pointing in the direction of one of the incident wavefronts will be orthogonal to the noise subspace,

$$\mathbf{v}_m^H \mathbf{a}(\omega,\theta) = 0, \quad m = P, P+1, \ldots, M-1, \tag{5.1}$$

where $\theta=\theta_0, \theta_1, \ldots, \theta_{P-1}$. For narrowband signals the matrix **A** is given by

$$\mathbf{A} = \left[\mathbf{a}\left(\omega_0 \frac{d}{c}\sin\theta_0\right), \mathbf{a}\left(\omega_1 \frac{d}{c}\sin\theta_1\right), \ldots, \mathbf{a}\left(\omega_{P-1}\frac{d}{c}\sin\theta_{P-1}\right)\right], \tag{5.2}$$

where we take a wavefront with a center frequency ω_p to be incident at an angle θ_p. Further, we assume that the center frequencies of the narrowband signals are known. We define a quadratic function involving a steering vector and noise subspace,

$$S_{\text{Music}}(\omega,\theta) = \frac{1}{\mathbf{a}^H(\omega,\theta)\mathbf{v}_\eta \mathbf{v}_\eta^H \mathbf{a}(\omega,\theta)}. \tag{5.3}$$

$S_{\text{Music}}(\omega,\theta)$, also known as the eigenvector spectrum, will show sharp peaks whenever $\theta=\theta_0, \theta_1, \ldots, \theta_{P-1}$. The subscript Music stands for Multiple Signal Classification. This acronym was coined by Schmidt [1] who discovered the subspace algorithm. At about the same time, but independently, Bienvenu and Kopp [2] proposed a similar algorithm. Pisarenko [3] had previously published a subspace-based algorithm in the context of harmonic analysis of time series. Note that, although we refer to $S_{\text{Music}}(\omega,\theta)$ as spectrum, it does not have the units of power; hence, it is not a true spectrum. Let us express the steering vector in terms of the product of frequency and time delay $\tau=(d/c)\sin\theta$,

$$\mathbf{a}(\omega\tau)=\text{col}\{1, e^{j\omega\tau}, e^{j2\omega\tau}, \ldots, e^{j(M-1)\omega\tau}\}.$$

The peaks of $S_{\text{Music}}(\omega,\theta)$ will now be at $\omega\tau=\omega_0\tau_0, \omega_1\tau_1, \ldots, \omega_{P-1}\tau_{P-1}$ and given the frequencies we can estimate the delays, but there is one difficulty. It is possible that two wavefronts with different center frequencies may arrive at such angles that $\omega_p(d/c)\sin\theta_p=\omega_{p'}(d/c)\sin\theta_{p'}$, in which case the two direction vectors will become identical, causing a loss of the full column rank property of the matrix **A**. When the frequencies are unknown, evidently, the delays or DOAs cannot be uniquely estimated.

We now turn to the signal subspace spanned by the eigenvectors corresponding to eigenvalues $\lambda_m=\alpha_m+\sigma_\eta^2,\ m=0,1,\ldots,P-1$. We shall show that the signal subspace is the same as the space spanned by the columns of the matrix **A**. Since **A** is a full rank matrix its polar decomposition gives

$$\mathbf{A}=\mathbf{T}\mathbf{G}, \tag{5.4}$$

where **G** is a full rank $P\times P$ matrix and **T** is a $M\times P$ matrix satisfying the following property,

$$\mathbf{T}^H\mathbf{T}=\mathbf{I}\ (P\times P \text{ unit matrix}).$$

In Equation 4.14b we shall replace **A** by its polar decomposition (Equation 5.4)

$$\mathbf{v}_m^H\mathbf{A}\mathbf{C}_0\mathbf{A}^H\mathbf{v}_m=\alpha_m, \tag{5.5a}$$

or in matrix form

$$\mathbf{v}_s^H\mathbf{T}[\mathbf{G}\mathbf{C}_0\mathbf{G}^H]\mathbf{T}^H\mathbf{v}_s=\text{diag}\{\alpha_m,\ m=0,1,\ldots,P-1\}. \tag{5.5b}$$

Let $\mathbf{T}=\mathbf{v}_s$, which is consistent with the assumed properties of **T**. Equation 5.5b now reduces to

$$\mathbf{G}\mathbf{C}_0\mathbf{G}^H=\text{diag}\{\alpha_m,\ m=0,1,\ldots,P-1\}, \tag{5.6a}$$

and

$$\mathbf{A}=\mathbf{v}_s\mathbf{G}. \tag{5.6b}$$

We can estimate **G** from Equation 5.6b by premultiplying both sides by $\mathbf{v}_s^H$ and obtain

$$\mathbf{G}=\mathbf{v}_s^H\mathbf{A}. \tag{5.6c}$$

Thus, the space spanned by **A** and $\mathbf{v}_s^H$ is identical. Further, by eliminating **G** from Equation 5.6b and c we obtain an interesting result

$$\mathbf{A}=\mathbf{v}_s\mathbf{v}_s^H\mathbf{A}. \tag{5.6d}$$

Let us define a complementary orthogonal space $(\mathbf{I}-\mathbf{v}_s\mathbf{v}_s^H)$, which will also be orthogonal to $\mathbf{A}$; therefore, $\mathbf{A}^H(\mathbf{I}-\mathbf{v}_s\mathbf{v}_s^H)\mathbf{A}=0$. An equivalent definition of the Music spectrum may be given using $(\mathbf{I}-\mathbf{v}_s\mathbf{v}_s^H)$,

$$S_{\text{Music}}(\omega,\theta)=\frac{1}{\mathbf{a}^H(\omega,\theta)(\mathbf{I}-\mathbf{v}_s\mathbf{v}_s^H)\mathbf{a}(\omega,\theta)}. \tag{5.7}$$

5.1.1.1 Signal Eigenvalues and Source Power

The signal eigenvalues and source power are related, although the relationship is rather involved except for the single source case. Let us first consider a single source case. The covariance matrix is given by $\mathbf{C}_f=\mathbf{a}_0 s_0\mathbf{a}_0^H$ which we use in Equation 5.5a and obtain

$$s_0=\frac{\alpha_0}{|\mathbf{v}_0^H\mathbf{a}_0|^2}. \tag{5.8}$$

Since $\mathbf{v}_0$ and $\mathbf{a}_0$ span the same space, $\mathbf{v}_0=g_0\mathbf{a}_0$, where g_0 is a constant that may be obtained by requiring that $\mathbf{v}_0^H\mathbf{v}_0=1$. We obtain $g_0=1/\sqrt{M}$. Equation 5.8 reduces to $s_0=\alpha_0/M$. Next, we consider the two source case. The covariance function is given by

$$\mathbf{C}_f=[\mathbf{a}_0,\mathbf{a}_1]\begin{bmatrix} s_0 & 0 \\ 0 & s_1 \end{bmatrix}[\mathbf{a}_0,\mathbf{a}_1]^H,$$

and using Equation 5.5a we obtain

$$\mathbf{v}_s^H\mathbf{A}\mathbf{C}_0\mathbf{A}^H\mathbf{v}_s=\text{diag}\{\alpha_0,\alpha_1\}, \tag{5.9a}$$

$$\mathbf{C}_0=[\mathbf{v}_s^H\mathbf{A}]^{-1}\text{diag}\{\alpha_0,\alpha_1\}[\mathbf{A}^H\mathbf{v}_s]^{-1}. \tag{5.9b}$$

Equation 5.9b has been used to estimate the powers of two equal power (signal-to-noise ratio; SNR=1.0) wavefronts incident on a 16-sensor uniform linear array (ULA) and the results are tabulated in Table 5.1. The estimation is error-free right down to a half-a-degree separation; however, this good performance deteriorates in the presence of model and estimation errors.

5.1.1.2 Aliasing

In Chapter 4, Section 4.1 (p.211), we pointed out that the sensor spacing in a ULA must be less than half the wavelength so that there is no aliasing in the frequency-wavenumber spectrum. The same requirement exists in all high-resolution methods, namely the Capon spectrum, maximum entropy

TABLE 5.1

Estimation of power given the azimuths of two equal power wavefronts incident on a 16-sensor ULA

Azimuths	Power #1 Source	Power #2 Source
30°, 40°	1.00	1.00
30°, 35°	1.00	1.00
30°, 33°	1.00	1.00
30°, 31°	1.00	1.00
30°, 30.5°	1.00	1.00

Note: Power estimates are error-free even when the wavefronts are closely spaced.

spectrum, and Music spectrum. The basic reason for aliasing lies in the fact that the direction vector is periodic. Consider any one column of the matrix **A**,

$$\mathbf{a}_p(\omega\tau) = \text{col}\left\{1, e^{-j\omega_p(d/c)\sin\theta_p}, e^{-j2\omega_p(d/c)\sin\theta_p}, \ldots, e^{-j(M-1)\omega_p(d/c)\sin\theta_p}\right\}.$$

Now let $d=(\lambda_p/\sin\theta_p)-\delta_p$ where $0\le\delta_p\le(\lambda_p/2\sin\theta_p)$, then $\mathbf{a}_p(\omega\tau)$ becomes

$$\begin{aligned}\mathbf{a}_p\left(\omega_p\frac{d}{c}\sin\theta_p\right) &= \text{col}\left\{1, e^{-j2\pi(d/\lambda_p)\sin\theta_p}, e^{-j4\pi(d/\lambda_p)\sin_p}, \ldots, e^{-j(M-1)2\pi(d/\lambda_p)\sin\theta_p}\right\}\\ &= \text{col}\left\{1, e^{j2\pi(\delta_p/\lambda_p)\sin\theta_p}, e^{j4\pi(\delta_p/\lambda_p)\sin\theta_p}, \ldots, e^{j(M-1)2\pi(\delta_p/\lambda_p)\sin\theta_p}\right\}\\ &= \mathbf{a}_p\left(\omega_p\frac{-\delta_p}{c}\sin\theta_p\right).\end{aligned}$$

We can find an angle $\hat{\theta}_p$ such that $(d/\lambda_p)\sin\hat{\theta}_p = -(\delta_p/\lambda_p)\sin\theta_p$ and hence

$$\mathbf{a}_p\left(\omega_p\frac{d}{c}\sin\theta_p\right) = \mathbf{a}_p\left(\omega_p\frac{d}{c}\sin\hat{\theta}_p\right).$$

An aliased spectral peak would appear at $\hat{\theta}_p$, which is related to θ_p,

$$\hat{\theta}_p = \sin^{-1}\left(\frac{-\delta_p}{d}\sin\theta_p\right) \tag{5.10}$$

As an example, consider an array with sensor spacing $(d/\lambda_p)=1$ and a wavefront incident at an angle $\theta_p=60°$. For this choice of array and wave parameters $\delta_p=0.1547$, and from Equation 5.10 we get the angle where the aliased peak will be located, $\hat{\theta}_p = -7.6993°$. The wavenumber spectrum computed by all four methods is shown in Figure 5.1.

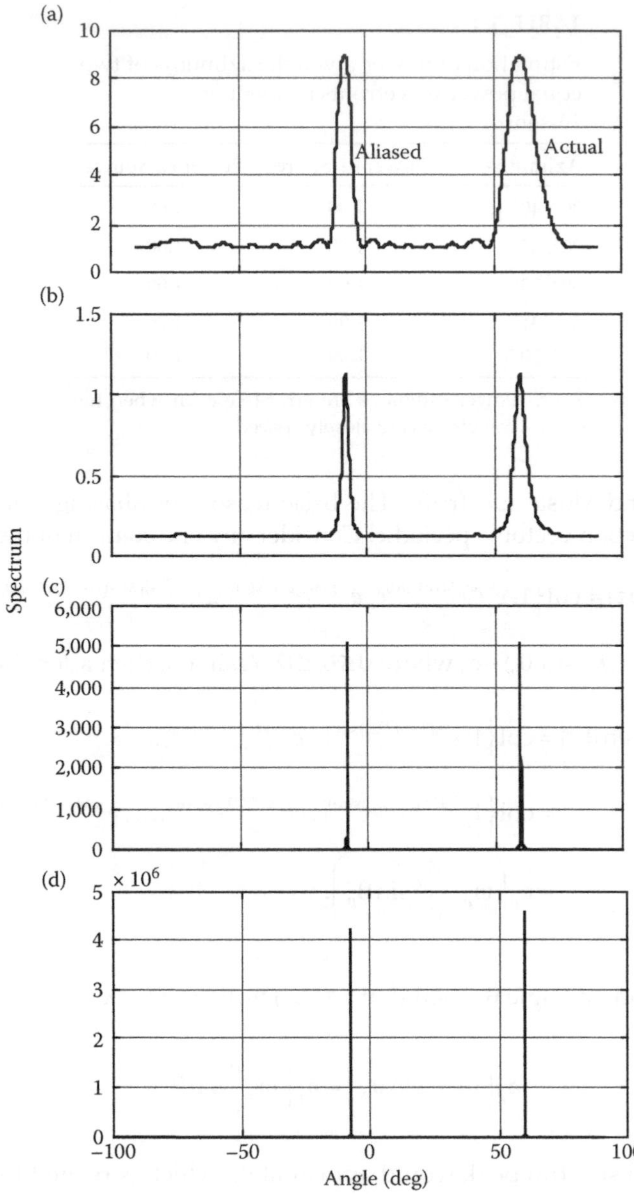

FIGURE 5.1
The aliasing effect due to under sampling of the wavefield ($d = \lambda$). All four methods of spectrum estimation have been used. (a) Bartlett spectrum, (b) Capon spectrum, (c) Maximum entropy spectrum and (d) Music spectrum. While the actual peak is at 60° the aliased peak appears at −7.69°.

Aliasing is on account of periodicity of a direction vector, which in turn is caused by periodicity present in a ULA. Thus, to avoid aliasing, it would be necessary to break this periodicity; for example, we may space the sensors nonuniformly. In a circular array, though sensors are uniformly spaced (e.g., uniform circular array (UCA), the time delays are nonuniform; therefore a UCA will yield an alias-free spectrum [4]. This is demonstrated in Figure 5.2, where we consider a wavefront that is incident at 60° (with respect to the *x*-axis) on a circular array consisting of 16 sensors uniformly spread over a circle of radius 8λ. The Capon spectrum is shown for this case. The aliasing phenomenon is not encountered in random arrays where the sensors are spaced at random intervals. But as shown in Chapter 2, the random array possesses a highly nonlinear phase response.

5.1.2 Correlated Sources

So far, we have assumed that the source matrix $\mathbf{S}_0$ is a full rank matrix. When sources are fully uncorrelated, $\mathbf{S}_0$ is a diagonal matrix with the nonzero elements representing the power of the sources. The source matrix is naturally full rank. Let us now consider a situation where the sources are partially or fully correlated. We model the source matrix as

$$\mathbf{S}_0 = \mathbf{s}\rho\mathbf{s}^H, \tag{5.11}$$

where

$$\mathbf{s} = \mathrm{diag}\left\{\sqrt{s_0}, \sqrt{s_1}, \ldots, \sqrt{s_{P-1}}\right\},$$

$s_0, s_1, \ldots, s_{P-1}$ represent the power of P sources and ρ is the coherence matrix whose (m,n)th element represents the normalized coherence between the mth and nth sources. The signal eigenvalues of the spectral matrix for $P=2$ are given by [5]

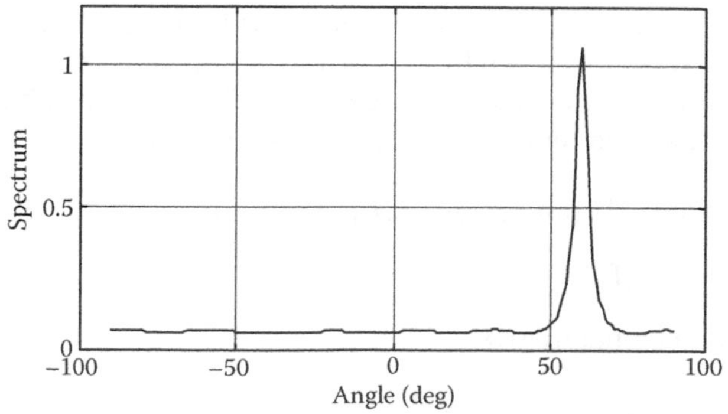

FIGURE 5.2
No aliasing effect is seen with a circular array. A 16-sensor UCA with radius=8λ (sensor spacing=3.12λ) is used. A plane wavefront is incident on the array at 60°.

$$\lambda_0=\frac{M}{2}(s_0+s_1)+M\sqrt{s_0s_1}\,\text{Re}\{\rho_{12}\}\psi$$
$$+\frac{M}{2}\left\{\left[(s_0+s_1)+2\sqrt{s_0s_1}\,\text{Re}\{\rho_{12}\}\psi\right]^2-4s_0s_1(1-|\psi|^2)(1-|\rho_{12}|^2)\right\}^{1/2}+\sigma_\eta^2$$
$$\lambda_1=\frac{M}{2}(s_0+s_1)+M\sqrt{s_0s_1}\,\text{Re}\{\rho_{12}\}\psi$$
$$-\frac{M}{2}\left\{\left[(s_0+s_1)+2\sqrt{s_0s_1}\,\text{Re}\{\rho_{12}\}\psi\right]^2-4s_0s_1(1-|\psi|^2)(1-|\rho_{12}|^2)\right\}^{1/2}+\sigma_\eta^2,$$

(5.12a)

where

$$\psi(M)=\frac{\sin\left(\pi\frac{d}{\lambda}M(\sin\theta_0-\sin\theta_1)\right)}{M\sin\left(\pi\frac{d}{\lambda}(\sin\theta_0-\sin\theta_1)\right)}.$$

The sum of the signal eigenvalues, that is,

$$(\lambda_0-\sigma_\eta^2)+(\lambda_1-\sigma_\eta^2)=M(s_0+s_1)+2M\sqrt{s_0s_1}\,\text{Re}\{\rho_{12}\}\psi(M), \qquad (5.12b)$$

represents coherent addition of power. For uncorrelated sources $\rho_{12}=0$, Equation 5.12a reduces to

$$\lambda_0=\frac{M}{2}(s_0+s_1)+\frac{M}{2}\left\{(s_0+s_1)^2-4s_0s_1(1-|\psi(M)|^2)\right\}^{1/2}+\sigma_\eta^2$$
$$\lambda_1=\frac{M}{2}(s_0+s_1)-\frac{M}{2}\left\{(s_0+s_1)^2-4s_0s_1(1-|\psi(M)|^2)\right\}^{1/2}+\sigma_\eta^2. \qquad (5.13)$$

Also, note that when the sources are in the same direction $\lambda_0=M(s_0+s_1)$ and $\lambda_1=0$.

The source spectral matrix may be modified by spatial smoothing of the array outputs. This is achieved by averaging the spectral matrices of subarray outputs over all possible subarrays (see Figure 5.3). The ith subarray (size μ) signal vector at a fixed temporal frequency is given by

$$\mathbf{F}_i=\text{col}\{F_i(\omega),F_{i+1}(\omega),\ldots,F_{i+\mu-1}(\omega)\}\quad 0\le i\le M-\mu+1$$
$$=\mathbf{I}_{i,\mu}\mathbf{F},$$

where $\mathbf{I}_{i,\mu}$ is a diagonal matrix,

$$\mathbf{I}_{i,\mu}=\text{diag}\left\{\underset{0\text{ to }i-1}{0,\ldots0,}\quad\underset{i\text{ to }i+\mu-1,}{1,\ldots1,}\quad\underset{+\mu\text{ to }M}{0,\ldots0,}\right\}$$
$$\mathbf{F}=\text{col}\{F_0(\omega),\ldots,F_{M-1}(\omega)\}.$$

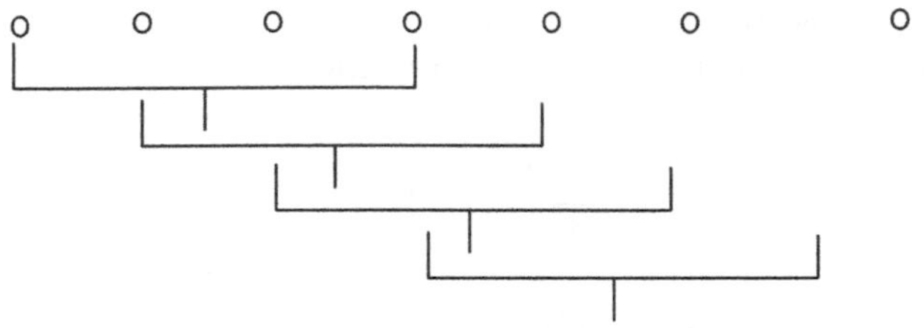

FIGURE 5.3
Overlapping subarrays are formed as shown. Each subarray has four sensors. It shares three sensors with its immediate neighbors.

The spectral matrix of the ith subarray is now given by $\mathbf{S}_{i,\mu} = E\{\mathbf{F}_i\mathbf{F}_i^H\} = \mathbf{I}_{i,\mu}\mathbf{S}\,\mathbf{I}_{i,\mu}^H$. We shall now average all subarray spectral matrices

$$\begin{aligned}\bar{\mathbf{S}} &= \frac{1}{M-\mu+1}\sum_{i=0}^{M-\mu+1}\mathbf{S}_{i,\mu} = \frac{1}{M-\mu+1}\sum_{i=0}^{M-\mu+1}\mathbf{I}_{i,\mu}\mathbf{S}\,\mathbf{I}_{i,\mu}^H \\ &= \frac{1}{M-\mu+1}\sum_{i=0}^{M-\mu+1}\mathbf{I}_{i,\mu}\mathbf{A}(\omega)\mathbf{S}_0(\omega)\mathbf{A}^H\,\mathbf{I}_{i,\mu}^H + \sigma_\eta^2\mathbf{I},\end{aligned} \tag{5.14a}$$

where we have used the spectral matrix of the signal model plane waves in the presence of white noise (Equation 4.12b). We can show that

$$\begin{aligned}\mathbf{I}_{i,\mu}\mathbf{A}(\omega) &= \mathbf{I}_{i,\mu}\left[\mathbf{a}_0, \mathbf{a}_1, \ldots, \mathbf{a}_{P-1}\right] \\ &= \left[\hat{\mathbf{a}}_0, \hat{\mathbf{a}}_1, \ldots, \hat{\mathbf{a}}_{P-1}\right]\mathrm{diag}\left\{e^{-j2\pi i(d/c)\sin\theta_0}, \ldots, e^{-j2\pi i(d/c)\sin\theta_{P-1}}\right\},\end{aligned} \tag{5.14b}$$

where

$$\hat{\mathbf{a}}_i(\omega\tau) = \mathrm{col}\{1, e^{-j\omega(d/c)\sin\theta_i}, e^{-j2\omega(d/c)\sin\theta_i}, \ldots, e^{-j(\mu-1)\omega(d/c)\sin\theta_i}\}.$$

We shall use Equation 5.14b in Equation 5.14a and obtain

$$\bar{\mathbf{S}} = \hat{\mathbf{A}}\left[\frac{1}{M-\mu+1}\sum_{i=0}^{M-\mu+1}\phi_i\mathbf{S}_0(\omega)\phi_i^H\right]\hat{\mathbf{A}}^H + \sigma_\eta^2\mathbf{I}, \tag{5.15a}$$

where

$$\hat{\mathbf{A}} = \left[\hat{\mathbf{a}}_0, \hat{\mathbf{a}}_1, \ldots, \hat{\mathbf{a}}_{P-1}\right],$$

$$\phi_i = \mathrm{diag}\left\{e^{-j2\pi i(d/c)\sin\theta_0}, e^{-j2\pi i(d/c)\sin\theta_1}, \ldots, e^{-j2\pi i(d/c)\sin\theta_{P-1}}\right\}.$$

The quantity inside the square brackets in Equation 5.15a may be computed by actual multiplication followed by summation,

$$\begin{aligned}[\ ]_{mn} &= \sqrt{s_m s_n}\rho_{mn}\frac{1}{M-\mu+1}\sum_{i=0}^{M-\mu+1} e^{-j2\pi(d/c)i(\sin\theta_m-\sin\theta_n)} \\ &= \sqrt{s_m s_n}\rho_{mn}\psi(M-\mu+1)e^{-j\pi(d/c)(M-\mu)(\sin\theta_m-\sin\theta_n)}.\end{aligned} \tag{5.15b}$$

From Equation 5.15b, it follows that the coherence after spatial smoothing may be written as

$$\bar{\rho}_{mn} = \rho_{mn}\psi(M-\mu+1)e^{-j\pi(d/c)(M-\mu)(\sin\theta_m-\sin\theta_n)}. \tag{5.16}$$

The magnitude of the off-diagonal terms in the coherence matrix ρ (see Equation 5.11), is reduced by a factor $\psi(M-\mu+1)$, which is indeed small ($\ll 1$) for large $(M-\mu+1)$. This contributes to the increase of the rank of the smoothed coherence matrix. It is shown in Ref. [6] that ρ becomes full rank when $M-\mu+1 \geq P$.

Let us now examine the effect of the spatial smoothing on the eigenvalues for the two source case. From Equation 5.12a

$$\begin{aligned}\bar{\lambda}_0 &= \frac{\mu}{2}(s_0+s_1)+\mu\sqrt{s_0 s_1}\,\mathrm{Re}\{\tilde{\rho}_{01}\}\psi(\mu)\psi(M-\mu+1) \\ &\quad +\frac{\mu}{2}\left\{\begin{matrix}\left[(s_0+s_1)+2\sqrt{s_0 s_1}\,\mathrm{Re}\{\tilde{\rho}_{01}\}\psi(\mu)\psi(M-\mu+1)\right]^2 \\ -4s_0 s_1(1-|\psi(\mu)|^2)(1-|\rho_{01}\psi(M-\mu+1)|^2)\end{matrix}\right\}^{1/2} + \sigma_\eta^2 \\ \bar{\lambda}_1 &= \frac{\mu}{2}(s_0+s_1)+\mu\sqrt{s_0 s_1}\,\mathrm{Re}\{\tilde{\rho}_{01}\}\psi(\mu)\psi(M-\mu+1) \\ &\quad -\frac{\mu}{2}\left\{\begin{matrix}\left[(s_0+s_1)+2\sqrt{s_0 s_1}\,\mathrm{Re}\{\tilde{\rho}_{01}\}\psi(\mu)\psi(M-\mu+1)\right]^2 \\ -4s_0 s_1(1-|\psi(\mu)|^2)(1-|\rho_{01}\psi(M-\mu+1)|^2)\end{matrix}\right\}^{1/2} + \sigma_\eta^2,\end{aligned} \tag{5.17}$$

where $\tilde{\rho}_{01} = \rho_{01}e^{j\pi(d/\lambda)(M-\mu)(\sin\theta_0-\sin\theta_1)}$. The sum of the signal eigenvalues is given by

$$(\bar{\lambda}_0-\sigma_\eta^2)+(\bar{\lambda}_1-\sigma_\eta^2) = M(s_0+s_1)+2M\sqrt{s_0 s_1}\,\mathrm{Re}\{\tilde{\rho}_{01}\}\psi(\mu)\psi(M-\mu+1).$$

If we select μ such that

$$\pi\frac{d}{\lambda}(M-\mu+1)|(\sin\theta_0-\sin\theta_1)| \approx \pi,$$

then $\psi(M-\mu+1)\approx 0$, $\bar{\lambda}_0 = Ms_0$, and $\bar{\lambda}_1 = Ms_1$ and the desired subarray size would be

$$\mu \approx (M+1) - \frac{2}{|(\sin\theta_0 - \sin\theta_1)|}.$$

Let $|(\sin\theta_0 - \sin\theta_1)| \approx 1/\mu$. Then the desired subarray size is

$$\mu \approx \frac{1}{3}(M+1). \tag{5.18}$$

This result is very close to that given in Ref. [7]. We have evaluated $\bar{\lambda}_1/\sigma_\eta^2$ as a function of the subarray size for two perfectly correlated equal power waves incident at angles θ_0 and θ_1 such that $\sin\theta_0 - \sin\theta_1 = 1/\mu$ on a ULA with 64 sensors and array SNR=10. The results are shown in Figure 5.4.

5.1.3 Direct Estimation

In MUSIC it is required to scan the entire angular range of $\pm\pi/2$. Since the spectral peaks are extremely fine the scanning must be done at a very fine interval; otherwise there is a high risk of missing the peak altogether. This is a serious drawback of the subspace methods based on scanning. Alternatively,

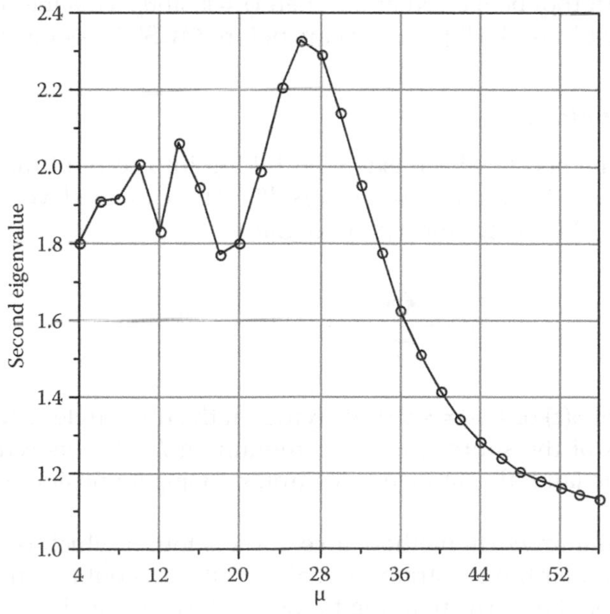

FIGURE 5.4
The second eigenvalue relative to noise variance is shown as a function of the subarray size (μ).

there are direct methods that enable us to estimate the DOA directly. We shall describe three such methods, namely, Pisarenko's method, minimum norm, and root Music and show how they are inter-related.

5.1.3.1 Pisarenko's Method

Let the array size be $M=P+1$, where P stands for the number of uncorrelated sources. The noise subspace is now spanned by a single eigenvector, $\mathbf{v}_{M-1}$. Hence, the direction vector, $\mathbf{a}$, will be orthogonal to $\mathbf{v}_{M-1}$,

$$\mathbf{v}_{M-1}^H \mathbf{a} = 0. \tag{5.19}$$

The direction vector of the pth source may be expressed as

$$\begin{aligned}\mathbf{a}_p &= \mathrm{col}\left\{1, e^{-j2\pi(d/\lambda)\sin\theta_p}, \ldots, e^{-j2\pi(M-1)(d/\lambda)\sin\theta_p}\right\} \\ &= \mathrm{col}\left\{1, z_p, z_p^2, \ldots, z_p^{M-1}\right\}, \quad p = 0, 1, \ldots, P-1,\end{aligned}$$

where $z_p = e^{-j2\pi(d/\lambda)\sin\theta_p}$. Consider a polynomial

$$\mathbf{v}_{M-1}^H \left[1, z, z^2, \ldots, z^{M-1}\right]^T = 0, \tag{5.20}$$

whose roots are indeed $z_0, z_1, \ldots, z_{p-1}$. In the complex plane, all roots lie on a unit circle. The angular coordinate of a root is related to the DOA. For example, let the pth root be located at φ_p. Then $\varphi_p = \pi \sin\theta_p$. This method was first suggested by Pisarenko [3], many years before MUSIC was invented!

5.1.3.2 Root Music

Pisarenko's method has been extended taking into account the noise space spanned by two or more eigenvectors [8]. The extended version is often known as root Music. Define a polynomial,

$$S(z) = \sum_{m=P}^{M-1} \mathbf{v}_m^H \left[1, z, z^2, \ldots., z^{M-1}\right]^T. \tag{5.21}$$

The roots of $S(z)$ or $D(z)=S(z)S(1/z)$ lying on the unit circle will correspond to the DOAs of the sources, and the remaining M-P roots will fall inside the unit circle (and also at inverse complex conjugate positions outside the circle).

In the minimum norm method, a vector $\mathbf{a}$ is found which is a solution of $\mathbf{a}^H \mathbf{S}_f \mathbf{a} = \min$ under the constraint that $\mathbf{a}^H \mathbf{a} = 1$. It turns out that the solution is an eigenvector of $\mathbf{S}_f$ corresponding to the smallest eigenvalue.

Thus, the minimum norm method belongs to the same class of direct methods of DOA estimation initiated by Pisarenko [3]. An example of DOA obtained by computing zeros on the unit circle under the ideal condition of no errors in the spectral matrix is shown in Table 5.2. Two equal power signal wavefronts are assumed to be incident on an eight-sensor ULA at angles 10° and 14°. The DOA estimates from the error-free spectral matrix are exact (columns one and two in Table 5.2), but in the presence of even a small error in the estimate of the spectral matrix, a significant error in DOA estimation may be encountered (see columns three and four in Table 5.2).

The zeros of the polynomial defined in any one of the above methods are important from the point of DOA estimation. In particular, the zeros that fall on the unit circle or close to it represent the DOAs of the incident wavefronts. These zeros are often called signal zeros and the remaining zeros, located deep inside the unit circle, are called the noise zeros. In MUSIC, a peak in the spectrum represents the DOA of the wavefront. But the height of the peak is greatly influenced by the position of the signal zero. The peak is infinite when the zero is right on the unit circle, but it rapidly diminishes when the zero moves away from the unit circle. The peak may be completely lost, particularly in the presence of another signal zero in the neighborhood but closer to the unit circle. Hence, it is possible that, while the spectral peaks remain unresolved, the signal zeros are well separated. The shift of the signal zero may be caused by errors in the estimation of the spectral matrix from finite data. Let a signal zero at z_i be displaced to $\hat{z}_i$. The displacement, both radial and angular, is given by

$$\Delta z_i \approx (z_i - \hat{z}_i) \approx \delta r \, e^{j(u+\delta u)},$$

where we have assumed that $\Delta u \ll 1$. Perturbation analysis reported in Ref. [9] for time series shows that the mean square error in δr and δu are related

TABLE 5.2

Direct estimation of DOA by computing signal zeros of a polynomial given by equation 5.20 or 5.21

		Estimation Error Variance=0.01	
Radial Coordinate	**Angular Coordinate (in radians)**	**Radial**	**Angular**
33.3162	0.9139	2.4854	−1.8515
2.2755	2.5400	1.5593	−0.7223
1.5793	2.9561	1.3047	−3.0944
0.8500	−1.7578	0.7244	−2.2490
0.8809	−0.9386	0.9336	1.6835
1.0000	**0.5455**	**1.0050**	**0.5647**
1.0000	**0.7600**	**1.0057**	**0.7864**

Note: The results shown in columns one and two are for the error-free spectral matrix and those in columns three and four are for a spectral matrix with random errors (zero mean and 0.01 variance).

$$\frac{E\{|\delta r_i|^2\}}{E\{|\delta\theta_i|^2\}} = 2N\left(\frac{\cos(\theta_i)}{\frac{\omega d}{c}}\right)^2 \le \frac{2N}{\pi^2}, \tag{5.22}$$

where N stand for the number of time samples (or snapshots). From Equation 5.22 it follows that $E\{|\Delta r_i|^2\} \gg E\{|\Delta\theta_i|^2\}$, particularly when the wavefronts are incident close to broadside. On account of the above finding, the magnitude of a spectral peak in MUSIC is likely to be more adversely affected by the errors due to finite data.

5.1.3.3 Subspace Rotation

The direct estimation of DOA discussed in the previous section requires a ULA. This itself is a restriction in many practical situations. A new principle of DOA estimation, known as subspace rotation, exploits a property of a dipole sensor array, where all dipoles are held in the same direction, which is the subspace spanned by the direction vectors pertaining to the upper sensors in the dipole array and those pertaining to lower sensors are related through a rotation. The principle is explained in Chapter 2, Section 2.2. The basic starting equation is Equation 2.57, which we reproduce here for convenience,

$$\mathbf{S}_{\tilde{f}} = \begin{bmatrix} \mathbf{A} \\ \mathbf{A\Gamma} \end{bmatrix} \mathbf{S}_f \begin{bmatrix} \mathbf{A} \\ \mathbf{A\Gamma} \end{bmatrix}^H + \begin{bmatrix} \mathbf{I}\sigma_{\eta_1}^2 & 0 \\ 0 & \mathbf{I}\sigma_{\eta_2}^2 \end{bmatrix},$$

where $\mathbf{S}_{\tilde{f}}$ is the spectral matrix of the dipole array sensors,

$$\mathbf{S}_{\tilde{f}} = \begin{bmatrix} \mathbf{S}_{f_1f_1} & \mathbf{S}_{f_1f_2} \\ \mathbf{S}_{f_2f_1} & \mathbf{S}_{f_2f_2} \end{bmatrix} = \begin{bmatrix} \mathbf{AS}_f\mathbf{A}^{\mathrm{H}} & \mathbf{AS}_f\mathbf{\Gamma}^{\mathrm{H}}\mathbf{A}^{\mathrm{H}} \\ \mathbf{A\Gamma S}_f\mathbf{A}^{\mathrm{H}} & \mathbf{AS}_f\mathbf{A}^{\mathrm{H}} \end{bmatrix}.$$

Note that $\mathbf{S}_{f_1f_1}$ is the spectral matrix of the upper sensor outputs and $\mathbf{S}_{f_2f_2}$ is the spectral matrix of the lower sensor outputs; $\mathbf{S}_{f_1f_2}$ is the cross-spectral matrix between the upper sensor and lower sensor outputs. From the eigenvalues of $\mathbf{S}_{f_1f_1}$ we can estimate the noise variance $\sigma_{\eta_1}^2$ (and $\sigma_{\eta_2}^2$ from $\mathbf{S}_{f_2f_2}$) and subtract it from the spectral matrix. Define $\mathbf{S}_{f_1f_1}^0 = \mathbf{S}_{f_1f_1} - \sigma_{\eta_1}^2$. Consider the generalized eigenvector (GEV) problem of the matrix pencil $\{\mathbf{S}_{f_1f_1}^0, \mathbf{S}_{f_1f_2}\}$,

$$\mathbf{S}_{f_1f_1}^0\mathbf{v} = \gamma\mathbf{S}_{f_1f_2}\mathbf{v}, \tag{5.23}$$

where γ is a generalized eigenvalue and $\mathbf{v}$ is a corresponding eigenvector. Consider the following:

$$\begin{aligned}\mathbf{v}^H\left[\mathbf{S}^0_{f_1f_1}-\gamma\mathbf{S}_{f_1f_2}\right]\mathbf{v}&=\mathbf{v}^H\left[\mathbf{A}\mathbf{S}_f\mathbf{A}^H-\gamma\mathbf{A}\mathbf{S}_f\boldsymbol{\Gamma}^H\mathbf{A}^H\right]\mathbf{v}\\&=\mathbf{v}^H\mathbf{A}\mathbf{S}_f\left[\mathbf{I}-\gamma\boldsymbol{\Gamma}^H\right]\mathbf{A}^H\mathbf{v}\\&=\mathbf{v}^H\mathbf{A}\mathbf{S}_f\left[\mathbf{I}-\gamma\boldsymbol{\Gamma}^H\right]\mathbf{A}^H\mathbf{v}\\&=\mathbf{v}^H\mathbf{Q}\mathbf{v}=0.\end{aligned}\tag{5.24}$$

Since $\mathbf{S}_f$ is full rank as the sources are assumed to be uncorrelated and $\mathbf{A}$ is assumed to be full rank, then, for the right-hand side of Equation 5.24 to be true, we must have $\gamma=e^{-j(\omega/c)\Delta\cdot\delta\Delta i}$. The rank of $\mathbf{Q}$ will be reduced by one and a vector $\mathbf{v}$ can be found to satisfy Equation 5.24. Thus, the generalized eigenvalues of the matrix pencil $\left\{\mathbf{S}^0_{f_1f_1},\mathbf{S}_{f_1f_2}\right\}$ are

$$\gamma_i=e^{-j(\omega/c)\Delta\cdot\delta i},\quad i=0,1,\ldots,P-1,$$

from which we can estimate θ_i, $i=0, 1,\ldots, P-1$. This method of DOA estimation is known as estimation of signal parameters via rotation invariance technique (ESPRIT) [10]. An example of DOA obtained via ESPRIT is in Table 5.3. The eigenvalues of the pencil matrix (Equation 5.23) under the ideal condition of no errors in the spectral matrix as well as with errors in the spectral matrix are shown in Table 5.3. Two equal power signal wavefronts are assumed to be incident on an eight-sensor ULA at angles 10° and 14°. The eigenvalues estimated from the error-free spectral matrix are exact (columns one and two in Table 5.3), but in the presence of a small error in the estimation of the spectral matrix, a significant error in DOA estimation may be encountered (see columns three and four in Table 5.3).

TABLE 5.3

Generalized eigenvalues of pencil matrix (equation 5.23)

		Estimation Error Variance=0.01	
Radial Coordinate	Angular Coordinate (in radians)	Radial	Angular
4.6642	1.8540	3.4640	−2.1554
2.0198	0.7276	2.5202	0.4402
0.8182	−2.3272	1.7050	−0.4834
0.2108	−2.6779	0.5738	−1.6783
0.5224	−0.3753	0.4586	1.6290
0.5193	1.1199	0.6753	0.9060
1.0000	**0.5455**	**0.9969**	**0.5533**
1.0000	**0.7600**	**1.0087**	**0.7720**

The GEV $\mathbf{v}_i$, corresponding to eigenvalue γ_i, possesses an interesting property. Let us write Equation 5.24 in expanded form

$$[\mathbf{v}_i^H\mathbf{a}_0,\ldots,\mathbf{v}_i^H\mathbf{a}_{i-1},\mathbf{v}_i^H\mathbf{a}_{i+1},\ldots,\mathbf{v}_i^H\mathbf{a}_{P-1}]\hat{\mathbf{S}}_f$$

$$\begin{bmatrix} 1-\gamma_i e^{j(\omega d/c)\sin\theta_0} & \cdots & 0 \\ 0 & \cdots & 1-\gamma_i e^{j(\omega d/c)\sin\theta_{P-1}} \end{bmatrix} \begin{bmatrix} \mathbf{v}_i^H\mathbf{a}_0 \\ \vdots \\ \mathbf{v}_i^H\mathbf{a}_{i-1} \\ \mathbf{v}_i^H\mathbf{a}_{i+1} \\ \vdots \\ \mathbf{v}_i^H\mathbf{a}_{P-1} \end{bmatrix} = 0,$$

where $\hat{\mathbf{S}}_f$ represents the source matrix where the ith column and the ith row are deleted. Since $\gamma_i \neq e^{-j(\omega 0/c)\Delta\cdot\delta k}$, $i\neq k$, the diagonal matrix will be full rank and also $\hat{\mathbf{S}}_f$ is full rank by assumption, we must have $\mathbf{v}_i^H\mathbf{a}_{k,}=0$ for all $k\neq i$. Thus, the generalized signal eigenvector of the pencil matrix $\{\mathbf{S}^0_{f1f1}, \mathbf{S}_{f1f2}\}(t)$ is orthogonal to all direction vectors except the ith direction vector. We shall exploit this property for signal separation and estimation in Chapter 6.

5.1.4 Diffused Source

A point source, on account of local scattering, or fluctuating medium, or reflections from an uneven surface, may appear as a diffused source, that is, the main source surrounded by many secondary sources. A sensor array will receive a large number of rays, all of which would have actually started from the same point (main source), but have traveled different paths due to scattering, or refraction, or reflection. Such a model was considered in Chapter 1, Section 1.5. Here we shall explore the possibilities of localizing such a diffused source, in particular, a diffused source due to local scattering (see Figure 1.29). The model of the diffused source signal is given by Equation 1.65

$$\mathbf{f}(t)=\tilde{\mathbf{a}}s(t)+\eta(t), \tag{5.25}$$

where $\tilde{\mathbf{a}}=\sum_{l=0}^{L-1}\alpha_l e^{-j\omega_c\delta t_l}\mathbf{a}(\theta+\delta\theta_l)$ where α_l, δt_l, and $\delta\theta_l$ are for the lth ray, complex amplitude, time delay, and DOA with respect to a direct ray from the main source, respectively. Assume that there are L rays reaching the sensor array. For the direct ray we shall assume that $\alpha_0=1$, $\delta t_0=0$ and $\delta\theta_0=0$. The covariance matrix of the array (ULA) output is given by (derived later in Chapter 7, Section 7.3.2)

$$\mathbf{C}_f \approx \sigma_s^2 L\sigma_\alpha^2 \mathbf{D}(\theta)\mathbf{Q}\mathbf{D}^H(\theta)+\sigma_\eta^2\mathbf{I} \tag{5.26}$$

where we have assumed that the scatterers are uniformly distributed in the angular range $\pm\Delta$. The **Q** matrix for the assumed uniform distribution is given by

$$\{\mathbf{Q}\}_{mn} = \frac{\sin 2\pi \frac{d}{\lambda}(m-n)\Delta\cos\theta}{2\pi\frac{d}{\lambda}(m-n)\Delta\cos\theta},$$

and

$$\mathbf{D} = \text{diag}[1, e^{j(2\pi d/\lambda)\sin\theta}, \ldots, e^{j(2\pi(M-1)d/\lambda)\sin\theta}].$$

The eigen-decomposition of **Q** shows some interesting properties [11]. These are r, where $r=[2(d/\lambda)M\Delta\cos\theta]$, significant eigenvalues close to unity ($[x]$ stands for the largest integer greater than x). The remaining $(M\text{-}r)$ eigenvalues are insignificant. Let $\mathbf{v}_\eta$ be a matrix whose columns are $M\text{-}r$ eigenvectors corresponding to $M\text{-}r$ insignificant eigenvalues (cf. the noise subspace in the Music algorithm). Now it follows that

$$\mathbf{v}_\eta^H\left[\mathbf{c}_f - \sigma_\eta^2\mathbf{I}\right]\mathbf{v}_\eta = \mathbf{0} \qquad \text{(a matrix } M\text{-}r \times M\text{-}r \text{ zeros).} \tag{5.27}$$

Using Equation 5.26 in Equation 5.27 we obtain

$$\sigma_s^2 L\sigma_\alpha^2 \sum_{m=0}^{r-1} \lambda_m \mathbf{v}_\eta^H \mathbf{D}(\theta)\mathbf{e}_m\mathbf{e}_m^H\mathbf{D}^H(\theta)\mathbf{v}_\eta = 0 \tag{5.28}$$

where we have used eigen-decomposition of

$$\mathbf{Q} \approx \sum_{m=0}^{r-1} \lambda_m \mathbf{e}_m \mathbf{e}_m^H.$$

Here λ_m (≈ 1) are significant eigenvalues of **Q**. Since σ_{f0}^2, and $\lambda_m > 0$ for all $m<r$, the following must hold good for all $m<r$:

$$\left|\mathbf{v}_\eta^H \mathbf{D}(\theta)\mathbf{e}_m\right|^2 = 0 \tag{5.29}$$

The azimuth of the center of the cluster can be estimated using the orthogonality property demonstrated in Equation 5.29. The eigenvectors $\mathbf{e}_m$, discrete prolate spheroidal sequence (DPSS), are obtained by eigen-decomposition of the **Q** matrix. But in **Q** there is an unknown parameter pertaining to the width of the cluster, namely, $\Delta\cos\theta$ which has to be estimated. In Chapter 7, Section 7.3.2 we will show that the rank of the covariance matrix (noise free) is closely related to this unknown parameter.

5.1.5 Adaptive Subspace

Adaptive methods for the estimation of the eigenvector corresponding to the minimum (or maximum) eigenvalue have been described by many researchers [12–14]. The method is based on inverse power iteration [15]

$$\begin{aligned}\hat{\mathbf{v}}^0 &= [1,0,\ldots,0]^T \\ \tilde{\mathbf{v}}^{k+1} &= \mathbf{S}_f^{-1}(\omega)\hat{\mathbf{v}}^k \\ \hat{\mathbf{v}}^{k+1} &= \frac{\tilde{\mathbf{v}}^{k+1}}{\sqrt{\tilde{\mathbf{v}}^{k+1^T}\tilde{\mathbf{v}}^{k+1}}}\end{aligned} \tag{5.30}$$

where $\hat{\mathbf{v}}^k$ is an eigenvector at kth iteration and the last equation is meant for normalization. To estimate the eigenvector corresponding to the largest eigenvalue, the algorithm shown in Equation 5.30 may be used with the difference that the spectral matrix, instead of its inverse, is used.

An attempt to extend the inverse power iteration method to the estimation of the entire subspace (signal or noise subspace) has also been reported [13]. All the above methods require computation of the gradient of a cost function that is minimized. The gradient computation is slow; furthermore, in the presence of noise, it is unstable. Additionally, the rate of convergence of a gradient method is known to be very slow [14]. The alternate approach is to use the well-known power method for estimation of an eigenvector corresponding to the minimum (or maximum) eigenvalue [15]. The power method is easy to understand and easy to implement on a computer. We shall briefly describe the power method for estimation of the entire signal or noise subspace. Let $\hat{\mathbf{v}}_m^N$, $m=1,2,\ldots,P$ be the eigenvectors based on the data until time N, that is, the eigenvectors of the spectral matrix, $\hat{\mathbf{S}}_f^N(\omega)$. When a new sample arrives, the eigenvectors are updated using the following recursive equation followed by Gram–Schmidt orthonormalization (GSO),

$$\begin{aligned}\hat{\mathbf{S}}_f^{N+1}\hat{\mathbf{v}}_m^N &= \mathbf{g}_m^{N+1} \\ \hat{\mathbf{v}}_m^{N+1} &= \mathrm{GSO}\{\mathbf{g}_m^{N+1}\}, \quad m=1,2,\ldots,P.\end{aligned} \tag{5.31}$$

The choice of initial eigenvectors to start the recursion is important for rapid convergence. A suggested choice in Ref. [14] is to orthonormalize P initial frequency snapshots and use them as the initial eigenvectors. The starting value of the spectral matrix is chosen as

$$\hat{\mathbf{S}}_f^P = (\mathbf{I} + \mathbf{F}_0\mathbf{F}_0^H)$$

The GSO process involves finding p orthonormal vectors that are linearly equivalent to data vectors [50]. Let $\mathbf{F}_1, \mathbf{F}_2, \ldots, \mathbf{F}_p$ be a set of frequency snapshots and $\mathbf{I}_1, \mathbf{I}_2, \ldots, \mathbf{I}_p$ be the set of linearly equivalent orthonormal vectors which satisfy the following conditions

$$\begin{aligned} &\mathbf{I}_k \perp \mathbf{F}_r \quad &\text{for} \quad &r<k, \\ &|\mathbf{I}_k|^2 = 1 \quad &\text{for all} \quad &k. \end{aligned} \tag{5.32}$$

The Gram–Schmidt method determines the orthonormal vectors by solving a system of linear equations [50, p. 206]

$$\begin{aligned} \mathbf{I}_1 &= \gamma_1^1\mathbf{F}_1 \\ \mathbf{I}_2 &= \gamma_1^2\mathbf{F}_1 + \gamma_2^2\mathbf{F}_2 \\ \mathbf{I}_3 &= \gamma_1^3\mathbf{F}_1 + \gamma_2^3\mathbf{F}_2 + \gamma_3^3\mathbf{F}_3 \\ &\cdots \\ \mathbf{I}_P &= \gamma_1^P\mathbf{F}_1 + \gamma_2^P\mathbf{F}_2 + \cdots + \gamma_P^P\mathbf{F}_P. \end{aligned} \tag{5.33}$$

5.2 Subspace Methods (Broadband)

Since the location parameters are independent of frequency, we ought to get identical location estimates at different frequencies. But, on account of errors in the estimation of the spectral matrix, the location estimates are likely to be different at different frequencies. The errors in the estimation of a spectral matrix are likely to be uncorrelated random variables. By suitably combining the spectral matrix estimates obtained at different frequencies, it is hoped that the error in the final location estimate can be reduced. This is the motivation for broadband processing. In the process of combining all spectral matrices, it is important that the basic structure of the spectral matrices should not be destroyed. For example, a spectral matrix at all signal frequencies is of rank one when there is only one source or rank two when there are two uncorrelated sources, and so on. Consider a ULA and a single source. The spectral matrix at frequency ω is given by $\mathbf{S}(\omega) = \mathbf{a}(\omega,\theta)S_0(\omega)\mathbf{a}^H(\omega,\theta) + \sigma_\eta^2\mathbf{I}$. Notice that the frequency dependence of direction vectors is explicitly shown. Let us consider a generalized linear transformation of the form $\mathbf{T}(\omega)\mathbf{S}(\omega)\mathbf{T}^H(\omega)$, where $\mathbf{T}(\omega)$ is a transformation matrix, yet to be chosen. The transformation matrix must map the direction vector into another direction vector at a chosen frequency and the background noise continues to remain white. The transformed spectral matrix is given by

$$\mathbf{T}(\omega)\mathbf{S}(\omega)\mathbf{T}^H(\omega)=\mathbf{T}(\omega)\mathbf{a}(\omega,\theta)S_0(\omega)\mathbf{a}^H(\omega,\theta)\mathbf{T}^H(\omega) \\ +\sigma_\eta^2\mathbf{T}(\omega)\mathbf{T}^H(\omega) \tag{5.34}$$

where the transformation matrix T(ω) must possess the following properties:

$$\mathbf{T}(\omega)\mathbf{a}(\omega,\theta)=\mathbf{a}(\omega_0,\theta) \\ \mathbf{T}(\omega)\mathbf{T}^H(\omega)=\mathbf{I} \tag{5.35}$$

where ω_0 is a selected frequency. Using Equation 5.35 in Equation 5.34 and averaging over a band of frequencies we obtain

$$\begin{aligned}\bar{\mathbf{S}}(\omega_0)&=\sum_{i\in bw}\mathbf{T}(\omega_i)\mathbf{S}(\omega_i)\mathbf{T}^H(\omega_i)\\ &=\mathbf{a}(\omega_0,\theta)\sum_{i\in bw}S_0(\omega_i)\mathbf{a}^H(\omega_0,\theta)+\sigma_\eta^2\mathbf{I}\end{aligned} \tag{5.36}$$

where bw stands for the signal bandwidth. In Equation 5.36, we observe that the signal power spread over a band of frequencies has been focused at one frequency, namely, ω_0. This idea of focusing of energy has been actively pursued in Refs. [16,17] for DOA estimation.

In the direction vector there is a parameter, namely, sensor spacing, which may be changed according to the frequency such that $\omega d/c$ remains constant. This will require the sensor spacing at frequency ω_i should be equal to $d_i=(\omega_0 d_0/\omega_i)$, $i\in bw$. In practice, the required change in the physical separation is difficult to achieve, but resampling of the wavefield through interpolation is possible. This approach to focusing was suggested in Ref. [18]. In Chapter 2, we introduced the spatio-temporal covariance matrix (STCM), which contains all spatio-temporal information present in a wavefield. It has been extensively used for source localization [19]. We shall probe into the signal and noise subspace structure of STCM of ULA as well as UCA and show how the structure can be exploited for source localization.

5.2.1 Wideband Focusing

Consider P uncorrelated point sources in a far field and a ULA for DOA estimation. As in Equation 4.12b the spectral matrix of an array signal may be expressed as $\mathbf{S}_f(\omega)=\mathbf{A}(\omega,\theta)\mathbf{S}_0(\omega)\mathbf{A}^H(\omega,\theta)$, where the columns of $\mathbf{A}(\omega,\theta)$ matrix are the direction vectors. We seek a transformation matrix $\mathbf{T}(\omega)$, which will map the direction vectors at frequency ω into direction vectors at the preselected frequency ω_0 as in Equation 5.35. There is no unique solution but a least squares solution is possible,

$$\mathbf{T}(\omega) \approx \mathbf{A}(\omega_0,\theta)\left[\mathbf{A}^H(\omega,\theta)\mathbf{A}(\omega,\theta)\right]^{-1}\mathbf{A}^H(\omega,\theta) \tag{5.37}$$

The transformation matrix given by Equation 5.37 depends upon the unknown azimuth information. However, it is claimed in Ref. [20] that approximate estimates obtained through beamformation or any other simple approach are adequate. As an example, consider a single source case. For this,

$$\mathbf{A}(\omega,\theta) = \text{col}\left\{1, e^{-j\omega(d/c)\sin\theta}, \ldots, e^{-j\omega(M-1)(d/c)\sin\theta}\right\}$$

and the transformation matrix given by Equation 5.37 is equal to

$$\mathbf{T}(\omega) \approx \frac{1}{N}\mathbf{A}(\omega_0,\theta)\mathbf{A}^H(\omega,\theta).$$

To show that this is not a unique answer, consider the following transformation matrix,

$$\mathbf{T}(\omega) = \text{diag}\left\{1, e^{-j(\omega_0-\omega)(d/c)\sin\theta}, \ldots, e^{-j(\omega_0-\omega)(M-1)(d/c)\sin\theta}\right\} \tag{5.38a}$$

Clearly, using Equation 5.38a we can also achieve the desired transformation. In Equation 5.38a let $\omega_0=0$; the transformation matrix becomes

$$\mathbf{T}(\omega) = \text{diag}\left\{1, e^{j\omega(d/c)\sin\theta}, \ldots, e^{j\omega(M-1)(d/c)\sin\theta}\right\}, \tag{5.38b}$$

which we shall use as a filter on the array output. The filtered output is given by

$$\begin{aligned}\mathbf{f}_1(t) &= \frac{1}{2\pi}\int_{-\infty}^{\infty}\mathbf{T}(\omega)\mathbf{F}(\omega)e^{j\omega t}d\omega \\ &= \text{col}\left\{f(t), f\left(t-\frac{d}{c}\sin\theta\right), \ldots, f\left(t-\frac{d}{c}(M-1)\sin\theta\right)\right\},\end{aligned} \tag{5.39}$$

which indeed is the same as the progressively delayed array output in beamformation processing. Let us next compute the covariance function of the filtered output,

$$
\begin{aligned}
\mathbf{c}_{f1} &= E\{\mathbf{f}_1(t)\mathbf{f}_1^H(t)\} \\
&= E\left\{ \begin{matrix} \left[f\left(t - m\frac{d}{c}\sin\theta\right), \quad m = 0,1,\ldots,M-1 \right] \\ \left[f\left(t - m\frac{d}{c}\sin\theta\right), \quad m = 0,1,\ldots,M-1 \right]^H \end{matrix} \right\} \\
&= \mathbf{c}_f\left((m-n)\frac{d}{c}(\sin\theta_0 - \sin\theta)\right) = [\mathbf{c}_f]_{mn},
\end{aligned}
$$

where $\mathbf{c}_f$ is the covariance matrix of the signal emitted by the source; $\mathbf{c}_{f_1}$ is known as the steered covariance matrix [21]. When $\theta=\theta_0$ all elements of $\mathbf{c}_{f_1}$ are equal to a constant equal to the total signal power, that is, $(1/2\pi)\int_{-\infty}^{\infty} S_f(\omega)d\omega$. Thus, in beamformation we seem to focus all signal power to zero frequency. When there is more than one source we should compute a series of steered covariance matrices over a range of azimuth angles. Whenever a steering angle matches with one of the DOA, the steered covariance matrix will display a strong 'dc' term, equal to the power of the source.

The main drawback of wideband focusing is the requirement that the DOAs of incident wavefronts must be known, at least approximately, but the resulting estimate is likely to have large bias and variance [20]. To overcome this drawback an alternate approach has been suggested in Ref. [22]. Let **A** and **B** be two square matrices and consider a two-sided transformation, $\mathbf{TBT}^H$, which is closest to **A**. It is shown in Ref. [22] that this can be achieved if $\mathbf{T} = \mathbf{v}_A\mathbf{v}_B^H$, where $\mathbf{v}_A$ is a matrix whose columns are eigenvectors of **A**, and $\mathbf{v}_B$ is similarly the eigenvector matrix of **B**. Now let **A** and **B** be the spectral matrices at two frequencies, $\mathbf{A}=\mathbf{S}_f(\omega_0)$ and $\mathbf{B}=\mathbf{S}_f(\omega_i)$.

To transform $\mathbf{S}_f(\omega_i)$ into $\mathbf{S}_f(\omega_0)$ the desired matrix is

$$
\mathbf{T}(\omega_0,\omega_i) = \mathbf{v}_s(\omega_0)\mathbf{v}_s^H(\omega_i), \tag{5.40}
$$

where $\mathbf{v}_s$ is a matrix whose columns are signal eigenvectors. It is easy to show that $\mathbf{T}(\omega_0,\omega_i)\mathbf{T}^H(\omega_0,\omega_i)=\mathbf{I}$. Applying the transformation on a spectral matrix at frequency ω_i we get

$$
\begin{aligned}
&\mathbf{T}(\omega_0,\omega_i)\mathbf{S}_f(\omega_i)\mathbf{T}^H(\omega_0,\omega_i) \\
&= \mathbf{v}_s(\omega_0)\mathbf{v}_s^H(\omega_i)\mathbf{S}_f(\omega_i)\mathbf{v}_s(\omega_i)\mathbf{v}_s^H(\omega_0) \\
&= \mathbf{v}_s(\omega_0)\boldsymbol{\lambda}(\omega_i)\mathbf{v}_s^H(\omega_0),
\end{aligned}
$$

where

$$\lambda(\omega_i)=\text{diag}\{\lambda_0(\omega_i),\lambda_1(\omega_i),\ldots,\lambda_{P-1}(\omega_i),0,\ldots,0\},$$

are the eigenvalues of the noise-free spectral matrix, $\mathbf{S}_f(\omega_i)$. Next, we average all transformed matrices and show that

$$\begin{aligned}\overline{\mathbf{S}}_f(\omega_0)&=\frac{1}{N}\sum_{i=1}^{N}\mathbf{T}(\omega_0,\omega_i)\mathbf{S}_f(\omega_i)\mathbf{T}^H(\omega_0,\omega_i)\\&=\mathbf{v}_s(\omega_0)\frac{1}{N}\sum_{i=1}^{N}\lambda(\omega_i)\mathbf{v}_s^H(\omega_0)\\&=\mathbf{v}_s(\omega_0)\overline{\boldsymbol{\lambda}}\mathbf{v}_s^H(\omega_0),\end{aligned} \tag{5.41}$$

$\overline{\mathbf{S}}_f(\omega_0)$ is the focused spectral matrix whose eigenvectors are the same as those of $\mathbf{S}_f(\omega_0)$ (before focusing), but its eigenvalues are equal to the averaged eigenvalues of all spectral matrices taken over the frequency band. We may now use the focused spectral matrix for DOA estimation.

5.2.2 Spatial Resampling

The basic idea is to resample the wavefield sensed by a fixed ULA so as to create a virtual ULA with a different sensor spacing depending upon the frequency. The spectral matrices computed at different frequencies are then simply averaged. Let us take a specific case of two wideband uncorrelated sources in the far field. The spectral matrix is given by

$$\mathbf{S}_f(\omega)=\begin{bmatrix}\mathbf{a}(\omega\frac{d}{c}\sin\theta_0)S_0(\omega)\mathbf{a}^H(\omega\frac{d}{c}\sin\theta_0)+\\ \mathbf{a}(\omega\frac{d}{c}\sin\theta_1)S_1(\omega)\mathbf{a}^H(\omega\frac{d}{c}\sin\theta_1)\end{bmatrix}+\sigma_\eta^2\mathbf{I}$$

where $S_0(\omega)$ and $S_1(\omega)$ are spectra of the first and the second source, respectively. We compute the spectral matrix at two different frequencies, ω_0 and ω_1, with different sensor spacings, d_0 and d_1, where

$$d_1=d_0\frac{\omega_0}{\omega_1},$$

and form a sum.

$$\mathbf{S}_f(\omega_0)+\mathbf{S}_f(\omega_1)=\left\{\begin{array}{l}\mathbf{a}(\omega_0\frac{d_0}{c}\sin\theta_0)S_0(\omega_0)\mathbf{a}^H(\omega_0\frac{d_0}{c}\sin\theta_0)\\ +\mathbf{a}(\omega_0\frac{d_0}{c_0}\sin\theta_1)S_1(\omega_0)\mathbf{a}^H(\omega_0\frac{d_0}{c}\sin\theta_1)\end{array}\right\}+\sigma_\eta^2\mathbf{I}$$

$$+\left\{\begin{array}{l}\mathbf{a}\left(\omega_1\frac{d_1}{c}\sin\theta_0\right)S_0(\omega_1)\mathbf{a}^H\left(\omega_1\frac{d_1}{c}\sin\theta_0\right)\\ +\mathbf{a}\left(\omega_1\frac{d_1}{c}\sin\theta_1\right)S_1(\omega_1)\mathbf{a}^H\left(\omega_1\frac{d_1}{c}\sin\theta_1\right)\end{array}\right\}+\sigma_\eta^2\mathbf{I}$$

$$=\mathbf{a}\left(\omega_0\frac{d_0}{c}\sin\theta_0\right)[S_0(\omega_0)+S_0(\omega_i)]\mathbf{a}^H\left(\omega_0\frac{d_0}{c}\sin\theta_0\right)$$

$$+\mathbf{a}\left(\omega_0\frac{d_0}{c}\sin\theta_1\right)[S_1(\omega_0)+S_1(\omega_i)]\mathbf{a}^H\left(\omega_0\frac{d_0}{c}\sin\theta_1\right)+2\sigma_\eta^2\mathbf{I} \quad (5.42)$$

Let us assume that N virtual arrays, each with different sensor spacing, have been created by a resampling process. We can generalize Equation 5.42,

$$\frac{1}{N}\sum_{i=0}^{N-1}\mathbf{S}_f(\omega_i)=\mathbf{a}\left(\omega_0\frac{d_0}{c}\sin\theta_0\right)\frac{1}{N}\sum_{i=0}^{N-1}\mathbf{S}_0(\omega_i)\mathbf{a}^H\left(\omega_0\frac{d_0}{c}\sin\theta_0\right)$$

$$+\mathbf{a}\left(\omega_0\frac{d_0}{c}\sin\theta_1\right)\frac{1}{N}\sum_{i=0}^{N-1}\mathbf{S}_1(\omega_i)\mathbf{a}^H\left(\omega_0\frac{d_0}{c}\sin\theta_1\right)+\sigma_\eta^2\mathbf{I}. \quad (5.43)$$

The above procedure for focusing may be extended to P sources; some of them may be correlated. While the idea of resampling is conceptually simple, its implementation would require us to perform interpolation of the wavefield in between physical sensors. Fortunately, this is possible because in a homogeneous medium the wavefield is limited to a spatial frequency range, $u^2+v^2\le\omega^2/c^2$ (see Chapter 1, Section 1.2). In two dimensions, the spatial frequency range is $-(\omega/c)\le u\le\omega/c$; hence, the wavefield is spatially bandlimited. This fact has been exploited in Ref. [21] for resampling.

The interpolation filter is a lowpass filter with its passband given by $-(\omega/c)\le u\le\omega/c$. Consider a ULA with sensor spacing equal to d_0 ($d_0=\lambda/2$). The maximum temporal frequency will be $(\omega_{max})=\pi(c/d_0)$. The lowpass filter is given by

$$h(m)=\frac{\sin\omega_{\max}\dfrac{(x-md_0)}{c}}{\omega_{\max}\dfrac{(x-md_0)}{c}}, \quad m=0,\pm 1,\ldots,\pm\infty \tag{5.44}$$

where x denotes the point where interpolation is desired, for example, $x=m'd_k$ where $d_k=(d_0\omega_{\max}/\omega_k)$.

5.2.3 STCM

In Chapter 2, we introduced the concept of extended direction vector (Chapter 2, Section 2.4) and STCM using the extended direction vector and the source spectrum (Equation 2.80), which we reproduce here for convenience,

$$\mathbf{C}_{\text{STCM}}=E\{\mathbf{f}_{\text{stacked}}(t)\mathbf{f}_{\text{stacked}}^H(t)\}=\frac{1}{2\pi}\int_{-\infty}^{\infty}\mathbf{h}(\omega,\varphi_0)\mathbf{h}^H(\omega,\varphi_0)S_{f0}(\omega)d\omega.$$

In this section, we shall show how the STCM can be used for the DOA estimation of a broadband source. For this, it is necessary that the eigenstructure, in particular, the rank of STCM, is ascertained. This has been done for a ULA in Ref. [23] and for a UCA in Ref. [4,24]. We shall assume that the source spectrum is a smooth function and that it may be approximated by a piecewise constant function

$$S_{f0}(\omega)\approx\sum_{l=-L}^{l=L}S_{0l}\text{rect}(\omega-\Delta\omega l), \tag{5.45}$$

where

$$\text{rect}(\omega-\Delta\omega l)=\begin{cases}\dfrac{1}{\Delta\omega} & \text{for}\quad \Delta\omega\left(l-\dfrac{1}{2}\right)\le\omega\le\Delta\omega\left(l+\dfrac{1}{2}\right),\\ 0 & \text{otherwise}\end{cases}$$

and S_{0l}, $(S_{0,l}=S_{0,-l})$ is the average spectrum in the lth frequency bin, $\Delta\omega l$. Using Equation 5.45 in Equation 2.80 we obtain

$$\mathbf{C}_{\text{STCM}}=\frac{1}{2\pi}\sum_{l=-L}^{l=L}S_{0l}\frac{1}{\Delta\omega}\int_{-\Delta\omega(l-1/2)}^{\Delta\omega(l-1/2)}\mathbf{h}(\omega,\varphi_0)\mathbf{h}^H(\omega,\varphi_0)d\omega, \tag{5.46}$$

where we have divided the frequency interval $\pm\pi$ into $2L+1$ non-overlapping frequency bins $(=2\pi/\Delta\omega)$. Using the definition of the extended direction vector given in Equation 2.80 in the integral in Equation 5.46, we obtain the following result:

$$\frac{1}{2\pi\Delta\omega}\int_{-\Delta\omega(l-\frac{1}{2})}^{\Delta\omega(l-\frac{1}{2})} \mathbf{h}(\omega,\varphi_0)\mathbf{h}^H(\omega,\varphi_0)d\omega = \mathbf{g}_l\mathbf{Q}\mathbf{g}_l^H$$

where

$$\mathbf{g}_l = \text{diag}\left\{\begin{matrix} e^{-j\omega\tau_0}, e^{-j\omega\tau_1}, \ldots, e^{-j\omega\tau_{M-1}}; e^{-j\omega(\Delta t+\tau_0)}, e^{-j\omega(\Delta t+\tau_1)}, \ldots, \\ e^{-j\omega(\Delta t+\tau_{M-1})}; \ldots; e^{-j\omega((N-1)\Delta t+\tau_0)}, e^{-j\omega((N-1)\Delta t+\tau_1)}, \ldots, \\ e^{-j\omega((N-1)\Delta t+\tau_{M-1})} \end{matrix}\right\}\Bigg|_{\omega=\Delta\omega l}$$

and

$$[\mathbf{Q}]_{\alpha,\alpha'} = \text{sin}\,c\left[((n-n')\Delta t + \tau_m - \tau_{m'})\frac{\Delta\omega}{2}\right],$$

where $\alpha = m + n \times M$, $\alpha' = m' + n' \times M$, $n, n' = 0, 1, \ldots, N-1$ and $m, m' = 0, 1, \ldots, M-1$. Note that the time delays $(\tau_0, \tau_1, \ldots, \tau_{M-1})$ depend upon the array geometry and, of course, on the azimuth; for example, for a linear array, the time delays are $(0, (d/c)\sin\varphi, \ldots, (M-1)(d/c)\sin\varphi)$ and for a circular array the time delays are $(\delta\cos\varphi, \delta\cos((2\pi/M)-\varphi), \ldots, \delta\cos((2\pi/M)(M-1)-\varphi))$, where $\delta = a/c$ and a is the radius of the circular array. Equation 5.46 may be expressed as

$$\mathbf{C}_{\text{STCM}} = \sum_{l=-L}^{l=L} S_{0l}\mathbf{g}_l\mathbf{Q}\mathbf{g}_l^H. \tag{5.47}$$

The rank of $\mathbf{C}_{\text{STCM}}$ is determined by the rank of matrix $\mathbf{Q}$. It follows from the results in Refs. [11,24] that 99.99% energy is contained in the first $[((N-1)\Delta t + (2a/c))(\Delta\omega/2\pi)+1]$ eigenvalues of $\mathbf{Q}$, where $[x]$ denotes the next integer greater than x. A comparison of the theoretical and numerically determined rank of $\mathbf{C}_{\text{STCM}}$ is given in Figure 5.5 for a circular array. For a linear array, the corresponding number is given by $\left[((N-1)\Delta t + ((M-1)d\cos\varphi/c)(\Delta\omega/2\pi)+1\right]$ [19]. Note that the dimension of the signal subspace is approximately equal to the time bandwidth product.

When the size of the observation space, that is, dimensions of $\mathbf{C}_{\text{STCM}}$, is larger than the rank of $\mathbf{C}_{\text{STCM}}$, $NM > [((N-1)\Delta t + (2a/c))(\Delta\omega/2\pi)+1]$, there exists a null subspace, a subspace of the observation space, of dimension equal to

$$\text{Dim}\{\mathbf{v}_{\text{null}}\} = NM - \left[\left((N-1)\Delta t + \frac{2a}{c}\right)\frac{\Delta\omega}{2\pi} + 1\right],$$

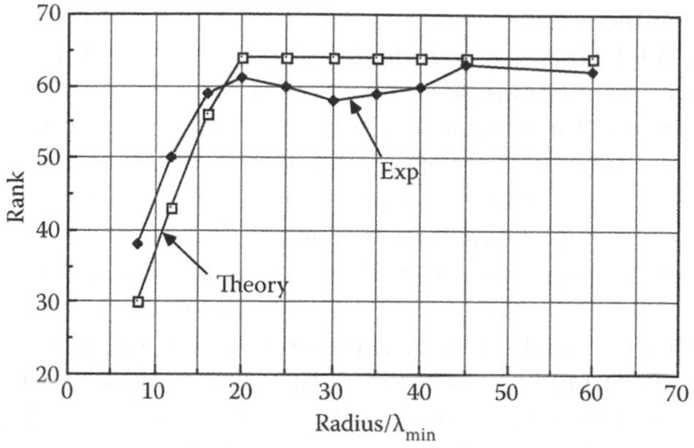

FIGURE 5.5
The effective rank of signal-only STCM as a function of the radius of a circular array. The drop in the rank at radius $30\lambda_{min}$ is due to a transition from smooth spectrum to line spectrum. We have considered a 16-sensor equispaced circular array with other parameters $N=4$, $B=0.8$ (normalized bandwidth), and one source. (From C. U. Padmini and P. S. Naidu, *Signal Process.*, vol. 37, pp. 243–254, 1994. With permission.)

where $\mathbf{v}_{null}$ eigenvectors correspond to the insignificant eigenvalues. Note that the null subspace thus defined will not be an ideal null space with zero power; some residual power (<0.01%) will be present.

Now, consider the following quadratic,

$$\begin{aligned}\mathbf{v}_i^H \mathbf{C}_{STCM} \mathbf{v}_i &= \frac{1}{2\pi}\int_{-\infty}^{\infty} \mathbf{v}_i^H \mathbf{h}(\omega,\varphi_0)\mathbf{h}^H(\omega,\varphi_0)\mathbf{v}_i S_{f0}(\omega)d\omega \\ &= \frac{1}{2\pi}\int_{-\infty}^{\infty} |\mathbf{v}_i^H \mathbf{h}(\omega,\varphi_0)|^2 S_{f0}(\omega)d\omega \approx 0.\end{aligned} \tag{5.48}$$

Assume that $S_{f0}(\omega)>0$ over some frequency band. For Equation 5.48 to hold good we must have in that frequency band

$$|\mathbf{v}_i^H \mathbf{h}(\omega,\varphi_0)|^2 \approx 0 \quad i \in \text{null space.} \tag{5.49}$$

Equation 5.49 is the basic result used for estimation of φ_0. For the purpose of locating the null we can define a parametric spectrum as in narrowband MUSIC.

5.2.4 Number of Sources

We have so far assumed that the number of sources is known *a priori* or can be estimated from knowledge of the significant and the repeated eigenvalues of the spectral matrix. The need for this information arises in the estimation of signal and noise subspaces. In time series analysis, the equivalent problem is estimation of the number of sinusoids or estimation of the order of a time series model. A lot of effort has gone into this problem in time series analysis [25]. In this section, we shall describe a test known as the sphericity test and its modern versions for estimating the number of repeated eigenvalues. Let P plane wavefronts be incident on an M sensor array. A covariance matrix of size $M \times M (M > P)$ has been computed using a finite number of N snapshots. Let $\hat{\lambda}_1 > \hat{\lambda}_2 > \hat{\lambda}_3 > \cdots > \hat{\lambda}_M$ be the eigenvalues of the estimated spectral matrix. We begin with the hypothesis that there are p wavefronts; then a subset of smaller eigenvalues, $\hat{\lambda}_{p+1} > \hat{\lambda}_{p+2} > \cdots > \hat{\lambda}_M$, is equal (or repeated). The null hypothesis, that the smallest eigenvalues has a multiplicity of M-p, is tested starting at $p=0$ until the test fails. Mauchley [26] suggested a test known as the sphericity test to verify the equality of all eigenvalues of an estimated covariance matrix. Define a likelihood ratio statistic

$$\Gamma(\hat{\lambda}_{p+1}, \hat{\lambda}_{p+2}, \ldots, \hat{\lambda}_M) = \ln \left[\frac{\left(\frac{1}{M-p} \sum_{i=p+1}^{M} \hat{\lambda}_i \right)^{M-p}}{\prod_{i=p+1}^{M} \hat{\lambda}_i} \right]^N, \tag{5.50a}$$

where N stands for the number of snapshots. The log likelihood ratio is then compared with a threshold γ. Whenever

$$\Gamma(\hat{\lambda}_{p+1}, \hat{\lambda}_{p+2}, \ldots, \hat{\lambda}_M) > \gamma,$$

the test is said to have failed. Modified forms of the sphericity tests have been suggested in Refs. [27, 28].

The choice of γ is subjective. Alternate approaches that do not require a subjective judgment have been proposed [29]. The number of wavefronts is given by that value of p in the range 0 to $M-1$ where, either

$$\Gamma(\hat{\lambda}_{p+1}, \hat{\lambda}_{p+2}, \ldots, \hat{\lambda}_M) + p(2M-p), \tag{5.50b}$$

or

$$\Gamma(\hat{\lambda}_{p+1}, \hat{\lambda}_{p+2}, \ldots, \hat{\lambda}_M) + \frac{p}{2}(2M-p)\ln N, \tag{5.50c}$$

is minimum. The first form is known as Akaike information criterion (AIC) and the second form is known as minimum description length (MDL). The expression consists of two parts; the first part is the log likelihood ratio that decreases monotonically and the second part is a penalty term that increases monotonically as p increases from 0 to $M-1$. The penalty term is equal to

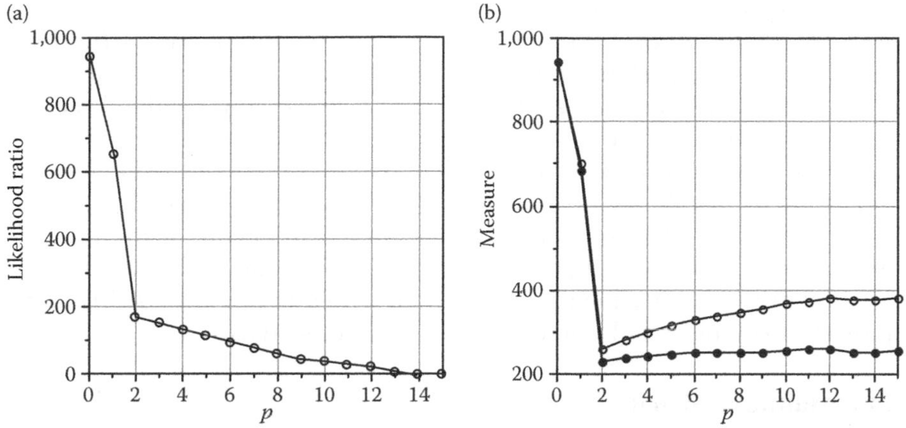

FIGURE 5.6
(a) Log likelihood ratio as a function of p, number of assumed wavefronts, (b) AIC (filled circles) and MDL (empty circles) as a function of p. Note the minimum at $p=2$. Twenty frequency snapshots. Sixteen-sensor ULA with $\lambda/2$ spacing. Two uncorrelated wavefronts are incident at angles 24° and 32°. SNR=10 dB per source.

free adjusted parameters. The log likelihood ratio plot around $p=P$ undergoes a rapid change of slope, which is responsible for the occurrence of a minimum at $p=P$. This is demonstrated in Figure 5.6.

The performance of AIC and MDL criteria has been studied in the context of detection of plane waves incident on a linear array in Ref. [30]. They define a probability of error as

$$prob.of\ error = prob(p_{\min} < P|H_P) + prob(p_{\min} > P|H_P)$$

where $p_{\min}$ stands for the position of the minimum of Equation 5.50 and H_p denotes the hypothesis that the true number of signal sources is P. The probability of error for the MDL criterion monotonically goes to zero for a large number of snapshots or high SNR. In contrast, the probability of error for the AIC criterion tends to a small finite value. However, for a small number of snapshots or low SNR the performance of AIC is better than that of MDL.

5.3 Coded Signals

In communication, a message-bearing signal is specifically designed to carry maximum information with minimum degradation. In analog communication, amplitude-modulated and frequency-modulated (FM) signals are often used, while in digital communication each pulse representing a bit is modulated with a sine wave (frequency shift keying) or with a pseudorandom sequence (spread spectrum). From the signal processing point, as these signals belong to a class of cyclostationary processes whose covariance function is periodic, but the background noise is simple ordinary stochastic process, these differences have been exploited in array processing for the direction arrival estimation. Sometimes, it is necessary to localize an active source for improved communication, as in the dense wireless telephone user environment. Here we would like to emphasize how a specially coded signal leads us to newer approaches to localization. We shall demonstrate with three different types of coded signals, namely, (i) multitone signal where the tones are spaced at prescribed frequency intervals but with random phases, (ii) binary phase shift keying signal, and (iii) cyclostationary signals.

5.3.1 Multitone Signal

Consider a circular boundary array (see Chapter 2) of M equispaced, omnidirectional, wideband sensors (Figure 5.7) with a target that is surrounded by the array. A source is assumed at a point with polar coordinates (r,φ) and it radiates a signal that is a sum of harmonically related random sinusoids,

$$f(t)=\sum_{k=0}^{Q-1}\alpha_k e^{-j(\omega_0+k\Delta\omega)t}, \tag{5.51}$$

where α_k is the complex amplitude of kth sinusoid. The output of the ith sensor is given by $f_i(t)=f(t+\Delta\tau_i)+\eta_i(t)$, where $\Delta\tau_i$ is time delay at the ith sensor with respect to a fictitious sensor at the center of the array. Each sensor

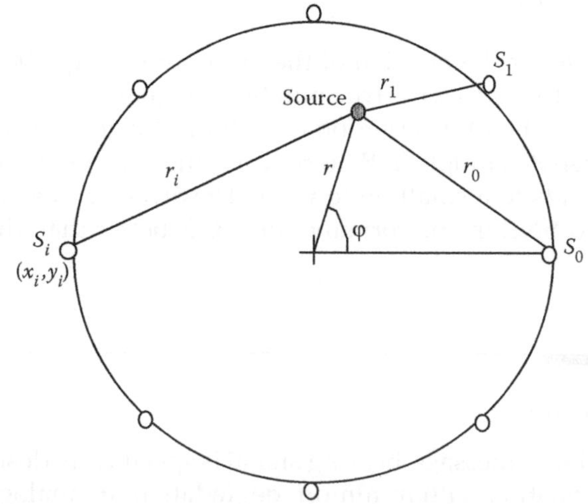

FIGURE 5.7
A circular array of sensors and a source inside the circle. The range (r) and azimuth (φ) of the source and all time delays are measured with respect to the center of the array as shown.

output is tapped at N time instants and arranged in a vector form, $\mathbf{f}_i=\mathrm{col}\{f_i(0), f_i(1),\ldots, f_i(N-1\}$.

The output data vector of the ith sensor may be written in a compact form:

$$\mathbf{f}_i=a(r_i)\mathbf{H}\mathbf{A}_i\,\varepsilon+\eta_i, \tag{5.52}$$

where

$$\mathbf{H}=[\mathbf{h}_0, \mathbf{h}_1,\ldots, \mathbf{h}_{Q-1}],$$

$$\mathbf{h}_k=\mathrm{col}\{1, e^{j\omega_k},\ldots, e^{j(N-1)\omega_k}\},$$

$$\omega_k=\omega_0+k\Delta\omega$$

$$\mathbf{A}_i=\mathrm{diag}\{e^{j\omega_0\Delta\tau_i}, e^{j\omega_1\Delta\tau_i},\ldots, e^{j\omega_{Q-1}\Delta\tau_i}\},$$

$$\varepsilon=\mathrm{col}[\alpha_0, \alpha_1,\ldots, \alpha_{Q-1}].$$

Next, we shall stack up all data vectors into one large vector $\mathbf{F}$ of dimension $MN\times 1$. It may be expressed as

$$\mathbf{F}=\mathbf{TDE}+\eta, \tag{5.53}$$

where

$$\mathbf{T}=\mathrm{diag}\{\mathbf{H}, \mathbf{H},\ldots, \mathbf{H}\}, (NM\times MQ),$$

$$\mathbf{D}=\mathrm{diag}\{a(r_0)\mathbf{A}_0, a(r_1)\mathbf{A}_1,\ldots, a(r_{M-1})\mathbf{A}_{M-1}\}, (MQ\times MQ)$$

$$\mathbf{E}=\left[\varepsilon^T, \varepsilon^T, \varepsilon^T,\ldots\right]^T, (MQ\times 1)$$

The covariance matrix (STCM) of $\mathbf{F}$ is given by

$$\mathbf{C}_f=\mathbf{TD}\Gamma_0\mathbf{D}^H\mathbf{T}^H+\sigma_\eta^2\mathbf{I}$$

where $\Gamma_0=E\{\mathbf{EE}^H\}=\mathbf{1}\otimes\Gamma$ where $\mathbf{1}$ is a square matrix of size $M\times M$ whose elements are all equal to 1, $\Gamma=\mathrm{diag}\{\gamma_0, \gamma_1,\ldots, \gamma_{Q-1}\}$ and $\gamma_0, \gamma_1,\ldots, \gamma_{Q-1}$ are powers of the random sinusoids. Symbol $\otimes$ stands for Kronecker product. We will assume hereafter that the noise variance, σ_η^2, is known or has been estimated from the array output when there is no signal transmission or by averaging the noise eigenvalues of the covariance matrix and that it has been subtracted from the covariance matrix.

Let us consider the structure of the mth column of the STCM. By straightforward multiplication it can be shown that the mth column of the covariance matrix is given by

$$\mathbf{c}_m=\left[\mathbf{c}_m^{0T}, \mathbf{c}_m^{1T},\ldots, \mathbf{c}_m^{Q-1^T}\right]^T, \tag{5.54}$$

where $\mathbf{c}_m^i = a(r_i)a(r_0)\mathbf{H}\mathbf{A}_i\Gamma\mathbf{A}_0^H\mathbf{H}^H\mathbf{u}_m$ and $\mathbf{u}_m$ is a $M\times 1$ vector consisting of all zeros except at mth location where there is one. Note $\mathbf{A}_i\Gamma\mathbf{A}_0^H$ is a diagonal matrix given by

$$\mathbf{A}_i\Gamma\mathbf{A}_0^H = \text{diag}\{\gamma_0 e^{j\omega_0\mu_i}, \gamma_1 e^{j\omega_1\mu_i}, \ldots, \gamma_{Q-1}e^{j\omega_{Q-1}\mu_i}\}, \tag{5.55}$$

where $\mu_i = \Delta\tau_i - \Delta\tau_0$. Further, we note that

$$\mathbf{H}^H\mathbf{u}_m = \text{col}\{e^{-jm\omega_0}, e^{-jm\omega_1}, \ldots, e^{-jm\omega_{Q-1}}\}. \tag{5.56}$$

Using Equations 5.55 and 5.56 in Equation 5.54, we obtain

$$\begin{aligned}\mathbf{c}_m^i &= a(r_i)a(r_0)\\ &\times \mathbf{H}\,\text{col}\left\{\gamma_0 e^{j\omega_0(\mu_i - m)}, \gamma_1 e^{j\omega_1(\mu_i - m)}, \ldots, \gamma_{Q-1}e^{j\omega_{Q-1}(\mu_i - m)}\right\}.\end{aligned} \tag{5.57}$$

Multiply both sides of Equation 5.57 by $\mathbf{H}^\# = (\mathbf{H}^H\mathbf{H})^{-1}\mathbf{H}^H$, the pseudo-inverse of $\mathbf{H}$, and obtain

$$\mathbf{H}^\#\mathbf{c}_m^i = a(r_i)a(r_0)\text{col}\left\{\gamma_0 e^{j\omega_0(\mu_i - m)}, \gamma_1 e^{j\omega_1(\mu_i - m)}, \ldots, \gamma_{Q-1}e^{j\omega_{Q-1}(\mu_i - m)}\right\}.$$

We now define a matrix $\mathbf{D}$ whose columns are $\mathbf{H}^\#\mathbf{c}_m^i$, $i = 0, 1, 2, \ldots, M-1$,

$$\begin{aligned}\mathbf{D} &= \left\{\mathbf{H}^\#\mathbf{c}_m^0, \mathbf{H}^\#\mathbf{c}_m^1, \ldots, \mathbf{H}^\#\mathbf{c}_m^{M-1}\right\}\\ &= a^2(r_0)\text{diag}\left\{\gamma_0 e^{-j\omega_0 m}, \gamma_1 e^{-j\omega_1 m}, \ldots, \gamma_{Q-1}e^{-j\omega_{Q-1}m}\right\}\mathbf{G}\end{aligned} \tag{5.58}$$

where

$$\mathbf{G} = \begin{bmatrix} 1, b_1 e^{j\omega_0\mu_1}, & \ldots b_{M-1}e^{j\omega_0\mu_{M-1}} \\ 1, b_1 e^{j\omega_1\mu_1}, & \ldots b_{M-1}e^{j\omega_1\mu_{M-1}} \\ & \ldots \\ & \ldots \\ 1, b_1 e^{j\omega_{Q-1}\mu_1}, & \ldots b_{M-1}e^{j\omega_{Q-1}\mu_{M-1}} \end{bmatrix}$$

and $b_i = a(r_i)/a(r_0)$, $i = 0, 1, \ldots, M-1$. Since the location information is present in $\mu_0, \mu_1, \ldots, \mu_{M-1}$, our aim naturally is to estimate these from the columns of $\mathbf{G}$. Each column may be considered as a complex sinusoid whose frequency can be estimated, provided $|\Delta\omega\mu_i| \le \pi$. For a boundary array, specifically a circular array, this limitation can be overcome. In Chapter 2, we have given an algorithm to estimate time delays or the location of the source from the phase estimates.

As a numerical example, we have considered two sources, each emitting eight random tones. The frequencies emitted by the first source are (0, 200, 400, 600, 800, 1,000, 1,200, 1,400 Hz) and those by the second are (1,600, 1,800, 2,000, 2,200, 2,400, 2,600, 2,800, 3,000 Hz). An eight-sensor UCA of radius 100 m surrounds both sources. Sixteen delayed snapshots (taps) were used, making the size of STCM as 128×128. The results are displayed in Table 5.4.

5.3.2 Binary Phase Shift Keying Signal

Binary phase shift keying signal consists of square pulses of fixed width with amplitude given by a Bernoulli random variable, an outcome of a coin-tossing experiment where head equals +1 and tail equals −1. A typical BPSK sequence is shown in Figure 5.8. The analytic representation of BPSK signal is given by

$$c_0(t) = \sum_{n=0}^{L-1} c_{0,n} h_{T_c}(t - nT_c), \tag{5.59a}$$

where $c_{0,n}$, $n=0, 1, \ldots, L-1$ are Bernoulli random variables and $h_{T_c}(t)$ is a rectangular function of width T_c. One of the most useful properties of a BPSK sequence is its narrow autocorrelation function with sidelobes whose variance is less than the inverse of the sequence length. We can easily generate many uncorrelated BPSK sequences. BPSK sequences have been used for time delay measurement [31], digital communication [32], and mobile communication [33] where it is of interest to exploit the spatial diversity for the purpose of increasing the capacity. Here we shall briefly show how the azimuth of a source emitting a known BPSK signal may be estimated.

TABLE 5.4

Localization of two sources by a circular (boundary) array of eight sensors (SNR = 10 dB)

Source #1		Source #2	
Range	Azimuth	Range	Azimuth
24.99 (25.00)	50.01 (50.00)	30.00 (30.00)	50.02 (50.00)
89.99 (90.00)	90.00 (90.00)	70.01 (70.00)	−90.00 (−90.00)
9.99 (10.00)	29.99 (30.00)	50.02 (50.00)	150.02 (150.00)
50.01 (50.1)	−60.00 (−60.00)	49.99 (50.00)	−60.00 (−60.00)

Note: The STCM was averaged over 100 independent estimates. The wave speed was assumed as 1,500 m/sec. The numbers inside brackets represent the true values (range in meters and azimuth in degrees).

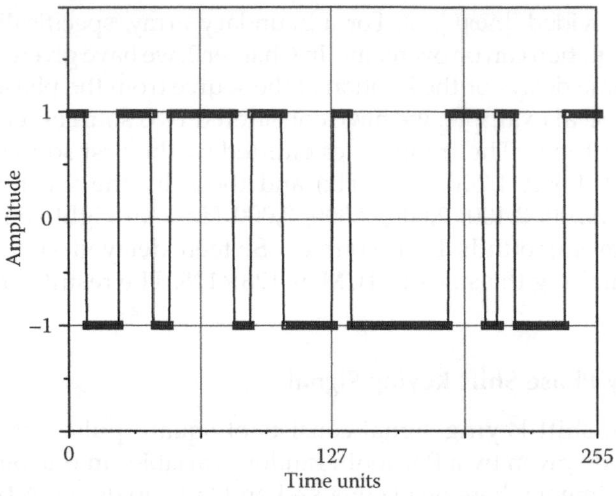

FIGURE 5.8
BPSK waveform. A train of rectangular pulses with amplitude alternating between+1 and −1.

The signal model considered here is as in Ref. [34]. There are Q uncorrelated sources emitting data bits that are encoded with a user-specific random sequence, a BPSK sequence, for example.

$$f_m(t) = \sum_{k=0}^{Q} p_k a_m(\theta_k) \sum_{l} b_{k,l} c_k(t - lT_s - \tau_k) + \eta_m(t) \quad m = 0, 1, 2, \ldots, M-1, \tag{5.59b}$$

where:

$a_m(\theta_k)$ = response of mth sensor to a signal coming from kth user,
p_k = signal amplitude of kth user,
$b_{k,l}$ = data stream from kth user,
$c_k(t)$ = random code of kth user,
T_S = bit duration,
τ_k = delay of a signal from kth user,
$\eta_m(t)$ = noise at mth sensor,
Q = number of users.

There are two different approaches to DOA estimation with the BPSK-coded signal. In the first approach the usual covariance matrix is computed and in the second approach, because the code used by the user of interest is known, the received signal is first cross-correlated with that code. The output of the cross-correlator is then used to compute the covariance matrix. Both approaches yield similar results; though in the second approach the interference from the users of no interest is reduced by a factor proportional to the code length.

5.3.2.1 Precorrelation Covariance Matrix

The outputs of the sensor array, after removing the carrier, are arranged in a vector form,

$$\mathbf{f}(t) = \sum_{k=0}^{Q} p_k \mathbf{a}(\theta_k) \sum_{l} b_{k,l} c_k (t - lT_s - \tau_k) + \eta(t), \tag{5.60a}$$

where $\mathbf{f}(t)$, $\mathbf{a}(\theta_k)$, and $\eta(t)$ are $M \times 1$ vectors. The precorrelation covariance matrix is given by

$$\mathbf{C}_f = E\{\mathbf{f}(t)\mathbf{f}^H(t)\}. \tag{5.60b}$$

Using Equation 5.60a in Equation 5.60b we obtain

$$\mathbf{C}_f = \left\{ \begin{array}{l} \sigma_\eta^2 \mathbf{I} + \sum_{k=0}^{Q} \sum_{k'=0}^{Q} p_k p_{k'} \mathbf{a}(\theta_k) \mathbf{a}^H(\theta_{k'}) \\ \left\{ \sum_{l} \sum_{l'} b_{k,l} c_k (t - lT_s - \tau_k) b_{k',l'} c_{k'} (t - l'T_s - \tau_{k'}) \right\} \end{array} \right\}. \tag{5.60c}$$

We shall assume that the data bits coming from different users are independent and codes assigned to different users are uncorrelated. As a result, Equation 5.60c reduces to

$$\mathbf{C}_f = \left\{ \sum_{k=0}^{Q} p_k^2 \mathbf{a}(\theta_k) \mathbf{a}^H(\theta_k) \left\{ \sum_{l} b_{k,l}^2 c_k^2 (t - lT_s - \tau_k) \right\} \right\} + \sigma_\eta^2 \mathbf{I}.$$

But the quantity inside the inner curly brackets is always equal to 1; hence we obtain

$$\mathbf{C}_f = \left\{ \sum_{k=0}^{Q} p_k^2 \mathbf{a}(\theta_k) \mathbf{a}^H(\theta_k) \right\} + \sigma_\eta^2 \mathbf{I},$$

which may be expressed in a standard form (Equation 4.12b) as

$$\begin{aligned} \mathbf{C}_f &= [\mathbf{a}(\theta_0), \mathbf{a}(\theta_1), \ldots, \mathbf{a}(\theta_{Q-1})] \operatorname{diag}\{p_0^2, p_1^2, \ldots, p_{Q-1}^2\} \\ &\quad [\mathbf{a}(\theta_0), \mathbf{a}(\theta_1), \ldots, \mathbf{a}(\theta_{Q-1})]^H + \sigma_\eta^2 \mathbf{I} \\ &= \mathbf{A} \operatorname{diag}\{p_0^2, p_1^2, \ldots, p_{Q-1}^2\} \mathbf{A}^H + \sigma_\eta^2 \mathbf{I}. \end{aligned} \tag{5.60d}$$

5.3.2.2 Postcorrelation Covariance Matrix

The outputs of the antenna array, after removing the carrier, are correlated with the desired user code assuming synchronization has been achieved. The postcorrelation array signal vector corresponding to the lth bit is given by

$$\mathbf{g}_0(l) = \frac{1}{T_s} \int_{lT_s}^{(l+1)T_s} \mathbf{f}(t) c_0(t - lT_s) dt, \tag{5.61}$$

where $\mathbf{f}(t)$ stands for array signal vector without carrier. Using the signal model given by Equation 5.60a, Equation 5.61 may be written as

$$\begin{aligned} \mathbf{g}_0(l) = {} & \frac{1}{T_s} p_0 \mathbf{a}(\theta_0) \int_{lT_s}^{(l+1)T_s} \left[\sum_{j=-\infty}^{\infty} b_{0j} c_0(t - jT_s) \right] c_0(t - lT_s) dt \\ & + \frac{1}{T_s} \sum_{k=1}^{Q-1} p_k \mathbf{a}(\theta_k) \int_{lT_s}^{(l+1)T_s} \left[\sum_{-\infty}^{\infty} b_{kj} c_k(t - jT_s + \tau_k) \right] c_0(t - lT_s) dt \\ & + \frac{1}{T_s} \int_{lT_s}^{(l+1)T_s} \eta(t) c_0(t - lT_s) dt, \end{aligned} \tag{5.62}$$

where we have set $\tau_0 = 0$. In Equation 5.62 the first term, equal to b_{0k}, is the signal term. The second term represents the interference. It is evaluated as follows: The integral is split into two parts, lT_s to $lT_s + \tau_k$, and $lT_s + \tau_k$ to $(l+1)T_s$. After a change of variable it reduces to

$$\begin{aligned} & \frac{1}{T_s} \sum_{k=1}^{Q-1} p_k \mathbf{a}(\theta_k) \left[b_{kl-1} \int_0^{\tau_k} c_k(t - T_s - \tau_k) c_0(t) dt + b_{kl} \int_0^{\tau_k} c_k(t - \tau_k) c_0(t) dt \right] \\ & = \mathbf{b}_l^1 + \mathbf{b}_l^2. \end{aligned} \tag{5.63}$$

Finally, the noise term reduces to

$$\eta_l = \frac{1}{T_s} \int_0^{T_s} \eta(t + lT_s) c_0(t) dt.$$

The postcorrelation covariance matrix of the array signal is given by

$$\begin{aligned}\mathbf{C}_{g_0g_0} &= E\{\mathbf{g}_0(l)\mathbf{g}_0^H(l)\} \\ &= p_0^2\mathbf{a}(\theta_0)\mathbf{a}^H(\theta_0) + E\{\mathbf{b}_l^1\mathbf{b}_l^{1H}\} + E\{\mathbf{b}_l^2\mathbf{b}_l^{2H}\} + E\{\eta_l\eta_l^H\}.\end{aligned} \tag{5.64a}$$

All cross-terms vanish. First, we shall evaluate $E\{\mathbf{b}_l^1\mathbf{b}_l^{1H}\}$.

$$E\{\mathbf{b}_l^1\mathbf{b}_l^{1H}\} = \frac{1}{T_s^2}\sum_{k=1}^{Q} p_k^2\mathbf{a}(\theta_k)\mathbf{a}(\theta_k)^H\rho_1, \tag{5.64b}$$

where

$$\rho_1 = E\left\{\left[\int_0^{\tau_k} c_k(t+T_s-\tau_k)c_0(t)dt\right]^2\right\},$$

and the sources are assumed to be uncorrelated. Using the representation of BPSK signal (Equation 5.59a) of length L, we obtain

$$\rho_1 = E\left\{\left[\int_0^{\tau_k}\sum_{m_0}^{L-1}\sum_{m_k}^{L-1} c_{0,m_0}c_{k,m_k}h_{T_c}(t-m_0T_c)h_{T_c}(t+T_s-m_kT_c-\tau_k)dt\right]^2\right\},$$

which, because c_{0,m_0} and c_{k,m_k} are uncorrelated random variables, reduces to

$$\rho_1 = \frac{1}{T_s}\int_0^{T_s}\sum_{m_0}^{L-1}\sum_{m_k}^{L-1}\left[\int_0^{\tau_k} \begin{matrix} h_{T_c}(t-m_0T_c) \\ h_{T_c}(t+T_s-m_kT_c-\tau_k)dt\end{matrix}\right]^2 d\tau_k, \tag{5.65a}$$

where τ_k, the arrival time of the signal from the kth user, is assumed to be uniformly distributed over 0 to T_s, the symbol duration. The integral over τ_k may be expressed as a sum over L integrals, one for each chip. When h_{T_c} is a rectangular function, the integral with respect to τ_k in Equation 5.65a can be evaluated in a closed form. After some algebraic manipulations we obtain

$$\rho_1 = \frac{L^2}{3T_s}T_c^3. \tag{5.65b}$$

Substituting Equation 5.65b in Equation 5.64b we obtain

$$E\{\mathbf{b}_l^1\mathbf{b}_l^{1H}\} = \frac{1}{3L}\sum_{k=1}^{Q} p_k^2 \mathbf{a}(\theta_k)\mathbf{a}(\theta_k)^H. \tag{5.66a}$$

Evaluation of $E\{\mathbf{b}_l^2\mathbf{b}_l^{2H}\}$ proceeds on the same lines as above. In fact, the result is identical, that is,

$$E\{\mathbf{b}_l^2\mathbf{b}_l^{2H}\} = \frac{1}{3L}\sum_{k=1}^{Q} p_k^2 \mathbf{a}(\theta_k)\mathbf{a}(\theta_k)^H = E\{\mathbf{b}_l^1\mathbf{b}_l^{1H}\}. \tag{5.66b}$$

Finally, we shall evaluate the noise term in Equation 5.62.

$$\begin{aligned} E\{\eta_l\eta_l^H\} &= \frac{1}{T_s^2}E\left\{\int_0^{T_s}\int_0^{T_s}\eta(t_1+lT_s)\eta^H(t_2+lT_s)c_0(t_1)c_0(t_2)dt_1dt_2\right\} \\ &= \frac{1}{T_s^2}\sum_{m_1=0}^{L-1}\int_0^{T_c}\int_0^{T_c} \begin{array}{l} E\{\eta(t_1+lT_c)\eta^H(t_2+lT_c)\} \\ h_{T_c}(t_1-m_1T_c)h_{T_c}(t_2-m_1T_c)dt_1dt_2 \end{array} \\ &= \frac{\sigma_\eta^2}{L}\mathbf{I}, \end{aligned} \tag{5.67}$$

where we assume that the noise is spatially uncorrelated but temporally correlated over the chip width T_c. Using Equations 5.65b, 5.66 and 5.67 in Equation 5.64, the postcorrelation covariance matrix may be expressed as

$$\mathbf{C}_{g_0g_0} = p_0^2\mathbf{a}(\theta_0)\mathbf{a}^H(\theta_0) + \frac{2}{3L}\sum_{k=1}^{Q} p_k^2\mathbf{a}(\theta_k)\mathbf{a}^H(\theta_k) + \frac{\sigma_\eta^2}{L}\mathbf{I}, \tag{5.68}$$

which may be expressed in a standard form (Equation 4.12b)

$$\begin{aligned} \mathbf{C}_{g_0g_0} &= [\mathbf{a}(\theta_0), \mathbf{a}(\theta_1), \ldots, \mathbf{a}(\theta_{Q-1})]\,\mathrm{diag}\left\{p_0^2, \frac{2}{3L}p_1^2, \ldots, \frac{2}{3L}p_{Q-1}^2\right\} \\ &\quad [\mathbf{a}(\theta_0), \mathbf{a}(\theta_1), \ldots, \mathbf{a}(\theta_{Q-1})]^H + \frac{\sigma_\eta^2}{L}\mathbf{I} \\ &= \mathbf{A}\,\mathrm{diag}\left\{p_0^2, \frac{2}{3L}p_1^2, \ldots, \frac{2}{3L}p_{Q-1}^2\right\}\mathbf{A}^H + \frac{\sigma_\eta^2}{L}\mathbf{I}. \end{aligned} \tag{5.69}$$

The DOA may be estimated following the subspace approach described earlier in Sections 5.1 and 5.2.

5.3.3 Cyclostationary Signals

A signal defined in Equation 5.59 and other similar communication signals possess an important property, namely, its temporal covariance function is periodic. This is a result of transmission of symbols at a fixed rate and also due to the use of a periodic carrier. Indeed, the periodicity in the covariance function of the signal, also known as cyclic frequency α, is equal to lf_b, where l is an integer and f_b is baud rate, that is, the number of symbols transmitted per second (for signal defined in Equation 5.59 baud rate, $f_b=1/T_c$). The baud rate is a unique property associated with each source and it is not affected by propagation. Since the baud rate of a signal of interest (SOI) is known *a priori* and it is different from that of the other interfering signals, it is possible to distinguish the SOI from the interference and the system noise whose covariance function is known to be aperiodic. In this section, we shall describe how a subspace method, i.e., MUSIC, may be devised by exploiting the property of cyclostationarity. The background information on the cyclostationary process will not be covered here as such material is already available in a book [35] and in a popular exposition [36]. However, we shall introduce enough material that is essential for the understanding of its application to DOA estimation. Let

$$\mathbf{f}(t)=\left\{f_0(t),\, f_0\left(t-\frac{d}{c}\sin\theta_0\right),\dots,\, f_0\left(t-(M-1)\frac{d}{c}\sin\theta_0\right)\right\}$$

be the output of a M sensor array on which a cyclostationary signal $f_0(t)$ is incident with an angle of incidence θ_0. We define frequency-shifted versions of $\mathbf{f}(t)$

$$\begin{aligned}\mathbf{f}_+(t)&=\mathbf{f}(t)e^{j\pi\alpha t}\\ \mathbf{f}_-(t)&=\mathbf{f}(t)e^{-j\pi\alpha t},\end{aligned} \tag{5.70}$$

and cyclic covariance matrix

$$\begin{aligned}\mathbf{C}_f^{\alpha}(\tau)&=\frac{1}{T}\sum_{t=-(T/2)}^{T/2} E\left\{\mathbf{f}_-\left(t+\frac{\tau}{2}\right)\mathbf{f}_+^H\left(t-\frac{\tau}{2}\right)\right\}\\ &=\frac{1}{T}\sum_{t=-(T/2)}^{T/2} E\left\{\mathbf{f}\left(t+\frac{\tau}{2}\right)\mathbf{f}^H\left(t-\frac{\tau}{2}\right)\right\}e^{-j2\pi\alpha t}.\\ &\quad T\to\infty\end{aligned}$$

Let us show how to evaluate the (k,l)th element of the matrix $\mathbf{C}_f^{\alpha}(\tau)$.

$$\left[\mathbf{C}_f^{\alpha}(\tau)\right]_{kl}=\frac{1}{T}\sum_{t=-(T/2)}^{T/2}E\left\{f_0\left(t+k\frac{d}{c}\sin\theta_0+\frac{\tau}{2}\right)f_0^*\left(t+l\frac{d}{c}\sin\theta_0-\frac{\tau}{2}\right)e^{-j2\pi\alpha t}\right\}$$

$$=\frac{1}{T}\sum_{t=-(T/2)}^{T/2}E\left\{f_0\left(t+k\frac{d}{c}\sin\theta_0+\frac{\tau}{2}\right)f_0^*\left(t+l\frac{d}{c}\sin\theta_0-\frac{\tau}{2}\right)e^{-j2\pi\alpha t}\right\},$$

$$=\frac{1}{T}\sum_{t=-(T/2)}^{T/2}E\left\{\begin{array}{l}f_0\left(t+\frac{k+l}{2}\frac{d}{c}\sin\theta_0+\frac{\tau}{2}+\frac{k-l}{2}\frac{d}{c}\sin\theta_0\right)\\ f_0^*\left(t+\frac{k+l}{2}\frac{d}{c}\sin\theta_0-\frac{\tau}{2}-\frac{k-l}{2}\frac{d}{c}\sin\theta_0\right)e^{-j2\pi\alpha t}\end{array}\right\}$$

$$T\rightarrow\infty$$

$$=c_{f_0}^{\alpha}\left(\tau+(k-l)\frac{d}{c}\sin\theta_0\right)e^{j\pi\alpha(k+l)(d/c)\sin\theta_0}. \tag{5.71a}$$

When a source is narrowband with center frequency ω_c so that following approximation holds good, $f_0(t+\tau)\approx f_0(t)e^{j\omega_c\tau}$ (see Equation 4.15b), the (k,l)th element of the matrix $\mathbf{C}_f^{\alpha}(\tau)$ may be approximated as

$$\left[\mathbf{C}_f^{\alpha}(\tau)\right]_{kl}\approx c_{f_0}^{\alpha}(\tau)e^{j\omega_c(k-l)(d/c)\sin\theta_0}e^{j\pi\alpha(k+l)(d/c)\sin\theta_0},$$

which may be further approximated as $\left[\mathbf{C}_f^{\alpha}(\tau)\right]_{kl}\approx c_{f_0}^{\alpha}(\tau)e^{j\omega_c(k-l)(d/c)\sin\theta_0}$. In deriving this result, we have approximated $2\pi(c/d)\approx\omega_c$ and used the fact that the cycle frequency is a multiple of the carrier frequency ω_c. Finally, we can express the cyclic covariance matrix as

$$\mathbf{C}_f^{\alpha}(\tau)\approx\mathbf{a}_0\mathbf{a}_0^H c_{f_0}^{\alpha}(\tau). \tag{5.71b}$$

The above relation is quite similar to Equation 4.12b, which was the starting point in the subspace algorithm, for example, in MUSIC. Naturally, based on Equation 5.71b, a subspace algorithm known as Cyclic MUSIC has been proposed in Ref. [37]. Although, in deriving Equation 5.71b we have assumed a single source, it holds good even in the presence of multiple signals with *different* cyclic frequencies and any type of stationary noise.

Let us consider only the diagonal terms of the cyclic covariance matrix. From Equation 5.71a the diagonal terms, k=l, are given by

$$\left[\mathbf{C}_f^{\alpha}(\tau)\right]_{k=l}=c_{f_0}^{\alpha}(\tau)e^{j2\pi\alpha l(d/c)\sin\theta_0}, \tag{5.72a}$$

which we shall express in a matrix form. Let

$$\tilde{\mathbf{c}}_f^{\alpha}(\tau) = \text{col}\left\{\left[\mathbf{c}_f^{\alpha}(\tau)\right]_{k=l=0}, \left[\mathbf{c}_f^{\alpha}(\tau)\right]_{k=l=1}, \ldots, \left[\mathbf{c}_f^{\alpha}(\tau)\right]_{k=l=M-1}\right\},$$

$$\mathbf{a}(\alpha,\theta_0) = \text{col}\left\{1, e^{j2\pi\alpha(d/c)\sin\theta_0}, \ldots, e^{j2\pi\alpha(M-1)(d/c)\sin\theta_0}\right\},$$

and using these vectors, Equation 5.72a may be written as

$$\tilde{\mathbf{c}}_f^{\alpha}(\tau) = c_{f0}^{\alpha}(\tau)\mathbf{a}(\alpha,\theta_0), \tag{5.72b}$$

and for P uncorrelated sources, but with the same cyclic frequency, we obtain

$$\tilde{\mathbf{c}}_f^{\alpha}(\tau) = \left\{\mathbf{a}(\alpha,\theta_0), \mathbf{a}(\alpha,\theta_1), \ldots, \mathbf{a}(\alpha,\theta_{P-1})\right\}\begin{bmatrix} c_{f0}^{\alpha}(\tau) \\ \cdots \\ c_{f_{P-1}}^{\alpha}(\tau) \end{bmatrix}. \tag{5.72c}$$

Equation 5.72c is evaluated over a set of values for $\tau=0, 1, 2,\ldots, N-1$ times sampling interval. All these quantities may be arranged in a matrix form,

$$\begin{aligned}&[\tilde{\mathbf{c}}_f^{\alpha}(0), \tilde{\mathbf{c}}_f^{\alpha}(1), \ldots, \tilde{\mathbf{c}}_f^{\alpha}(N-1)] \\ &= \left\{\mathbf{a}(\alpha,\theta_0), \mathbf{a}(\alpha,\theta_1), \ldots, \mathbf{a}(\alpha,\theta_{P-1})\right\}\begin{bmatrix} c_{f0}^{\alpha}(0) c_{f0}^{\alpha}(1) & \cdots & c_{f0}^{\alpha}(N-1) \\ \vdots & & \vdots \\ c_{f_{P-1}}^{\alpha}(0)\, c_{f_{P-1}}^{\alpha}(1) & \cdots & c_{f_{P-1}}^{\alpha}(N-1) \end{bmatrix}.\end{aligned} \tag{5.72d}$$

In a compact form we express Equation 5.72d in matrix form $\tilde{\mathbf{C}}_f^{\alpha} = \mathbf{A}\tilde{\mathbf{C}}^{\alpha}$, where $\tilde{\mathbf{C}}_f^{\alpha}$ is now a data matrix and $\tilde{\mathbf{C}}^{\alpha}$ a source matrix. We can write Equation 5.72d in a standard form (Equation 4.12b) for application of subspace method (e.g., MUSIC)

$$\tilde{\mathbf{C}}_f^{\alpha}\tilde{\mathbf{C}}_f^{\alpha H} = \mathbf{A}\tilde{\mathbf{C}}^{\alpha}\tilde{\mathbf{C}}^{\alpha H}\mathbf{A}^{H}. \tag{5.72e}$$

Note that in deriving Equation 5.72 we have not used the narrowband approximation that was earlier used in deriving Equation 5.71b. Another useful feature is that, since cross-cyclic covariance terms are not required, there is no need for interconnecting different sensors in a large array such as in a sensor array network. The computer simulation results reported in Ref. [38] showed improved results obtained by using Equation 5.72c over Equation 5.71b except when the bandwidth is only a tiny fraction (<0.05) of the carrier frequency. In Equation 5.72b or Equation 5.72c, the noise term is absent as it is not a cyclostationary process. This is a very important feature of any system using the cyclostationary signals. Whether the noise is correlated or white, its cyclostationary autocorrelation function at any cyclic frequency vanishes.

Another important feature of great significance is that the condition of two sources being mutually cyclically uncorrelated is weaker than the condition of mutually uncorrelated. Two cyclostationary signals $f(t)$ and $g(t)$ are said to be uncorrelated if their cross-cyclic covariance function defined as

$$c_{fg}^{\alpha}(\tau) = \frac{1}{T} \sum_{t=-(T/2)}^{T/2} E\left\{ f^{*}\left(t - \frac{\tau}{2}\right) g\left(t + \frac{\tau}{2}\right)\right\} e^{-j2\pi\alpha t}, \qquad (5.73)$$

$$T \to \infty$$

vanishes for all τ, and α is equal to the cyclic frequency either of $f(t)$ or $g(t)$. To demonstrate this property, let us consider $f(t)=s(t)\cos(\omega_1 t+\theta_1)$ and $g(t)=s(t)\cos(\omega_2 t+\theta_2)$ and substituting in Equation 5.73 we obtain

$$c_{fg}^{\alpha}(\tau) = c_s(\tau)\frac{1}{2T} \sum_{t=-(T/2)}^{T/2} \left[\begin{array}{l} \cos\left((\omega_1+\omega_2)t - (\omega_1-\omega_2)\frac{\tau}{2} + \theta_1 + \theta_2\right) \\ +\cos\left((\omega_1-\omega_2)t - (\omega_1+\omega_2)\frac{\tau}{2} + \theta_1 - \theta_2\right) \end{array}\right] e^{-j2\pi\alpha t},$$

$$T \to \infty$$

which becomes zero except when $2\pi\alpha=\pm(\omega_1+\omega_2)$ or $=\pm(\omega_1-\omega_2)$. Thus, two sources transmitting the same message but with different carriers become cyclically uncorrelated. Generalizing the above result, P sources transmitting even the same message but with different carrier frequencies may become *cyclically uncorrelated*. Further, we notice that even when the carrier frequencies are the same, the sources can become cyclically uncorrelated unless we choose $2\pi\alpha=\pm 2\omega_1$ or 0. As a result, it is possible through an appropriate choice of α to selectively cancel all signals of no interest. As a numerical example, we computed the cyclic cross-correlation function between two sources radiating the same stochastic signal with different carrier frequencies. In Figure 5.9a, the cyclic cross-correlation function is shown as a function of the cyclic frequency. The carrier frequencies are 0.15 and 0.20 Hz. (The Nyquist frequency is 0.5 Hz and 1,024 time samples were used.) Note the peaks at the sum and difference frequencies. In Figure 5.9b, the cyclic cross-correlation function is shown as a function of the carrier frequency of the second source.

5.3.4 Constant Modulus (CM) Signals

The temporal structure relates to the properties of the signal, and includes modulation format, pulse-shaping function, and symbol constellation. Constant modulus (CM) and finite alphabet (FA) are some typical temporal structures. The transmitted signal has a constant envelope waveform. Analytically it implies that

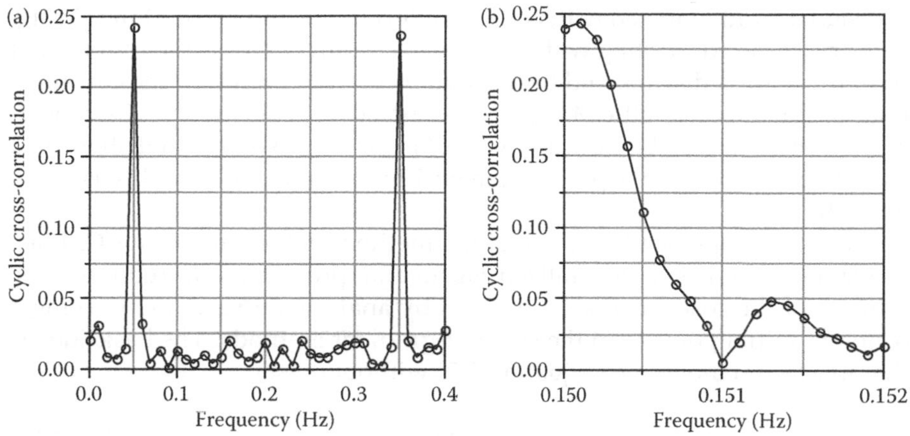

FIGURE 5.9
Cyclic cross-correlation function at zero lag as a function of (a) cyclic frequency and (b) carrier frequency in Hertz.

$$|f(t)| = \text{constant},$$

or $f(t)$ is of the form

$$f(t) = \exp(j\phi(t)),$$

where $\phi(t)$ is any function. A simple example of CM waveform is the well-known FM signal,

$$f(t) = \exp(-(\omega_0 t - \alpha t^2)).$$

In digital communication, $\phi(t)$ is a step function, for example, in binary phase shift keying, $\phi(k) = (0, \pi)$ (see Equation 5.59a for BPSK signal waveform) and in quadrature phase shift keying (QPSK) $\phi(k) = (\pi/4, 3/4\pi, 5/4\pi, 7/4\pi)$. Note that $|f(k)| = 1$ in both cases. Its own replicas due to multipath propagation, interference from other sources, and ubiquitous noise often contaminate the signal that reaches the receiver. One consequence of this is that the received signal no longer has the CM property. We can, however, restore this property with a suitably designed filter acting on the received signal and thus recover the transmitted CM signal. This approach was first suggested by Sato [58] in the context of blind equalization. A similar approach has also been tried for blind DOA estimation [59,60]. The approach consists of stochastic gradient-descent optimizer of cost function, which is given as

$$J = E\left\{\left[|\mathbf{w}^H \mathbf{f}(t)|^2 - 1\right]^2\right\}, \tag{5.74}$$

where $\mathbf{f}(t)$ is the array output and $\mathbf{w}$ is the desired filter whose output magnitude is as close as possible to one. The array output consists of Q uncorrelated

CM signals contaminated with non-CM interference and noise. Only CM signals are extracted, starting with the strongest source. This is then subtracted from the observed signal and the residual is used to minimize the cost function. In this manner, all CM signals are extracted sequentially. The convergence may, however, be slow and irregular for weak sources and short data lengths. Some of these aspects of the optimization process are discussed in Ref. [51].

In another approach [52], an analytic method (deterministic) for CM factorization, based on a generalized eigenvalue problem, is introduced. The minima of the cost functions are found by analysis rather than by trying a different initial point as in the steepest-gradient methods. The basic idea is that in a signal model (see Equation 5.75 without noise)

$$\mathbf{X}=\mathbf{AS}, \tag{5.75}$$

where $\mathbf{X}$ is a $M\times N$ data matrix consisting of N snapshot vectors $(M\times 1)$, $\mathbf{A}$ is a matrix of Q direction vectors $(M\times 1)$, and $\mathbf{S}$ is a signal matrix $(Q\times N)$. We assume that both $\mathbf{A}$ and $\mathbf{S}$ are full column rank matrices. The rank (row rank) of the data matrix is $Q\leq \text{M}$. On account of Equation 5.75, Q row vectors of the signal matrix span the row space of the data matrix. Let $\mathbf{s}$ be one such signal vector, then the problem of finding $\mathbf{s}$, given the data model, is equivalent to finding all linearly independent signals that satisfy

$$\begin{aligned} &\mathbf{s}\in \text{row }(\mathbf{X}) \\ &\mathbf{s}\in \text{CM}. \end{aligned} \tag{5.76}$$

Since the desired signal is in the row space of the data matrix, it may be expressed as a linear combination of minimal basis vectors of the row span of $\mathbf{X}$. The basis vectors are the right singular vectors of $\mathbf{X}$ corresponding to the nonzero singular values of $\mathbf{X}$. Let the rows of $\hat{\mathbf{V}}$ form an orthonormal basis of row span of $\mathbf{X}$. The first condition in Equation 5.76 is then equivalent to

$$\mathbf{s}=\mathbf{w}\hat{\mathbf{V}}, \tag{5.77}$$

where $\mathbf{w}$ $(Q\times 1)$ are weighting coefficients. Further, since $\mathbf{s}$ is a CM signal, we have an additional constraint that

$$|\{\mathbf{s}\}_k|^2=1,\ k=1,\ldots,N. \tag{5.78}$$

Combining Equation 5.77 and Equation 5.78 we obtain

$$\mathbf{w}\mathbf{v}_k\mathbf{v}_k^H\mathbf{w}^H=1, \quad k=1,\ldots,N, \tag{5.79}$$

where $\mathbf{v}_k$ is kth column of $\hat{\mathbf{V}}$. There are N simultaneous quadratic equations in Equation 5.79 to solve for $\mathbf{w}$.

5.4 Array Calibration

Throughout, we have implicitly assumed that all sensors are ideal, that is, they are omnidirectional point sensors with constant response. Further, the shape of the array is fully known. If we are dealing with a ULA, we assume that all sensors are on a perfectly straight line and sensors are equispaced. Any deviations from the assumptions can cause considerable loss of performance, particularly in the use of subspace methods [39]. One way out of this limitation is to actually measure the properties of the array, including its shape, and use this information in the design of the processor. Often this requires a source whose location is known. The properties of the array can be measured, a process known as array calibration. We shall describe two types of calibration. In the first instance, we shall give a method of computing the amplitude and the phase variations. Next, we shall describe a method for the shape estimation, which is vital in the use of flexible arrays.

5.4.1 Amplitude and Phase Variation of a Sensor

From Equation 2.17e, the direction vector may be expressed as a product of two components

$$\mathbf{a}(\theta_0) = \boldsymbol{\alpha}(\theta_0)\boldsymbol{\phi}(\theta_0),$$

where

$$\boldsymbol{\alpha}(\theta_0) = \text{diag}\{\alpha_0(\theta_0), \alpha_1(\theta_0), \ldots, \alpha_{M-1}(\theta_0)\}$$

$$\boldsymbol{\phi}(\theta_0) = \text{col}\{1, e^{-j\omega(d/c)\sin\theta_0}, e^{-j\omega(2d/c)\sin\theta_0}, \ldots, e^{-j\omega((M-1)d/c)\sin\theta_0}\}.$$

The purpose of array calibration is to estimate $\boldsymbol{\alpha}(\theta_0)$ from actual array observations. We preclude the possibility of direct *in-situ* measurement of sensor sensitivity (hardware calibration). Recall the relation we had derived between the direction vectors and eigenvectors of $\mathbf{v}_s\mathbf{v}_s^H$ (Equation 5.6d). Consider a single source whose DOA is known. For single source (Equation 5.6d) may be expressed as

$$\boldsymbol{\alpha}(\theta_0)\boldsymbol{\phi}(\theta_0) = \mathbf{v}_s\mathbf{v}_s^H\boldsymbol{\alpha}(\theta_0)\boldsymbol{\phi}(\theta_0). \tag{5.80}$$

Clearly $\boldsymbol{\alpha}(\theta_0)\boldsymbol{\phi}(\theta_0)$ is an eigenvector of $\mathbf{v}_s\mathbf{v}_s^H$ and the corresponding eigenvalue is 1. Since the DOA of the source is known, we can compute $\boldsymbol{\phi}(\theta_0)$ and demodulate the eigenvector. We can thus estimate $\boldsymbol{\alpha}(\theta_0)$ but for a complex constant. Assume that $\boldsymbol{\alpha}(\theta_0)$ is independent of θ_0. We shall express Equation 5.80 in a different form. Note that

$$\begin{aligned}\boldsymbol{\alpha}\boldsymbol{\phi}(\theta_0) &= \mathrm{diag}\{\alpha_0, \alpha_1 e^{-j\omega(d/c)\sin\theta_0}, \ldots, \alpha_{M-1}e^{-j\omega((M-1)d/c)\sin\theta_0}\} \\ &= \mathrm{diag}\{1, e^{-j\omega(d/c)\sin\theta_0}, \ldots, e^{-j\omega((M-1)d/c)\sin\theta_0}\}\mathrm{col}\{\alpha_0, \ldots, \alpha_{M-1}\} \\ &= \boldsymbol{\phi}_d(\theta_0)\boldsymbol{\alpha}_c.\end{aligned}$$

Using the above result in Equation 5.80 we obtain

$$\boldsymbol{\alpha}_c = \boldsymbol{\phi}_d^H(\theta_0)\mathbf{v}_s\mathbf{v}_s^H\boldsymbol{\phi}_d(\theta_0)\boldsymbol{\alpha}_c. \tag{5.81a}$$

When there are P sources with known DOAs, θ_p, $p=0, 1, \ldots, P-1$, Equation 5.81a may be expressed as $\boldsymbol{\alpha}_c = \mathbf{Q}\boldsymbol{\alpha}_c$ where

$$\mathbf{Q} = \sum_{p=0}^{P-1} \boldsymbol{\phi}_d^H(\theta_p)\mathbf{v}_s\mathbf{v}_s^H\boldsymbol{\phi}_d(\theta_p), \tag{5.81b}$$

and $\boldsymbol{\alpha}_c$ is the eigenvector of $\mathbf{Q}$ corresponding to its largest eigenvalue equal to P [40].

5.4.2 Mutual Coupling

When the output of a sensor in an array is influenced by neighboring sensors, often by immediate neighbors, through reflection or scattering of the incident wavefield, the sensor is said to be coupled with its neighbors. The output voltage of each sensor is a sum of the primary voltage due to the incident wavefield, plus the contributions from various coupling sources from each of its neighbors. The coupling effect may be modeled through a matrix operation,

$$\mathbf{a}(\theta) = \mathbf{C}\boldsymbol{\phi}(\theta), \tag{5.82}$$

where $\mathbf{C}$ is a $M \times M$ complex matrix often known as a mutual coupling matrix (MCM), whose elements on the main diagonal are equal to one and the magnitude of the off-diagonal elements lie between 1 and 0. $\mathbf{C}$ matrix is independent of θ. For ULA or UCA, the MCM assumes a particularly simple form. For ULA, the MCM is a banded symmetric Toeplitz matrix and for UCA, the MCM is a banded circulant Toeplitz matrix. As an illustration, we show MCM for an eight-element ULA in Figure 5.10a and for an eight-element UCA in Figure 5.10b, where a sensor is coupled to two immediate neighbors.

(a)

$$\begin{bmatrix} 1 & c_1 & c_2 & 0 & 0 & 0 & 0 & 0 \\ c_1 & 1 & c_1 & c_2 & 0 & 0 & 0 & 0 \\ c_2 & c_1 & 1 & c_1 & c_2 & 0 & 0 & 0 \\ 0 & c_2 & c_1 & 1 & c_1 & c_2 & 0 & 0 \\ \vdots & & & & & & & \\ 0 & 0 & 0 & 0 & 0 & c_2 & c_1 & 1 \end{bmatrix}$$

(b)

$$\begin{bmatrix} 1 & c_1 & c_2 & 0 & 0 & 0 & c_2 & c_1 \\ c_1 & 1 & c_1 & c_2 & 0 & 0 & 0 & c_2 \\ c_2 & c_1 & 1 & c_1 & c_2 & 0 & 0 & 0 \\ 0 & c_2 & c_1 & 1 & c_1 & c_2 & 0 & 0 \\ \vdots & & & & & & & \\ c_1 & c_2 & 0 & 0 & 0 & c_2 & c_1 & 1 \end{bmatrix}$$

FIGURE 5.10
Mutual coupling matrix for ULA (a) and for UCA (b).

In the MUSIC algorithm the direction vector, which is now composed of three components, namely, $\boldsymbol{\alpha}$ (diagonal matrix representing gain-phase variations), $\mathbf{C}$ (MCM), and propagation vector $\boldsymbol{\phi}$, is orthogonal to the noise subspace v_η (see Equation 5.82)

$$v_\eta^H \mathbf{a} = v_\eta^H \boldsymbol{\alpha} \mathbf{C} \boldsymbol{\phi} = 0,$$

or

$$\mathbf{a}^H v_\eta v_\eta^H \mathbf{a} = \boldsymbol{\phi}^H \mathbf{C}^H \boldsymbol{\alpha}^H v_\eta \mathbf{v}_\eta^H \boldsymbol{\alpha} \mathbf{C} \boldsymbol{\phi} = 0. \tag{5.83}$$

In practice, one needs to minimize the quadratic on the left-hand side of Equation 5.83 with respect to DOAs, gain-phase, and mutual coupling parameters. This is what Friedlander [53] proposed to carry out sequentially, first assuming that gain-phase and mutual coupling parameters are known, DOAs are estimated using the MUSIC algorithm. In the second step, using the estimated DOAs, the quadratic in Equation 5.83 is minimized with respect to gain-phase parameters assuming the mutual coupling parameters are known. In the third step using the estimated DOAs and gain-phase parameters, the quadratic is minimized with respect to mutual coupling parameters. The minimization is carried on repeatedly until the quadratic reaches a value below a preset number. It is reported in Ref. [53] that the error in the estimated parameters decreases rapidly in the first few iterations, but from then on the decrease is slow. To achieve a few percentage error, one would need a few tens of iterations.

The iterative scheme has been avoided in a subspace-based method [54]. We assume that gain-phase parameters are known or equal to unity. The method is based on an important result that we can express $\mathbf{C}\phi$ as $T[\phi]\mathbf{c}$, where $T[\phi]$ is a transformation of the column ϕ into a matrix $(M \times L)$ and $\mathbf{c}$ is a vector $(L \times 1)$ given by

$$[\mathbf{c}]_i = [\mathbf{C}]_{1i}, \quad i = 1, 2, \ldots, L,$$

where L is the highest superdiagonal of $\mathbf{C}$ that is different from zero. It is also known as degrees of freedom of MCM. The transformation utilizes the symmetric Toeplitz structure of matrix $\mathbf{C}$ (associated with ULA). The transformation for ULA is outlined below:

$$\mathbf{T}=\mathbf{T}_1+\mathbf{T}_2, \tag{5.84}$$

where

$$[\mathbf{T}_1\}_{lm}=\begin{cases}[\phi]_{l+m-1} & \text{for } \ l+m\le M+1,\\ 0 & \text{otherwise,}\end{cases}$$

and

$$[\mathbf{T}_2\}_{lm}=\begin{cases}[\phi]_{l+m-1} & l\ge m\ge 2\\ 0 & \text{otherwise,}\end{cases}$$

(see Ref. [13] for details). Using the above transformation in Equation 5.83 we obtain a useful result

$$\mathbf{c}^H T[\phi]^H \mathbf{v}_\eta \mathbf{v}_\eta^H T[\phi]\mathbf{c}=\mathbf{c}^H\mathbf{Q}\mathbf{c}=0, \tag{5.85}$$

where the matrix $\mathbf{Q}$ $(L\times L)$ is independent of the mutual coupling parameters. It is argued that, since $\mathbf{c}\neq 0$, the quadratic will vanish only when the rank of $\mathbf{Q}$ drops or becomes singular [54]. This will happen at the true DOAs. Then the eigenvector corresponding to the null eigenvalue (or the least eigenvalue) is an estimate of $\mathbf{c}$. When there are P distinct sources there will be P independent estimates of $\mathbf{c}$, which may be averaged to yield a better (lower variance) estimate of the coupling parameters. However, there is a possibility of ambiguous DOA estimates, particularly for large L, which represents strong mutual coupling [55].

In this example, we would like to bring out the serious consequence of the mutual coupling in a small circular array (ULA) of radius equal to 0.2λ and eight sensors. The mutual coupling between the nearest neighbors is 0.1 and that with the next sensor is 0.01. Thus, $c_1=0.1$ and $c_2=0.01$. There are three uncorrelated equal power sources at DOA: $-67°$, $31°$, and $120°$. First, we show the effect of just ignoring the presence of mutual coupling. The MUSIC spectrum obtained by accepting the mutual coupling is shown in Figure 5.11b and that obtained by ignoring the fact that the mutual coupling exists is shown in Figure 5.11a. In both computations, we have assumed exact covariance matrix, which is required in the MUSIC algorithm. In Figure 5.11b, we encounter many spurious peaks over and above peaks at right DOA. The spurious peaks are the direct result of ignoring the mutual coupling.

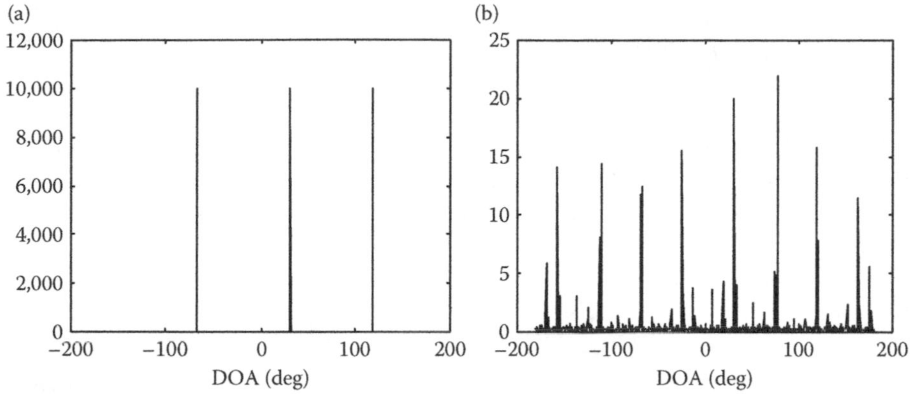

FIGURE 5.11
The effect of mutual coupling on MUSIC spectrum. In (a) the coupling is accounted for, but in (b) coupling is ignored.

5.4.3 Shape Estimation

In any practical towed array system used in sonar or seismic exploration the position of sensors is continuously monitored. Additionally, it is also possible to estimate the shape from the array output. Here, from the point of array processing, our natural interest is in the second approach. Consider a ULA that has been deformed either on account of towing or ocean currents. Assume that we have a known source radiating a narrowband signal and the background noise is spatially uncorrelated. We have shown in Equation 5.4 the eigenvector of the spatial covariance matrix corresponding to the largest eigenvalue is proportional to the direction vector. In fact, the relation is given by $\mathbf{v}_0=(1/\sqrt{M})\mathbf{a}_0$. For deformed array, the direction vector may be obtained from Equation 2.34c

$$\mathbf{a}(\theta_0,\varphi_0)=\text{col}\left\{1,\begin{matrix}e^{-j\omega_c(d/c)[\gamma_1\sin\theta_0\sin\varphi_0+\varepsilon_1\sin\theta_0\cos\varphi_0+\xi_1\sin\theta_0]},\ldots,\\ e^{-j\omega_c(d/c)[\gamma_{M-1}\sin\theta_0\sin\varphi_0+\varepsilon_{M-1}\sin\theta_0\cos\varphi_0+\xi_{M-1}\sin\theta_0]}\end{matrix}\right\} \tag{5.86}$$

Therefore, we can relate the eigenvector to the direction vector

$$\mathbf{v}_0=\frac{1}{\sqrt{M}}\text{col}\left\{1,\begin{matrix}e^{-j\omega_c(d/c)[\gamma_1\sin\theta_0\sin\varphi_0+\varepsilon_1\sin\theta_0\cos\varphi_0+\xi_1\sin\theta_0]},\ldots,\\ e^{-j\omega_c(d/c)[\gamma_{M-1}\sin\theta_0\sin\varphi_0+\varepsilon_{M-1}\sin\theta_0\cos\varphi_0+\xi_{M-1}\sin\theta_0]}.\end{matrix}\right\} \tag{5.87a}$$

Note that since the source is known, its azimuth and elevation are known *a priori*. The unknowns are γ_m, ε_m, $m=0, 1,\ldots, M-1$. (Note ξ_m, $m=0, 1,\ldots, M-1$ are dependent on γ_m, ε_m.) To solve for a pair of unknowns we would need one more source, say, with different azimuth and elevation, (θ_1,ϕ_1). Once again, the largest eigenvector may be related to the direction vector of the second source.

$$\mathbf{v}_1 = \frac{1}{\sqrt{M}}\text{col}\left\{ \begin{array}{l} e^{-j\omega_c(d/c)[\gamma_1 \sin\theta_1 \sin\varphi_1 + \varepsilon_1 \sin\theta_1 \cos\varphi_1 + \xi_1 \sin\theta_1]}, \ldots, \\ e^{-j\omega_c(d/v)[\gamma_{M-1} \sin\theta_1 \sin\varphi_1 + \varepsilon_{M-1} \sin\theta_1 \cos\varphi_1 + \xi_{M-1} \sin\theta_1]}. \end{array} \right\} \tag{5.87b}$$

We shall, for the sake of simplicity, assume that the deformed array is in the x-y plane and also place the calibrating source in the x-y plane. Then, a single source is enough for estimation of γ_m and ε_m, which now take a form

$$\begin{aligned} \gamma_m &= \sum_{i=0}^{m} \cos\alpha_i, \quad m = 0, 1, \ldots, M-1 \\ \varepsilon_m &= \sum_{i=0}^{m} \sin\alpha_i, \quad m = 0, 1, \ldots, M-1. \end{aligned} \tag{5.88}$$

Using Equation 5.88 in Equation 5.87b we obtain the following basic result

$$\begin{aligned} &\frac{\lambda_c}{2\pi}[\angle\{\mathbf{v}_0\}_m - \angle\{\mathbf{v}_0\}_{m-1}] + \lambda_c n \\ &= d\cos\alpha_m \cos\varphi_0 + d\sin\alpha_m \sin\varphi_0 = \Delta x_m \cos\varphi_0 + \Delta y_m \sin\varphi_0, \end{aligned} \tag{5.89}$$

where n is an unknown integer and Δx_m and Δy_m are x,y coordinates of mth sensor relative to m-1st sensor. The ambiguity, arising out of the unknown integer, may be resolved through geometrical considerations. Note that the mth sensor must be on a circle of radius d centered at m-1st sensor. Further, the position of the sensor must satisfy Equation 5.89, which, incidentally, is an equation of a straight line [41]. For illustration let the line corresponding to $n=1$ intersect the circle at two points p_1 and p_2 (see Figure 5.12). The sensor can either be at p_1 or p_2. This ambiguity is resolved by choosing a point that results in minimum array distortion [42].

5.5 Source in Bounded Space

When a source is in a bounded space a sensor array will receive signals reflected from the boundaries enclosing the space, for example, radar returns from a low-flying object [43], an acoustic source in shallow water [44], a speaker in a room [45]. In all these cases, the reflections are strongly correlated with the direct signal and come from a set of predictable directions. The complexity of the multipath structure increases with the increasing number of reflecting surfaces, as illustrated in Figure 5.13. In this section, we shall briefly consider two simple cases involving one reflecting surface (a low-flying object) and two reflecting surfaces (an acoustic source in shallow water) shown in Figure 5.13a and b.

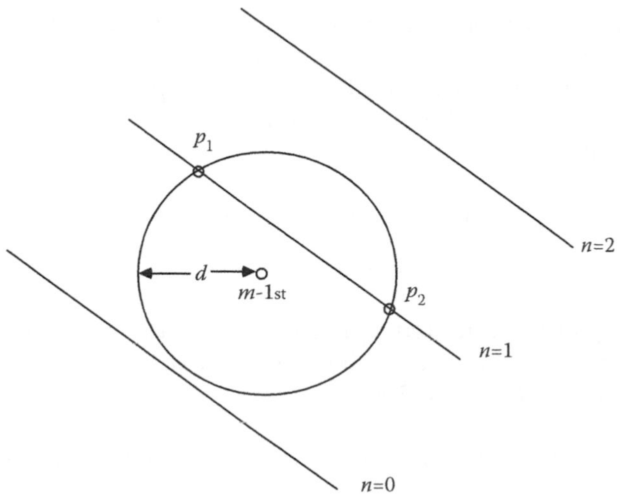

FIGURE 5.12
The ambiguity in Equation 5.77 is resolved by requiring that the line it represents must intersect the circle of radius d drawn at m-1st sensor. There are two intersection points. The sensor may be at any one of the two intersections.

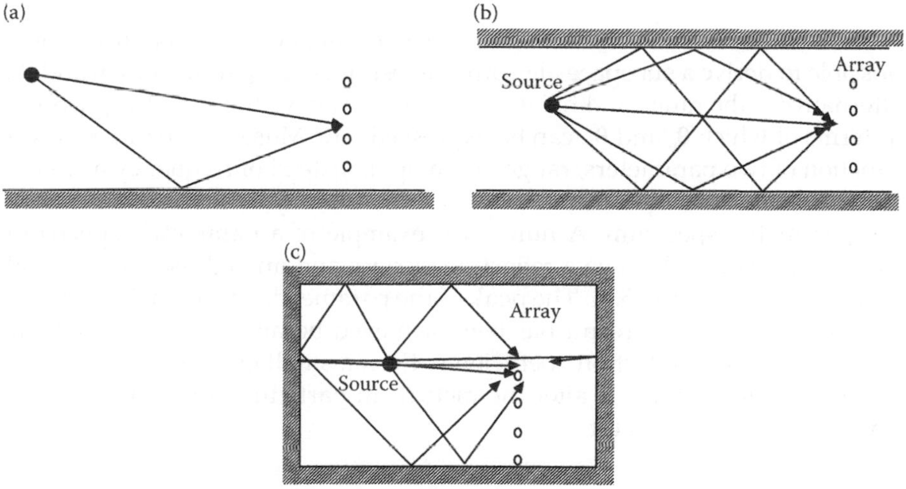

FIGURE 5.13
Three examples of multipath structure with increasing complexity. Assuming the boundaries are well defined, we can predict the multipaths given the source and array locations.

5.5.1 Single Reflecting Surface

Assume that the source emits a stationary stochastic signal. A vertical sensor array is used to receive the signal radiated by the source. The array output in the frequency domain may be written in terms of the radiated signal as follows:

$$
\begin{aligned}
d\mathbf{F}(\omega) &= \mathbf{a}_0 dF_0(\omega) + w_1 e^{-j\omega\tau_1}\mathbf{a}_1 dF_0(\omega) + d\mathbf{N}(\omega) \\
&= [\mathbf{a}_0, \mathbf{a}_1][1, w_1 e^{-j\omega\tau_1}]^T dF_0(\omega) + d\mathbf{N}(\omega),
\end{aligned}
\tag{5.90}
$$

where $dF_0(\omega)$ is the differential of the generalized Fourier transform of stochastic signal emitted by the source, $d\mathbf{N}(\omega)$ is the differential of the generalized Fourier transform of the background noise presumed to be spatially and temporally white, and $\mathbf{a}_0$ and $\mathbf{a}_1$ are the direction vectors of direct and reflected signals, respectively,

$$\mathbf{a}_0 = \text{col}\{1, e^{-j\omega(d/c)\sin\theta_0}, \ldots, e^{-j\omega(M-1)(d/c)\sin\theta_0}\}$$

$$\mathbf{a}_1 = \text{col}\{1, e^{-j\omega(d/c)\sin\theta_1}, \ldots, e^{-j\omega(M-1)(d/c)\sin\theta_1}\},$$

τ_1 is the delay of the reflected signal relative to the direct arrival and w_1 stands for reflection coefficient. The spectral matrix of the array output is easily derived from Equation 5.90

$$\mathbf{S}_f(\omega) = [\mathbf{a}_0, \mathbf{a}_1][1, w_1 e^{-j\omega\tau_1}]^T [1, w_1 e^{-j\omega\tau_1}][\mathbf{a}_0, \mathbf{a}_1]^H S_0(\omega) + \sigma_\eta^2 \mathbf{I}. \tag{5.91}$$

Define $\tilde{\mathbf{a}} = [\mathbf{a}_0, \mathbf{a}_1][1, w_1 e^{-j\omega\tau_1}]^T = \mathbf{a}_0 + w_1 e^{-j\omega\tau_1}\mathbf{a}_1$ and rewrite Equation 5.81 as

$$\mathbf{S}_f(\omega) = \tilde{\mathbf{a}} S_0(\omega) \tilde{\mathbf{a}}^H + \sigma_\eta^2 \mathbf{I}. \tag{5.92}$$

Note that Equation 5.92 is of the same form as Equation 4.12b. Naturally, it is possible to derive a subspace algorithm to estimate the parameters θ_0 and θ_1; alternatively, the range and height of the source above the reflecting surface, in terms of which θ_0 and θ_1, can be expressed. The Music spectrum is now a function of two parameters, range and height, instead of frequency as in the conventional Music spectrum. For this reason it may be worthwhile to call it a parametric spectrum. A numerical example of a parametric spectrum for a source situated above a reflecting surface and an eight-sensor vertical ULA is given in Figure 5.14. The peak of the parametric spectrum is correctly located, but this good result has been achieved because we have used the exact value of the reflection coefficients. Even a small error, on the order of 1%, appears to completely alter the picture; in particular, the range estimation becomes very difficult.

5.5.2 Two Reflecting Surfaces

To keep the analysis reasonably simple, we shall confine ourselves to two parallel reflecting surfaces and a vertical ULA. A broadband signal is assumed, as simple frequency averaging seems to yield good results (see Figure 5.15). Consider a uniform channel of depth H meters where a vertical array of equispaced (spacing $d_0 < \lambda/2$) sensors is located at depth z_R (depth is measured from the surface of the channel to the top sensor of the array) and a source of radiation at a depth z_S and is horizontally separated by a distance R_0 from the array. In Chapter 1 (page 26), we have noted that depending upon

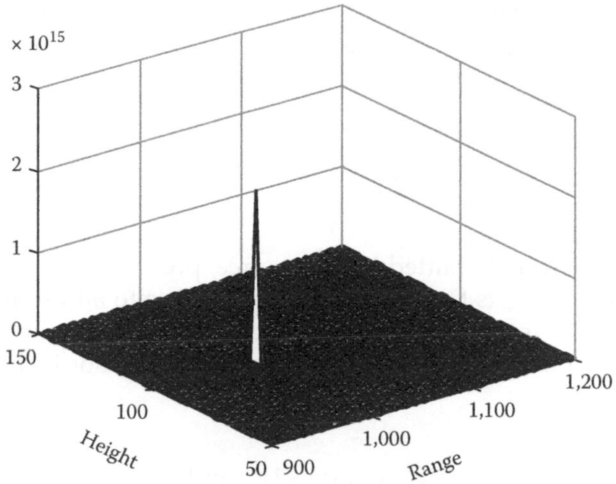

FIGURE 5.14
Parametric spectrum for a source situated above a reflecting surface (100λ) and 1,000λ away from an eight element vertical ULA. The array center is at 16λ above the reflecting surface.

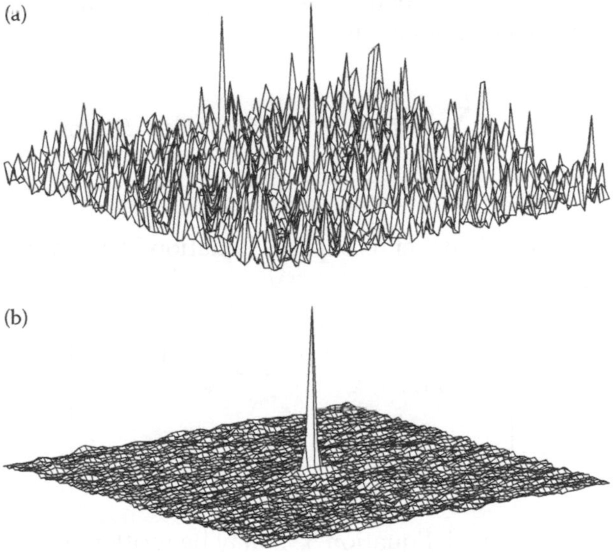

FIGURE 5.15
(a) Parametric spectrum at 1,600 Hz is shown as a function of range and depth. A broadband acoustic source is assumed at 2,000 m (range) and 25 m (depth) in a channel of depth 100 m. Twenty frequency snapshots were used to compute the spectral matrix. Next, the parametric spectra computed at 50 frequencies, equispaced in the band 1,000–2,000 Hz, were averaged. The averaged spectrum is shown in (b). (From P. S. Naidu and H. Uday Shankar, *Sig. Process.*, vol. 72, pp. 107–116, 1999. [47]. With permission.)

the range and the channel characteristics, the array would be able to see a certain number of significant images, say P. The waveform received from P images at mth sensor ($m=0$ is the top sensor) is given by

$$f_m(t)=\sum_{p=0}^{P-1}\frac{\alpha_p}{R_p}f_0(t-\tau_{pm})+\eta_m(t), \tag{5.93}$$

where $f_0(t)$ is the signal emitted by the source, presumed to be a stationary stochastic process, τ_{pm} is time delay from pth image to mth sensor, R_p is the distance to the pth image, and α_p is the reflection coefficient for the ray arriving from the pth image. Noise received by the mth sensor is $\eta_m(t)$. It is easy to show that

$$\tau_{pm}=\tau_p+(m-1)\frac{d}{c}\sin\theta_p \tag{5.94}$$

where τ_p is travel time from pth image to the top sensor, θ_p is the azimuth angle (with respect to the horizontal plane) of the wave vector from pth image, and c is sound speed in sea water. This angle can be computed from the geometry of the image structure, as shown in Figure 5.14. Using the spectral representation of the stationary stochastic process, Equation 5.93 may be written in the frequency domain as

$$dF_m(\omega)=\sum_{p=0}^{P-1}w_p dF_0(\omega)e^{j(m-1)2\pi(d/\lambda)\sin\theta_p}+dN_m(\omega), \tag{5.95}$$

where $w_p=(\alpha_p/R_p)e^{j\omega\tau_p}$ [44]. Let us express Equation 5.95 in a compact matrix notation by defining

$$\mathbf{A}=\begin{bmatrix} 1 & \cdots & 1 \\ e^{j\omega(d/c)\sin\theta_0} & \cdots & e^{j\omega(d/c)\sin\theta_{P-1}} \\ \vdots & & \vdots \\ e^{j\omega(d/c)(M-1)\sin\theta_0} & \cdots & e^{j\omega(d/c)(M-1)\sin\theta_{P-1}} \end{bmatrix},$$

and $\mathbf{w}=\text{col}[w_0, w_1, \ldots, w_{P-1}]$. Equation 5.95 may be written in a matrix form,

$$d\mathbf{F}(\omega)=\mathbf{Aw}dF_0(\omega)+d\mathbf{N}(\omega). \tag{5.96}$$

The spectral matrix is obtained from Equation 5.96 by squaring and averaging

$$\mathbf{S}_f(\omega)=\mathbf{Aww}^H\mathbf{A}^H S_{f0}(\omega)+\sigma_\eta^2\mathbf{I}, \tag{5.97}$$

where $S_{f0}(\omega)$ (scalar) is the spectrum of the source radiation, and σ_η^2 is the variance of the background noise. Define a vector $\tilde{\mathbf{A}} = \mathbf{Aw}$ and rewrite Equation 5.97 as

$$\mathbf{S}_f(\omega) = \tilde{\mathbf{A}} S_{f0}(\omega) \tilde{\mathbf{A}}^H + \sigma_\eta^2 \mathbf{I}. \tag{5.98}$$

Note that the structure of the spectral matrix in Equation 5.98 is the same as in Equation 4.12b. This enables us to develop a subspace algorithm to estimate unknown parameters, range, and depth of the source (also azimuth, if a horizontal array is used) [44].

Source localization in a bounded space is prone to errors in the assumed model as well as the estimation errors due to finite available data. The geometry of the channel needs to be specified with an accuracy of a fraction of wavelength and the wavespeed must be known accurately. In practice, these requirements are often difficult to satisfy. However, it is possible, at the cost of increased array length, to achieve good results when the channel is only partially known [46]. Good results have also been obtained with limited data but using a broadband signal, as demonstrated in Figure 5.15 [47].

5.6 Azimuth/Elevation Estimation

In three-dimensional space, an incident wavefront will have two DOA parameters, namely, azimuth and elevation (see Figure 2.16). We require for the estimation of two DOA parameters, a two-dimensional (2D) array of sensors arranged in a plane s. Some of the commonly employed 2D array geometries are shown in Figure 2.15. A uniform planar array (UPA) consisting of M^2 sensors is an obvious extension of ULA having M sensors in one dimension (1D). Thus, the cost of array hardware increases exponentially, which one can afford to employ only in military application and to some extent in oil exploration. In other less-demanding applications, many types of sparse 2D array geometries are available. We shall focus on these simple array geometries for azimuth and elevation estimation. In particular, we shall consider a circular array and a cross-array.

5.6.1 Circular Array

In Chapter 2, we have already discussed circular array and have derived an expression connecting array signal spectrum matrix to the source spectrum matrix and 2D direction vectors (Equation 2.49a). This has an interesting structure that can be expressed as a product of a vector that is function of azimuth only and a diagonal matrix whose elements are Bessel functions of elevation only. We shall now exploit this structure for application of ESPRIT, which is based on spatial invariance property. Consider any one-direction vector, say, of ith source.

$$\mathbf{a}_i(\theta_i,\varphi_i)=\text{col}\left\{J_0\left(\frac{\omega a}{c}\sin\theta_i\right), J_1\left(\frac{\omega a}{c}\sin\theta_i\right)\right. \\ \left. e^{-j(\varphi_i-(\pi/2))},\ldots, J_{r-1}\left(\frac{\omega a}{c}\sin\theta_i\right)e^{-j(r-1)(\varphi_i-(\pi/2))}\right\} \tag{5.99}$$

We form three subvectors by selecting upper, middle, and lower $r-2$ terms, as shown below:

$$\mathbf{a}_i^-(\theta_i,\varphi_i)=\text{col}\left\{J_0\left(\frac{\omega a}{c}\sin\theta_i\right),\ldots,J_{r-3}\left(\frac{\omega a}{c}\sin\theta_i\right)e^{-j(r-3)(\varphi_i-(\pi/2))}\right\}, \tag{5.100a}$$

$$\mathbf{a}_i^0(\theta_i,\varphi_i)=\text{col}\left\{J_1\left(\frac{\omega a}{c}\sin\theta_i\right)e^{-j(\varphi_i-(\pi/2))},\ldots,J_{r-2}\left(\frac{\omega a}{c}\sin\theta_i\right)e^{-j(r-2)(\varphi_i-(\pi/2))}\right\} \tag{5.100b}$$

$$\mathbf{a}_i^+(\theta_i,\varphi_i)=\text{col}\left\{J_2\left(\frac{\omega a}{c}\sin\theta_i\right)e^{-j2(\varphi_i-(\pi/2))},\ldots,J_{r-1}\left(\frac{\omega a}{c}\sin\theta_i\right)e^{-j(r-1)(\varphi_i-(\pi/2))}\right\}. \tag{5.100c}$$

Now we invoke the recurrence relation of Bessel functions stated in Chapter 2, Section 2.2 and reproduced for convenience

$$J_{k+1}(x)=\frac{2k}{x}J_k(x)-J_{k-1}(x),$$

which we rewrite as

$$J_k(x)=\frac{x}{2k}[J_{k+1}(x)+J_{k-1}(x)],$$

and use in Equation 5.100b for $k=1, 2,\ldots, r-2$. It is easy to show that $\mathbf{a}_i^0(\theta_i,\varphi_i)$ is a linear combination of $\mathbf{a}_i^-(\theta_i,\varphi_i)$ and $\mathbf{a}_i^+(\theta_i,\varphi_i)$, which are given in Equations 5.100a and 5.100c.

$$\mathbf{a}_i^0(\theta_i,\varphi_i)=\gamma\left[\mathbf{a}_i^-(\theta_i,\varphi_i)\frac{\frac{\omega a}{c}\sin\theta_i}{2}e^{-j(\varphi_i-(\pi/2))}+\mathbf{a}_i^+(\theta_i,\varphi_i)\frac{\frac{\omega a}{c}\sin\theta_i}{2}e^{+j(\varphi_i-(\pi/2))}\right], \tag{5.101}$$

where $\gamma=\text{diag}\{1, 1/2,\ldots, 1/r-2\}$.

Let $\mathbf{A}=[\mathbf{a}_0\ \mathbf{a}_1\ \dots\ \mathbf{a}_{P-1}]$ be the direction matrix. Note that $\mathbf{a}_i$ stands for $\mathbf{D}_i\mathbf{Z}_i$ in Equation 2.50. We shall define a set of direction matrices through a selection process illustrated in Equation 5.100, that is,

$$\mathbf{A}^0=[\mathbf{a}_0^0\ \mathbf{a}_1^0\ \dots\ \mathbf{a}_{P-1}^0]$$

$$\mathbf{A}^-=[\mathbf{a}_0^-\ \mathbf{a}_1^-\ \dots\ \mathbf{a}_{P-1}^-]$$

$$\mathbf{A}^+=[\mathbf{a}_0^+\ \mathbf{a}_1^+\ \dots\ \mathbf{a}_{P-1}^+].$$

Using the above matrices, we can express Equation 5.101 in a compact matrix form

$$\mathbf{A}^0=\gamma[\mathbf{A}^-\psi+\mathbf{A}^+\psi^H], \tag{5.102}$$

where

$$\psi=\operatorname{diag}\left\{\frac{\frac{\omega a}{c}\sin\theta_0}{2}e^{-j(\varphi_0-(\pi/2))},\ \frac{\frac{\omega a}{c}\sin\theta_1}{2}e^{-j(\varphi_1-(\pi/2))},\dots,\ \frac{\frac{\omega a}{c}\sin\theta_{P-1}}{2}e^{-j(\varphi_{P-1}-(\pi/2))}\right\}.$$

Recall that the direction vectors and eigenvectors spanning the signal subspace of a spectral matrix are related as in Equation 5.6b. Multiply Equation 5.102 from the right with a unitary matrix (but unknown) $\mathbf{G}$ and use Equation 5.6b. We obtain the following key result

$$\mathbf{v}_s^0=\gamma\left[\mathbf{v}_s^-\overline{\psi}+\mathbf{v}_s^+\overline{\psi}^H\right] \tag{5.103}$$

where $\mathbf{v}_s^0$, $\mathbf{v}_s^-$, and $\mathbf{v}_s^+$ are defined analogously as $\mathbf{A}^0$, $\mathbf{A}^-$, and $\mathbf{A}^\#$ but using the signal eigenvectors of the spectral matrix $\overline{\psi}=\mathbf{G}\psi\mathbf{G}^H$.

Finally, to estimate the unknown DOAs we need to solve Equation 5.103, which we express in a matrix form,

$$\underset{(r-2)\times P}{\mathbf{v}_s^0}=\underset{(r-2)\times 2P}{\left[\gamma\mathbf{v}_s^-\ \gamma\mathbf{v}_s^+\right]}\underset{2P\times P}{\begin{bmatrix}\overline{\psi}\\ \overline{\psi}^H\end{bmatrix}}+\eta, \tag{5.104}$$

where η represents the noise component in the total system. The least squares solution of Equation 5.104 is given by

$$\begin{bmatrix}\overline{\psi}\\ \overline{\psi}^H\end{bmatrix}=\left[(\gamma\mathbf{v}_s^-\ \gamma\mathbf{v}_s^+)^H(\gamma\mathbf{v}_s^-\ \gamma\mathbf{v}_s^+)\right]^{-1}\left[\gamma\mathbf{v}_s^-\ \gamma\mathbf{v}_s^+\right]^H\mathbf{v}_s^0. \tag{5.105}$$

Equation 5.105 straight away yields azimuth and elevation. From the eigenvalues of $\bar{\Psi}$ we obtain the azimuth as the angle of the complex eigenvalue and the elevation from its magnitude. Note that azimuth and elevation angles are properly paired, unlike in ESPRIT application to planar array where extra effort is required for pairing [56].

5.6.2 L-Shaped Array of Crossed Dipoles

We now consider an L-shaped array with crossed electric dipoles arranged uniformly on x and y arms, as shown in Figure 5.16. In Chapter 2, we had considered ULA of crossed dipoles to estimate 1D DOA (i.e., azimuth). Here we shall show how to estimate both azimuth and elevation, and also pair them correctly. The dipoles are assumed to be small in size ($<10\,\lambda$) and spaced at half wavelength. There are M crossed dipoles on each arm. Let P diversely polarized EM waves, not necessarily incoherent, be incident on the array.

The outputs of dipoles on the x-arm are: $\mathbf{u}_x$, the electric field along the x-axis, and $\mathbf{v}_x$, the electric field along the y-axis. Similarly, the outputs of dipoles on the y-arm are: $\mathbf{u}_y$, the electric field along the x-axis, and $\mathbf{v}_y$, the electric field along the y-axis. Each is a vector ($M\times 1$) representing a single snapshot. Following Ref. [57] we express the above quantities as:

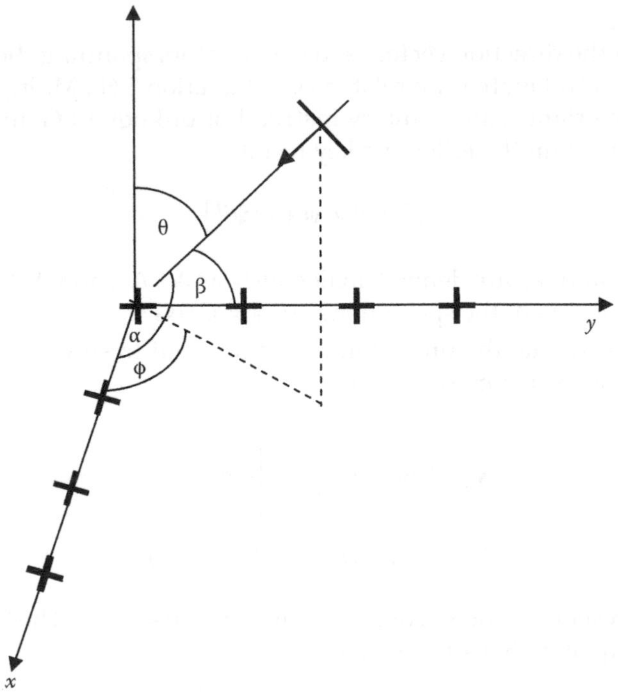

FIGURE 5.16
L-shaped array of crossed electric dipoles. The two arms share a common dipole at the center. There are M dipoles in each arm but there are 4 M-2 outputs in all.

$$\begin{bmatrix} \mathbf{u}_x(n) \\ \mathbf{v}_x(n) \end{bmatrix} = l_0 \sum_{p=0}^{P-1} s_p(n)\mathbf{a}(\alpha_p) \otimes \begin{bmatrix} \cos\theta_p \cos\phi_p \sin\gamma_{pp} e^{j\kappa_p} - \sin\phi_p \cos\gamma_p \\ \cos\theta_p \sin\phi_p \sin\gamma_{pp} e^{j\kappa_p} + \cos\phi_p \cos\gamma_p \end{bmatrix}, \tag{5.106a}$$

where

$$\mathbf{a}(\alpha_p) = [1, e^{-j(2\pi\Delta/\lambda_0)\cos(\alpha_p)}, e^{-j(4\pi\Delta/\lambda_0)\cos(\alpha_p)}, \ldots, e^{(2(M-1)\pi\Delta/\lambda_0)\cos(\alpha_p)}],$$

$S_p(n)$ is the output of pth source at sample instant n, l_0 is the dipole size ($<\lambda_0$), and Δ is the spacing between dipoles. For y-arm, we get similar expressions,

$$\begin{bmatrix} \mathbf{u}_y(n) \\ \mathbf{v}_y(n) \end{bmatrix} = l_0 \sum_{p=0}^{P-1} s_p(n)\mathbf{a}(\beta_p) \otimes \begin{bmatrix} \cos\theta_p \cos\phi_p \sin\gamma_{pp} e^{j\kappa_p} - \sin\phi_p \cos\gamma_p \\ \cos\theta_p \sin\phi_p \sin\gamma_{pp} e^{j\kappa_p} + \cos\phi_p \cos\gamma_p \end{bmatrix}, \tag{5.106b}$$

where

$$\mathbf{a}(\beta_p) = [1, e^{-j(2\pi\Delta/\lambda_0)\cos(\beta_p)}, e^{-j(4\pi\Delta/\lambda_0)\cos(\beta_p)}, \ldots, e^{(2(M-1)\pi\Delta/\lambda_0)\cos(\beta_p)}].$$

Note that α_p and β_p, as shown in Figure 5.16, stand for angles between wave vector and x- and y-arms, respectively.

5.6.3 ESPRIT Algorithm

To apply the ESPRIT algorithm, we need to exploit the spatial invariance property of the data matrices defined below:

$$\mathbf{H}_1 = \begin{bmatrix} \mathbf{u}_x^1(0)\ \mathbf{u}_x^1(1) \ldots \mathbf{u}_x^1(N-1) \\ \mathbf{v}_x^1(0)\ \mathbf{v}_x^1(1) \ldots \mathbf{v}_x^1(N-1) \end{bmatrix}$$

$$\mathbf{H}_2 = \begin{bmatrix} \mathbf{u}_x^2(0)\ \mathbf{u}_x^2(1) \ldots \mathbf{u}_x^2(N-1) \\ \mathbf{v}_x^2(0)\ \mathbf{v}_x^2(1) \ldots \mathbf{v}_x^2(N-1) \end{bmatrix},$$

$$2L \times N$$

where

$$\mathbf{u}_x^1(n) = [u_{0,x}(n), u_{1,x}(n) \ldots u_{L-1,x}(n)]^T,$$

$$\mathbf{u}_x^2(n) = [u_{1,x}(n), u_{2,x}(n) \ldots u_{L,x}(n)]^T,$$

$\mathbf{v}_x^1(n)$ and $\mathbf{v}_x^2(n)$ are similarly defined and L is a constant $P<L<M$.

$$\mathbf{H}_1 = l_0\mathbf{A}_\alpha \circ \begin{bmatrix} \cos\theta_p \cos\phi_p \sin\gamma_{pp} e^{j\kappa_p} - \sin\phi_p \cos\gamma_p \\ \cos\theta_p \sin\phi_p \sin\gamma_{pp} e^{j\kappa_p} + \cos\phi_p \cos\gamma_p \end{bmatrix} \mathbf{S}, \tag{5.107a}$$

$$\underset{2L\times N}{\mathbf{H}_2} = l_0\underset{2L\times P}{\mathbf{A}_\alpha \circ \begin{bmatrix} \cos\theta_p \cos\phi_p \sin\gamma_{pp} e^{j\kappa_p} - \sin\phi_p \cos\gamma_p \\ \cos\theta_p \sin\phi_p \sin\gamma_{pp} e^{j\kappa_p} + \cos\phi_p \cos\gamma_p \end{bmatrix}} \underset{P\times P}{\alpha} \underset{P\times N}{\mathbf{S}}, \tag{5.107b}$$

where $\circ$ is Khattri–Rao product,

$$\alpha = \text{diag}\left\{e^{-j(2\pi\Delta/\lambda_0)\cos(\alpha_0)}, e^{-j(2\pi\Delta/\lambda_0)\cos(\alpha_1)}, \ldots, e^{-j(2\pi\Delta/\lambda_0)\cos(\alpha_{P-1})}\right\},$$

and

$$\mathbf{S} = \begin{bmatrix} s_0(0) & s_0(1) & \cdots & s_0(N-1) \\ s_1(0) & s_1(1) & \cdots & s_1(N-1) \\ \vdots & \vdots & & \vdots \\ s_{P-1}(0) & s_{P-1}(1) & \cdots & s_{P-1}(N-1) \end{bmatrix}.$$

From the structure of Equation 5.107 we conclude that α is the generalized eigenvalue matrix of $\mathbf{H}_1$ and $\mathbf{H}_2$. Note that $\mathbf{H}_1$ and $\mathbf{H}_2$ are rank P matrices. The exact same procedure is followed to estimate β's from the y-arm.

5.6.4 Pairing Algorithm

Equation 5.106a is split into two equations, one for the u component and the other for the v component of the electric field from the x-arm.

$$[\mathbf{u}_x(0)\,\mathbf{u}_x(1)\ldots\mathbf{u}_x(N-1)] = l_0[\mathbf{a}(\alpha_0)\,\mathbf{a}(\alpha_1)\ldots\mathbf{a}(\alpha_{P-1})]\text{diag}\{q_x^0\, q_x^1\ldots q^{P-1}\}\mathbf{S}, \tag{5.108a}$$

$$[\mathbf{v}_x(0)\,\mathbf{v}_x(1)\ldots\mathbf{v}_x(N-1)] = l_0[\mathbf{a}(\alpha_0)\,\mathbf{a}(\alpha_1)\ldots\mathbf{a}(\alpha_{P-1})]\text{diag}\{r_x^0\, r_x^1\ldots r^{P-1}\}\mathbf{S}, \tag{5.108b}$$

where

$$q^p = \cos\theta_p \cos\phi_p \sin\gamma_p e^{j\kappa_p} - \sin\phi_p \cos\gamma_p,$$

$$r^p = \cos\theta_p \sin\phi_p \sin\gamma_p e^{j\kappa_p} + \cos\phi_p \cos\gamma_p.$$

We shall express Equation 5.108 in short matrix form

$$\mathbf{U}_x = l_0 \mathbf{A}_\alpha \mathbf{QS}$$
$$\mathbf{V}_x = l_0 \mathbf{A}_\alpha \mathbf{RS}. \tag{5.109}$$

We like to estimate **QS** and **RS** from Equation 5.109

$$l_0 \mathbf{QS} = \mathbf{A}_\alpha^{\#} \mathbf{U}_x$$
$$l_0 \mathbf{RS} = \mathbf{A}_\alpha^{\#} \mathbf{V}_x, \tag{5.110}$$

where # stands for pseudoinverse. Let $\mathbf{1}=\text{col}\{1\ 1\ \dots\ 1\}$ vector of size $(N\times 1)$. Postmultiply on both sides of Equation 5.110 and obtain

$$l_0 \mathbf{QS1} = l_0 \mathbf{Q}\left[\sum_{n=0}^{N-1} s_0(n) \sum_{n=0}^{N-1} s_1(n) \dots \sum_{n=0}^{N-1} s_{P-1}(n)\right]^T = \mathbf{A}_\alpha^{\#} \mathbf{U}_x \mathbf{1}, \tag{5.111a}$$

$$l_0 \mathbf{RS1} = l_0 \mathbf{R}\left[\sum_{n=0}^{N-1} s_0(n) \sum_{n=0}^{N-1} s_1(n) \dots \sum_{n=0}^{N-1} s_{P-1}(n)\right]^T = \mathbf{A}_\alpha^{\#} \mathbf{V}_x \mathbf{1}. \tag{5.111b}$$

Element-by-element division of two vectors given by Equation 5.111a and b we obtain

$$\left[\frac{q^0}{r^0} \frac{q^1}{r^1} \dots \frac{q^{P-1}}{r^{P-1}}\right]^T = \mathbf{A}_\alpha^{\#} \mathbf{U}_x \mathbf{1} ./ \mathbf{A}_\alpha^{\#} \mathbf{V}_x \mathbf{1}, \tag{5.112a}$$

For the y-arm, carrying out the above steps we arrive at a parallel result

$$\left[\frac{q^0}{r^0} \frac{q^1}{r^1} \dots \frac{q^{P-1}}{r^{P-1}}\right]^T = \mathbf{A}_\beta^{\#} \mathbf{U}_y \mathbf{1} ./ \mathbf{A}_\beta^{\#} \mathbf{V}_y \mathbf{1}, \tag{5.112b}$$

where ./ stands for element-by-element division (MATLAB® convention). From Equation 5.112a and b we observe that

$$\mathbf{A}_\alpha^{\#} \mathbf{U}_x \mathbf{1} ./ \mathbf{A}_\alpha^{\#} \mathbf{V}_x \mathbf{1} = \mathbf{A}_\beta^{\#} \mathbf{U}_y \mathbf{1} ./ \mathbf{A}_\beta^{\#} \mathbf{V}_y \mathbf{1}. \tag{5.113}$$

Equation 5.113 provides a basis for pairing [57]. The order in which α's occur in $\mathbf{A}_\alpha$ and β's in $\mathbf{A}_\beta$ must be identical. To pair the DOAs, compute the left-hand side of Equation 5.113 for one fixed order of α's and compute the right-hand side for all possible combinations of β's. ($P!$ different combinations of β's.) Whenever the right-hand side is closest (minimum mean square difference) to the left-hand side, we will have achieved the right pairing.

5.7 Exercises

(1) The line represented by Equation 5.89 intersects the circle at two points p_1 and p_2 (see Figure 5.12). Let two adjacent sensors be on the x-axis. Show that one of the points will be at the intersection of the x-axis and circle. Where will the second point be?

(2) In Chapter 4 (Equation 4.13a) it was shown that $\mathbf{A}^H\mathbf{Q}=\mathbf{0}$, where $\mathbf{Q}$ was defined in terms of partitions of the spectral matrix. Is $\mathbf{Q}$ itself the noise subspace? Remember that in obtaining $\mathbf{Q}$ no eigendecomposition was required.

(3) Show that, taking into account the variation in the sensitivity of the sensors, Equation 5.72b takes the form

$$\tilde{\mathbf{c}}_f^{\alpha}(\tau) = c_f^{\alpha}(\tau)\alpha_2(\theta_0)\mathbf{a}(\alpha,\theta_0),$$

where

$$\alpha_2(\theta_0)=\mathrm{diag}\{|\alpha_0(\theta_0)|^2,\ |\alpha_1(\theta_0)|^2,\ \ldots,\ |\alpha_{M-1}(\theta_0)|^2\}.$$

Interestingly, phase variations themselves do not affect the relation given in Equation 5.71b.

(4) How would you cancel the interference coming from the users of no interest in Equations 5.60d and 5.69? Can you also cancel the term due to the noise?

(5) In Section 5.1.2, we have shown how to restore the rank of a rank-deficient spectral matrix by smoothing the spectral matrices of subarrays. Show that this can also be achieved by smoothing of the signal subspace eigenvectors of the spectral matrix of the full array. This approach is suggested in Ref. [48] but on a covariance matrix.

(6) Let $\mathbf{J}$ be an exchange matrix that collects all odd and even elements of a vector; for example, a 4×4 exchange matrix is

$$\mathbf{J}=\begin{bmatrix}1&0&0&0\\0&0&1&0\\0&1&0&0\\0&0&0&1\end{bmatrix}.$$

Consider a ULA with M sensors and P uncorrelated wavefronts that are incident on the array. Let $\mathbf{v}_s$ be a matrix whose columns are the signal eigenvectors of the array spectral matrix. Show that $\mathbf{J}\mathbf{v}_s=[\mathbf{A}\ \mathbf{A}\Gamma]\mathbf{G}^{-1}$, where matrices $\mathbf{A}$ $((M/2)\times P)$ and $\mathbf{G}$ $(P\times P)$ are as defined in Equation 5.6c and the $\Gamma(P\times P)$ is as defined in Equation 2.57.

This result provides an alternate approach to the ESPRIT algorithm described in Section 5.1. It can be used to extend the concept of subspace rotation to multiple subarrays as described in Ref. [49].

References

1. R. O. Schmidt, Multiple emitter location and signal parameter estimation, in *Proceedings of RADC Spectral Estimation Workshop*, Griffiths AFB, New York, pp. 243–258, Reprinted in *IEEE Trans.*, AP-34, pp. 276–280, 1986.
2. G. Bienvenu and L. Kopp, Principe de la goniometri passive adaptive, *Proc. Colloque*, CRESTI, Nice (France), pp. 106/1–10, 1979.
3. V. F. Pisarenko, The retrieval of harmonics from covariance function, *Geophys. J. R. Astron. Soc.*, vol. 33, pp. 347–366, 1973.
4. C. U. Padmini and P. S. Naidu, Circular array and estimation of direction of arrival of a broadband source, *Signal Process.*, vol. 37, pp. 243–254, 1994.
5. A. B. Gershman and V. T. Ermolaev, *Electron. Lett.*, vol. 28, pp. 1114–1115, 1992.
6. S. U. Pillai, *Array Signal Processing*, Springer-Verlag, New York, 1989.
7. A. B. Gershman and V. T. Ermolaev, Optimal subarray size for spatial smoothing, *IEEE SP Lett.*, vol. 2, pp. 28–30, 1995.
8. A. J. Barabell, Improving the resolution performance of eigenstructure based direction finding algorithm, *IEEE Proc. ICASSP*, pp. 336–339, 1983.
9. B. D. Rao, Performance analysis of root Music, *IEEE Trans.*, ASSP-37, pp. 1939–1049, 1989.
10. R. Roy, A. Paulraj, and T. Kailath, Direction of arrival estimation by subspace rotation method – ESPRIT, *IEEE Proc. ICASSP*, Tokyo, pp. 2495–2498, 1986.
11. D. Slepian, H. D. Pollack, and H. J. Landau, Prolate spheroidal wave function, Fourier analysis and uncertainty, *Bell Sys. Tech. J.*, vol. 40, pp. 43–84, 1961.
12. U. Reddy, B. Egart, and T. Kailath, Least squares type algorithm for adaptive implementation of Pisarenko's harmonic retrieval method, *IEEE Trans.*, ASSP-30, pp. 339–405, 1982.
13. P. Yang and M. Kaveh, Adaptive eigensubspace algorithm for direction or frequency estimation and tracking, *IEEE Trans.*, ASSP-36, pp. 241–251, 1988.
14. J. Karhunen, Adaptive algorithms for estimation of eigenvectors of correlation type matrices, *IEEE Proc.*, ICASSP-84, pp. 14.6.1–4, 1984.
15. G. Stewart, *Introduction to Matrix Computations*, Academic Press, New York, 1973.
16. H. Wang and M. Kaveh, Coherent signal subspace processing for the detection and estimation of multiple wideband sources, *IEEE*, ASSP-30, pp. 823–831, 1985.
17. H. Wang and M. Kaveh, Focussing matrices for coherent signal subspace processing, *IEEE*, ASSP-36, pp. 1272–1281, 1988.
18. J. Krolick and D. N. Swinger, Focussed wideband array processing via spatial sampling, *IEEE*, ASSP-38, pp. 356–360, 1990.
19. K. M. Buckly and L. Griffiths, Broadband signal subspace spatial spectrum (BASS-ALE) estimation, *IEEE*, ASSP-36, pp. 953–964, 1988.

20. D. N. Swinger and J. Krolik, Source location bias in the coherently focused high resolution broadband beamformation, *IEEE*, ASSP-37, pp. 143–144, 1989.
21. J. Krolik and D. Swingler, Multiple broadband source location using steered covariance matrices, *IEEE Trans.*, ASSP-37, pp. 1481–1494, 1989.
22. S. Valae and P. Kabal, Wideband array processing using a two sided correlation transformation, *IEEE Trans.*, SP-43, pp. 160–172, 1995.
23. K. M. Buckley and L. Griffiths, Eigenstructure based broadband source localization, *IEEE*, ICASSP-86, pp. 1867–1872, 1986.
24. C. Usha Padmini, Wideband source localization, ME dissertation, ECE Dept, IISc, Bangalore, India, 1991.
25. P. S. Naidu, *Modern Spectrum Analysis of Time Series*, CRC Press, Boca Raton, FL, 1996.
26. J. W. Mauchley, Significance test for sphericity of a normal *n*-variate distribution, *Ann. Math. Stat.*, vol. 11, pp. 204–209, 1940.
27. W. S. Ligget Jr., Passive sonar: Fitting models to multiple time series, in *Signal Processing: Proc. NATO ASI Signal Processing*, J. W. R. Griffiths and P. L. Stocklin (Eds), Academic, New York, pp. 327–345, 1973.
28. D. B. Williams and D. H. Johnson, Using the sphericity test for source detection with narrow band passive array, *IEEE Trans.*, ASSP-38, pp. 2008–2014, 1990.
29. M. Wax and T. Kailath, Detection of signals by information theoretic criteria, *IEEE Trans.*, ASSP-33, pp. 387–392, 1985.
30. Q.-T. Zhang, K. M. Wong, P. C. Yip, and J. P. Reilly, Statistical analysis of the performance of information theoretic criteria in the detection of the number of signals in array processing, *IEEE Trans.*, ASSP-39, pp. 1557–1567, 1989.
31. J. P. Henrioux and G. Jourdain, Use of large bandwidth-duration binary phase shift keying signals for bearing measurements in active sonar classification, *JASA*, vol. 97, pp. 1737–1746, 1995.
32. R. L. Pickholz, D. L. Schilling, and L. B. Milstein, Theory of spread spectrum communications – A tutorial, *IEEE Trans.*, COM-30, pp. 855–884, 1982.
33. R. L. Pickholz, L. B. Milstein, and D. L. Schilling, Spread spectrum for mobile communications, *IEEE Trans. Vehicular Technol.*, vol. 40, pp. 313–322, 1991.
34. A. F. Naguib, A. Paulraj, and T. Kailath, Capacity improvement with base station antenna array in cellular CDMA, *IEEE Trans.*, vol. 43, pp. 691–698, 1994.
35. W. A. Gardner, *Introduction to Random Processes with Application to Signals and Systems*, Macmillan, New York, 1985.
36. W. A. Gardner, Exploitation of spectral redundancy in cyclostationarity signals, *IEEE Signal Proc. Mag.*, April, pp. 14–36, 1991.
37. S. V. Schell, R. A. Calabretta, W. A. Gardner, and B. G. Agee, Cyclic MUSIC algorithm for signal selective direction finding, *Proc. ICASSP89*, vol. 4, pp. 2278–2281, 1989.
38. G. Xu and T. Kailath, Direction of arrival estimation via exploitation of cyclostationarity – A combination of temporal and spatial processing, *IEEE Trans.*, SP-40, pp. 1775–1786, 1992.
39. R. T. Compton, Pointing accuracy and dynamic range in steered beam adaptive array, *IEEE Trans.*, AES-16, pp. 280–287, 1980.
40. V. C. Soon, L. Tong, Y. F. Huang, and R. Liu, A subspace method for estimating sensor gains and phases, *IEEE Trans.*, SP-42, pp. 973–976, 1994.
41. J. J. Smith, Y. H. Leung, and A. Cantoni, The partitioned eigenvector method for towed array shape estimation, *IEEE Trans.*, SP-44, pp. 2273–2283, 1996.

42. D. A. Gray, W. O. Wolfe, and J. L. Riley, An eigenvector method for estimating the positions of the elements of an array of receivers, in *Proc. ASSPA 89, Signal Processing Theories, Implementation, and Applications*, R. F. Barrett (Ed.), Adelaide, Australia, pp. 391–393, 1989.
43. S. B. Kesler, J. Kesler, and G. Levita, Resolution of coherent target in the near field of an array antenna, *IEEE Proc.*, ICASSP-86, pp. 1937–1940, 1986.
44. P. S. Naidu, A new method of acoustic source localization in shallow sea, *Proc. Inst. Acoust.*, vol. 18, part 5, pp. 39–48, 1996.
45. H. F. Silverman, Some analysis of microphone arrays for speech data acquisition, *IEEE Trans.*, ASSP-35, pp. 1699–1712, 1987.
46. P. S. Naidu and T. Ganesan, Source localization in a partially known shallow water channel, *J. Acoust. Soc. Am.*, vol. 98, pp. 2554–2559, 1995.
47. P. S. Naidu and H. Uday Shankar, Broadband source localization in shallow water, *Signal Processing*, vol. 72, pp. 107–116, 1999.
48. S. S. Reddi and A. B. Gershman, An alternative approach to coherent source location problem, *IEEE Trans.*, SP-59, pp. 221–233, 1997.
49. A. L. Swindlehurst, B. Ottersten, R. Roy, and T. Kailath, Multiple invariance ESPRIT, *IEEE Trans.*, SP-40, pp. 867–881, 1992.
50. A. Papoulis, *Probability, Random Variables and Stochastic Processes*, Third edition, McGraw-Hill Book Co., New York, 1991.
51. D. Liu and L. Tong, An analysis of constant modulus algorithm for array signal processing, *Signal Process.*, vol. 73, pp. 81–104, 1999.
52. A.-J. van der Veen and A. Paulraj, An analytic constant modulus algorithm, *IEEE Trans. Signal Process.*, vol. 44, pp. 1136–1155, 1996.
53. B. Friedlander and A. J. Weiss, Direction finding in the presence of mutual coupling, *IEEE Trans. Antenn. Propag.*, vol. 39, pp. 273–284, 1991.
54. C. Qi, Y. Wang, Y. Zhang, and H. Chen, DOA estimation and self-calibration algorithm for uniform circular array, *Electron. Lett.*, vol. 51, pp. 1–2, 2005.
55. D.-Y. Gao, B.-H. Wang, and Y. Guo, Comments on "Blind calibration and DOA estimation with uniform circular arrays in the presence of mutual coupling", *IEEE Antennas Wireless Propagat. Lett.*, vol. 5, pp. 566–569, 2006.
56. A.-J. van der Veen, P. B. Ober, and E. F. Deprettere, Azimuth and elevation computation in high resolution DOA estimation, *IEEE Trans. Signal Processing*, vol. 40, pp. 1828–1832, 1992.
57. J. E. Fernandez del Rio and M. F. Catedra-Perez, The matrix pencil method for two-dimensional direction of arrival estimation employing an L-shaped array, *IEEE Trans. Antenn. Propag.*, vol. 45, pp. 1093–1094, 1997.
58. Y. Sato, A method of self-recovering equalization for multilevel amplitude-modulation, *IEEE Trans. Comms.*, pp. 680–682, 1975.
59. J. R. Treichler and B. G. Agee, A new approach to multipath correction of constant modulus signals, *IEEE Trans.*, ASSP-31, pp. 459–471, 1983.
60. J. R. Treichler, and M. G. Larimore, New processing techniques based on the constant modulus adaptive algorithm, *IEEE Trans.*, ASSP-33, pp. 420–431, 1985.

6

Source Estimation

The temporal signals radiated by sources are separated on the principle of non-overlapping or partially overlapping temporal spectral characteristics of the signals. A perfect separation is possible only when the signal spectra are non-overlapping. The spatio-temporal signals possess an additional degree of variability, namely, the spatial spectrum. The differences in the spatial spectra can be used, in addition to the differences in the temporal spectra, for the purpose of signal separation. The signals coming from widely different directions will have non-overlapping spatial spectra and therefore they can be separated using an array of sensors. However, when the signal sources are quite close, perfect separation is not possible. There will be some cross-talk. We shall evaluate the Wiener filter, which was derived in Chapter 3, from the point of cross-talk power in relation to the total signal power. Suppression of unwanted signal or interference is achieved by placing a null or a collection of nulls in the spatial frequency band occupied by the interference. The effectiveness of nulls is enhanced when additional constraints are placed on the filter; for example, the filter response is unity in the direction of useful signal. This leads to the well-known Capon filter, which is also known as the minimum variance filter. It is found that the Capon filter is quite effective when the signal and the interference sources are highly directional. The filter will automatically place a null wherever there is a strong interference. Finally, when the direction of interference is known *a priori*, it is possible to devise a filter that will place a sharp null at the spatial frequency corresponding to the direction of arrival (DOA) of the interference. The null can be steered to any desired position, depending upon how the interference is changing its direction. Thus, null steering can be effectively used to suppress slowly varying interference. In Section 6.4, we take up the problem of source estimation in space-time domain leading to wideband adaptive beamformation, and in Section 6.5 we consider frequency invariant (FI) beamformation often used in the estimation of speech signal.

6.1 Wiener Filters

In Chapter 3, we have derived Wiener filters for extraction of a wavefront incident at M sensor uniform linear array (ULA) and uniform circular array (UCA). It may be recalled that the Wiener filter minimizes the mean square error (MSE) between the filter output and the desired signal. The signal and noise are assumed to be stationary stochastic processes. The basic equation (Equation 3.38) is reproduced here for convenience,

$$\mathbf{S}_f(\omega)\mathbf{H}(\omega) = \mathbf{S}_0(\omega),$$

where $\mathbf{S}_f(\omega)$ is the spectral matrix of the array output,

$$\mathbf{S}_f(\omega) = \begin{bmatrix} S_{0,0}(\omega) & S_{0,1}(\omega) & \cdots & S_{0,M-1}(\omega) \\ S_{1,0}(\omega) & S_{1,1}(\omega) & \cdots & S_{1,M-1}(\omega) \\ \vdots & \vdots & & \vdots \\ \vdots & \vdots & & \vdots \\ \vdots & \vdots & & \vdots \\ S_{M-1,0}(\omega) & S_{M-1,1}(\omega) & \cdots & S_{M-1,M-1}(\omega) \end{bmatrix},$$

and $\mathbf{S}_0(\omega)$ is the cross-spectral vector between the desired output and the array output. We had derived a specific result for a single wavefront with white background noise. It was shown that the filter function is given by

$$\mathbf{H}_w(\omega,\theta) = Q\mathbf{a}_0(\omega,\theta_0), \tag{6.1a}$$

where

$$Q = \frac{\dfrac{S_0(\omega)}{\sigma_\eta^2}}{1 + M\dfrac{S_0(\omega)}{\sigma_\eta^2}},$$

$\mathbf{a}_0(\omega, \theta_0)$ is a steering vector and $\mathbf{S}_0(\omega)$ is the spectrum of the incident signal.

6.1.1 Filter Output

We shall now use this filter on the Fourier transformed array output, $\hat{\mathbf{F}}(\omega)$

$$\begin{aligned} \hat{\mathbf{F}}(\omega) &= \mathbf{H}_W^H(\omega, \theta_0)\mathbf{F}(\omega) \\ &= Q\mathbf{a}_0^H(\omega, \theta_0)\left[\mathbf{a}_0(\omega, \theta_0)\Xi_0(\omega) + \mathbf{N}(\omega)\right] \\ &= \Xi_0(\omega) + Q\mathbf{a}_0^H(\omega, \theta_0)\mathbf{N}(\omega). \end{aligned} \tag{6.1b}$$

Note that the signal component in the output remains undisturbed and the noise variance is reduced by a factor of $1/M$. The filter response to an incident plane wave coming from a different direction with a direction vector $\mathbf{a}(\omega, \theta)$ is given by $Q\mathbf{a}_0^H(\omega, \theta_0)\mathbf{a}(\omega, \theta)$. The response of the Wiener filter will be compared later with that of the Capon filter in Figure 6.4.

6.1.1.1 Array Gain

Array gain is defined as a ratio of the output signal-to-noise ratio (SNR) to the input SNR. For the purpose of illustration, let us consider a single source in the presence of white noise. The spectral matrix of the array output (Equation 4.12b), which we shall write in a slightly different form showing the signal power explicitly

$$\mathbf{S}_f(\omega) = \sigma_s^2 \mathbf{a}_0(\omega, \theta_0)\tilde{S}_0(\omega)\mathbf{a}_0^H(\omega, \theta_0) + \sigma_\eta^2 \mathbf{I},$$

where $\tilde{S}_0(\omega)$ is a normalized spectrum. The input SNR is evidently equal to σ_s^2/σ_η^2. We assume that the Wiener filter (Equation 6.1a) is used to extract the signal whose DOA is known *a priori*. The output power is given by

$$\mathbf{H}_W^H(\omega,\theta_0)\mathbf{S}_f(\omega)\mathbf{H}_W(\omega,\theta_0) = \sigma_s^2 Q^2 \left|\mathbf{a}_0(\omega,\theta_0)\right|^2 \tilde{S}_0(\omega)\left|\mathbf{a}_0(\omega,\theta_0)\right|^2 + \sigma_\eta^2 Q^2 \left|\mathbf{a}_0(\omega,\theta_0)\right|^2.$$

It is easy to show the output SNR as $(\sigma_s^2/\sigma_\eta^2)\left|\mathbf{a}_0(\omega,\theta_0)\right|^2$. By definition, the array gain is equal to $|\mathbf{a}_0(\omega, \theta_0)|^2$ which, for ideal sensor array with omnidirectional and unit response sensors, is equal to M (the number of sensors).

6.1.2 Two Source Case

Let us now consider two wavefronts incident on an array (ULA) in the presence of white background noise. The sources are assumed to be uncorrelated between themselves as well as with the background noise. The aim is to extract a first source signal while suppressing the signal from the second source as well as the background noise. Corresponding to this model, the spectral matrix and the cross-spectral vector in the Wiener filter equation take the following form,

$$\begin{aligned} \mathbf{S}(\omega) &= \left\{\mathbf{a}_0(\omega,\theta_0)\mathbf{a}_0^H(\omega,\theta_0)S_0(\omega) + \mathbf{a}_1(\omega,\theta_1)\mathbf{a}_1^H(\omega,\theta_1)S_1(\omega) + \sigma_\eta^2\mathbf{I}\right\} \\ \mathbf{S}_0(\omega) &= \mathbf{a}_0(\omega,\theta_0)S_0(\omega), \end{aligned} \tag{6.2}$$

where

$$\mathbf{a}_0(\omega,\theta_0) = \mathrm{col}[1, e^{-j(\omega/c)d\sin\theta_0}, e^{-j(\omega/c)2d\sin\theta_0}, \ldots, e^{-j(\omega/c)(M-1)d\sin\theta_0}],$$

and

$$\mathbf{a}_1(\omega,\theta_1) = \mathrm{col}[1, e^{-j(\omega/c)d\sin\theta_1}, e^{-j(\omega/c)2d\sin\theta_1}, \ldots, e^{-j(\omega/c)(M-1)d\sin\theta_1}],$$

where θ_0 and θ_1 are DOA angles of the first and the second wavefront, respectively. The filter response for extraction of the first source signal is given by

$$\mathbf{H}_W(\omega,\theta_0) = \left[\mathbf{a}_0(\omega,\theta_0)\mathbf{a}_0^H(\omega,\theta_0) + \mathbf{a}_1(\omega,\theta_1)\mathbf{a}_1^H(\omega,\theta_1)\frac{S_1(\omega)}{S_0(\omega)} + \frac{\sigma_\eta^2}{S_0(\omega)}\mathbf{I}\right]^{-1} \mathbf{a}_0(\omega,\theta_0). \tag{6.3}$$

The inverse of the quantity inside the square brackets in Equation 6.3 has been derived in Chapter 4, Section 4.3. Using that result we obtain

$$\mathbf{H}_W(\omega,\theta_0)=\left[\mathbf{V}^{-1}-\frac{S_1(\omega)}{S_0(\omega)}\frac{\mathbf{V}^{-1}\mathbf{a}_1(\omega,\theta_1)\mathbf{a}_1^H(\omega,\theta_1))\mathbf{V}^{-1}}{1+\frac{S_1(\omega)}{S_0(\omega)}\mathbf{a}_1^H(\omega,\theta_1)\mathbf{V}^{-1}\mathbf{a}_1(\omega)}\right]\mathbf{a}_0(\omega,\theta_0), \tag{6.4a}$$

where

$$\mathbf{V}^{-1}=\frac{S_0(\omega)}{\sigma_\eta^2}\left[\mathbf{I}-Q\mathbf{a}_0(\omega,\theta_0)\mathbf{a}_0^H(\omega,\theta_0)\right], \tag{6.4b}$$

where

$$Q=\frac{\frac{\mathbf{S}_0(\omega)}{\sigma_\eta^2}}{1+M\frac{\mathbf{S}_0(\omega)}{\sigma_\eta^2}}.$$

Though the filter is tuned to receive the first wavefront, some amount of energy from the second wavefront will leak into the filter output. This is known as cross-talk. Ideally, the cross-talk should be zero. Let $\mathbf{a}_1(\omega)\Xi_1(\omega)$ be the Fourier transform of the signal emitted by the second source. The output of the filter tuned to the first source is given by

$$\text{Output}=\left\{\mathbf{H}_W^H(\omega,\theta_0)\mathbf{a}_0(\omega,\theta_0)\Xi_0(\omega)+\mathbf{H}_W^H(\omega,\theta_0)\mathbf{a}_1(\omega,\theta_1)\Xi_1(\omega)+\mathbf{H}_W^H(\omega,\theta_0)\mathbf{N}(\omega)\right\}. \tag{6.5a}$$

Further, to evaluate different terms in Equation 6.5a we need the following results:

$$\mathbf{a}_0^H(\omega,\theta_0)\mathbf{V}^{-1}\mathbf{a}_1(\omega,\theta_1)=\frac{S_0(\omega)}{\sigma_\eta^2}\frac{\mathbf{a}_0^H(\omega,\theta_0)\mathbf{a}_1(\omega,\theta_1)}{1+M\frac{S_0(\omega)}{\sigma_\eta^2}}, \tag{6.5b}$$

and

$$\mathbf{a}_1^H(\omega,\theta_1)\mathbf{V}^{-1}\mathbf{a}_1(\omega,\theta_1)=\frac{S_0(\omega)}{\sigma_\eta^2}\left[M-Q\left|\mathbf{a}_1^H(\omega,\theta_1)\mathbf{a}_0(\omega,\theta_0)\right|^2\right]. \tag{6.5c}$$

Using Equations 6.5b and 6.5c in Equation 6.5a we obtain

$$\mathbf{H}_W^H(\omega,\theta_0)\mathbf{a}_0(\omega,\theta_0)\Xi_0(\omega)$$

$$=MQ\Xi_0(\omega)-\frac{S_1(\omega)}{S_0(\omega)}\frac{Q^2\left|\mathbf{a}_1^H(\omega,\theta_1)\mathbf{a}_0(\omega,\theta_0)\right|^2\Xi_0(\omega)}{1+\frac{S_1(\omega)}{\sigma_\eta^2}\left[M-Q\left|\mathbf{a}_1^H(\omega,\theta_1)\mathbf{a}_0(\omega,\theta_0)\right|^2\right]}, \tag{6.6a}$$

$$\mathbf{H}_W^H(\omega,\theta_0)\mathbf{a}_1(\omega,\theta_1)\Xi_1(\omega) = Q\frac{\mathbf{a}_0^H(\omega,\theta_0)\mathbf{a}_1(\omega,\theta_1)\Xi_1(\omega)}{1+\frac{S_1(\omega)}{\sigma_\eta^2}\left[M - Q\left|\mathbf{a}_1^H(\omega,\theta_1)\mathbf{a}_0(\omega,\theta_0)\right|^2\right]}. \tag{6.6b}$$

We shall assume that the array signal-to-noise ratio (aSNR) is much greater than one, that is, $MS_0(\omega)/\sigma_\eta^2 \gg 1$. Using this approximation in Equation 6.6 the array output power may be approximated as

$$\begin{aligned}\text{Output power} \approx S_0(\omega) + \left\{\begin{array}{l} \left(\frac{S_1(\omega)}{S_0(\omega)}\right)^2 \times \frac{\frac{\left|\mathbf{a}_1^H(\omega,\theta_1)\mathbf{a}_0(\omega,\theta_0)\right|^4}{M^4}S_0(\omega)}{\left[1+\frac{S_1(\omega)}{\sigma_\eta^2}M\left[1-\frac{\left|\mathbf{a}_1^H(\omega,\theta_1)\mathbf{a}_0(\omega,\theta_0)\right|^2}{M^2}\right]\right]^2} \\ -2\frac{\frac{\left|\mathbf{a}_1^H(\omega,\theta_1)\mathbf{a}_0(\omega,\theta_0)\right|^2}{M^2}S_1(\omega)}{1+\frac{S_1(\omega)}{\sigma_\eta^2}M\left[1-\frac{\left|\mathbf{a}_1^H(\omega,\theta_1)\mathbf{a}_0(\omega,\theta_0)\right|^2}{M^2}\right]} \\ +\frac{\frac{\left|\mathbf{a}_0^H(\omega,\theta_0)\mathbf{a}_1(\omega,\theta_1)\right|^2}{M^2}S_1(\omega)}{\left[1+\frac{S_1(\omega)}{\sigma_\eta^2}M\left[1-\frac{\left|\mathbf{a}_1^H(\omega,\theta_1)\mathbf{a}_0(\omega,\theta_0)\right|^2}{M^2}\right]\right]^2} \\ +\mathbf{H}_W^H(\omega,\theta_0)\mathbf{H}_W(\omega,\theta_0)\sigma_\eta^2(\omega) \end{array}\right\}.\end{aligned} \tag{6.7}$$

The first term is the desired signal power. The remaining terms represent the interference. Of these three, the magnitude of the first is much lower than that of the second and third terms. The magnitude of the first is proportional to $\left|\mathbf{a}_1^H(\omega,\theta_1)\mathbf{a}_0(\omega,\theta_0)\right|^4/M^4$, while the magnitude of the second and third terms is proportional to $\left|\mathbf{a}_1^H(\omega,\theta_1)\mathbf{a}_0(\omega,\theta_0)\right|^2/M^2$. Hence, we drop the first term from the interference expression. The cross-talk, defined as a ratio of the power leaked from the second source and the actual power in the second source, is given by

$$\text{Cross talk} \approx \frac{\left|\mathbf{a}_0^H(\omega,\theta_0)\mathbf{a}_1(\omega,\theta_1)\right|^2}{M^2}\left\{1-\frac{\left[\frac{S_1(\omega)}{\sigma_\eta^2}M\left(1-\frac{\left|\mathbf{a}_0^H(\omega,\theta_0)\mathbf{a}_1(\omega,\theta_1)\right|^2}{M^2}\right)\right]^2}{\left[1+\frac{S_1(\omega)}{\sigma_\eta^2}M\left(1-\frac{\left|\mathbf{a}_0^H(\omega,\theta_0)\mathbf{a}_1(\omega,\theta_1)\right|^2}{M^2}\right)\right]^2}\right\}. \tag{6.8}$$

Note that $|\mathbf{a}_1^H(\omega,\theta_1)\mathbf{a}_0(\omega,\theta_0)|^2/M^2$ represents the square of cosine of the angle between two direction vectors $\mathbf{a}_0(\omega, \theta_0)$ and $\mathbf{a}_1(\omega, \theta_1)$. When $\mathbf{a}_0(\omega, \theta_0)=\mathbf{a}_1(\omega, \theta_1)$, $|\mathbf{a}_1^H(\omega, \theta_1)\mathbf{a}_0(\omega, \theta_0)|^2/M^2 = 1$ and the cross-talk $=1$, as expected. But when $\mathbf{a}_0(\omega, \theta_0) \perp \mathbf{a}_1(\omega, \theta_1)$, $|\mathbf{a}_0^H(\omega, \theta_0)\mathbf{a}_1(\omega, \theta_1)|^2/M^2 = 0$ and the cross-talk is zero. Aside from these two extreme situations, the cross-talk may be reduced to zero if $MS_1(\omega)/\sigma_\eta^2 \gg 1$, that is, the aSNR for the second source is very large. The cross-talk is shown in Figure 6.1. The cross-talk also depends upon the DOA of the first course. For off-broadside DOA the cross-talk is greater, that is, it persists over a wider angular width. This is shown in Figure 6.1 where all other parameters are the same as in Figure 6.2.

6.1.3 Linear Least Squares Estimate (LLSE)

As previously stated the signal model is

$$\mathbf{f}(t) = [\mathbf{a}_0, \mathbf{a}_1, \ldots, \mathbf{a}_{P-1}] \begin{bmatrix} \xi_0(t) \\ \xi_1(t) \\ \xi_{P-1}(t) \end{bmatrix} + \eta(t),$$
$$= \mathbf{A}\xi(t) + \eta(t), \qquad t = 0, 1, \ldots, N \tag{6.9}$$

where $\mathbf{a}_0, \mathbf{a}_1, \ldots, \mathbf{a}_{P-1}$ are steering vectors of P sources. It is assumed that $\mathbf{A}$ is known and $\xi(t)$ is unknown but deterministic. The least squares estimate is obtained by minimizing with respect to $\xi(t)$

$$\|\mathbf{f}(t) - \mathbf{A}\xi(t)\|^2 = \min. \tag{6.10}$$

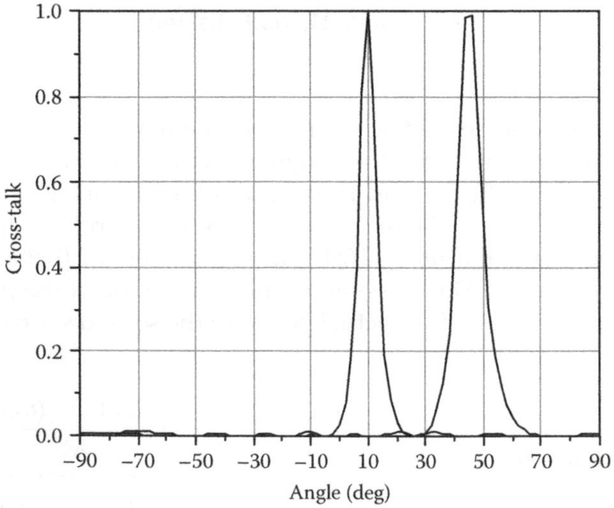

FIGURE 6.1
Cross-talk as a function of angular distance between two sources for two different off-broadside DOA of the first source.

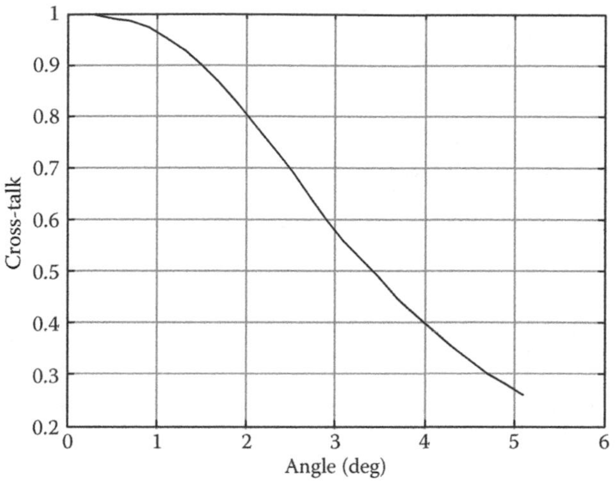

FIGURE 6.2
Wiener filter cross-talk as a function of angular distance between two sources. The first source is on broadside (DOA=0°). Array signal-to-noise ratio (aSNR)=10 and an eight-sensor ULA is assumed.

Differentiating Equation 6.10 with respect to $\xi(t)$ and setting the derivative to zero, we obtain

$$\begin{aligned}\hat{\xi}(t) &= \left[\mathbf{A}^H\mathbf{A}\right]^{-1}\mathbf{A}^H\mathbf{f}(t)\\ &= \xi(t) + \left[\mathbf{A}^H\mathbf{A}\right]^{-1}\mathbf{A}^H\eta(t).\end{aligned} \tag{6.11}$$

We observe that the signal term is extracted without any distortion, but the noise term, given by $[\mathbf{A}^H\mathbf{A}]^{-1}\mathbf{A}^H\eta(t)$, behaves differently; for example, the noise becomes correlated. The output noise covariance matrix is given by

$$\mathbf{C}_{\hat{\eta}} = \left[\mathbf{A}^H\mathbf{A}\right]^{-1}\sigma_{\eta}^2. \tag{6.12}$$

When $\mathbf{A}^H\mathbf{A}$ is singular or close to being singular, that is, with a large eigenvalue spread, the noise in the output may become amplified. Consider a case of two sources with direction vectors $\mathbf{a}_0$ and $\mathbf{a}_1$ and corresponding to DOAs θ_0 and θ_1, respectively. It is easy to show that

$$\det\left[\mathbf{A}^H\mathbf{A}\right] = M^2\left(1 - \left|\frac{\mathbf{a}_0^H\mathbf{a}_1}{M}\right|^2\right). \tag{6.13}$$

Using Equation 6.13 in Equation 6.12, we obtain the variance of the noise in all array outputs

$$\sigma_{\hat{\eta}}^2 = \frac{1}{M\left(1 - \left|\frac{\mathbf{a}_0^H\mathbf{a}_1}{M}\right|^2\right)}\sigma_{\eta}^2. \tag{6.14}$$

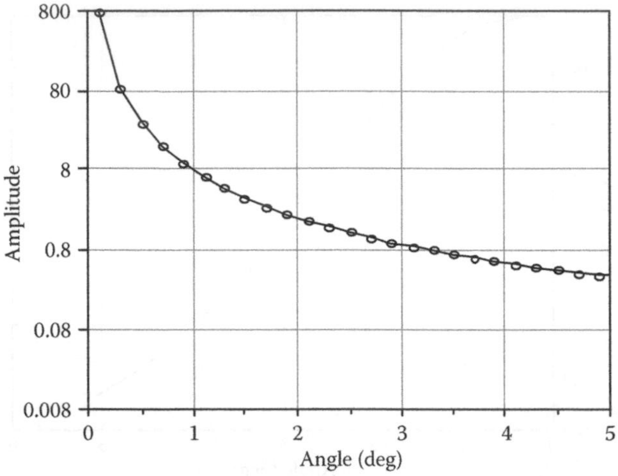

FIGURE 6.3
The noise amplification factor as a function of DOA of source #2. Source #1 is held fixed at 0° and source #2 is moved. Array size $M=8$.

In Figure 6.3, we have plotted the noise amplification factor, the factor multiplying σ_η^2, as a function of the angular separation between two sources. From the plot, we note that the noise is amplified only when the angular separation is a fraction of a degree.

In the stochastic model, the signal waveform is (stationary) stochastic and the linear least squares estimate (LLSE) turns out to be the same as the Wiener filter, which is easily derived as (Section 6.6, Exercise 1),

$$\mathbf{H}_W(\omega) = \mathbf{S}_f^{-1}(\omega)\mathbf{A}\mathbf{S}_0(\omega),$$

where $\mathbf{A}=[\mathbf{a}_0, \mathbf{a}_1, \ldots, \mathbf{a}_{P-1}]$. In the Wiener filter, we require spectral matrix and cross-spectral vectors; both of these are obtained by statistical averaging. Such a filter will naturally be applicable to an ensemble of time series having the same second order statistical properties. In the deterministic approach, the LLSE filter is adapted to a particular data set. It is claimed in Ref. [1] that the stochastic LLSE results in a lower mse than the deterministic LLSE.

6.1.3.1 Circular Array

The result on cross-talk (Equation 6.8) is of a general nature, valid for any array geometry. The direction vector needs to be appropriately defined. We now consider a UCA for which the direction vector is given by Equation 2.51 where we let $\theta=90°$, so that the source is in the same plane as the circular array,

$$\mathbf{a}(\varphi)=\text{col}[e^{-j(\omega a/c)\cos(\varphi)}, e^{-j(\omega a/c)\cos((2\pi/M)-\varphi)}, \ldots, e^{-j(\omega a/c)\cos((2\pi(M-1)/M)-\varphi)}]. \quad (6.15)$$

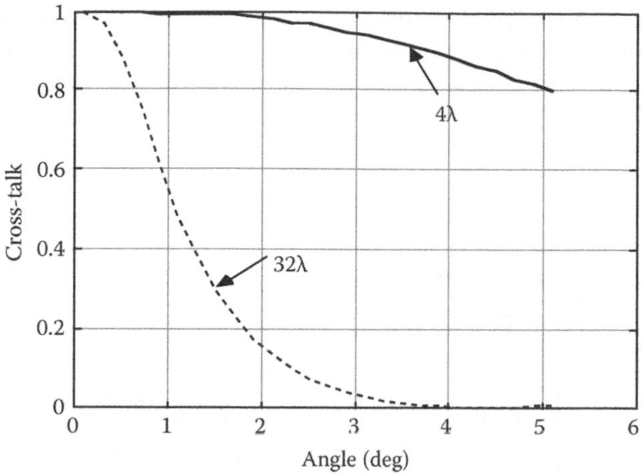

FIGURE 6.4
Cross-talk as a function of angular separation of two sources. Eight sensors are uniformly spaced over a circle of radius 4λ (solid curve) and 32λ (dashed curve).

The interest is to find out how a circular array fares in comparison with a linear array with respect to cross-talk capability. For a given number of sensors (say, M) the maximum aperture of a ULA is fixed at $(M-1)\lambda/2$, but the aperture of a circular array can be very large, at least in principle. Since the array aperture is the main factor deciding the cross-talk, it is expected that the circular array ought to perform better in terms of lower cross-talk. Using the same number of sensors, cross-talk may be reduced when arranged over a large circle. This is shown in Figure 6.4 for an eight-sensor UCA. The cross-talk has been reduced considerably when the array radius is increased from four to thirty-two wavelengths. A linear array of eight sensors will have an aperture of 3.5λ. A UCA having an aperture of 3.5λ was found to show much higher cross-talk than that of the ULA shown in Figure 6.1. Thus, the performance of a UCA is significantly better only when the radius is increased considerably.

6.1.4 Effects of Errors in DOA

Waveform estimation requires a knowledge of the DOA of wavefronts. But the DOA estimates are subject to errors largely on account of the finite data length used in their estimates. In this section, we investigate the effects of errors in DOA estimates on interference from signals of no interest. A comprehensive analysis of the effects of model errors including the errors in DOA may be found in Ref. [2]. Let $\hat{\theta}_p=\theta_p+\Delta\theta_p$, $p=0, 1,\ldots, P-1$ be the estimated DOAs, where θ_p is correct DOA and $\Delta\theta_p$ is estimation error. We shall rewrite Equation 6.11, showing explicitly the dependence on the estimated DOAs

$$\hat{\xi}(t)=\left[\mathbf{A}^H(\hat{\theta})\mathbf{A}(\hat{\theta})\right]^{-1}\mathbf{A}^H(\hat{\theta})\mathbf{f}(t). \tag{6.16}$$

Let us express $\mathbf{A}(\hat{\theta}) = [\mathbf{B}\ \mathbf{c}]$, where $\mathbf{B}$ is a matrix whose columns are the direction vectors of all interfering sources (signals of no interest) and $\mathbf{c}$ is a vector representing the direction vector of the signal of interest (SOI). $\left[\mathbf{A}^H(\hat{\theta})\mathbf{A}(\hat{\theta})\right]^{-1}$ may be simplified [2] as

$$\begin{aligned}
\left[\mathbf{A}^H(\hat{\theta})\mathbf{A}(\hat{\theta})\right]^{-1} &= \left[[\mathbf{B}\ \mathbf{c}]^H[\mathbf{B}\ \mathbf{c}]\right]^{-1} \\
&= \begin{bmatrix} \mathbf{B}^H\mathbf{B} & \mathbf{B}^H\mathbf{c} \\ \mathbf{c}^H\mathbf{B} & \mathbf{c}^H\mathbf{c} \end{bmatrix}^{-1} \\
&= \begin{bmatrix} \left[\mathbf{B}^H\mathbf{B} - \dfrac{\mathbf{B}^H\mathbf{c}\mathbf{c}^H\mathbf{B}}{\|\mathbf{c}\|^2}\right]^{-1} & -\left[\mathbf{B}^H\mathbf{B} - \dfrac{\mathbf{B}^H\mathbf{c}\mathbf{c}^H\mathbf{B}}{\|\mathbf{c}\|^2}\right]^{-1}\dfrac{\mathbf{B}^H\mathbf{c}}{\|\mathbf{c}\|^2} \\ -\left[\mathbf{c}^H\mathbf{c} - \mathbf{c}^H\mathbf{P}\mathbf{c}\right]^{-1}\mathbf{c}^H\mathbf{B}_1 & \left[\mathbf{c}^H\mathbf{c} - \mathbf{c}^H\mathbf{P}\mathbf{c}\right]^{-1} \end{bmatrix} \\
&= \frac{1}{\mathbf{c}^H(\mathbf{I}-\mathbf{P})\mathbf{c}} \begin{bmatrix} (\mathbf{B}^H\mathbf{B})^{-1}\mathbf{c}^H(\mathbf{I}-\mathbf{P})\mathbf{c} + \mathbf{B}_1\mathbf{c}\mathbf{c}^H\mathbf{B}_1^H & -\mathbf{B}_1\mathbf{c} \\ -\mathbf{c}^H\mathbf{B}_1^H & 1 \end{bmatrix},
\end{aligned} \tag{6.17}$$

where $\mathbf{P} = \mathbf{B}(\mathbf{B}^H\mathbf{B})^{-1}\mathbf{B}^H$ and $\mathbf{B}_1 = (\mathbf{B}^H\mathbf{B})^{-1}\mathbf{B}^H$. Using Equation 6.17 in Equation 6.16 and simplifying we obtain

$$\begin{aligned}
&\left[\mathbf{A}^H(\hat{\theta})\mathbf{A}(\hat{\theta})\right]^{-1}\mathbf{A}^H(\hat{\theta}) \\
&= \frac{1}{\mathbf{c}^H(\mathbf{I}-\mathbf{P})\mathbf{c}} \begin{bmatrix} (\mathbf{B}^H\mathbf{B})^{-1}\mathbf{c}^H(\mathbf{I}-\mathbf{P})\mathbf{c} + \mathbf{B}_1\mathbf{c}\mathbf{c}^H\mathbf{B}_1^H & -\mathbf{B}_1\mathbf{c} \\ -\mathbf{B}_1\mathbf{c} & 1 \end{bmatrix} \begin{bmatrix} \mathbf{B}^H \\ \mathbf{c}^H \end{bmatrix} \\
&= \frac{1}{\mathbf{c}^H(\mathbf{I}-\mathbf{P})\mathbf{c}} \begin{bmatrix} \mathbf{B}_1\left[\mathbf{c}^H(\mathbf{I}-\mathbf{P})\mathbf{c}\mathbf{I} - \mathbf{c}\mathbf{c}^H(\mathbf{I}-\mathbf{P})\right] \\ \mathbf{c}^H(\mathbf{I}-\mathbf{P}) \end{bmatrix}.
\end{aligned} \tag{6.18}$$

Using Equation 6.18 in Equation 6.16, an estimate of the SOI is obtained

$$\hat{\xi}_{\text{soi}}(t) = \frac{\mathbf{c}^H(\mathbf{I}-\mathbf{P})}{\mathbf{c}^H(\mathbf{I}-\mathbf{P})\mathbf{c}}\mathbf{f}(t) = \mathbf{d}^H(\hat{\theta})\mathbf{f}(t). \tag{6.19}$$

We now expand $\mathbf{d}^H(\hat{\theta})$ in a Taylor's series and retain the first derivative term only, $\mathbf{d}(\hat{\theta}) \approx \mathrm{d}(\theta) + \mathrm{d}_1(\theta)\Delta\theta$, where $\mathbf{d}_1(\theta)$ is the first derivative of $\mathbf{d}(\theta)$ with respect to θ. The estimated SOI reduces to

$$\begin{aligned}
\hat{\xi}_{\text{soi}}(t) &= \mathbf{d}^H(\theta)\mathbf{A}\xi(t) + \Delta\theta^H\mathbf{d}_1^H(\theta)\mathbf{A}\xi(t) + \mathbf{d}(\hat{\theta})\eta(t) \\
&= \xi_{\text{soi}}(t) + \Delta\theta^H\mathbf{d}_1^H(\theta)\mathbf{A}\xi(t) + \mathbf{d}(\hat{\theta})\eta(t).
\end{aligned} \tag{6.20}$$

The interference term in Equation 6.20 is $\Delta\theta^H \mathbf{d}_1^H(\theta)\mathbf{A}\xi(t)$, which we shall rewrite showing the contribution of each DOA

$$\Delta\xi_{\text{soi}}(t) = \sum_{p=0}^{P-1} \left[\beta^H \xi(t)\right]_p \Delta\theta_p, \tag{6.21}$$

where $\Delta\xi_{\text{soi}}(t)$ stands for the error in $\hat{\xi}_{\text{soi}}(t)$, $\beta = [\mathbf{A}^H \mathbf{d}_1(\theta)]$ and $[.]_p$ stands for the pth element of a column vector.

6.2 Minimum Variance (Capon Method)

In Section 4.3, we have devised a filter (in the frequency domain), which minimizes the output power while maintaining unit response within an angular sector (extended source) whose axis is directed in the desired direction. This leads to the following equation,

$$\frac{1}{\lambda}\mathbf{H} = (\mathbf{S}_f^{-1}\Gamma)\mathbf{H}, \tag{6.22}$$

where

$$\Gamma = \frac{1}{\Delta\theta} \int_{\theta-(\Delta\theta/2)}^{\theta+(\Delta\theta/2)} \mathbf{a}(\omega,\phi)\mathbf{a}^H(\omega,\phi)d\phi, \tag{6.23}$$

where $\mathbf{a}(\omega,\phi)$ is the steering vector and $\lambda = \mathbf{H}^H \mathbf{S}_f \mathbf{H}$ is the output power. To further minimize the output power we must select a filter vector as the eigenvector of $(\mathbf{S}_f^{-1}\Gamma)$ corresponding to the largest eigenvalue; in which case the output power will be equal to the inverse of the largest eigenvalue of $(\mathbf{S}_f^{-1}\Gamma)$. A closed-form solution of Equation 6.22 can be obtained for $\Delta\theta = 0$ (point source). Equation 6.22 simplifies to

$$\frac{\mathbf{H}}{\mathbf{H}^H \mathbf{S}_f \mathbf{H}} = (\mathbf{S}_f^{-1}\mathbf{a}(\omega,\theta)\mathbf{a}^H(\omega,\theta)\mathbf{H}). \tag{6.24}$$

Multiply both sides of Equation 6.24 by $\mathbf{a}^H(\omega, \theta)$ and simplify the resulting expression,

$$\mathbf{H}^H \mathbf{S}_f \mathbf{H} = \frac{1}{\mathbf{a}^H(\omega,\theta)\mathbf{S}_f^{-1}\mathbf{a}(\omega,\theta)}. \tag{6.25}$$

By direct substitution in Equation 6.24 it may be verified that the solution is given by

$$\mathbf{H}_{\text{Cap}} = \frac{\mathbf{S}_f^{-1}\mathbf{a}(\omega,\theta)}{\mathbf{a}^H(\omega,\theta)\mathbf{S}_f^{-1}\mathbf{a}(\omega,\theta)}, \tag{6.26}$$

and

$$\text{Output power} = \frac{1}{\mathbf{a}^H(\omega,\theta)\mathbf{S}_f^{-1}\mathbf{a}(\omega,\theta)}.$$

The optimum filter given in Equation 6.26 is also known as the Applebaum filter [3].

6.2.1 Extended Source

For ULA the Γ matrix takes the following form

$$[\Gamma]_{m,n} = \sum_{k=0,2,4\ldots} \varepsilon_k J_k\left((m-n)\frac{\omega d}{c}\right)\cos(k\theta)\frac{\sin\left(k\frac{\Delta\theta}{2}\right)}{k\frac{\Delta\theta}{2}}$$

$$+ j\sum_{k=1,3,5\ldots} 2J_k\left((m-n)\frac{\omega d}{c}\right)\sin(k\theta)\frac{\sin\left(k\frac{\Delta\theta}{2}\right)}{k\frac{\Delta\theta}{2}},$$

where $\varepsilon_0=1$ and $\varepsilon_k=2$ for all k. To evaluate the integral in Equation 6.23, we have used the result derived in Equation 1.53b. For UCA the Γ matrix takes the following form:

$$\underset{(M\times M)}{\Gamma} = \underset{(M\times\alpha)}{\mathbf{W}} \; \underset{(\alpha\times\alpha)}{\Theta} \; \underset{(\alpha\times M)}{\mathbf{W}^H} \quad \alpha\to\infty$$

where

$$\{\mathbf{W}\}_{km} = e^{j(2\pi km/M)} \quad \begin{Bmatrix} m=0,1,\ldots,M-1 \\ k=0,\pm1,\ldots,\infty \end{Bmatrix},$$

$$[\Theta]_{k,l} = J_k\left(\frac{\omega a}{c}\right)J_l\left(\frac{\omega a}{c}\right)e^{-j((\pi/2)+\varphi_0)(k-l)}\operatorname{sinc}\left((k-l)\frac{\Delta\theta}{2}\right),$$

valid for all values of $\Delta\theta$. We have used a series expansion of the steering vector for a circular array (Equation 2.51).

6.2.2 Single Point Source Case

Let there be a single source emitting a plane wave with a direction vector $\mathbf{a}_0(\omega,\theta_0)$ in the presence of white noise. The spectral matrix is given by $\mathbf{S}_f = \mathbf{a}_0(\omega,\theta_0)\mathbf{a}_0^H(\omega,\theta_0)\mathbf{S}_\xi(\omega) + \sigma_\eta^2\mathbf{I}$, and its inverse is given by

$$\mathbf{S}_f^{-1} = \frac{1}{\sigma_\eta^2}\left[\mathbf{I} - Q\mathbf{a}_0(\omega,\theta_0)\mathbf{a}_0^H(\omega,\theta_0)\right], \tag{6.27}$$

where

$$Q = \frac{\dfrac{\mathbf{S}_\xi(\omega)}{\sigma_\eta^2}}{1 + M\dfrac{\mathbf{S}_\xi(\omega)}{\sigma_\eta^2}}.$$

For large aSNR, $(\mathbf{S}_\xi(\omega)/\sigma_\eta^2) \gg 1$, we may approximate $Q \approx (1/M)$. Using Equation 6.27 in Equation 6.26 we obtain

$$\mathbf{H}_{\text{Cap}} = \frac{1}{M}\left[\frac{\mathbf{a}(\omega,\theta) - Q\left[\mathbf{a}_0^H(\omega,\theta_0)\mathbf{a}(\omega,\theta)\right]\mathbf{a}_0(\omega,0_0)}{1 - Q\dfrac{\left|\mathbf{a}^H(\omega,\theta)\mathbf{a}_0(\omega 0_0)\right|^2}{M}}\right]. \tag{6.28}$$

Let us obtain the filter output for the signal model of a single source in the presence of white noise

$$\hat{F}(\omega) = \mathbf{H}_{\text{Cap}}^H\left[\mathbf{a}_0(\omega,\theta_0)\Xi_0(\omega) + \mathbf{N}(\omega)\right]. \tag{6.29a}$$

The signal term turns out to be

$$\mathbf{H}_{\text{Cap}}^H\mathbf{a}_0(\omega,\theta_0)\Xi_0(\omega) = \frac{\mathbf{a}^H(\omega,\theta)\mathbf{a}_0(\omega,\theta_0)}{M}\left[\frac{1 - QM}{1 - QM\dfrac{\left|\mathbf{a}^H(\omega,\theta)\mathbf{a}_0(\omega\theta_0)\right|^2}{M^2}}\right]\Xi_0(\omega) \tag{6.29b}$$

$$= \Xi_0(\omega) \qquad \text{when } \mathbf{a}^H(\omega,\theta) = \mathbf{a}_0(\omega,\theta_0),$$

and the noise term

$$\mathbf{H}_{\text{Cap}}^H\mathbf{N}(\omega) = \frac{1}{M}\left[\frac{\mathbf{a}(\omega,\theta) - Q\left[\mathbf{a}_0^H(\omega,\theta_0)\mathbf{a}(\omega,\theta)\right]\mathbf{a}_0(\omega,\theta_0)}{1 - QM\dfrac{\left|\mathbf{a}^H(\omega,\theta)\mathbf{a}_0(\omega\theta_0)\right|^2}{M^2}}\right]\mathbf{N}(\omega) \tag{6.29c}$$

$$= \frac{\mathbf{a}_0^H(\omega,\theta_0)}{M}\mathbf{N}(\omega) \qquad \text{when } \mathbf{a}^H(\omega,\theta) = \mathbf{a}_0(\omega,\theta_0).$$

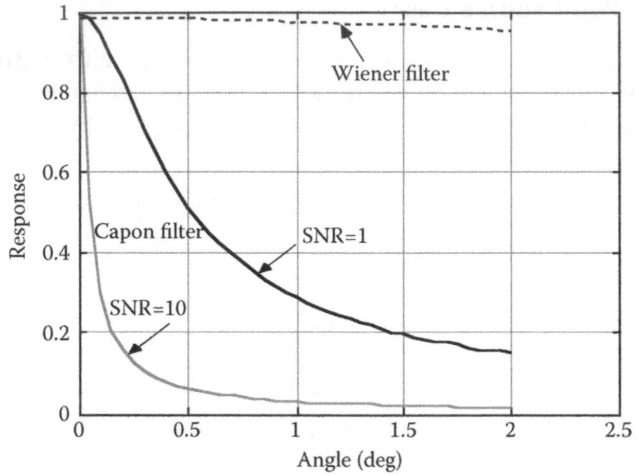

FIGURE 6.5
Response of Capon filter as a function of the angle of the steering vector. A ULA of eight elements was assumed. Equivalent Wiener filter response is also shown by the dashed curve.

The variance of the noise in the filter output (Equation 6.29c), when $\mathbf{a}^H(\omega, \theta)=\mathbf{a}_0(\omega, \theta_0)$, turns out to be (σ_N^2/M). The response of the filter was computed from Equation 6.29a for different steering directions. The DOA of the incident wavefront was assumed to be at 0°. A ULA with eight sensors was considered. The response of the filter depends upon the SNR. For SNR=10 (or aSNR=80) the response becomes extremely sharp, as demonstrated in Figure 6.5. For comparison, we have also plotted a Wiener filter response (dashed curve) where we have assumed the SNR equal to 10 while other parameters remain the same as for the Capon filter. The response of the Capon filter is far superior to that of an equivalent Wiener filter. The shape of the Capon filter is strongly dependent on SNR; on the other hand, the dependence of the Wiener filter on the SNR is merely to the extent of changing the scale factor. Even at a very low SNR, the Capon filter seems to outperform the Wiener filter.

6.2.3 Two Source Case

As in the case of the Wiener filter, we shall now consider the Capon filter specifically for two uncorrelated wavefronts in the presence of white noise. The spectral matrix is given by Equation 6.2 and its inverse is computed in the same manner as shown there. Using the expression for inverse of the spectral matrix in Equation 6.26, we obtain an expression for the Capon filter, $\mathbf{H}_{\text{Cap}}(\omega, \theta_0)=\text{num/den}$, where

$$\text{num} = \left[\mathbf{V}^{-1} - \frac{S_1(\omega)}{S_0(\omega)} \frac{\mathbf{V}^{-1}\mathbf{a}_1(\omega,\theta_1)\mathbf{a}_1^H(\omega,\theta_1)\mathbf{V}^{-1}}{1+\dfrac{S_1(\omega)}{S_0(\omega)}\mathbf{a}_1^H(\omega,\theta_1)\mathbf{V}^{-1}\mathbf{a}_1(\omega,\theta_1)}\right]\mathbf{a}_0(\omega,\theta_0), \tag{6.30a}$$

den $= \mathbf{a}_0^H(\omega,\theta_0)\cdot$num and $\mathbf{V}$ is defined in Equation 6.4b. The denominator may be further reduced to

$$\text{den}=\frac{S_0(\omega)}{\sigma_\eta^2}(1-QM)\frac{M\left(1+\frac{S_1(\omega)}{\sigma_\eta^2}\left[1-\frac{\left|\mathbf{a}_1^H(\omega,\theta_1)\mathbf{a}_0(\omega,\theta_0)\right|^2}{M^2}\right]\right)}{\left(1+\frac{S_1(\omega)}{\sigma_\eta^2}\left[1-QM\frac{\left|\mathbf{a}_1^H(\omega,\theta_1)\mathbf{a}_0(\omega,\theta_0)\right|^2}{M^2}\right]\right)}. \tag{6.30b}$$

The filter output may be shown to be

$$\text{Output}=\left\{\Xi_0(\omega)+\mathbf{H}_{\text{Cap}}^H(\omega,\theta_0)\mathbf{a}_1(\omega,\theta_1)\Xi_1(\omega)+\mathbf{H}_{\text{Cap}}^H(\omega,\theta_0)\mathbf{N}(\omega)\right\}. \tag{6.31}$$

Notice that the signal term is equal to the actual signal. But this was not the case for the Wiener filter, where we had to assume a large SNR in order to arrive at this result.

The contribution of the second source is represented by the second term in Equation 6.31. The cross-talk is then given by

$$\begin{aligned}\text{Cross talk} &= \mathbf{a}_1^H(\omega,\theta_1)\mathbf{H}_{\text{Cap}}(\omega,\theta_0)\mathbf{H}_{\text{Cap}}^H(\omega,\theta_0)\mathbf{a}_1(\omega,\theta_1)\\ &=\frac{\left|\frac{\mathbf{a}_0^H(\omega,\theta_1)\mathbf{a}_1(\omega,\theta_0)}{M}\right|^2}{\left\{1+\frac{S_1(\omega)}{\sigma_\eta^2}\left[1-\frac{\left|\mathbf{a}_1^H(\omega,\theta_1)\mathbf{a}_0(\omega,\theta_0)\right|^2}{M^2}\right]\right\}^2}.\end{aligned} \tag{6.32}$$

The cross-talk is plotted in Figure 6.6 for a ULA and in Figure 6.7 for a UCA. Compare these two figures with Figures 6.1 and 6.3, where we have plotted the cross-talk for the Wiener filter. Clearly, the Capon filter performs better for both types of array geometries. An expression for SNR, which agrees with Equation 6.32, is derived in Ref. [4].

Finally, we shall evaluate the noise term, that is, the leftover noise in the array output,

$$\begin{aligned}&\mathbf{H}_{\text{Cap}}^H(\omega,\theta_0)\mathbf{N}(\omega)\\ &=\frac{\mathbf{a}_0^H(\omega,\theta_0)\left[\mathbf{V}^{-1}-\frac{S_1(\omega)}{S_0(\omega)}\frac{\mathbf{V}^{-1}\mathbf{a}_1(\omega,\theta_1)\mathbf{a}_1^H(\omega,\theta_1)\mathbf{V}^{-1}}{1+\frac{S_1(\omega)}{S_0(\omega)}\mathbf{a}_1^H(\omega,\theta_1)\mathbf{V}^{-1}\mathbf{a}_1(\omega,\theta_1)}\right]\mathbf{N}}{\text{den}}\\ &=\frac{\mathbf{a}_0^H(\omega,\theta_0)\mathbf{V}^{-1}\mathbf{N}-\psi\mathbf{a}_1^H(\omega,\theta_1)\mathbf{V}^{-1}\mathbf{N}}{\text{den}},\end{aligned} \tag{6.33}$$

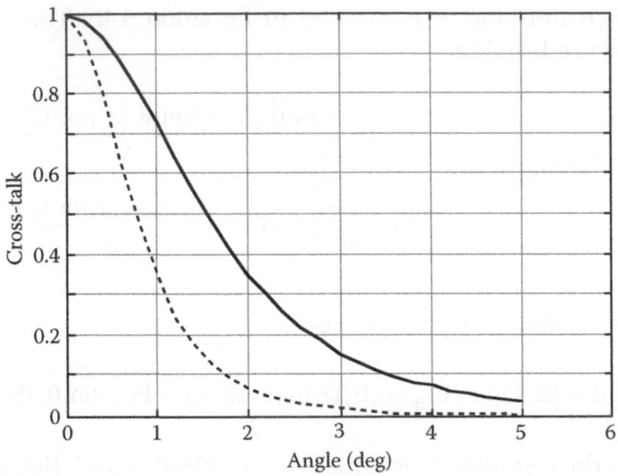

FIGURE 6.6
Capon filter: cross-talk as a function of angular distance between two sources. The first source is on broadside (DOA=0°). Array signal-to-noise ratio (aSNR)=10. Continuous curve: eight-sensor ULA and dashed curve: 16-sensor ULA.

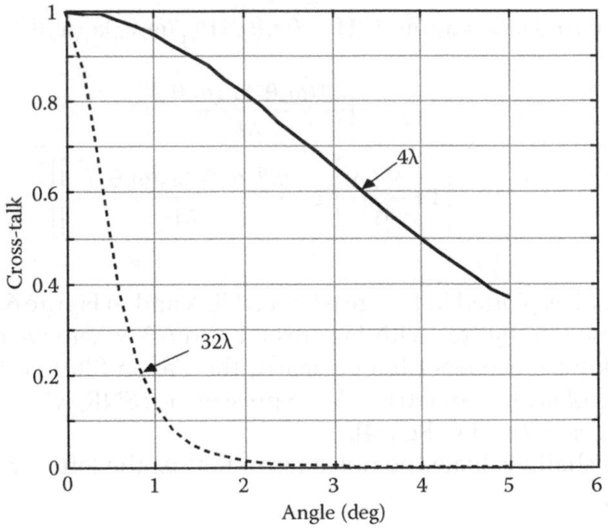

FIGURE 6.7
Capon filter: cross-talk as a function of angular separation of two sources. Eight sensors are uniformly spaced over a circle of radius 4λ (solid curve) and 32λ (dashed curve).

where the denominator term, den, is given in Equation 6.30b and

$$\psi = \frac{\dfrac{S_1(\omega)}{S_0(\omega)}\mathbf{a}_0^H(\omega,\theta_0)\mathbf{V}^{-1}\mathbf{a}_1(\omega,\theta_1)}{1+\dfrac{S_1(\omega)}{S_0(\omega)}\mathbf{a}_1^H(\omega,\theta_1)\mathbf{V}^{-1}\mathbf{a}_1(\omega,\theta_1)}.$$

The variance of the leftover noise may be computed from Equation 6.33. We obtain after simplification

$$\mathrm{Var}\left\{\mathbf{H}_{\mathrm{Cap}}^{H}(\omega,\theta_0)\mathbf{N}(\omega)\right\} = \frac{\sigma_\eta^2}{M}\left[1-\psi\frac{\mathbf{a}_1^H(\omega,\theta_1)\mathbf{a}_0(\omega,\theta_0)}{M}\right]$$

$$= \frac{\sigma_\eta^2}{M}\left[\frac{1+\dfrac{S_1(\omega)}{\sigma_\eta^2}M\left(1-\left|\dfrac{\mathbf{a}_1^H(\omega,\theta_1)\mathbf{a}_0(\omega,\theta_0)}{M}\right|^2\right)}{1+\dfrac{S_1(\omega)}{\sigma_\eta^2}M\left(1-QM\left|\dfrac{\mathbf{a}_1^H(\omega,\theta_1)\mathbf{a}_0(\omega,\theta_0)}{M}\right|^2\right)}\right]. \tag{6.34a}$$

When the array SNR is large, then $QM \approx 1$ and Equation 6.34 reduces to a simple form

$$\mathrm{Var}\left\{\mathbf{H}_{\mathrm{Cap}}^{H}(\omega,\theta_0)\mathbf{N}(\omega)\right\} \approx \frac{\sigma_\eta^2}{M}. \tag{6.34b}$$

6.3 Adaptive Beamformation

In adaptive beamformation, the array processor is so designed that it receives a signal coming from a desired direction and it *automatically* suppresses signals (that is, interference) coming from all other directions. This is achieved by means of a filter that adapts itself to the incoming signal and interference. Suppression of interference is achieved through predictive cancellation; an example of this approach was described in Chapter 2 in connection with noise cancellation. The weight coefficients are upgraded through an algorithm, such as LMS or one of its kind. There are excellent texts on the topic of adaptive signal processing that include adaptive beamformation [5, 6]. We do not intend to cover this topic in any detail. The aim here is to briefly discuss a few selected topics, namely, null steering, adaptive interference canceller, and adaptive Capon filter.

6.3.1 Null Steering

A basic step in beamformation is the weighted and delayed sum of the array outputs

$$\hat{f}(t) = \sum_{m=0}^{M-1} a_m f_m(t-m\tau), \tag{6.35a}$$

where τ is delay per sensor and a_m is a real weight coefficient. In the temporal frequency domain, Equation 6.35a may be expressed as

$$\begin{aligned}\hat{F}(\omega) &= \sum_{m=0}^{P-1}(a_m e^{-jm\tau\omega})F(md,\omega) \\ &= \frac{1}{2\pi}\int_{-\pi}^{\pi} A(u)F^*(u,\omega)du,\end{aligned} \tag{6.35b}$$

where

$$A(u) = \sum_{m=0}^{P-1} \tilde{a}_m e^{-jmu}$$

$$\tilde{a}_m = a_m e^{-jm\tau\omega},$$

$$F(u,\omega) = \sum_{m=0}^{P-1} F(md,\omega)e^{-jmu}.$$

Through a process of selecting the complex weight coefficients, we can assign the desired property to the response function of the weight function; for example, $A(u)$ may be required to have a deep null in some desired direction and unit response in all directions. Let us express $A(u)$ as a z-transform of $\tilde{a}_m$,

$$A(z) = \sum_{m=0}^{P-1} \tilde{a}_m z^{-1},$$

where $z=e^{ju}$. We can now design a notch filter in the z-plane. A simple approach is to place a zero on the unit circle and two poles, one on either side of the zero. Let the zero be at $z_0 = e^{j\pi\sin\theta_0}$ and the poles at $z_+ = re^{j\pi\sin(\theta_0+(\Delta\theta/2))}$ and $z_- = re^{j\pi\sin(\theta_0-(\Delta\theta/2))}$ (see Figure 6.8) where $0.9 \le r \le 0.99$. We have assumed that $d=0.5\lambda$.

$$A(z) = \frac{(z-z_0)}{(z-z_+)(z-z_-)}. \tag{6.36}$$

The response of the notch filter given by Equation 6.36 is shown in Figure 6.9. See Section 6.6, Exercise 3 for another type of null steering filter independent frequency.

6.3.2 Beam-Steered Adaptive Array

In the Capon filter (Equation 6.26) let us replace the inverse of the spectral matrix by its eigenvalue eigenvector representation,

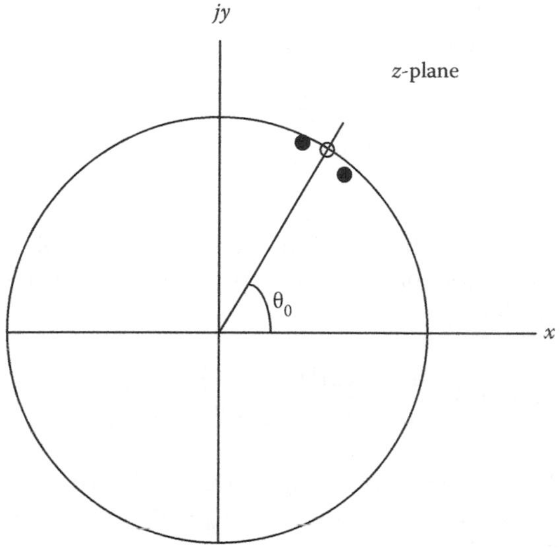

FIGURE 6.8
Position of a null and two poles in its immediate neighborhood but slightly inside the unit circle.

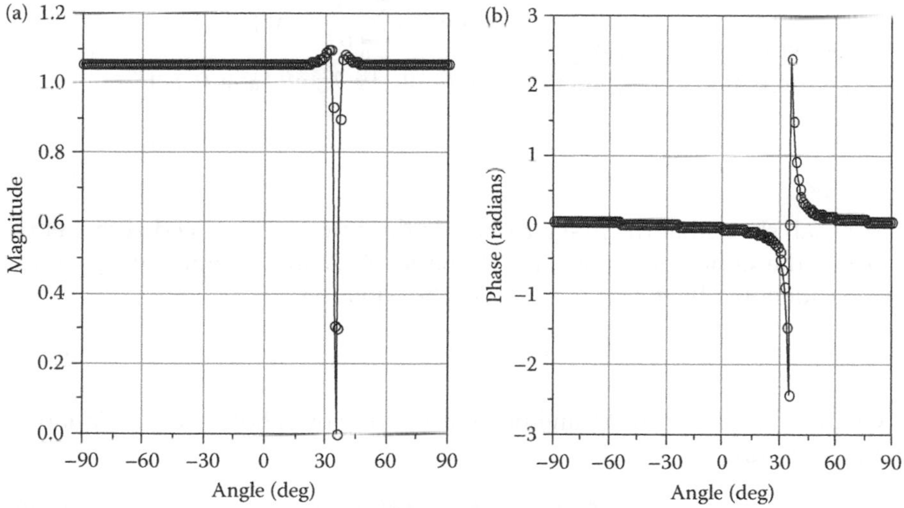

FIGURE 6.9
Wavenumber response of the notch filter (a) amplitude response and (b) phase response. Assumed parameters are $r=0.95$, $\theta_0=35°$, and $\Delta\theta=3°$.

$$
\begin{aligned}
\mathbf{H}_{\text{Cap}} &= \frac{\mathbf{S}_f^{-1}\mathbf{a}(\omega,\theta)}{\mathbf{a}^H(\omega,\theta)\mathbf{S}_f^{-1}\mathbf{a}(\omega,\theta)} \\
&= \frac{\left\{\mathbf{v}_s\lambda_s^{-1}\mathbf{v}_s^H + \mathbf{v}_\eta\lambda_\eta^{-1}\mathbf{v}_\eta^H\right\}\mathbf{a}(\omega,\theta)}{\mathbf{a}^H(\omega,\theta)\left\{\mathbf{v}_s\lambda_s^{-1}\mathbf{v}_s^H + \mathbf{v}_\eta\lambda_\eta^{-1}\mathbf{v}_\eta^H\right\}\mathbf{a}(\omega,\theta)}.
\end{aligned}
\tag{6.37a}
$$

We assume that the look direction vector $\mathbf{a}(\omega, \theta)$ lies in the signal subspace; hence $\mathbf{a}(\omega, \theta) \perp \mathbf{v}_\eta$. Hence, Equation 6.37a simplifies to

$$\mathbf{H}_{\text{Cap}} = \frac{\{\mathbf{v}_s \lambda_s^{-1} \mathbf{v}_s^H\} \mathbf{a}(\omega,\theta)}{\mathbf{a}^H(\omega,\theta)\{\mathbf{v}_s \lambda_s^{-1} \mathbf{v}_s^H\} \mathbf{a}(\omega,\theta)}. \tag{6.37b}$$

The Capon filter has an interesting property, that is, when the look direction coincides with one of the signal directions, for example, $\mathbf{a}(\omega, \theta) = \mathbf{a}(\omega, \theta_0)$, it is approximately orthogonal to all direction vectors, $\mathbf{a}(\omega, \theta_m)$, $m = 1,2,\ldots,P-1$. To show this, consider

$$\mathbf{H}_{\text{Cap}}^H \mathbf{a}(\omega,\theta_m) = \frac{\mathbf{a}^H(\omega,\theta_0)\{\mathbf{v}_s \lambda_s^{-1} \mathbf{v}_s^H\} \mathbf{a}(\omega,\theta_m)}{\mathbf{a}^H(\omega,\theta_0)\{\mathbf{v}_s \lambda_s^{-1} \mathbf{v}_s^H\} \mathbf{a}(\omega,\theta_0)}, \qquad m \neq 0. \tag{6.38a}$$

In Equation 6.38a the inverse of the signal eigenvalues may be approximated, for large SNR or large array size, by $\lambda_m^{-1} \approx \alpha_m^{-1} - (\sigma_\eta^2/\alpha_m^2)$. Therefore, Equation 6.38a may be expressed as

$$\mathbf{H}_{\text{Cap}}^H \mathbf{a}(\omega,\theta_m)$$

$$\approx \left\{ \frac{\mathbf{a}^H(\omega,\theta_0)\{\mathbf{v}_s \boldsymbol{\alpha}_s^{-1} \mathbf{v}_s^H\} \mathbf{a}(\omega,\theta_m)}{\mathbf{a}^H(\omega,\theta_0)\{\mathbf{v}_s \boldsymbol{\lambda}_s^{-1} \mathbf{v}_s^H\} \mathbf{a}(\omega,\theta_0)} - \frac{\mathbf{a}^H(\omega,\theta_0) \sum_{m=0}^{P-1} \frac{\sigma_\eta^2}{\alpha_m^2} \mathbf{v}_m \mathbf{v}_m^H \mathbf{a}(\omega,\theta_m)}{\mathbf{a}^H(\omega,\theta_0)\{\mathbf{v}_s \boldsymbol{\lambda}_s^{-1} \mathbf{v}_s^H\} \mathbf{a}(\omega,\theta_0)} \right\}, \qquad m \neq 0. \tag{6.38b}$$

We shall show that the first term inside the brackets is indeed zero. Let us consider the signal term alone from Equation 4.12b and compute the pseudo-inverse (denoted by #) on both sides of the equation

$$\mathbf{S}_f^{\#} = \mathbf{A}^{\#} \mathbf{S}_0^{-1} \mathbf{A}^{\#H}, \tag{6.39}$$

On premultiplying by $\mathbf{A}^H$ and postmultiplying by $\mathbf{A}$ both sides of Equation 6.39, we obtain

$$\mathbf{A}^H \mathbf{S}_f^{\#} \mathbf{A} = \mathbf{A}^H \mathbf{A}^{\#} \mathbf{S}_0^{-1} \mathbf{A}^{\#H} \mathbf{A} = \mathbf{S}_0^{-1}. \tag{6.40}$$

When the sources are *uncorrelated*, $\mathbf{S}_0$ as well as $\mathbf{S}_0^{-1}$ will be diagonal matrices. It, therefore, follows that all cross-product terms in $\mathbf{A}^H \mathbf{S}_f^{\#} \mathbf{A}$ must be equal to zero. The numerator of the first term is one such cross-product term. The second term in Equation 6.38b will be small when the aSNR is high. Thus, we obtain the following approximate result

$$\mathbf{H}_{\text{Cap}}^H \mathbf{a}(\omega,\theta_m) \approx 0 \qquad m = 1, 2, \ldots, P-1. \tag{6.41}$$

A modified Capon filter is defined as

$$\tilde{\mathbf{H}}_{\mathrm{Cap}} = \frac{\mathbf{S}_s^{\#}(\omega)\mathbf{a}(\omega,\theta)}{\mathbf{a}^H(\omega,\theta)\mathbf{S}_s^{\#}(\omega)\mathbf{a}(\omega,\theta)}, \tag{6.42}$$

where $\mathbf{S}_s^{\#}(\omega)$ is the pseudoinverse of the signal-only spectral matrix. It is shown in Ref. [7] that the modified Capon filter is robust against look direction errors. In Chapter 5, Section 5.1, it was shown that the generalized eigenvector corresponding to a given DOA is orthogonal to all other direction vectors.

6.3.3 Adaptive Capon Filter

In Chapter 3, we have derived two different filters to extract a wavefront coming from a specified direction. It may be recalled that the Wiener filter minimizes the mse between the filter output and the desired signal, and the Capon filter minimizes the output power while maintaining unit response in the desired direction. The Wiener filter is given by

$$\mathbf{H}_W(\omega) = \mathbf{S}_f^{-1}(\omega)\mathbf{S}_0(\omega),$$

and the Capon filter is given by

$$\mathbf{H}_{\mathrm{Cap}}(\omega) = \frac{\mathbf{S}_f^{-1}(\omega)\mathbf{a}(\omega,\theta)}{\mathbf{a}^H(\omega,\theta)\mathbf{S}_f^{-1}(\omega)\mathbf{a}(\omega,\theta)},$$

where $\mathbf{a}(\omega, \theta)$ is the direction vector in the desired direction. Note that $\mathbf{S}_0(\omega)=\mathbf{a}(\omega, \theta)S_0(\omega, \theta)$, where $\mathbf{S}_0(\omega)$ is power from the desired direction. In the Wiener filter this power is assumed to be a constant, but in the Capon filter the power is estimated ($= (1/\mathbf{a}^H(\omega,\theta)\mathbf{S}_f^{-1}(\omega)\mathbf{a}(\omega,\theta))$) and it is a function of the desired direction. This subtle difference is probably responsible for the improved performance of the Capon filter.

To implement the Wiener or Capon filter, we need the inverse of the spectral matrix $\mathbf{S}_f^{-1}(\omega)$. In practical terms, with the arrival of new data in the form of new time or frequency snapshot, we should be able to improve upon the available estimate of the spectral matrix and its inverse. In Chapter 4, we have shown how to recursively estimate these quantities. We shall rewrite Equation 4.62a and c in a more general fashion

$$\hat{\mathbf{S}}_f^{N+1}(\omega) = \mu\hat{\mathbf{S}}_f^{N}(\omega) + (1-\mu)\mathbf{F}_{N+1}(\omega)\mathbf{F}_{N+1}^H(\omega), \tag{6.43a}$$

$$\left[\hat{\mathbf{S}}_f^{N+1}(\omega)\right]^{-1} = \frac{1}{\mu}\left[\hat{\mathbf{S}}_f^{N}(\omega)\right]^{-1} - \frac{(1-\mu)\mathbf{z}\mathbf{z}^H}{\mu\left((1-\mu)\mathbf{F}_{N+1}^H(\omega)\left[\hat{\mathbf{S}}_f^{N}(\omega)\right]^{-1}\mathbf{F}_{N+1}(\omega)+\mu\right)}, \tag{6.43b}$$

where $\mathbf{z}=\left[\hat{\mathbf{S}}_f^N(\omega)\right]^{-1}\mathbf{F}_{N+1}(\omega)$ and $\mu \in (0, 1)$, which is a free parameter to be chosen depending upon how fast the estimated spectral matrix changes from snapshot to snapshot. For stationary process, where the change is small, $\mu \approx 1$. Using Equation 6.43a and b, we can recursively compute the Wiener and Capon filters. We will do this for the Capon filter. Multiply both sides of Equation 6.43b by

$$\frac{\mathbf{a}(\omega,\theta)}{\mathbf{a}^H(\omega,\theta)\left[\hat{\mathbf{S}}_f^N(\omega)\right]^{-1}\mathbf{a}(\omega,\theta)}.$$

Noting the definition of the Capon filter and assuming that the power from the desired direction does not change much from snapshot to snapshot, we obtain

$$\mathbf{H}_{\text{Cap}}^{N+1}=\frac{1}{\mu}\mathbf{H}_{\text{Cap}}^{N}-\beta\mathbf{z}\mathbf{z}^H\frac{\mathbf{a}(\omega,\theta)}{\mathbf{a}^H(\omega,\theta)\left[\hat{\mathbf{S}}_f^N(\omega)\right]^{-1}\mathbf{a}(\omega,\theta)}, \tag{6.43c}$$

where

$$\beta=\frac{(1-\mu)}{\mu\left((1-\mu)\mathbf{F}_{N+1}^H(\omega)\left[\hat{\mathbf{S}}_f^N(\omega)\right]^{-1}\mathbf{F}_{N+1}(\omega)+\mu\right)}.$$

But, we can show that

$$\mathbf{z}^H\frac{\mathbf{a}(\omega,\theta)}{\mathbf{a}^H(\omega,\theta)\left[\hat{\mathbf{S}}_f^N(\omega)\right]^{-1}\mathbf{a}(\omega,\theta)}=\mathbf{F}_{N+1}^H(\omega)\mathbf{H}_{\text{Cap}}^N,$$

and reduce Equation 6.43c to

$$\mathbf{H}_{\text{Cap}}^{N+1}=\frac{1}{\mu}\mathbf{H}_{\text{Cap}}^{N}-\beta\mathbf{z}\mathbf{F}_{N+1}^H(\omega)\mathbf{H}_{\text{Cap}}^N. \tag{6.43d}$$

It is shown in Ref. [8] that Equation 6.43d yields a stable estimate of the Capon filter.

6.4 Wideband Adaptive Beamformation

In this section, we shall look into space-time filter, which avoids the need for prior Fourier transformation of array signal, and then apply the frequency domain filter at each frequency. The final step is to compute the inverse Fourier transform.

6.4.1 Constrained Least Mean Square (LMS) Problem

A broadband wavefront is incident at broadside on a ULA. The background random noise and other interferences are assumed to be uncorrelated with the signal. The array output is sampled as described for broadband signal in Chapter 2, Section 2.4. A large data vector is constructed as in Equation 2.76. The length of the data vector is $M.N$, where M stands for the number of sensors and N stands for the number of delayed time samples. Let $\mathbf{w}$ be a $(MN\times 1)$ weight vector defined as

$$\mathbf{w}=\{w_{m.N+n}, \quad m=0,1,\ldots,M-1, \quad n=0,1,\ldots,N\}, \tag{6.44a}$$

which may be expressed in terms of a set of spatial filters acting on temporal snapshots,

$$\mathbf{w}=\left[\mathbf{w}_0^T, \mathbf{w}_1^T, \ldots, \mathbf{w}_{N-1}^T\right]^T, \tag{6.44b}$$

where $\mathbf{w}_n$ is a filter operating on nth snapshot. The output of the array at time t is given by

$$\hat{f}(t)=\mathbf{w}^T\mathbf{f}_{MN}(t), \tag{6.45}$$

where $\mathbf{f}_{MN}(t)$ is the data vector. From Equation 6.45 the expected output power is obtained

$$\begin{aligned} E\left\{\hat{f}^2(t)\right\} &= \mathbf{w}^T E\left\{\mathbf{f}_{MN}(t)\mathbf{f}_{MN}^T(t)\right\}\mathbf{w} \\ &= \mathbf{w}^T\mathbf{C}_{\mathrm{STCM}}\mathbf{w}, \end{aligned} \tag{6.46}$$

where $\mathbf{C}_{\mathrm{STCM}}$ is spatio-temporal covariance matrix as defined in Equation 2.77. Further the weight vector is so constrained that

$$\sum_{m=0}^{M-1} w_{mN+n}=h_n, \tag{6.47a}$$

where h_n, $n=0,1,\ldots,N-1$, are prescribed filter coefficients that define the frequency response of the beamformer output in the look direction. Equation 6.47a may be expressed in a matrix form as

$$\Gamma^{\mathrm{T}}\mathbf{w}=\mathbf{h}, \tag{6.47b}$$

where $\Gamma=\mathbf{I}_N\otimes 1_M$, where $\mathbf{I}_N$ is $N\times N$ diagonal matrix and $\mathbf{1}_M$ is $(M\times 1)$ unit column vector. Γ will be a rectangular matrix of size $(MN\times N)$. An example of Γ for $M=4$ and $N=2$ is given as

$$\underset{(8\times 2)}{\Gamma} = \begin{bmatrix} 1 & 0 \\ 1 & 0 \\ 1 & 0 \\ 1 & 0 \\ 0 & 1 \\ 0 & 1 \\ 0 & 1 \\ 0 & 1 \end{bmatrix},$$

h is a vector of filter coefficients. Equation 6.47a can also be expressed as

$$\mathbf{1}_M^T \mathbf{w}_n = h_n. \tag{6.47c}$$

Since the look direction response is fixed by the N constraints given by Equation 6.47b, the minimization of power in the nonlook direction is the same as minimization of the total power. Thus, we seek a weight vector that minimizes Equation 6.46 subject to constraints given by Equation 6.47b. This is the constrained least mean square (LMS) algorithm

$$\text{minimize} \quad \mathbf{w}^T \mathbf{C}_{\text{STCM}} \mathbf{w}$$

$$\text{subject to } \Gamma^T \mathbf{w} = \mathbf{h}. \tag{6.48}$$

The cost to be minimized is

$$\text{cost function} = \mathbf{w}^T \mathbf{C}_{\text{STCM}} \mathbf{w} + \lambda^T (\Gamma^T \mathbf{w} - \mathbf{h}), \tag{6.49}$$

where λ is Lagrange multipliers ($N \times 1$ vector). Taking a gradient of Equation 6.49 with respect to **w** and setting it to zero, we get

$$\mathbf{C}_{\text{STCM}} \mathbf{w} + \Gamma \lambda = 0. \tag{6.50a}$$

In terms of the Lagrange multiplier, the optimum weight vector is given by

$$\mathbf{w}_{\text{opt}} = -\mathbf{C}_{\text{STCM}}^{-1} \Gamma \lambda. \tag{6.50b}$$

To obtain the Lagrange multiplier we substitute Equation 6.50b into the second equation of Equation 6.48

$$\lambda = -\left[\Gamma^T \mathbf{C}_{\text{STCM}}^{-1} \Gamma\right]^{-1} \mathbf{h}.$$

Using this result in Equation 6.50 we get an expression for the optimum weight vector,

$$\mathbf{w}_{\text{opt}} = \mathbf{C}_{\text{STCM}}^{-1}\Gamma\left[\Gamma^{\text{T}}\mathbf{C}_{\text{STCM}}^{-1}\Gamma\right]^{-1}\mathbf{h}, \tag{6.51}$$

where we shall assume that $\left[\Gamma^{\text{T}}\mathbf{C}_{\text{STCM}}^{-1}\Gamma\right]^{-1}$ exists. This follows from the fact that $\mathbf{C}_{\text{STCM}}$, being a covariance matrix, is positive definite and that Γ has full column rank. In $M\times N$ dimensional space, the set of constraint equations define a hyper plane and the perpendicular distance to this hyper plane is $\mathbf{w}_0=\Gamma[\Gamma^T\Gamma]^{-1}\mathbf{h}$, which is indeed a minimum norm solution of the constraint equation (see Figure 6.10a). It may also be considered as an optimum weight vector, a solution of Equation 6.51 when $\mathbf{C}_{\text{STCM}}$ is a purely diagonal matrix, a situation where there is no propagating signal but only spatially and temporally uncorrelated noise. The output equal-power surfaces, given by Equation 6.46, are shown as contours in Figure 6.10b. The error power increases away from the center of the space. The surface that is tangential to the constraint plane is the desired minimum power surface, and the tangential point of contact is the required solution. The vector to the tangential point of contact is the optimum filter vector, denoted by $\mathbf{w}_{\text{opt}}$ in Figure 6.10b.

6.4.2 Beam Pattern

It is instructive to understand the beam pattern of the optimum weight vector. The frequency response of the weight vector is given by

$$H(\omega,\theta) = \sum_{n=0}^{N-1} \mathbf{E}^H(\omega,\theta)\mathbf{D}(\omega)\mathbf{w}_n \exp(-j\omega n\Delta t), \tag{6.52a}$$

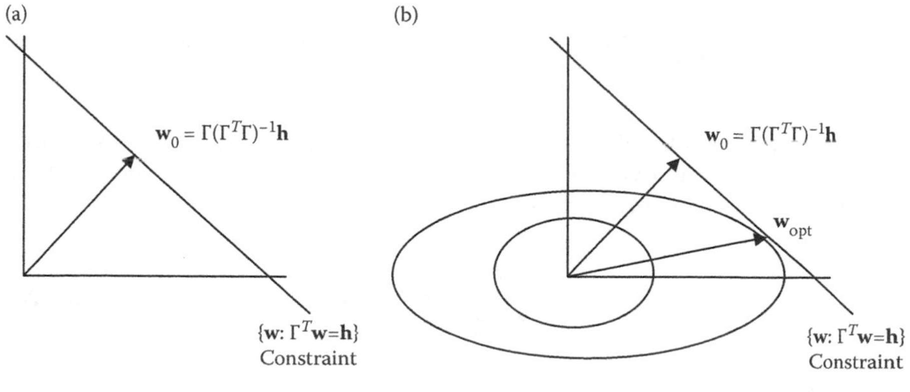

FIGURE 6.10
(a) MN dimensional weight vector space and (MN–N) dimensional hyper plane defined by N constraints (b) Contours of constant power. The optimum solution is where the hyper plane is tangential to a contour.

where

$$\mathbf{E}(\omega) = \mathrm{col}\left\{e^{-j\omega\tau'_0}, e^{-j\omega\tau'_1}, \ldots, e^{-j\omega\tau'_{M-1}}\right\},$$

and

$$\mathbf{D}(\omega) = \mathrm{diag}\left\{e^{-j\omega\tau_0}, e^{-j\omega\tau_1}, \ldots, e^{-j\omega\tau_{M-1}}\right\},$$

where $\tau'_0, \tau'_1, \ldots, \tau'_{M-1}$ are relative propagation delays at different sensors and $\tau_0, \tau_1, \ldots, \tau_{M-1}$ are the delays applied to the array outputs. Note that the propagation delay at mth sensor is given by $\tau'_m = md(\sin\theta/c)$. The array response in the look direction ($\theta=\theta_0$) may be obtained using Equations 6.52a and 6.47c as

$$\begin{aligned} H(\omega,\theta_0) &= \sum_{n=0}^{N-1} \mathbf{1}_M^T \mathbf{w}_n \exp(-j\omega n\Delta t) \\ &= \sum_{n=0}^{N-1} h_n \exp(-j\omega n\Delta t) = H(\omega). \end{aligned}$$

The beam pattern is equal to the magnitude squared response function (Equation 6.52a) integrated over a frequency band ($\omega_l-\omega_u$),

$$B(\theta) = \left[\frac{1}{2\pi}\int_{-\omega_u}^{\omega_u} \left|H(\omega,\theta)\right|^2 d\omega - \frac{1}{2\pi}\int_{-\omega_l}^{\omega_l} \left|H(\omega,\theta)\right|^2 d\omega\right], \tag{6.52b}$$

where

$$\left|H(\omega,\theta)\right|^2 = \sum_{n'=0}^{N-1}\sum_{n=0}^{N-1} \mathbf{w}_{n'}^H \mathbf{D}^H(\omega)\mathbf{E}(\omega,\theta)\mathbf{E}^H(\omega,\theta)\mathbf{D}(\omega)\mathbf{w}_n \exp(-j\omega(n-n')\Delta t).$$

The integration operation may be taken inside the summation signs and then it is carried out over each element of the matrix product. The result of integration may be expressed as

$$B(\theta) = \sum_{n'}^{N-1}\sum_{n=0}^{N-1} \mathbf{w}_{n'}^H \mathbf{Q}^{n'n} \mathbf{w}_n, \tag{6.52c}$$

where $\mathbf{Q}^{n'n}$ is a $M\times M$ square matrix whose (m, m') th element is given by

$$\left[\mathbf{Q}^{n,n'}\right]_{m,m'} = \frac{1}{2\pi}[\omega_u \sin c(\omega_u \hat{\tau}_{m,m'}^{n,n'}) - \omega_l \sin c(\omega_l \hat{\tau}_{m,m'}^{n,n'})],$$

where

$$\hat{\tau}_{m,m'}^{n,n'} = (\tau'_{m'} - \tau'_m) - (\tau_{m'} - \tau_m) + (n - n')\Delta t,$$

and

$$\sin c(\omega\tau) = \frac{\sin(\omega\tau)}{(\omega\tau)}.$$

Finally, Equation 6.52c may be expressed in a matrix form

$$B(\phi) = \mathbf{w}^H \mathbf{Q} \mathbf{w}, \tag{6.52d}$$

where $\mathbf{Q}$ is a $(MN \times MN)$ square matrix

$$\mathbf{Q} = \begin{bmatrix} \mathbf{Q}^{0,0} & \mathbf{Q}^{0,1} & \cdots & \mathbf{Q}^{0,N-1} \\ \mathbf{Q}^{1,0} & \mathbf{Q}^{1,1} & \cdots & \mathbf{Q}^{1,N\ 1} \\ \vdots & \vdots & & \vdots \\ \mathbf{Q}^{N-1,0} & \mathbf{Q}^{N-1,1} & \cdots & \mathbf{Q}^{N-1,N-1} \end{bmatrix}.$$

It may be noted that since

$$[\mathbf{Q}^{n,n'}]_{m,m'} = [\mathbf{Q}^{n',n}]_{m',m},$$

$\mathbf{Q} = \mathbf{Q}^T$ where superscript T stands for transpose.

6.4.3 Frost's Adaptive Algorithm

For computation of the optimum weight vector, we must know $\mathbf{C}_{\text{STCM}}$ and be able to compute its inverse. The former requires a large amount of data and hence large delay, while the latter requires computational complexity to the order of cube of the number of weights. Frost [18] presented an adaptive approach to overcome the above-mentioned difficulties. The cost function (Equation 6.49) is minimized by going down the steepest gradient, starting from the initial point,

$$\mathbf{w}(0) = \Gamma[\Gamma^T \Gamma]^{-1} \mathbf{h}.$$

The step size is proportional to the magnitude of the gradient (see Equation 6.50a) and scaled by a constant μ. After k iterations, the new weight vector will be given by

$$\mathbf{w}(k+1) = \mathbf{w}(k) - \mu[\mathbf{C}_{\text{STCM}}\, \mathbf{w}(k) + \Gamma\lambda(k)]. \tag{6.53}$$

At each step the constraints (second equation in Equation 6.48) must be satisfied. This requirement enables us to solve for the Lagrange multipliers. On substituting the value of the Lagrange multiplier thus obtained into Equation 6.53, we obtain

$$\mathbf{w}(k+1)=\mathbf{w}(k)-\mu[\mathbf{I}-\Gamma[\Gamma^T\Gamma]^{-1}\Gamma^T]\,\mathbf{C}_{\text{STCM}}\,\mathbf{w}(k)+\Gamma[\Gamma^T\Gamma]^{-1}(\mathbf{h}-\Gamma^T\mathbf{w}(k)). \tag{6.54a}$$

Further simplification of Equation 6.54a yields

$$\mathbf{w}(k+1)=[\mathbf{I}-\Gamma[\Gamma^T\Gamma]^{-1}\Gamma^T]\,\{\mathbf{w}(k)-\mu\mathbf{C}_{\text{STCM}}\,\mathbf{w}(k))\}+\Gamma[\Gamma^T\Gamma]^{-1}\,\mathbf{h}, \tag{6.54b}$$

or

$$\mathbf{w}(k+1)=\mathbf{P}\{\mathbf{w}(k)-\mu\mathbf{C}_{\text{STCM}}\,\mathbf{w}(k))\}+\mathbf{Q}, \tag{6.54c}$$

where

$$\mathbf{P}=[\mathbf{I}-\Gamma[\Gamma^T\Gamma]^{-1}\Gamma^T]\ (M.N\times M.N\text{ matrix}),$$

$$\mathbf{Q}=\Gamma[\Gamma^T\Gamma]^{-1}\mathbf{h}\ (MN\times 1\text{ vector}).$$

The starting value for recursion is $\mathbf{w}(0)=\mathbf{Q}$. It is easy to show the following relations

$$\begin{aligned}
&\mathbf{Q}=[\mathbf{I}-\mathbf{P}]\mathbf{w}_{\text{opt}},\\
&\mathbf{P}\mathbf{C}_{\text{STCM}}\mathbf{w}_{\text{opt}}=0,
\end{aligned} \tag{6.55}$$

$$\begin{aligned}
\mathbf{w}(k+1)-\mathbf{w}_{\text{opt}}&=\mathbf{P}\{\mathbf{w}(k)-\mu\mathbf{C}_{\text{STCM}}\mathbf{w}(k)\}-\mathbf{P}\mathbf{w}_{\text{opt}}\\
&=\mathbf{P}(\mathbf{w}(k)-\mathbf{w}_{\text{opt}})-\mu\mathbf{P}\mathbf{C}_{\text{STCM}}\mathbf{w}(k)-0\\
&=\mathbf{P}(\mathbf{w}(k)-\mathbf{w}_{\text{opt}})-\mu\mathbf{P}\mathbf{C}_{\text{STCM}}(\mathbf{w}(k)-\mathbf{w}_{\text{opt}}),
\end{aligned}$$

$$\delta\mathbf{w}(k+1)=\mathbf{P}\delta\mathbf{w}(k)-\mu\mathbf{P}\mathbf{C}_{\text{STCM}}\,\delta\mathbf{w}(k), \tag{6.56}$$

where $\delta\mathbf{w}(k)=(\mathbf{w}(k)-\mathbf{w}_{\text{opt}})$. Premultiplying Equation 6.56 with $\mathbf{P}$ and noting that $\mathbf{P}$ is an idempotent matrix (i.e., $\mathbf{P}^2=\mathbf{P}$), we obtain an interesting result, that is,

$$\mathbf{P}\delta\mathbf{w}(k+1)=\delta\mathbf{w}(k+1),$$

for all k. Using this result in Equation 6.56 we obtain

$$\delta\mathbf{w}(k+1)=[\mathbf{I}-\mu\mathbf{P}\mathbf{C}_{\text{STCM}}\mathbf{P}]\,\delta\mathbf{w}(k). \tag{6.57}$$

The solution of the above difference equation is simply

$$\delta\mathbf{w}(k)=[\mathbf{I}-\mu\mathbf{P}\mathbf{C}_{\text{STCM}}\mathbf{P}]^k\,\delta\mathbf{w}(0),$$

where $\delta\mathbf{w}(0)=\mathbf{Q}-\mathbf{w}_{\text{opt}}$. If μ is chosen such that

$$0 \le \mu \le \frac{1}{\sigma_{\max}},$$

then the length of $\delta\mathbf{w}(k)$ is bounded between two ever-decreasing geometric progressions

$$(1-\mu\sigma_{\max})^{k+1}\|\delta\mathbf{w}(0)\| \le \|\delta\mathbf{w}(k+1)\| \le (1-\mu\sigma_{\min})^{k+1}\|\delta\mathbf{w}(0)\|,$$

the maximum and the minimum eigenvalues of $\mathbf{PC}_{\text{STCM}}\mathbf{P}$ are $\sigma_{\max}$ and $\sigma_{\min}$, respectively. It, therefore, follows that as $k \to \infty$, $\|\delta\mathbf{w}(k)\| \to 0$ hence,

$$\mathbf{w}(k) \to \mathbf{w}_{\text{opt}}.$$

Equation 6.54 is the deterministic constrained LMS algorithm requiring the exact covariance matrix. But, a simple approximation of the covariance matrix is the outer product of the array output at kth step, which leads to the stochastic constrained LMS algorithm,

$$\begin{aligned}\tilde{\mathbf{w}}(k+1) &= \mathbf{P}\left\{\tilde{\mathbf{w}}(k)-\mu\mathbf{f}_{MN}(k)\mathbf{f}_{MN}^{T}(k)\tilde{\mathbf{w}}(k)\right\}+\mathbf{Q}\\ &= \mathbf{P}\left\{\tilde{\mathbf{w}}(k)-\mu\hat{f}(k)\mathbf{f}_{MN}(k)\right\}+\mathbf{Q},\end{aligned} \tag{6.58}$$

where $\hat{f}(k)$ (scalar) is the array output at step k. Note that the weight vector, denoted by $\tilde{\mathbf{w}}$, is now a random vector. Assuming that the successive snapshots are independent, the mean of $\tilde{\mathbf{w}}(k)$ can be shown to approach $\mathbf{w}_{\text{opt}}$ as $k \to \infty$. The proof follows the same line of argument as in the deterministic constrained LMS algorithm. In order that the snapshots are independent, the tap delay, Δt, must be sufficiently large and the array output is Gaussian. The beamformer structure is shown in Figure 6.11a; $\tau_1, \tau_2, \ldots, \tau_{M-1}$ are delays applied to each sensor output in order to orient the array to receive the SOI as if it is incident on broadside. The box representing the weight vector $\tilde{\mathbf{w}}$ actually stands for spatio-temporal filter. The internal structure of $\tilde{\mathbf{w}}$ is shown in Figure 6.11b, where Δt is the sampling interval. The LMS algorithm is used to update the coefficients following Equation 6.58.

6.4.4 Generalized Sidelobe Canceler (GSC)

It may be noted that Γ is of rank N with range space of dimension N. The projection of any weight vector on the range space of Γ, that is, $\Gamma(\Gamma^T\Gamma)^{-1}\Gamma^T\mathbf{w}$, is equal to the minimum norm solution, $\mathbf{w}_0=\Gamma(\Gamma^T\Gamma)^{-1}\mathbf{h}$. Similarly, the projection of a vector $\mathbf{w}_a$ onto the null space of Γ (that is, $[\mathbf{I}-\Gamma[\Gamma^T\Gamma]^{-1}\Gamma^T]=\mathbf{P}$) is $\mathbf{Pw}_a$. Formally, a weight vector is expressed as a sum of $\mathbf{w}_0$ and $\mathbf{Pw}_a$, known as GSC decomposition,

$$\mathbf{w}=\mathbf{w}_0-\mathbf{Pw}_a. \tag{6.59}$$

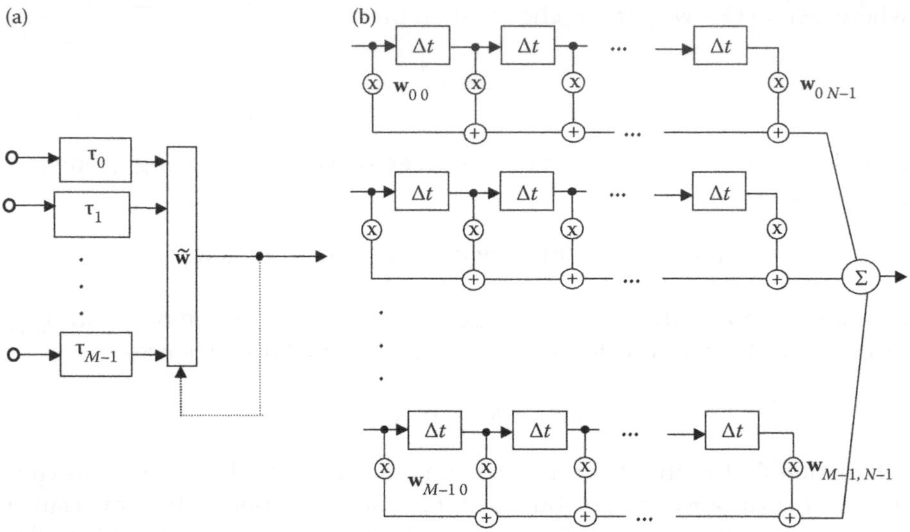

FIGURE 6.11
Linearly constrained adaptive beamformer. (a) Structure of the beamformer. (b) Internal structure of $\tilde{\mathbf{w}}$.

Equation 6.59 will satisfy the constraint part of Equation 6.48 for any $\mathbf{w}_a$. Hence, the constrained minimization (Equation 6.48) reduces to unconstrained minimization. Further, we shall show that $\mathbf{P}$ blocks the signal vector, that is, it blocks a signal vector given by,

$$\mathbf{s}(t) = \mathrm{col}\left\{s(t)\mathbf{1}_M^T, s(t-\Delta t)\mathbf{1}_M^T, s(t-2\Delta t)\mathbf{1}_M^T, \ldots, s(t-(N-1)\Delta t)\mathbf{1}_M^T\right\},$$

where $\mathbf{s}(t)$ is the signal waveform presumed to be incident at broadside. We note that the signal vector is a linear combination of column vectors of Γ, therefore it lies in the range space of Γ. It therefore follows that

$$\mathbf{P}^T\mathbf{s}(t) = \mathbf{0}. \tag{6.60}$$

Thus, $\mathbf{P}$ acts as a signal-blocking matrix.

The weight vector consists of two parts, namely, a data independent vector, $\mathbf{w}_0$, also known as a quiescent weight vector, and a data dependent vector, $\mathbf{w}_a$. Since it is adaptively estimated from the data, it is known as an adaptive weight vector. Based on the above model of the weight vector, the structure of the GSC is shown in Figure 6.12.

The weight vector structure, given by Equation 6.59, ensures that the constraint part of Equation 6.48 is automatically satisfied, thus the constrained minimization reduces to unconstrained minimization given by

$$\text{minimize: } (\mathbf{w}_0 - \mathbf{P}\mathbf{w}_a)^T \mathbf{C}_{\text{STCM}}(\mathbf{w}_0 - \mathbf{P}\mathbf{w}_a),$$

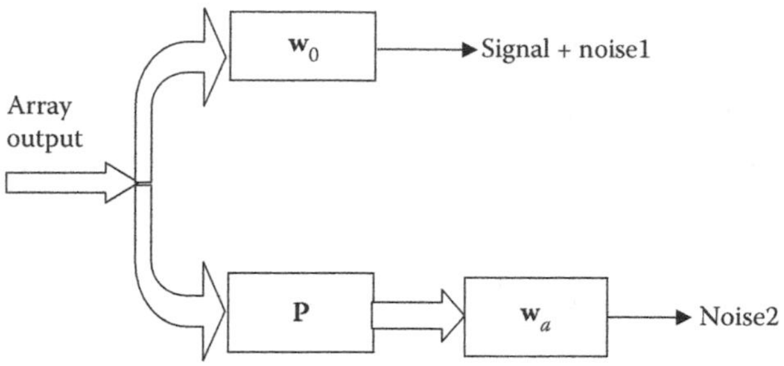

FIGURE 6.12
Structure of the generalized sidelobe canceler.

which further simplifies to

$$\begin{aligned}\text{minimize: } &= \mathbf{w}_0^T \mathbf{C}_{\text{STCM}} \mathbf{w}_0 + \mathbf{w}_a^T \mathbf{P}^T \mathbf{C}_{\text{Noise}} \mathbf{P} \mathbf{w}_a \\ &\quad - \mathbf{w}_a^T \mathbf{P}^T \mathbf{C}_{\text{Noise}} \mathbf{w}_0 - \mathbf{w}_0^T \mathbf{C}_{\text{Noise}} \mathbf{P} \mathbf{w}_a,\end{aligned} \tag{6.61}$$

where we have used the fact that $\mathbf{P}^T\mathbf{C}_{\text{Sig}}=0$. We have implicitly assumed that the signal and noise (including interference) are uncorrelated. Therefore, the covariance matrix of signal plus noise will be equal to the sum of signal and noise covariance matrices, that is,

$$\mathbf{C}_{\text{STCM}} = \mathbf{C}_{\text{Sig}} + \mathbf{C}_{\text{Noise}}.$$

By minimizing Equation 6.61 with respect to $\mathbf{w}_a$ we obtain

$$\mathbf{w}_a = (\mathbf{P}^T \mathbf{C}_{\text{Noise}} \mathbf{P})^{-1} \mathbf{P}^T \mathbf{C}_{\text{Noise}}\, \mathbf{w}_0. \tag{6.62}$$

In absence of white noise the rank of $\mathbf{P}$ is $(MN-N)$ but, the rank $\mathbf{C}_{\text{Noise}}$ is controlled by the bandwidth of the interference and the geometry of the array. It is generally much smaller than $(MN-N)$. The effective dimension of the weight vector will depend upon the rank of $\mathbf{P}^T\mathbf{C}_{\text{Noise}}\mathbf{P}$, $K \ll (N.M-N)$. This property has been exploited by introducing, after the blocking matrix, a transformation matrix consisting of significant eigenvectors of $\mathbf{P}^T\mathbf{C}_{\text{Noise}}\mathbf{P}$. This reduced dimension weight vector, often known as partially adaptive, has a performance quite close to that of a fully adaptive weight vector [9].

The GSC structure shown in Figure 6.12 has two branches. The output of the top branch consists of signal and Noise1, which is the noise and interference received by the array and modified by quiescent weight vector, $\mathbf{w}_0$. The output of the lower branch consists of Noise2, that is, noise and interference received by the array and modified by filter $\mathbf{P}$ and $\mathbf{w}_a$. The aim is to cancel out Noise1 by Noise2, which is made as close to Noise1 as possible by optimally selecting $\mathbf{w}_a$. The optimal value of $\mathbf{w}_a$, given by Equation 6.62, is the only data dependent vector. Unfortunately, it requires knowledge of the noise matrix $\mathbf{C}_{\text{Noise}}$, which is not readily available.

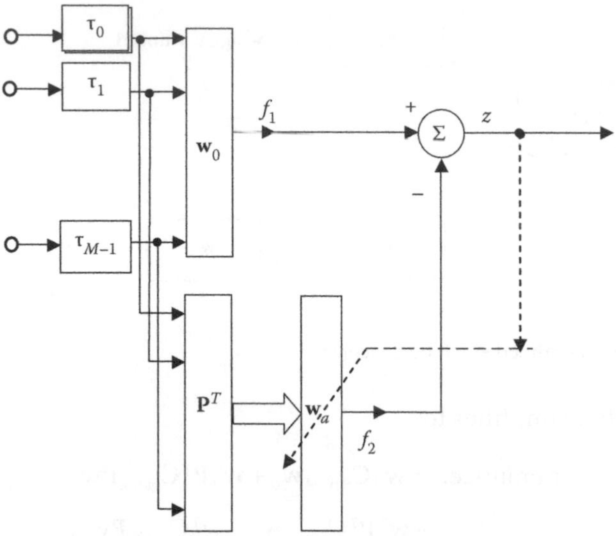

FIGURE 6.13
Adaptive implementation of the generalized sidelobe canceler. A uniform linear array is assumed and the direction of arrival of signal of interest is assumed known.

However, an adaptive approach (unconstrained LMS algorithm) for the estimation of $\mathbf{w}_a$ is possible. We shall now briefly describe this adaptive approach and follow it up with some illustrative examples.

An array of receivers, not necessarily a ULA, receives a SOI along with interference and white noise. The DOA of the SOI is assumed so that the delays at different sensors can be calculated and used to align the SOI for incidence on broadside. Signal output at different points in the circuit are marked in Figure 6.13. For example, $z(k)$ is defined as

$$z(k)=f_1(k)-f_2(k),$$

where $f_1(k)$ is the output of quiescent weight vector, $\mathbf{w}_0$ and $f_2(k)$ is the output of adaptive weight vector $\mathbf{w}_a$.

The LMS algorithm can be used to adapt the weight vector, $\mathbf{w}_a$, to the desired solution,

$$\mathbf{w}_a(k+1)=\mathbf{w}_a(k)+\mu z(k)\mathbf{f}'(k). \tag{6.63}$$

The step size μ controls the convergence rate and steady-state noise behavior. The convergence is assured if μ satisfies

$$\mu=\frac{\alpha}{\sum_{m=0}^{M-1}\sum_{n=0}^{N-1}|f(m,n)|^2},$$

where $0<\alpha<1$.

6.4.5 Griffiths' Adaptive Beamformer

The GSC shown in Figure 6.13 can be simplified by exploiting the fact that the quiescent weight vector and the signal-blocking matrix are data independent. Indeed the quiescent weight vector can be expressed as a uniform spatial weight vector followed by a finite impulse response (FIR) filter. This follows from the fact that the quiescent weight vector may be written as $\mathbf{w}_0=\Gamma\mathbf{h}/M$. The spatial filter is $1/M$ and the FIR filter is $\mathbf{h}$. The signal-blocking matrix may be expressed as

$$\begin{aligned}\mathbf{P}&=[\mathbf{I}-\Gamma[\Gamma^T\Gamma]^{-1}\Gamma^T]\\&=[\mathbf{I}-\Gamma\Gamma^T/M].\end{aligned}$$

It may be observed that $\Gamma\Gamma^T$ is an outer product of two block diagonal matrices whose diagonal entries are column vectors of ones and of length M. The product is also a block diagonal matrix, but with square matrices of size $M\times M$. Therefore, it follows that the signal-blocking matrix $\mathbf{P}$ will also be a block diagonal matrix. The rank of each block matrix on the diagonal is $M-1$. Furthermore, each block matrix acts as a signal-blocking matrix, that is, the entries in a column sum up to zero. Since the role of a signal-blocking matrix is simply to block a uniform SOI, it can be achieved in many ways while keeping the required rank condition and column entries summing up to zero. This, in fact, was pointed out by Griffiths' [10] and further investigated by Jablon [11]. Griffiths' alternate scheme for broadband adaptive beamforming is illustrated in Figure 6.14.

The signal-blocking matrix of size $(M\times M-1)$ may be computed intuitively, keeping in mind that column entries sum up to zero and the rank is $M-1$. Of the many matrices that can be generated with this property, two possibilities are shown below [10] for $M=4$.

$$\mathbf{B}_1=\begin{bmatrix}1 & 1 & -1\\ 1 & -1 & -1\\ -1 & -1 & 1\\ -1 & 1 & 1\end{bmatrix},$$

$$\mathbf{B}_2=\begin{bmatrix}1 & 0 & 0\\ -1 & 1 & 0\\ 0 & -1 & 1\\ 0 & 0 & -1\end{bmatrix}.$$

It is easy to generalize the above matrices for any M.

We consider an idealized (that is noise-free) situation. There are two uncorrelated sources with DOAs at 0.0° and 15° incident on an eight-element ULA. The SOI lasts over a duration of 512 sample points, but the interference lasts

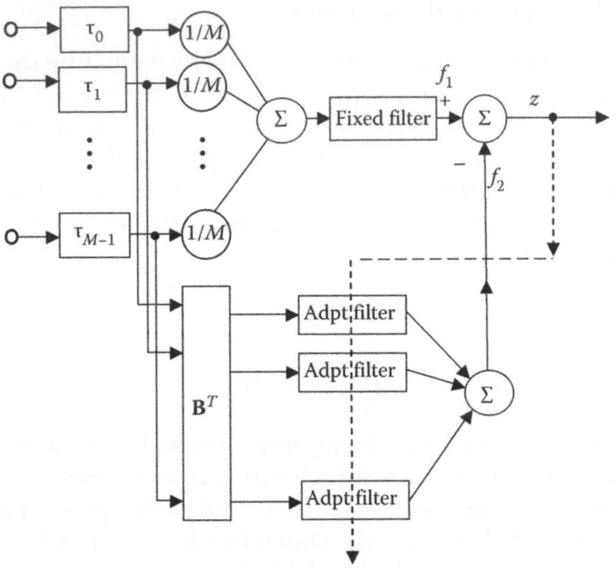

FIGURE 6.14
Alternate implementation of adaptive beamformer. Fixed filter is the constraint coefficients in Frost filter. The signal-blocking matrix **B** is intuitively designed. All entries in a column sum up to zero and the rank of **B** is $M-1$.

over the entire duration of 1,024 sample points. Both signals are broadband signals with a center frequency of 0.2 Hz and a bandwidth of 0.1 Hz. The output of the first sensor (reference sensor) is shown in plot #1 of Figure 6.15. The output of the delay-and-sum beamformer is shown in plot #2. The next two plots show the outputs of optimum adaptive broadband beamformers. For this we have used four taps. Plot #3 shows the output of the Frost beamformer, while plot #4 shows the output of the Griffiths' beamformer. Both optimum beamformers yield roughly similar outputs, though it was found that the mse is less in Griffiths' beamformer. The basic assumption of uncorrelated signal and interference is invalid in many real life situations, for example, when the SOI follows multipath propagation.

6.5 FI Beamforming

The response of a fixed length array is frequency-dependent except when the wavefront is incident at broadside. To emphasize this point, we have plotted the frequency response of an eight-element array when a broadband signal (±0.5 Hz) is incident at 10°, 30°, and 60° (see Figure 6.16). The response is frequency independent only when the wavefront is incident at 0°, that is, at broadside. There will be some signal shape distortion in all other cases. This is not acceptable when it is required to cancel broadband interference by using

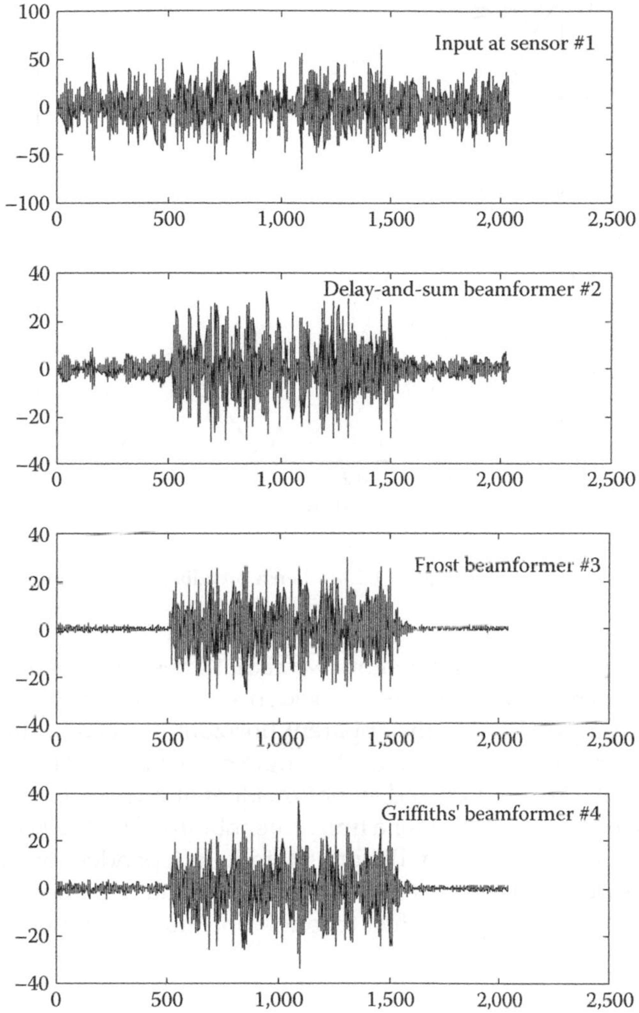

FIGURE 6.15
A numerical experiment to show the relative performance of broadband adaptive beamformers.

replicas of the interference signals received through different sidelobes [12]. Hence, there is a need for frequency independent beamforming so that the output of the array is a replica of incident broadband signal. Fan filter is a simple example of FI beamforming filter. A somewhat more involved approach is to devise space-dependent filters [13].

6.5.1 Fan Filter as FI Beamformer

We have already covered the topic of fan filters in Chapter 3, Section 3.1, where we have considered a uniform fan with a unit response inside the fan-shaped pass region and zero outside. There, we have pointed out that the maximum

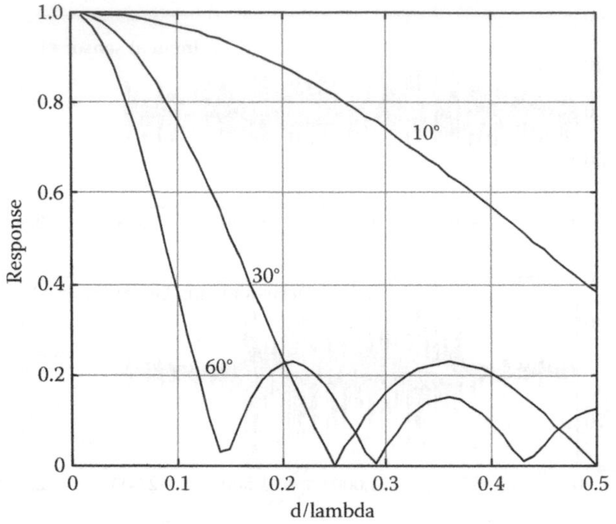

FIGURE 6.16
Frequency response of an eight-element ULA when a broadband signal (±0.5 Hz) is incident at 10°, 30°, and 60°.

size of the pass region is bounded by two radial lines whose slopes are equal to $\pm c$, which correspond to angles of incidence equal to $\pm\pi/2$. The slope of a radial line is proportional to the apparent horizontal speed of the wavefront or inversely proportional to sine of the angle of incidence. The frequency of an ideal fan filter to a broadband signal incident at any angle within $\pm\pi/2$ is independent of frequency, but for a finite filter obtained by truncating an ideal filter, the frequency response will not be totally independent of frequency, as shown in Figure 6.17. The ripples are indeed the result of truncation. It is possible to minimize this ripple effect through optimum filter design just as in the design of a lowpass filter.

6.5.2 Design of Fan Filter via Mapping

In Chapter 3, Section 3.2, we have described a method of mapping a 1D optimum filter into a 2D filter via McClellan's transformation. Different types of 2D filter contours can be generated using a specific transformation. We have noted that the transformation given by Equation 3.28b, reproduced here for convenience, maps a 1D filter into fan-shaped contours

$$\cos(\omega') = \frac{1}{2}[\cos(\omega) - \cos(u)],$$

where ω' is the frequency variable of a 1D filter, and ω and u are frequency variables (temporal and spatial) of a 2D filter. The contours generated for different values of ω' are shown in Figure 3.12, where note that the y-axis is for temporal frequency (ω). The contours are only approximately fan-shaped,

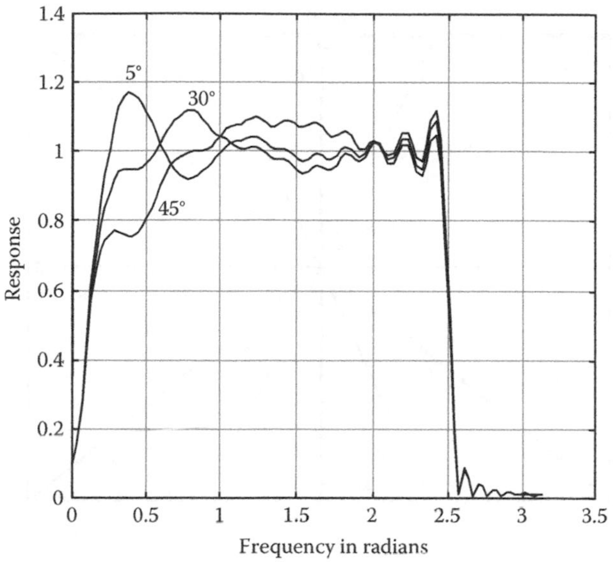

FIGURE 6.17
Frequency response of a 16-element ULA when a broadband signal (±0.5 Hz) is incident at 5°, 30°, and 45°. An ideal fan filter truncated to 16×64 spatio-temporal is used.

particularly poor where the desired passband is likely to be positioned. Segikuchi [12] introduced another type of transformation

$$\omega' = \pi \sin(\theta) = \frac{\pi c}{c_{\text{app}}} \frac{u}{\omega}, \tag{6.64}$$

where θ is the DOA and $c_{\text{app}} = d/\Delta t$ is the apparent speed of the propagating wave across the array.

In the above discussion, the array is steered to broadside (i.e., $\theta = 0°$). It is possible to steer the array to any desired direction, θ_0. The required transformation is given by

$$\omega' = \pi \left[\frac{c}{c_{\text{app}}} \frac{u}{\omega} - \sin \theta_0 \right]. \tag{6.65}$$

Next, a 1D FIR filter (not necessarily linear phase) may be mapped into a 2D filter using Equation 6.64 or Equation 6.65. The algorithmic steps are as follows:

The frequency response of a 1D prototype FIR filter is written as

$$H(\omega') = \sum_{n=0}^{N-1} h(n) \exp(-j\omega' n),$$

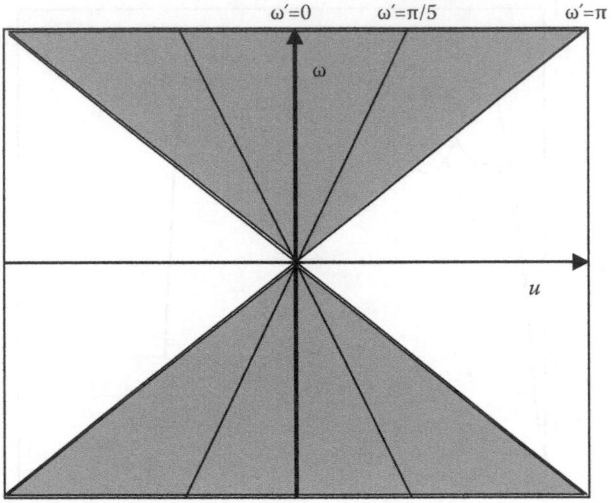

FIGURE 6.18
Fan-shaped contours produced by the mapping given in Equation 6.64. Note that the diagonals correspond to $\pm\theta=90°$ and ω-axis corresponds to $\theta=0°$. The propagating waves exist only in the shaded region (see Chapter 1).

where $h(n)$ is the impulse response. Apply the transformation given in Equation 6.65

$$H(\omega,u)=\sum_{n=0}^{N-1} h(n)\exp\left(-j\pi\left(\frac{c}{c_{\text{app}}}\frac{u}{\omega}-\sin\theta_0\right)n\right).$$

Apply the inverse Fourier transform to obtain 2D filter coefficients.

In Figure 6.18, $H(\omega, u)$ is defined in the shaded fan area. Outside the shaded fan, since the waves are only transient, that is, nonpropagating; we can assume any value for the transfer function. For the sake of continuity across the diagonal, we select a constant value equal to $\pi(1-\sin(\theta_0))$.

6.5.3 Frequency-Scaled Continuous Sensor

The array pattern is controlled by the array aperture, which, in case of a linear array, is the length of the array expressed in wavelengths. Consider an array where the sensors are spaced at half-wavelength corresponding to the upper frequency, ω_U, of a wideband signal whose spectrum is spread over a frequency band, ω_L to ω_U. Construct a subarray by picking alternate sensors, now spaced at half-wavelength corresponding to the frequency $\omega_U/2$. Notice that the new subarray will have the same aperture as the original one. We can carry on this procedure of constructing new subarrays having the same aperture but at frequencies $\omega_U/4$, $\omega_U/8$, $\omega_U/16$..., covering the entire bandwidth. These are known as harmonically nested subarrays [14]. They produce identical beam patterns at frequencies ω_U, $\omega_U/2$, $\omega_U/4$, $\omega_U/8$, $\omega_U/16$. When the upper frequency ω_U is very large, the sensor spacing in the array will be very small, which in the limiting case leads to a continuous sensor array.

Let $\rho(x,\omega)$ represent space (one dimensional) and frequency-dependent continuous gain or sensitivity distribution. A plane wave is incident on the continuous sensor. The output of such a sensor is given by

$$b(\theta,\omega)=\int_{-\infty}^{\infty}\rho(x,\omega)\exp\left(-j\omega\frac{x\sin(\theta)}{c}\right)dx, \tag{6.66}$$

where θ is the angle (measured with respect to broadside) of arrival and c is the propagation speed of the wavefield. Now we make an important assumption, namely, the sensitivity distribution has the following functional form [15]:

$$\rho(x,\,\omega)=\omega G(x\omega)\quad \forall\omega>0, \tag{6.67}$$

where $G(.)$ is the frequency-scaled arbitrary absolutely integrable complex function. Then, the sensor output takes on an interesting form

$$\begin{aligned} b(\theta,\omega)&=\int_{-\infty}^{\infty}\omega G(x\omega)\exp\left(-j\omega\frac{x\sin(\theta)}{c}\right)dx \\ &=\int_{-\infty}^{\infty}G(y)\exp-jy\left(\frac{\sin(\theta)}{c}\right)dy \\ &=\Gamma\left(\frac{\sin(\theta)}{c}\right). \end{aligned} \tag{6.68}$$

The sensor response becomes frequency independent and it is equal to the Fourier transform of the sensitivity distribution in the frequency range $\pm 1/c$. Outside this range, the Fourier transform of the sensitivity distribution can take on any value consistent with the requirement of square integrability.

The sensitivity distribution, $G(x\omega)$ for a fixed ω, $A_\omega(x)$, is known as an aperture function and for a fixed $x(\geq 0)$, $H_x(\omega)$, is known as a primary filter.

$$G(x\omega)=A_\omega(x)=H_x(\omega),\quad \forall x,\quad \omega>0. \tag{6.69}$$

The primary filter response takes the same shape as the aperture distribution, but ω and x cannot be freely interchanged as, while ω must be positive, x may take any value over $\pm\infty$. For $x>0$ the primary filter function is equal to the aperture distribution function where x and ω are interchanged,

$$H_{x^+}(\omega)=A_x(\omega),\quad x>0. \tag{6.70a}$$

But for $x<0$ we have

$$H_{x^-}(\omega)=A_{-x}(-\omega),\quad x<0. \tag{6.70b}$$

6.5.4 Discrete Approximation

Let $H_{x_1}(\omega)$ and $H_{x_2}(\omega)$ be the primary filters at x_1 and x_2 (>0), respectively. It is easy to show that

$$H_{x_2}(\omega)=H_{x_1}\left(\frac{x_2}{x_1}\omega\right), \tag{6.71a}$$

and, in general, the primary filter at x_i

$$H_{x_i}(\omega)=H_{x_1}\left(\frac{x_i}{x_1}\omega\right). \tag{6.71b}$$

Thus, a primary filter at any location is a scaled version of the filter at $x_1=1$. All primary filters are identical up to frequency dilation. The primary filter at $x=0$ is constant and hence infinite bandwidth. As the distance of a sensor increases, the bandwidth decreases linearly. As a result, the sensors may be arranged with linearly increasing spacing to form a nonuniform array. All sensors to the left of the ith sensor are spaced at intervals less than or equal to half-wavelength corresponding to the maximum frequency of ith primary filter. These sensors are said to be active and constitute an array, which does not cause any spatial aliasing. Let P be the aperture of the array measured in half-wavelength and λ_i be the wavelength corresponding to the maximum frequency of ith primary filter. We have the following relation

$$x_i = P\frac{\lambda_i}{2}. \tag{6.72a}$$

The maximum frequency of ith primary filter will be equal to $\omega_i=(Pc/x_i)\pi$. The spacing between consecutive sensors will satisfy a recursive relation

$$x_i = x_{i-1}+\frac{\lambda_i}{2}. \tag{6.72b}$$

By solving Equation 6.72a and b we obtain

$$x_i = \frac{P}{P-1}x_{i-1},$$

or

$$x_i = \alpha\left(\frac{P}{P-1}\right)^i, \tag{6.73}$$

where α is an unknown constant. Let the incident signal be bandlimited in the range ω_L to ω_U. The smallest wavelength is λ_U, hence the least sensor spacing is equal to $\lambda_U/2$. The dense part of the array may be replaced by a uniformly spaced linear array of length equal to $P(\lambda_U/2)$. Using this in Equation 6.73, we can solve for α. We obtain

$$\alpha = P\frac{\lambda_U}{2}\left(\frac{P}{P-1}\right)^{-P}.$$

The location of ith sensor for $i > P$ is given by

$$x_i = P\frac{\lambda_U}{2}\left(\frac{P}{P-1}\right)^{i-P} \qquad P < i < N-1, \tag{6.74a}$$

and the location of the last sensor is given by

$$x_{N-1} = P\frac{\lambda_L}{2}. \tag{6.74b}$$

Given P and the lower and upper frequency bounds on the signal spectrum, we can compute, using Equation 6.74b, the length of the array. Finding N, the number of sensors, is however a bit more involved. Equation 6.72 for $i=N-1$ becomes

$$x_{N-1} = x_{N-2} + \frac{\lambda_L}{2}. \tag{6.75}$$

Using Equation 6.74a and b in Equation 6.75 we obtain

$$P\frac{\lambda_L}{2} = P\frac{\lambda_U}{2}\left(\frac{P}{P-1}\right)^{N-2-P} + \frac{\lambda_L}{2}.$$

After some algebraic simplification of the above equation, we obtain an expression for N,

$$N = P+1+\left\lceil \frac{\log\left(\frac{\lambda_L}{\lambda_U}\right)}{\log\frac{P}{P-1}} \right\rceil, \tag{6.76}$$

where $\lceil . \rceil$ denotes the ceiling function.

Finally, in order that the impulse response function of a primary filter is a real function, we must redefine the primary filter by introducing the complex conjugate symmetry:

$$\begin{aligned} \hat{H}_x(\omega) &= \frac{1}{2}H_{x_1}(\omega) \qquad \omega > 0 \\ &= H_{x_1}(0) \qquad \omega = 0 \\ &= \frac{1}{2}H^*_{x_1}(-\omega) \qquad \omega < 0. \end{aligned} \tag{6.77}$$

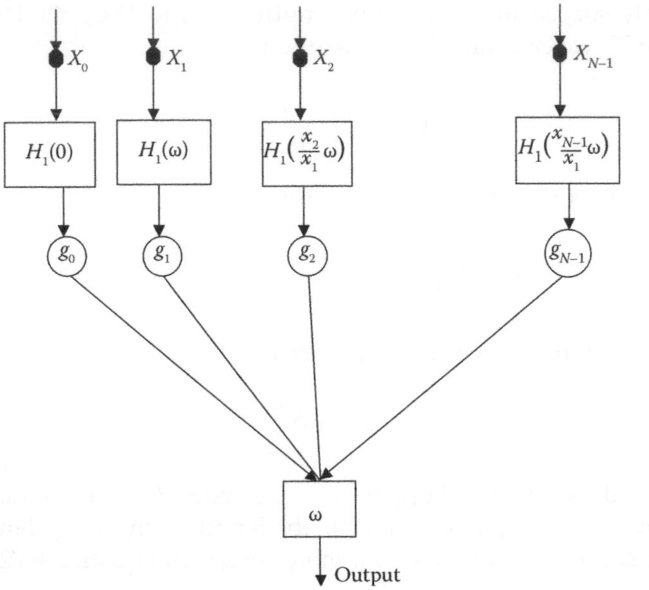

FIGURE 6.19
Realization of Equation 6.78.

6.5.4.1 FI Array

We have shown how a continuous sensor may be represented by a nonuniform array of omnidirectional point sensors whose outputs are filtered by a set of filters, which are simple dilations of a single frequency response function. A block diagram of broadband frequency independent array is shown in Figure 6.19, where coefficients $g_0, g_1, \ldots, g_{N-1}$ are frequency independent and are required to improve the approximation of an integral involving a continuous sensor by a sum of discrete sensor outputs, just as in the approximation of an integral by means of a sum of discrete samples. The trapezoidal rule yields

$$g_0 = (x_1 - x_0)/2,\ g_1 = (x_2 - x_0)/2,\ g_2 = (x_3 - x_1)/2,\ \ldots,\ g_{N-2} = (x_{N-1} - x_{N-3})/2,\ g_{N-1} = (x_{N-1} - x_{N-2})/2.$$

The integral (Equation 6.68) is now approximated by a discrete sum, where the integrant is evaluated at discrete sample points $0, x_1, x_2, \ldots, x_{N-1}$.

$$\begin{aligned} b(\theta,\omega) &= \int_{-\infty}^{\infty} \omega G(x\omega)\exp\left(-j\omega\frac{x\sin(\theta)}{c}\right)dx \\ &= \sum_{n=0}^{N-1} \omega\, g_n G(x_n\omega)\exp\left(-j\omega\frac{x_n\sin(\theta)}{c}\right) \qquad (6.78) \\ &= \omega\sum_{n=0}^{N-1} g_n H_1\left(\frac{x_n}{x_1}\omega\right)\exp\left(-j\omega\frac{x_n\sin(\theta)}{c}\right), \end{aligned}$$

where we have used Equations 6.69 and 6.71b. Realization of Equation 6.78 is illustrated in Figure 6.19, where $X_0, X_1, \ldots, X_{N-1}$ are samples of incident wavefront at different sensors. We shall now replace $H_1((x_n/x_1)\omega)$ in terms of filter coefficients,

$$H_1\left(\frac{x_n}{x_1}\omega\right)=\sum_{m=0}^{M-1}h(m)\exp\left(-j\frac{x_n}{x_1}\omega m\right)$$
$$=\mathbf{h}^H\phi_n,$$

where

$$\mathbf{h}=\{h(0), h(1),\ldots,h(M-1)\}^T$$

$$\phi_n=\left\{1, \exp\left(-j\frac{x_n}{x_1}\omega\right),\ldots,\exp\left(-j\frac{x_n}{x_1}\omega(M-1)\right)\right\}^T.$$

Equation 6.78 reduces to a compact form given in Equation 6.79 [16]. To avoid aliasing, the maximum frequency of a primary filter must be limited to a frequency appropriate at the sensor location (given by Equation 6.72a)

$$\begin{aligned}b(\theta,\omega)&=\omega\mathbf{h}^H\sum_{n=0}^{N-1}g_n\phi_n\exp\left(-j\omega\frac{x_n\sin(\theta)}{c}\right)\\&=\mathbf{h}^H\phi\mathbf{a}(\theta,\omega)\\&\approx b(\theta),\end{aligned}\tag{6.79}$$

where $\phi=\{\phi_0, \phi_1, \ldots, \phi_{N-1}\}$ is a $(M\times N)$ matrix and $\mathbf{a}(\theta, \omega)$ is an array response vector $(N\times 1)$, given by

$$\mathbf{a}(\theta,\omega)=\left\{g_0, g_1\exp\left(-j\omega\frac{x_1\sin(\theta)}{c}\right),\ldots,g_{N-1}\exp\left(-j\omega\frac{x_{N-1}\sin(\theta)}{c}\right)\right\}^T.$$

Consider an example of broadband array; $P=8$, $\lambda_U=5$, and $\lambda_L=50$ units. This corresponds to a frequency band, 0.02–0.2 Hz (normalizes frequency). The wave speed c is 1 (unit distance per sample interval) and the sampling rate is 1 sample/sec. Using Equation 6.76, it turns out that $N=25$ and the array length is 32 units. The sensor location and spacing are as shown in Table 6.1.

The primary filter at reference sensor ($x_{\text{ref}}=P\lambda_U/2$) is selected as a lowpass FIR filter having a passband ±0.2 Hz. The filter coefficients are: $h=[-0.0437$ -0.0001 0.0711 -0.0601 -0.0921 0.3015 0.6001 0.3015 -0.0921 -0.0601 0.0711 -0.0001 $-0.0437]$. The beam pattern given by Equation 6.79 was computed for 10 frequencies in the band 0.02–0.2 Hz. Care was taken to limit the passband

TABLE 6.1
Nonuniform linear array for broadband frequency invariant beamformation

Sensor	Sensor Location	Spacing
1	0	
2	0.5	0.5
3	1.0	0.5
4	1.5	0.5
5	2.0	0.5
6	2.5	0.5
7	3.0	0.5
8	3.5	0.5
9	4.0	0.5
10	4.57	0.57
11	5.22	0.65
12	5.97	0.75
13	6.82	0.85
14	7.80	0.98
15	8.91	1.11
16	10.18	1.27
17	11.64	1.46
18	13.30	1.66
19	15.20	1.90
20	17.38	2.18
21	19.85	2.47
22	22.69	2.74
23	25.94	3.25
24	29.64	3.70
25	33.88	4.24
26	38.72	4.84
27	40.00	1.28

Note: Sensor locations are normalized by λ.

of the primary filters according to the location of a sensor. The computed beam patterns are shown in Figure 6.20, where for comparison all 11 beam patterns are plotted in the same figure.

6.5.5 Frequency Independent Capon Beamformer

In the previous section, we have used a bandpass filter for FI beamformation. In this section, we shall show how to select filter coefficients to minimize interference while keeping unit response in the look direction, that is, a Capon filter. The output of FI array, which is shown in Figure 6.20, is given by

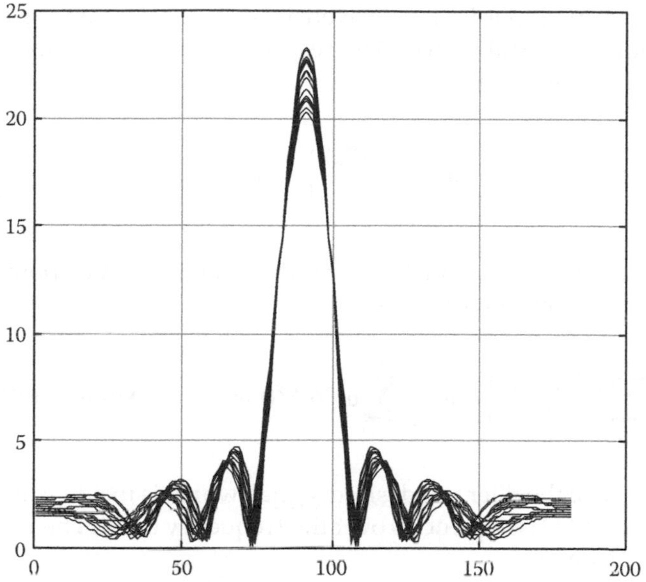

FIGURE 6.20
The beam pattern given by Equation 6.79 was computed for 10 frequencies in the band 0.02–0.2 Hz.

$$\begin{aligned} Z(\omega) &= \omega \sum_{n=0}^{N-1} g_n H_1\left(\frac{x_n}{x_{\text{ref}}}\omega\right) X_n(\omega) \\ &= \mathbf{h}^H \sum_{n=0}^{N-1} \omega g_n \phi_n X_n(\omega) \\ &= \mathbf{h}^H \omega [g_0\phi_0, g_1\phi_1, \ldots, g_{N-1}\phi_{N-1}][X_0(\omega), \ldots, X_{N-1}(\omega)]^T \\ &= \mathbf{h}^H \omega \phi \mathbf{X}(\omega). \end{aligned} \tag{6.80}$$

The average output power is given by

$$\begin{aligned} E\left\{|Z(\omega)|^2\right\} &= \mathbf{h}^H \omega^2 \phi E\left\{\mathbf{X}\mathbf{X}^H\right\} \phi^H \mathbf{h} \\ &= \mathbf{h}^H \omega^2 \phi \mathbf{S}_X \phi^H \mathbf{h}, \end{aligned} \tag{6.81}$$

where

$$\begin{aligned} \mathbf{S}_X(\omega) &= [\mathbf{a}(\theta_0,\omega), \mathbf{a}(\theta_1,\omega), \ldots, \mathbf{a}(\theta_P,\omega)]\mathbf{S}_0(\omega)[\mathbf{a}(\theta_0,\omega), \mathbf{a}(\theta_1,\omega), \ldots, \mathbf{a}(\theta_P,\omega)]^H \\ &= \mathbf{A}(\theta,\omega)\mathbf{S}(\omega)\mathbf{A}(\theta,\omega)^H. \end{aligned} \tag{6.82}$$

We have assumed that P plane wavefronts are incident on the array. $\mathbf{S}_0(\omega)$ is a $P \times P$ source spectral matrix. The response of the array (nonuniform) to the pth wavefront is given by

$$\mathbf{a}(\theta_p,\omega) = \left\{1, \exp\left(-j\omega \frac{x_1 \sin(\theta_p)}{c}\right), \ldots, \exp\left(-j\omega \frac{x_{N-1} \sin(\theta_p)}{c}\right)\right\}^T.$$

Using Equation 6.82 in Equation 6.81, the average output power over the frequency band may be written as

$$\frac{1}{L}\sum_{l=1}^{L} E\left\{|Z(\omega_l)|^2\right\} = \mathbf{h}^H \frac{1}{L}\sum_{l=1}^{L} \omega_l^{\,2}\phi\, \mathbf{A}(\boldsymbol{\theta},\omega_l)\mathbf{S}_0(\omega_l)\mathbf{A}(\theta,\omega_l)^H \phi^H \mathbf{h}. \quad (6.83a)$$

The response of the filter to a desired signal, which is incident at an angle θ_1, is set to one at L ($\leq M$) frequencies over the frequency band. The unit response condition is given by

$$\mathbf{C}^H \mathbf{h} = \mathbf{1}, \quad (6.83b)$$

where

$$\mathbf{C} = \left[\omega_1\phi(\omega_1)\mathbf{a}(\theta_1,\omega_1), \omega_2\phi(\omega_2)\mathbf{a}(\theta_1,\omega_2), \ldots, \omega_L\phi(\omega_L)\mathbf{a}(\theta_1,\omega_L)\right],$$
$$(M \times L)$$

and

$$\mathbf{1} = [1, 1, \ldots, 1]^H \; (L \times 1).$$

We need to minimize the output power given by Equation 6.83a subject to unit response in the look direction given by Equation 6.83b. The solution is given by [16]

$$\mathbf{h} = \hat{\mathbf{S}}^{-1}\mathbf{C}[\mathbf{C}^H \hat{\mathbf{S}}^{-1}\mathbf{C}]^{-1}\mathbf{1}, \quad (6.84)$$

where

$$\hat{\mathbf{S}} = \frac{1}{L}\sum_{l=1}^{L} \omega_l^2 \phi\, \mathbf{S}_x(\omega_l)\phi^H$$
$$= \frac{1}{L}\sum_{l=1}^{L} \omega_l^2 \phi\, \mathbf{A}(\theta,\omega_l)\mathbf{S}_0(\omega_l)\mathbf{A}(\theta,\omega_l)^H \phi^H.$$

6.6 Exercises

1. Consider P uncorrelated wavefronts incident on an M sensor ULA. Let the background noise be uncorrelated with all signals. Show that the Wiener filter in the frequency domain is given by

$$\mathbf{H}_W(\omega) = \mathbf{S}_f^{-1}(\omega)\mathbf{A}\mathbf{S}_0(\omega),$$

where $\mathbf{A}$ is a matrix whose columns are direction vectors of the wavefronts that are incident on the array.

2. Show that the Wiener filter (also Capon filter) can be written as

$$\mathbf{H}_W = \mathbf{v}_s \alpha_s^{-1} \mathbf{v}_s^H \mathbf{a}_0 S_0,$$

where $\mathbf{a}_0$ is the direction vector of the wavefront from the first source and S_0 is signal power. (Hint: Use Equation 3.38b and the property given in Equation 4.14b.) Such a filter is robust against calibration error (see Section 6.3).

3. In Sections 6.1 and 6.2 we have seen that in estimating a waveform in the presence of interference there is always some cross-talk, that is, leakage of power from the interfering signal to the desired signal. It is possible to devise a set of filters, $H_m(\omega)$, $m=0, 1,\ldots,M-1$, which will null the interference without distorting, that is, $\sum_{m=0}^{M-1} H_m(\omega) = 1$, the desired signal (but with no noise). The array has been steered to receive the desired signal. The DOA of the desired signal and that of the interference are known. Show that

$$H_m(\omega) = \frac{M - e^{j\omega\tau_m}\sum_{i=0}^{M-1} e^{-j\omega\tau_i}}{M^2 - \left|\sum_{i=0}^{M-1} e^{-j\omega\tau_i}\right|^2},$$

where τ_i is the time delay of the interference at ith sensor. The interference is nulled except when $\omega\tau_i$ is equal to an integer multiple of 2π [17].

4. Design a broadband frequency independent linear array. The desired array aperture is $P=10$. Other parameters are $f_U=0.2$ Hz and $f_L=0.1$ Hz. The primary filter at the reference sensor is an FIR filter. You may select the coefficients given on page 343. Compute the number and locations of all sensors in the array. Finally, compute the array pattern at different frequencies.

5. In (6.52d) the broadband beam pattern is expressed in a matrix form. Show that **Q** is a positive definite matrix. Let the weight vector **w** be further required to satisfy unit norm condition. Show that the beam pattern has upper and lower bounds given by the maximum and minimum eigenvalue of **Q**.

References

1. B. Otterstein, R. Roy, and T. Kailath, Signal waveform estimation in sensor array processing, *Asilomar Conf.*, pp. 787–791, 1989.
2. B. Friedlander and A. J. Weiss, Effects of model errors on waveform estimation using the MUSIC algorithm, *IEEE Trans.*, SP-42, pp. 147–155, 1994.
3. S. P. Applebaum, Adaptive array, *IEEE Trans.*, AP-24, pp. 585–598, 1976.
4. L. C. Godara, Error analysis of the optimal antennal array processors, *IEEE Trans.*, AES-22, pp. 395–409, 1986.
5. B. Widrow and S. D. Stearms, *Adaptive Signal Processing*, Prentice-Hall, Englewood Cliffs, NJ, 1985.
6. J. R. Treichler, C. R. Johnson Jr., and M. G. Larimore, *Theory and Design of Adaptive Filters*, Wiley Interscience, New York, 1987.
7. J. W. Kim and C. K. Un, A robust adaptive array based on signal subspace approach, *IEEE Trans.*, SP-41, pp. 3166–3171, 1993.
8. R. Schreiber, Implementation of adaptive array algorithms, *IEEE Trans.*, ASSP-34, pp. 1038–1045, 1986.
9. B. D. Van Veen, Eigenstructure based partially adaptive array design, *IEEE Trans. Antenn. Propag.*, vol. 36, pp. 357–362, 1988.
10. L. Griffiths' and C. W. Jim, An alternate approach to linearly constrained adaptive beamforming, *IEEE Trans. Antenn. Propag.*, vol. AP-30, pp. 27–34, 1982.
11. N. K. Jablon, Steady state analysis of the generalized sidelobe canceller by adaptive noise cancelling techniques, *IEEE Trans. Antenn. Propag.*, vol. AP-34, pp. 330–337, 1986.
12. T. Sekiguchi and Y. Karasawa, Wideband beamspace adaptive array utilizing FIR fan filters for multibeam forming, *IEEE Trans. Signal Process.*, vol. 48, pp. 277–284, 2000.
13. D. B. Ward, Z. Ding, and R. A. Kennedy, Broadband DOA estimation using frequency invariant beamforming, *IEEE Signal Process.*, vol. 46, pp. 1463–1469, 1998.
14. J. H. Doles III and F. D. Benedict, Broad-band array design using the asymptotic theory of unequally spaced arrays, *IEEE Trans. Antenn. Propag.*, vol. 36, pp. 27–33, 1988.
15. D. B. Ward, Theory and design of broadband sensor arrays with frequency invariant far-field beam patterns, *J. Acoust. Soc. Am.*, vol. 97, pp. 1023–1034, 1995.
16. D. B. Ward, Technique for broadband correlated interference rejection in microphone arrays, *IEEE Trans. Speech and Audio Proc.*, vol. 6, pp. 414–417, 1998.
17. M. T. Hanna and M. Simaan, Absolutely optimum array filters for sensor array, *IEEE Trans.*, ASSP-33, pp. 1380–1386, 1985.
18. O. L. Frost III, An algorithm for linearly constrained adaptive array processing, *Proc. IEEE*, vol. 60, pp. 926–935, 1972.

7

Multipath Channel

The medium, that is, a channel between transmitter and receiver, is often heterogeneous, which causes the signal to travel along different paths. Such a channel is known as a multipath channel. In this chapter, we shall study how to characterize a multipath channel given the input and output or just the output. A channel may support a finite number of discrete multipaths caused by large reflectors or boundaries. We will investigate such a discrete channel in Sections 7.1 and 7.2. With a single receiver, we shall receive overlapping echoes of the transmitted signal. In Section 7.1, we shall explore how to estimate the parameters of multipaths, namely delay and strength multipaths from a single sensor output; while in Section 7.2 we show how to use an array sensor to estimate the direction of arrival (DOA) of different multipaths. When there are many scatterers, a large number of closely spaced multipaths with small delay spread are produced. The discrete model is then no longer useful, as it is not possible to estimate so many parameters. In Section 7.3, a different approach is adopted to characterize a scatter channel. Finally, in Section 7.4, a channel, where there are a number of multipaths with large delay spread, is modeled as a finite impulse response (FIR) filter.

7.1 Overlapping Echoes

In many real-life situations, a signal (also interference) may reach a sensor via more than one path. The resulting signal will be an overlap of many copies of the transmitted signal. The different copies of the signal may differ in arrival time, amplitude, and DOA, but they remain highly correlated unless they have traveled over widely different paths through a random medium. For the time being, we shall not consider such a complex situation. We shall assume that all copies of the signal remain correlated. Even in this simplified model of multipath propagation, the waveform received by a sensor may have no resemblance with the signal emitted by the source. To appreciate the kind of deterioration a signal may suffer, let us consider a source emitting a signal described by a sinc function ($=\sin(t)/t$) and a multipath channel having four paths. The relative amplitudes and delays are assumed to be random numbers. The signal received at one sensor is given by

$$f(t)=\{f_0(t)-0.0562\,f_0(t-7.0119)+0.5135\,f_0(t+9.1032)+0.3967\,f_0(t+2.6245)\},$$

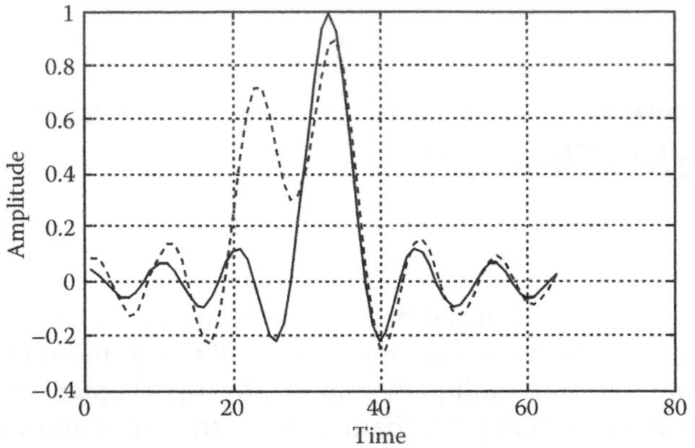

FIGURE 7.1
A source emits a waveform (sinc function) shown by a solid curve. The signal reaches a sensor via four paths. The sensor output is shown by a dashed curve.

where the signal emitted by the source is a sin *c* function,

$$f_0(t)=\sin c(0.2\pi(t-32)).$$

The sum waveform is shown in Figure 7.1

Generally, the broadband signals are more distorted than the narrowband signals. It is of practical interest to estimate amplitude gains and delays with or without knowledge of the transmitted signal from the output of a single sensor. Estimation of DOA will, however, require use of an array of sensors, which we shall consider later in this chapter.

7.1.1 Estimation of Delays and Attenuation

First, we shall consider a simple situation where a single sensor is used to receive a known signal that has traveled via multiple paths. We encounter such a situation, for example, in seismic mapping of layered strata. The received signal may be modeled by the following equation

$$f(t)=\sum_{i=1}^{d} m_i f_0(t-\tau_i)+\eta(t), \tag{7.1}$$

$f_0(t)$ is the transmitted signal, which is known. There are d multipaths and each multipath introduces a delay, τ_l and attenuation coefficients, m_i; $\eta(t)$ is background noise. The received signal is sampled at time instants $t_1, t_2, \ldots, t_N$ not necessarily at uniform intervals. We shall express Equation 7.1 in matrix form. For this let us introduce the following vectors and matrices:

$$\mathbf{f}_j=[f_j(t_1), f_j(t_2), \ldots, f_j(t_N)]^T,$$

$$\mathbf{A}(\tau)=\begin{bmatrix} f_0(t_1-\tau_1) & f_0(t_1-\tau_2) & \cdots & f_0(t_1-\tau_d) \\ f_0(t_2-\tau_1) & f_0(t_2-\tau_2) & \cdots & f_0(t_2-\tau_d) \\ \vdots & \vdots & & \vdots \\ f_0(t_N-\tau_1) & f_0(t_N-\tau_2) & \cdots & f_0(t_N-\tau_d) \end{bmatrix},$$

$$\mathbf{m}_j=[m_{1j}, m_{2j}, \ldots, m_{dj}]^T,$$

and

$$\eta_j=[\eta_j(t_1), \eta_j(t_2), \ldots, \eta_j(t_N)]^T.$$

The subscript j refers to jth transmission or experiment. It is further assumed that the delays remain unchanged from transmission to transmission, but the random amplitudes undergo a change as a result of Rayleigh fading. Equation 7.1 may be written in terms of the above matrices as

$$\mathbf{f}_j=\mathbf{A}\mathbf{m}_j+\eta_j. \tag{7.2}$$

Compare Equation 7.2 with Equation 2.18b, which was derived as a signal model for an array of sensors receiving P-plane wavefronts. There the columns of matrix **A** represented the direction vectors, which lie in an array manifold spanned by a direction vector as the DOA ranges over $\pm\pi/2$ (for uniform linear array [ULA]). In the present case, the columns of **A** matrix are obtained by sampling the transmitted signal, which has been delayed by an amount equal to the propagation delay of an echo. The columns of **A** matrix now lie in a temporal manifold spanned by a sampled signal vector with delay, which covers all possible propagation delays. The temporal manifold is generally known as the transmitted signal is known, but the rank of **A** is not guaranteed to be equal to d even though the delays are unequal. The rank of **A** depends upon the type of signal. For example, when the signal is constant over the duration of transmission, the rank will be equal to one, but when the signal is rapidly varying, the rank will be equal to d. Therefore, the ability to estimate the delays greatly hinges on the signal structure. Given that the matrix **A** is full rank we readily apply the MUSIC algorithm to estimate the time of arrival (TOA), in place of DOA. We demonstrate this through an example.

A broadband signal such as a linear frequency modulated (FM) signal appears to be ideal. We consider an FM signal given by

$$f_0(t)=\sin(\gamma\pi(t^2+t)),$$

where γ is a suitable constant, which controls the rate of frequency change. This signal is used to illuminate a target. The reflected signal consists of two multipaths with delays 23 and 25.5 units of time, which is also equal to the sampling interval. The reflected signals travel along different paths and hence they undergo independent amplitude modulation (real). We have

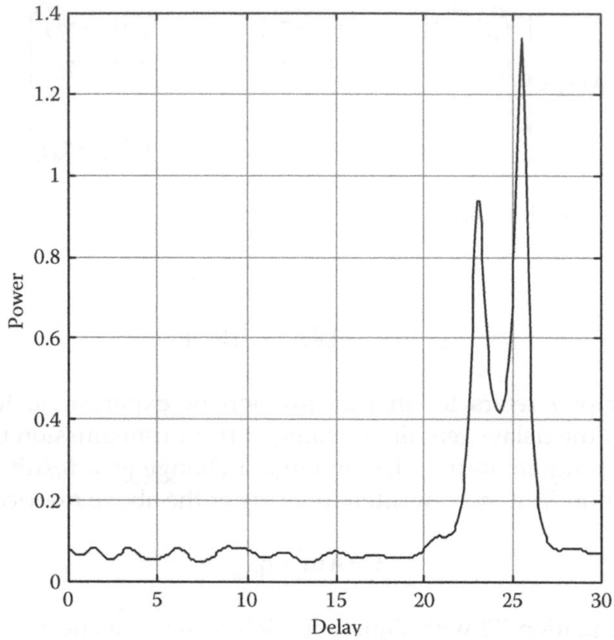

FIGURE 7.2
Application of MUSIC algorithm for estimation of time of arrival (TOA). The assumed delays are 23 and 25.5 sampling units.

added unit variance zero mean Gaussian background noise to the received signal. Note that there is a single sensor. The covariance matrix, which is required for the application of the MUSIC algorithm, was computed using the data from 64 independent experiments. In simulation, the independence was achieved by introducing a random phase in the FM signal. The results are shown in Figure 7.2.

Having obtained the delays, the **A** matrix is now fully known. We can now obtain the minimum mean square estimate (MMSE) of the amplitudes, *m*, from Equation 7.3, where the inverse of $\mathbf{A}^H\mathbf{A}$ exists as **A** is full (column) rank.

$$\hat{\mathbf{m}} = (\mathbf{A}^H\mathbf{A})^{-1}\mathbf{A}^H\mathbf{f}. \tag{7.3}$$

7.1.1.1 Capon's Minimum Variance Method

In Capon's method, a filter is sought that minimizes the filter power output subject to the constraint that it maintains unit response towards a multipath of prescribed TOA [1]. We shall work in the frequency domain. Equation 7.1 in frequency domain is given by

$$F(\omega) = \sum_{i=1}^{d} m_i F_0(\omega) e^{-j\omega\tau_i} + \mathrm{N}(\omega), \tag{7.4a}$$

where $F(\omega)$, $F_0(\omega)$, and $N(\omega)$ are, respectively, Fourier transform of the received signal, Fourier transform of the transmitted signal, and Fourier transform of background noise. Consider a set of discrete frequencies, not necessarily uniformly spaced $(\omega_0, \omega_1,\ldots, \omega_{M-1})$. We shall now express Equation 7.4a in a matrix form

$$\mathbf{F}=\mathbf{F}_0\mathbf{A}\mathbf{m}+\mathbf{N}, \tag{7.4b}$$

where

$$\mathrm{F}=[F(\omega_0), F(\omega_1),\ldots, F(\omega_{M-1})]^T,$$

$$\mathrm{F}_0=\mathrm{diag}\,[F_0\,(\omega_0), F_0\,(\omega_1),\ldots, F_0\,(\omega_{M-1})],$$

$$\mathbf{A}=\begin{bmatrix} e^{-j\omega_0\tau_1} & e^{-j\omega_0\tau_2} & \cdots & e^{-j\omega_0\tau_d} \\ e^{-j\omega_0\tau_1} & e^{-j\omega_1\tau_2} & \cdots & e^{-j\omega_1\tau_d} \\ \vdots & \vdots & & \vdots \\ e^{-j\omega_{M-1}\tau_1} & e^{-j\omega_{M-1}\tau_2} & \cdots & e^{-j\omega_{M-1}\tau_d} \end{bmatrix},$$

$$\mathbf{m}=[m_1, m_2,\ldots, m_{\mathrm{d}}]^T,$$

$$\mathbf{N}=[\mathrm{N}(\omega_0), \mathrm{N}(\omega_1),\ldots, \mathrm{N}(\omega_{M-1})]^T.$$

Let $\mathbf{w}=[w_0, w_1,\ldots, w_{M-1}]^T$ be a filter vector. The output power is given by

$$\mathbf{w}^H E\{\mathbf{F}\mathbf{F}^H\}\mathbf{w}=\mathbf{w}^H\mathbf{F}_0\mathbf{A}E\{\mathbf{m}\mathbf{m}^H\}\mathbf{A}^H\mathbf{F}_0^H+\mathbf{w}^H E\{\mathbf{N}\mathbf{N}^H\}\mathbf{w},$$

or

$$\mathbf{w}^H\mathbf{R}_f\mathbf{w}=\mathbf{w}^H\mathbf{F}_0\mathbf{A}\mathbf{R}_m\mathbf{A}^H\mathbf{F}_0^H+\mathbf{w}^H\mathbf{R}_\eta\mathbf{w}, \tag{7.5}$$

where

$$\mathbf{R}_f=E\,\{\mathbf{F}\mathbf{F}^H\},\ \mathbf{R}_m=E\,\{\mathbf{m}\mathbf{m}^H\}\text{ and }\mathbf{R}_\eta=E\,\{\mathbf{N}\mathbf{N}^H\}.$$

We wish to find the filter $\mathbf{w}$ that minimizes filter output power

$$\mathbf{w}^H\mathbf{R}_f\mathbf{w}=\min, \tag{7.6a}$$

subject to the constraint that the filter maintains unit response toward a multipath of prescribed delay,

$$\mathbf{w}^H\,\mathbf{F}_0\mathbf{a}(\tau)=1, \tag{7.6b}$$

where $\mathbf{a}(\tau)=\left[e^{-j\omega_0\tau}, e^{-j\omega_1\tau},\ldots,e^{-j\omega_M\tau}\right]^T$ and τ is the prescribed delay. This is achieved by minimizing the Lagrange function

$$\mathbf{w}^H\mathbf{R}_f\mathbf{w}+\lambda\,(1-\mathbf{w}^H\mathbf{F}_0\mathbf{a}(\tau))=\min,$$

where λ is a Lagrange constant. The solution is given by

$$\mathbf{w} = \frac{\mathbf{R}_f^{-1}\mathbf{F}_0\mathbf{a}(\tau)}{\mathbf{a}^H(\tau)\mathbf{F}_0^H\mathbf{R}_f^{-1}\mathbf{F}_0\mathbf{a}(\tau)}, \tag{7.7a}$$

$$P(\tau) = \frac{1}{\mathbf{a}^H(\tau)\mathbf{F}_0^H\mathbf{R}_f^{-1}\mathbf{F}_0\mathbf{a}(\tau)}, \tag{7.7b}$$

where $P(\tau)$ is the minimum power output of the filter as a function of the prescribed delay [1]. A plot of the power output as a function of delay shows peaks at true delays. We shall illustrate this property through an example. We repeat the experiment described on pages 350–351. In place of the MUSIC algorithm we have used Capon's minimum variance algorithm. The algorithm works in the frequency domain and we have used 16 uniformly spaced frequency samples to compute the $\mathbf{R}_f$ matrix used in Equation 7.7. The results are shown in Figure 7.3.

7.1.2 Correlated Echoes

Until now the underlying assumption has been that the attenuation coefficients of the echoes are uncorrelated random variables, which require a propagation medium that is both spatially and temporally variable as, for

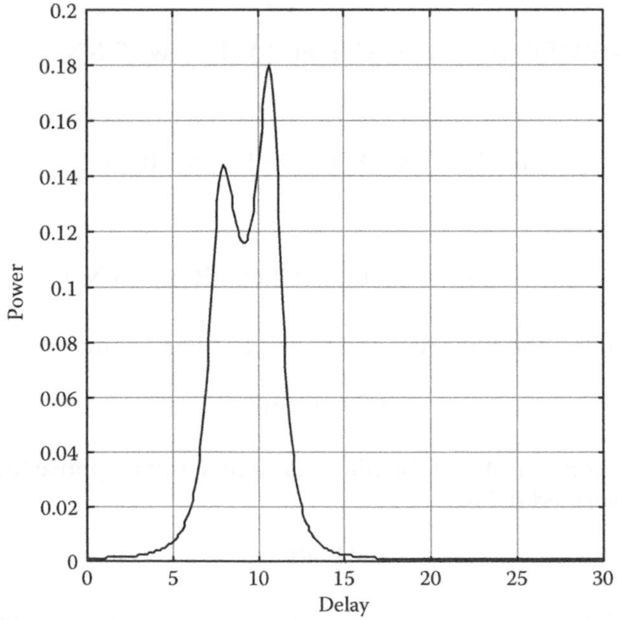

FIGURE 7.3
Application of Capon's algorithm for estimation of time of arrival (TOA). All parameters for simulation are as in the experiment described on pages 350–351. The assumed delays are 8.0 and 10.5 sampling units.

example, in ocean. Correlated echoes will result when the propagation medium is relatively stable. Then, we must treat both attenuation and delay as deterministic parameters and estimate them by minimizing the mean square error (mse) or maximizing the likelihood ratio, which is computationally intensive, as it requires iterative estimation of the delays. The maximum likelihood (ML) method is optimal in the sense that for large data the error approaches Cramer–Rao bounds.

Note that the cost function is a linear function of the attenuation coefficients but a nonlinear function of the delays. First, we need good initial estimates of the delays, which are then used to compute the attenuation coefficients. The cost function is then expressed as a function of the delays only. It has been observed that the cost function for real attenuation is highly oscillatory, which makes it difficult to locate the global minimum. On the other hand, when the attenuation is complex, the cost function is smooth, but the minimum is not located at true delay, leading to a biased estimate [2].

Simple but suboptimal methods have been devised for practical applications. In one such method, Equation 7.4 is expressed as a sum of complex exponentials (without noise term)

$$\frac{F(\omega)}{F_0(\omega)} = \sum_{i=1}^{d} m_i e^{-j\tau_i \omega}. \tag{7.8}$$

The left-hand side of Equation 7.8 is known and we can now fit d complex sinusoids using the well-known Prony's algorithm or its modern versions [1,2]. Briefly, we define forward and backward data matrices and linear prediction coefficient vector **h**,

$$\mathbf{z}_k^f = \begin{bmatrix} z(k) & z(k-1) & \cdots & z(k-L) \\ z(k+1) & z(k) & \cdots & z(k+1-L) \\ \vdots & \vdots & & \vdots \\ z(N-1) & z(N-2) & \cdots & z(N-1-L) \end{bmatrix},$$

$$\mathbf{z}_k^b = \begin{bmatrix} z^*(k) & z^*(k+1) & \cdots & z^*(k+L) \\ z^*(k+1) & z^*(k+2) & \cdots & z^*(k+1+L) \\ \vdots & \vdots & & \vdots \\ z^*(N-1-L) & z^*(N-2) & \cdots & z^*(N-1) \end{bmatrix},$$

where superscript f and b stand for forward and backward prediction, respectively, and (*) stands for complex conjugate. The prediction filter is of length $L+1$. Let $z(k)$, $k=0, 1, 2,\ldots, N-1$ be a ratio of finite discrete Fourier transform (FDFT) coefficients of the observed and the transmitted signals,

$$z(k) = \frac{F(k\Delta\omega)}{F_0(k\Delta\omega)} = \sum_{i=1}^{d} m_i e^{-jk\Delta\omega\tau_d} \quad \Delta\omega = \frac{2\pi}{N}.$$

We assume that $L>d$, number of sinusoids (but d is not known). As $z(k)$ is a sum of complex sinusoids, it is a solution of a linear prediction equation

$$z(k)+h_1z(k-1)+h_2z(k-2)+\ldots+h_Lz(k-L)=0,$$

which in matrix form becomes

$$\begin{bmatrix}\mathbf{z}_k^f\\ \mathbf{z}_k^b\end{bmatrix}\mathbf{h}=\mathbf{z}_k\mathbf{h}=\mathbf{0}, \tag{7.9a}$$

where the linear prediction coefficient vector $\mathbf{h}$ is given by

$$\mathbf{h}=[1h_1\ h_2\ldots h_L]^T.$$

We premultiply Equation 7.9a with Hermitian conjugate of the data matrix and obtain

$$\mathbf{z}_k^H\mathbf{z}_k\mathbf{h}=\mathbf{C}_z\mathbf{h}=\mathbf{0}, \tag{7.9b}$$

where $\mathbf{C}_z$ is a correlation matrix of size $(L+1)\times(L+1)$. The rank of the correlation matrix is equal to d [3]. This property enables us to estimate d, at least in a low-noise case. Having estimated d, we set $L=d$ and redefine the data matrix. Now, the correlation matrix $\mathbf{C}_z$ will be of size $(d+1)\times(d+1)$. The null space of the correlation matrix is spanned by a lone eigenvector corresponding to the null eigenvalue (in a no-noise case). The null eigenvector is proportional to the linear prediction vector, $\mathbf{h}$. Having obtained the prediction coefficients, the delays can be computed from the roots of a polynomial,

$$H(z)=1+h_1z^{-1}+h_2z^{-2}+\ldots+h_dz^{-d}=0. \tag{7.10}$$

Let z_i, $i=1, 2,\ldots, d$ be the roots of $H(z)$. The delays are given by

$$\tau_i=\frac{N}{2\pi}\angle z_i, \tag{7.11}$$

in units of the sampling interval.

Another example of a suboptimal but simple method is one where the delay is expressed as an FIR filter [4, 5]. We need to assume that the transmitted signal is bandlimited ($\pm\omega_b$), hence its discrete samples $x(n\Delta t)$ may be used to interpolate at any time instant.

$$x(t)=\sum_{n=-\infty}^{\infty}x(n\Delta t)\operatorname{sinc}(t-n\Delta t)\quad -\infty<t<\infty, \tag{7.12a}$$

where

$$\sin c(t) = \frac{\sin \omega_b t}{\omega_b t},$$

and $\Delta t = \pi/\omega_b$. For simplicity we shall assume that $\Delta t = 1$. Let $y(t) = x(t - \tau)$ and using Equation 7.12a we can write for $y(t)$

$$y(t) = \sum_{n=-\infty}^{\infty} x(n) \sin c(t - \tau - n). \tag{7.12b}$$

Let $\tau = (l + r)$, where l is an integer and r is a fraction, $0 < r < 1$. At $t = k$, Equation 7.12b reduces to

$$y(k) = \sum_{n=-\infty}^{\infty} x(n) \sin c(k - l - r - n). \tag{7.12c}$$

In Equation 7.12c we substitute $k - n$ by i and obtain

$$y(k) = \sum_{k-i=-\infty}^{\infty} x(k - i) \sin c(i - l - r),$$

or for finite k we have

$$y(k) = \sum_{i=-\infty}^{\infty} x(k - i) h(i), \tag{7.12d}$$

where

$$\begin{aligned} h(i) &= \sin c(i - l - r) \\ &= \frac{(-1)^{i-l}}{\pi(i - l - r)} \sin(r), \end{aligned}$$

which may be looked upon as a filter obtained by shifting the interpolation filter by an amount equal to $(l + r)$. This filter will have a peak at $i = (l + r)$.

We now consider a case where there are d delayed signals with different delays,

$$y(t) = \sum_{\delta=1}^{d} m_\delta x(t - \tau_\delta). \tag{7.13}$$

Similar to Equation 7.12d, which was derived for a single delayed signal, we obtain an equivalent result for multiple delayed signals

$$\begin{aligned} y(k) &= \sum_{\delta=1}^{d} \sum_{i=-\infty}^{\infty} x(k-i) m_\delta h_\delta(i) \\ &= \sum_{i=-\infty}^{\infty} x(k-i) h_d(i), \end{aligned} \tag{7.14a}$$

where

$$\begin{aligned} h_d(i) &= \sum_{\delta=1}^{d} m_\delta h_\delta(i) \\ h_\delta(i) &= \sin c(i-\tau_\delta), \end{aligned} \tag{7.14b}$$

where τ_δ is expressed in units of the sampling interval. We shall next adopt Equation 7.14a for FIR filter. Let $h_d(i) = 0$ for $|i| \geq P$, where P is a constant greater than maximum delay. We obtain

$$y(k) \approx \sum_{i=-P}^{P} x(k-i) h_d(i). \tag{7.15}$$

We shall express Equation 7.15 in matrix form. For this we define the following vectors and matrices

$$\mathbf{y} = [y(P)\ y(P+1), \ldots, y(N-P-1)]^T,$$

$$\mathbf{h} = [h_d(-P)\ h_d(-P+1), \ldots, h_d(0), \ldots, h_d(P)]^T,$$

$$\mathbf{X} = \begin{bmatrix} x(2P) & x(2P-1) & \cdots & x(1)x(0) \\ x(2P+1) & x(2P) & \cdots & x(2)x(1) \\ \vdots & \vdots & & \vdots \\ x(N-1) & & \cdots & x(N-2P-1) \end{bmatrix}.$$

We obtain

$$\mathbf{y} \approx \mathbf{X}\mathbf{h}. \tag{7.16}$$

The MMSE solution of **h** is given by

$$\mathbf{h} = \left[\mathbf{X}^H \mathbf{X}\right]^{-1} \mathbf{X}^H \mathbf{y}.$$

For invertability, we need to assume that $[\mathbf{X}^H\mathbf{X}]$ is full rank. The amplitudes and locations of the maxima are, respectively, the estimates of attenuations and delays of different multipaths.

An example described below illustrates the type of results one would expect. A linear FM signal is used as a broadband signal.

$$x(n) = \sin(\pi(\gamma n^2 + \gamma n)).$$

For $\gamma = 0.004$ the bandwidth is equal to 0.375 Hz and the center frequency is equal to 0.25 Hz. Note that the sampling interval is one and hence the highest frequency is 0.5 Hz. A direct arrival with zero delay and two delayed signals with delays equal to three and six units, respectively (one unit = one sampling interval) were assumed. The following parameters were assumed: N (signal length) = 128, $p = 8$ (length of interpolation filter = 17), and signal to noise ratio (SNR) = 0 dB. The attenuations suffered by the delayed signals are 1.0, 0.8, and 0.5. All paths are fully correlated. The estimated interpolation function is illustrated in Figure 7.4.

Noninteger delay causes broadening of the peak, which makes localization of the true peak difficult. An interpolation technique is suggested in Ref. [4]. The true peak will lie between $i = l$ and $i = l + 1$. The filter coefficients (see Equation 7.12d) evaluated at $i = l$ and $i = l + 1$ are

$$h(l) = \frac{\sin(r)}{\pi r} \quad \text{and} \quad h(l+1) = \frac{\sin(r)}{\pi(1-r)}.$$

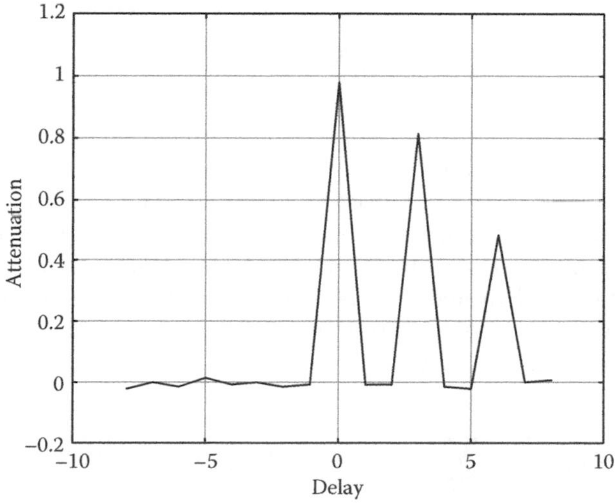

FIGURE 7.4
Estimated interpolation function. There are three peaks whose positions correctly estimate the delays and the amplitudes give an estimate the attenuations.

The fractional component may be obtained from the ratio of filter coefficients

$$\frac{h(l)}{h(l+1)} = \frac{(1-r)}{r} \quad \text{or} \quad r = \frac{h(l+1)}{h(l)+h(l+1)}.$$

When the bandwidth of the signal is inadequate, $[\mathbf{X}^H\mathbf{X}]$ becomes singular, which can be inverted only after some diagonal loading. But this causes broadening of each peak in the interpolation function, resulting in loss of resolution. When the bandwidth of the signal is small, a different approach has been used to minimize Equation 7.16 [5]. In this approach the mse

$$||\mathbf{y}-\mathbf{Xh}||^2 = \min,$$

is minimized under the constraint that (l_1 form)

$$||\mathbf{h}|| \le \varepsilon.$$

7.1.3 Unknown Echoes

We shall now consider a slightly more complex situation where the signal is unknown. The propagation model remains the same as in the previous section. Along with delay and amplitude, we like to estimate the signal waveform as well. This situation is typical in wireless communication in air or underwater and also in speech waveform detection in a closed space. For this problem, it is convenient to work in the frequency domain. We shall rewrite Equation 7.1 in the frequency domain,

$$F(\omega) = F_0(\omega)\sum_{i=1}^{d} m_i e^{-j\tau_i\omega} + \mathrm{N}(\omega), \tag{7.17}$$

where $F(\omega)$ and $F_0(\omega)$ are Fourier transforms of received and transmitted signals, respectively. The attenuation coefficients, m_i, $i=1,2,\ldots,d$ are uncorrelated or their correlation matrix is full rank. Let us now sample the received signal with sampling interval, say, unit time under the assumption that the transmitted signal is bandlimited, that is, $F_0(\omega)=0$ for $|\omega|>\pi$. We have N samples of the observed signal. We compute its FDFT. Equation 7.17 in terms of the FDFT coefficients can be written as

$$F\left(\frac{2\pi}{N}k\right) = F_0\left(\frac{2\pi}{N}k\right)\sum_{i=1}^{d} m_i e^{-j\tau_i(2\pi/N)k} + \mathrm{N}\left(\frac{2\pi}{N}k\right), \quad k=0,1,\ldots,N-1. \tag{7.18}$$

Since $f(t)$ is a real signal, the FDFT coefficients will have complex conjugate symmetry, therefore it is enough to consider the first half, that is, $N/2$ FDFT coefficients. Let us now represent Equation 7.18 in a compact matrix form. We define the following vectors and matrices:

$$\mathbf{f}=[F(0), F(1), F(2),\ldots, F(L-1)]^T,$$

$$\mathbf{F}_0=\text{diag}\,[F_0(0), F_0(1), F_0(2),\ldots, F_0(L-1)],$$

$$\eta=[\mathrm{N}(0), \mathrm{N}(1),\ldots, \mathrm{N}(L-1)]^T,$$

$$\mathbf{m}=[m(1), m(2),\ldots, m(d)]^T,$$

$$\mathbf{A}=\begin{bmatrix} 1 & 1 & 1 & \cdots & 1 \\ \phi_1 & \phi_2 & \phi_3 & \cdots & \phi_d \\ \phi_1^2 & \phi_2^2 & \phi_3^2 & \cdots & \phi_d^2 \\ \vdots & \vdots & \vdots & & \vdots \\ \phi_1^{L-1} & \phi_2^{L-1} & \phi_3^{L-1} & \cdots & \phi_d^{L-1} \end{bmatrix},$$

where

$$\phi_1 = e^{-j(2\pi/N)\tau_1}, \phi_2 = e^{-j(2\pi/N)\tau_2},\ldots, \phi_d = e^{-j(2\pi/N)\tau_d},$$

and L is even integer $\le N/2-1$. Equation 7.18 may be expressed in terms of the above vectors and matrices as

$$\mathbf{f}=\mathbf{F}_0\mathbf{Am}+\mathbf{N}. \tag{7.19}$$

We now compute the expected value of the outer product of the data vector, that is, we evaluate

$$\begin{aligned}\mathbf{Z} &= E\{\mathbf{ff}^H\} \\ &= \mathbf{F}_0\mathbf{A}E\{\mathbf{mm}'\}\mathbf{A}'\mathbf{F}_0 + \mathbf{I}\sigma_\eta^2.\end{aligned}$$

Let $\mathbf{V}_s$ represent the eigenvectors corresponding to the d dominant eigenvalues (signal eigenvalues) of $\mathbf{Z}$. Following Equation 5.6c, $\mathbf{V}_s$ may be related to $\mathbf{F}_0\mathbf{A}$ as

$$\mathbf{V}_s=\mathbf{F}_0\mathbf{AT}, \tag{7.20}$$

where $\mathbf{T}$ is $d\times d$ full rank matrix. Now we divide $\mathbf{V}_s$ into two matrices, $\mathbf{V}_s^e$ and $\mathbf{V}_s^o$, by collecting even and odd rows. Diagonal $\mathbf{F}_0$ matrix is also divided into two diagonal matrices containing the even and odd indexed elements,

$$\mathbf{F}_0^e = \text{diag}[F_0(0), F_0(2),\ldots,F_0(L-2)]$$

$$\mathbf{F}_0^o = \text{diag}[F_0(1), F_0(3),\ldots,F_0(L-1)].$$

A matrix is similarly divided into two matrices $\mathbf{A}^e$ and $\mathbf{A}^o$ containing the even and odd rows,

$$\mathbf{A}^e = \begin{bmatrix} 1 & 1 & 1 & \cdots & 1 \\ \phi_1^2 & \phi_2^2 & \phi_3^2 & \cdots & \phi_d^2 \\ \vdots & \vdots & \vdots & & \vdots \\ \phi_1^{L-2} & \phi_2^{L-2} & \phi_3^{L-2} & \cdots & \phi_d^{L-2} \end{bmatrix},$$

$$\mathbf{A}^o = \mathbf{A}^e\, \phi$$

$$\phi = \text{diag}\,[\phi_1, \phi_2, \ldots, \phi_d].$$

Equation 7.20 may be expressed in terms of the above defined matrices,

$$\begin{bmatrix} \mathbf{V}_s^e \\ \mathbf{V}_s^o \end{bmatrix} = \begin{bmatrix} \mathbf{F}_0^e \mathbf{A}_e \\ \mathbf{F}_0^o \mathbf{A}_e \phi \end{bmatrix} \mathbf{T}. \tag{7.21a}$$

At this stage, we introduce an important assumption [6] that Fourier coefficients of the signal are pairwise equal, that is, $F_0(0)=F_0\,(1)$, $F_0\,(2)=\mathrm{F}_0\,(3)$, etc. As claimed in Ref. [6], this can be achieved by zero-padding of $f(n)$ before FDFT computation. But it is found that the delay estimates are sensitive to even a slight mismatch between the odd and even coefficients. Alternatively, we can transform any (real) signal into a (complex) signal, which exactly satisfies the pairwise equality of FDFT coefficients. With this approximation, Equation 7.21a may now be expressed in a form that fits the ESPRIT model (Chapter 2, Section 2.2),

$$\begin{bmatrix} \mathbf{V}_s^e \\ \mathbf{V}_s^o \end{bmatrix} = \bar{\mathbf{F}}_0 \begin{bmatrix} \mathbf{A}^e \\ \mathbf{A}^e \phi \end{bmatrix} \mathbf{T}, \tag{7.21b}$$

where

$$\bar{\mathbf{F}}_0 = \frac{1}{2}(\mathbf{F}_0^e + \mathbf{F}_0^o).$$

Alternatively, as suggested in Ref. [6], we can obtain ϕ as the least squares or a total least squares estimate. For the least squares estimate, we write Equation 7.21b as two equations

$$\mathbf{V}_s^e = \bar{\mathbf{F}}_0 \mathbf{A}^e \mathbf{T}, \tag{7.22a}$$

$$\mathbf{V}_s^o = \bar{\mathbf{F}}_0 \mathbf{A}^e \phi \mathbf{T}. \tag{7.22b}$$

Since $\mathbf{T}$ is a $d \times d$ full rank matrix, hence its inverse exists. We can write Equation 7.22b as

$$\begin{aligned} \mathbf{V}_s^o &= \bar{\mathbf{F}}_0 \mathbf{A}^e \mathbf{T} \mathbf{T}^{-1} \phi \mathbf{T} \\ &= \mathbf{V}_s^e \mathbf{T}^{-1} \phi \mathbf{T} = \mathbf{V}_s^e \psi. \end{aligned} \tag{7.23a}$$

The least square solution of ψ is easily obtained as

$$\psi = \left[\mathbf{V}_s^{e\prime} \mathbf{V}_s^e \right]^{-1} \mathbf{V}_s^{e\prime} \mathbf{V}_s^o. \tag{7.23b}$$

Since ψ is full rank, its inverse exist, therefore we can rewrite Equation 7.23a as

$$\mathbf{V}_s^e = \mathbf{V}_s^o \psi^{-1}, \tag{7.24a}$$

whose least squares solution is given by

$$\psi^{-1} = \left[\mathbf{V}_s^{o\prime} \mathbf{V}_s^o \right]^{-1} \mathbf{V}_s^{o\prime} \mathbf{V}_s^e. \tag{7.24b}$$

When there is error in the estimation of signal subspace, the two solutions given by Equations 7.23b and 7.24b will not be identical. Note that, since the eigenvalues of ψ are identical to those of ϕ, we can estimate the delays by computing the eigenvalues of ψ. These eigenvalues, interestingly, are not influenced by the Fourier transform of the signal. However, the *pairwise equality* must hold good.

The next step after estimation of the delays is to estimate the transmitted signal itself. For this we consider Equations 7.21 and 7.23a. We have the following two equations where the unknowns are $\mathbf{T}$ and $\mathbf{F}_0$.

$$\begin{aligned} \mathbf{V}_s &= \mathbf{F}_0 \mathbf{A} \mathbf{T} \\ \psi &= \mathbf{T}^{-1} \phi \mathbf{T}. \end{aligned} \tag{7.25}$$

We need to eliminate $\mathbf{T}$ and solve for $\mathbf{F}_0$. We have earlier noted that the eigenvalues of ψ are identical to those of ϕ, which is a diagonal matrix. Hence, the eigenvalues of ψ are the diagonal elements of ϕ. Let $\mathbf{V}_\psi$ be eigenvectors of ψ. Then

$$\psi = \mathbf{V}_\psi \phi \mathbf{V}_\psi^H = (\mathbf{T}^{-1} \Delta^{-1}) \phi (\Delta \mathbf{T}),$$

where Δ is any diagonal scaling matrix. We can now equate

$$\mathbf{V}_\psi = (\mathbf{T}^{-1} \Delta^{-1}),$$

or

$$\mathbf{T} = \Delta^{-1}\mathbf{V}_{\psi}^{H}. \tag{7.26}$$

Substituting for **T** in the first equation of Equation 7.25 we obtain

$$\begin{aligned} \mathbf{V}_s &= \mathbf{F}_0\mathbf{A}\Delta^{-1}\mathbf{V}_{\psi}^{H} \\ \mathbf{V}_s\mathbf{V}_{\psi} &= \mathbf{F}_0\mathbf{A}\Delta^{-1}. \end{aligned} \tag{7.27}$$

Define a matrix **Q** as follows

$$\mathbf{Q} = \mathbf{V}_s\,\mathbf{V}_{\psi}./\,\mathbf{A} = \mathbf{F}_0\mathbf{A}\Delta^{-1}./\,\mathbf{A}, \tag{7.28}$$

where ./ stands for element by element division (MATLAB® convention). Observe that $\mathbf{V}_s\mathbf{V}_{\psi}$ and **A** are compatible matrices, hence element by element division is possible. The right-hand side of Equation 7.28 becomes

$$\mathbf{F}_0\mathbf{A}\Delta^{-1}./\mathbf{A} = \mathbf{F}_0\begin{bmatrix} 1 & \cdots & 1 \\ 1 & \cdots & 1 \\ \vdots & & \vdots \\ 1 & \cdots & 1 \end{bmatrix}\Delta^{-1} = \begin{bmatrix} F_0(0) \\ F_0(1) \\ \vdots \\ F_0(L) \end{bmatrix}\begin{bmatrix} \frac{1}{\Delta_1} & \frac{1}{\Delta_2} & \cdots & \frac{1}{\Delta_d} \end{bmatrix}.$$

Using the above result in Equation 7.28 we obtain

$$\mathbf{Q} = \mathbf{V}_s\mathbf{V}_{\psi}./\,\mathbf{A} = \mathbf{f}_0\delta^{T}, \tag{7.29}$$

where $\mathbf{f}_0 = [F_0(0), F_0(1), \ldots, F_0(L-1)]^T$ and $\delta = [1/\Delta_1, 1/\Delta_2, \ldots, 1/\Delta_2]^T$. Thus, **Q** is a rank one matrix. The Singular value decomposition (SVD) of **Q** will enable us to obtain $\mathbf{f}_0$ and δ as the left and right eigenvectors corresponding to the largest singular value. There is, however, an unknown scale factor, which scales up $\mathbf{f}_0$ and scales down δ without affecting Equation 7.29.

As an illustration, we shall consider a three multipath channel. A source emits a complex Gaussian signal at time instant $t=0$. All three echoes reach a single sensor after some delay and attenuation. The complex Gaussian signal is obtained from a real Gaussian signal by repeating every FDFT coefficient so that the odd and even coefficients are equal. The real and complex Gaussian signals are shown in Figure 7.5. The estimated and actual delays are shown in Table 7.1. While the estimated signal is shown in Figure 7.6, SNR=52 dB, 64 independent experiments, and time unit=1 sec.

7.1.4 Cluster of Sensors

In the previous section, it was necessary to assume some special properties for the unknown signal. Also, it was necessary to assume that the attenuation of multipaths are uncorrelated. Further, to a compute covariance matrix

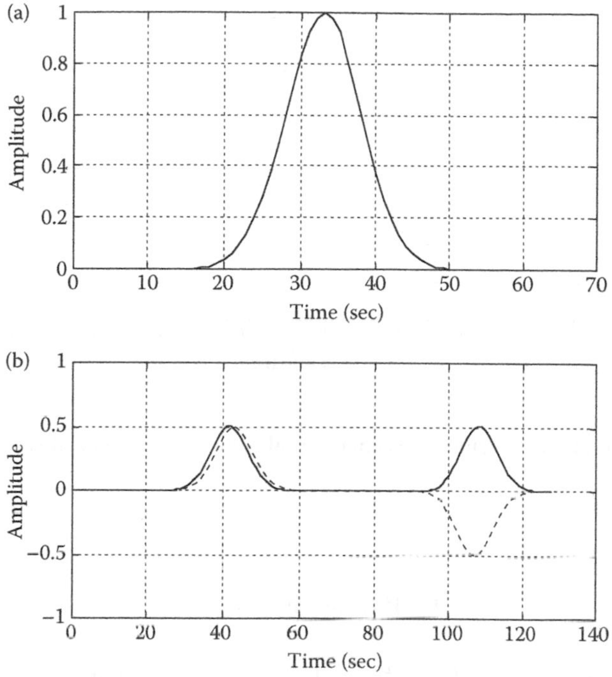

FIGURE 7.5
(a) Real Gaussian signal, (b) complex Gaussian signal. Real (solid) and imaginary (dashed) parts are shown.

TABLE 7.1
Estimated and actual delays in second

Estimated Delay	Actual Delay
2.00	2
9.95	10
11.99	12

it was required to carry out many independent experiments. Here we will get over all the above limitations. The central idea is to devise a cluster of sensors whose spatial response is a full rank matrix and independent of frequency. One way to achieve such a receiver is to form a closely packed (ideally colocated) cluster of randomly oriented direction sensitive sensors. Mutual coupling, if any, is assumed not to affect the full rank character of the spatial response matrix. Assume that there are M sensors in the cluster. Let $\mathbf{R}$ $(M \times d)$ represent the spatial response of the cluster where $[\mathbf{R}]_{ik}$ is the response of the ith sensor to the kth multipath. The output of each sensor is given by Equation 7.19. A matrix representation of all outputs of cluster array is given by

$$\mathbf{F}=\mathbf{F}_0\mathbf{A}\mathbf{B}^T+\mathbf{N}\ (L\times M), \tag{7.30}$$

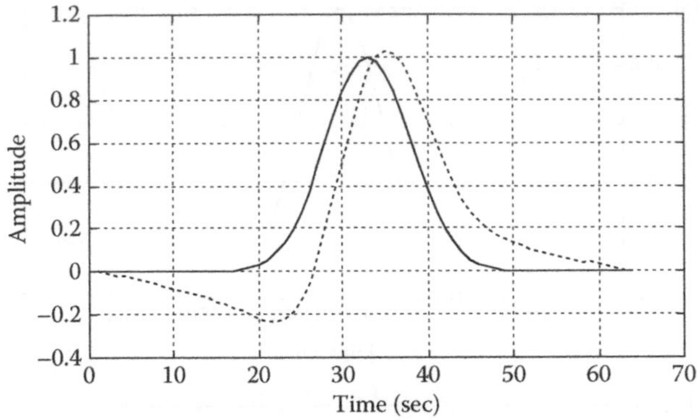

FIGURE 7.6
Estimated signal (dashed) against the actual signal (solid). The estimated signal is scaled up by a factor 35.

where

$$\mathbf{B}=\mathbf{R}\ \text{diag}\ \{\mathbf{m}\}\ (M\times d).$$

Form an outer product of the data matrix and evaluate the expected operation,

$$\begin{aligned}\mathbf{Z} = E\{\mathbf{FF}'\} &= \mathbf{F}_0\mathbf{AB}'\mathbf{BA}'\mathbf{F}_0' + E\{\mathbf{NN}'\} \\ &= \mathbf{F}_0\mathbf{A}\Gamma\mathbf{A}'\mathbf{F}_0' + \sigma_\eta^2\mathbf{I},\end{aligned} \tag{7.31}$$

where

$$\Gamma=\text{diag}\ \{\mathbf{m}^*\}\ \mathbf{R}'\mathbf{R}\ \text{diag}\ \{\mathbf{m}\},$$

Γ is a full rank matrix provided **R** is a full column rank. This may be achieved by designing a randomly oriented directional sensor cluster.

The eigenvector decomposition of **Z** will yield d significant eigenvalues and $L-d$ constant eigenvalues equal to the noise power. This property is analogous to that of a covariance matrix of ULA with d uncorrelated waves and white noise interference. Here, however, we do not require uncorrelated multipaths. Let $\mathbf{V}_s$ represent eigenvectors (signal subspace) corresponding to d significant eigenvalues. Analogous to Equation 7.7 we have

$$\mathbf{V}_s=\mathbf{F}_0\mathbf{AT}. \tag{7.32}$$

We define two matrices $\mathbf{V}_1$ and $\mathbf{V}_2$ as, respectively, the first and last $L-\delta$ rows of $\mathbf{V}_s$. Similarly, we define matrices $\mathbf{A}_1$ and $\mathbf{A}_2$ from **A**. Further, we define two diagonal matrices $\mathbf{F}_1$ and $\mathbf{F}_2$ by selecting the first and the last L–δ diagonal terms from $\mathbf{F}_0$ and then Equation 7.32 leads to a pair of equations,

$$\begin{aligned} \mathbf{V}_1 &= \mathbf{F}_1\mathbf{A}_1\mathbf{T} \\ \mathbf{V}_2 &= \mathbf{F}_2\mathbf{A}_2\mathbf{T}. \end{aligned} \tag{7.33}$$

It is easy to show that $\mathbf{A}_1$ and $\mathbf{A}_2$ are related as

$$\mathbf{A}_2 = \mathbf{A}_1\phi, \tag{7.34}$$

where ϕ is now given by

$$\phi = \mathrm{diag}\left[e^{-j2\pi\delta\tau_1/L}, e^{-j2\pi\delta\tau_2/L}, \ldots, e^{-j2\pi\delta\tau_d/L}\right].$$

Using Equation 7.34 in Equation 7.33 we obtain

$$\begin{aligned} \mathbf{V}_1 &= \mathbf{F}_1\mathbf{A}_1\mathbf{T} \\ \mathbf{V}_2 &- \mathbf{F}_2\mathbf{A}_1\psi\mathbf{T}, \end{aligned}$$

or

$$\begin{aligned} \mathbf{V}_2 &= \mathbf{F}_2\mathbf{A}_1\phi\mathbf{T} = \mathbf{F}_2\mathbf{A}_1\mathbf{T}\mathbf{T}^{-1}\phi\mathbf{T} = \mathbf{F}_2\mathbf{A}_1\mathbf{T}\psi \\ \mathbf{F}_1\mathbf{F}_2^{-1}\mathbf{V}_2 &= \mathbf{F}_1\mathbf{A}_1\mathbf{T}\psi = \mathbf{V}_1\psi. \end{aligned} \tag{7.35a}$$

Let $\Delta = \mathbf{F}_1\,\mathbf{F}_2^{-1}$ (a diagonal matrix) and rewrite Equation 7.35a as

$$\Delta\mathbf{V}_2 = \mathbf{V}_1\psi. \tag{7.35b}$$

When data is large and noise free, Equation 7.35b holds even without carrying out the expected operation used in Equation 7.18. In practice, however, in the presence of noise we will have to take recourse to least squares solution. We need to minimize

$$\underset{\Delta\psi}{\arg\min}\left\|\Delta\mathbf{V}_2 - \mathbf{V}_1\psi\right\|_F^2. \tag{7.36}$$

Holding Δ, **the** fixed solution of Equation 7.36 is given by

$$\hat{\psi} = (\mathbf{V}_1^*\mathbf{V}_1)^{-1}\mathbf{V}_1^*\Delta\mathbf{V}_2. \tag{7.37}$$

Using Equation 7.37 in Equation 7.36, the quantity to be minimized may be simplified so that the solution is simply an eigenvector corresponding to the smallest eigenvalue [7].

$$\begin{aligned}\left\|\Delta\mathbf{V}_2-\mathbf{V}_1(\mathbf{V}_1^H\mathbf{V}_1)^{-1}\mathbf{V}_1^H\Delta\mathbf{V}_2\right\|_F^2 &= \left\|\Delta\mathbf{V}_2-(\mathbf{I}-\mathbf{P}_{\mathbf{V}_1}^{\perp})\Delta\mathbf{V}_2\right\|_F^2\\ &= \left\|\Delta\mathbf{P}_{\mathbf{V}_1}^{\perp}\Delta\mathbf{V}_2\right\|_F^2 = Tr\left(\mathbf{V}_2^H\Delta^{\mathrm{H}}\mathbf{P}_{\mathbf{V}_1}^{\perp}\Delta\mathbf{V}_2\right) \qquad (7.38)\\ &= Tr\left(\Delta^{\mathrm{H}}\mathbf{P}_{\mathbf{V}_1}^{\perp}\Delta\mathbf{V}_2\mathbf{V}_2^H\right)\\ &= \lambda^H(\mathbf{P}_{\mathbf{V}_1}^{\perp}.\times\mathbf{V}_2\mathbf{V}_2^H)\lambda,\end{aligned}$$

where λ is the vector formed from the diagonal elements of Δ and $.\times$ stands for element-wise multiplication (MATLAB® convention). The last result in Equation 7.38 follows from lemma 1 in Ref. [7]. Note that

$$\mathbf{P}_{\mathbf{V}_1}^{\perp}=(\mathbf{I}-\mathbf{V}_1(\mathbf{V}_1^H\mathbf{V}_1)^{-1}\mathbf{V}_1^H),$$

is the projection matrix. The procedure for minimization of Equation 7.36 is, first to compute the eigenvector corresponding to the minimum eigenvalue of

$$(\mathbf{P}_{\mathbf{V}_1}^{\perp}.\times\mathbf{V}_2\mathbf{V}_2^H),$$

and then obtain Δ from the minimum eigenvector. Using this estimate in Equation 7.37, we compute an estimate of ψ.

The eigenvalues of ψ are proportional to the diagonal elements of ϕ. The constant of proportionality is however unknown as the eigenvector, λ, in Equation 7.25 can be determined only within an unknown complex constant of unit magnitude. This problem, however, can be resolved by normalizing the diagonal elements of ϕ with respect to the first element. This implies that all delays are relative, usually with respect to the first arrival.

It is not possible to estimate $\mathbf{F}_0$ from estimated Δ. We will go back to Equation 7.32, which we rewrite in a different form

$$\mathbf{V}_s\mathbf{T}^{-1}=\mathbf{F}_0\mathbf{A}.$$

We seek the least squares solution for $\mathbf{F}_0$ and $\mathbf{T}^{-1}$ by minimizing

$$\min_{\mathbf{F}_0\mathbf{T}}\left\|\mathbf{V}_s\mathbf{T}^{-1}-\mathbf{F}_0\mathbf{A}\right\|_F^2. \qquad (7.39)$$

Holding $\mathbf{F}_0$ fixed, we obtain the least squares solution for $\mathbf{T}^{-1}$ as

$$\mathbf{T}^{-1} = \left(\mathbf{V}_s^H \mathbf{V}_s\right)^{-1} \mathbf{V}_s^H \mathbf{F}_0 \mathbf{A},$$

which we use in Equation 7.39 and obtain

$$\begin{aligned}
\left\|\mathbf{V}_s \mathbf{T}^{-1} - \mathbf{F}_0\mathbf{A}\right\|_F^2 &= \left\|\mathbf{V}_s\left(\mathbf{V}_s^H\mathbf{V}_s\right)^{-1}\mathbf{V}_s^H\mathbf{F}_0\mathbf{A} - \mathbf{F}_0\mathbf{A}\right\|_F^2 \\
&= \left\|\mathbf{P}_s^{\perp}\mathbf{F}_0\mathbf{A}\right\|_F^2 = Tr\left(\mathbf{A}^H\mathbf{F}_0^H\mathbf{P}_s^{\perp}\mathbf{F}_0\mathbf{A}\right) \\
&= Tr\left(\mathbf{F}_0^H\mathbf{P}_s^{\perp}\mathbf{F}_0\mathbf{A}\mathbf{A}^H\right) = \mathbf{f}_0^H(\mathbf{P}_s^{\perp}.\times\mathbf{A}\mathbf{A}^H)\mathbf{f}_0.
\end{aligned} \tag{7.40}$$

The solution for Equation 7.39 is then given by the eigenvector of $\mathbf{P}_{v_s}^{\perp}.\times\mathbf{A}\mathbf{A}^H$ corresponding to its minimum eigenvalue. Note that $\mathbf{f}_0$ is a vector formed from the diagonal elements of $\mathbf{F}_0$ and

$$\mathbf{P}_s^{\perp} = \mathbf{I} - \mathbf{V}_s\left(\mathbf{V}_s^H\mathbf{V}_s\right)^{-1}\mathbf{V}_s^H,$$

is a projection matrix. Since eigenvectors are normalized (unit norm), the FDFT coefficients of the signal can be obtained within a scale factor. Therefore, the retrieved signal in time domain shall be scaled by the same factor.

This example illustrates the strength of sensor cluster array in determining delays and estimating the transmitted signal waveform. A cluster of eight randomly oriented directional sensors is assumed. The transmitter emits a BPSK signal, which is shown in Figure 7.7a. There are four multipaths arriving with delays (0, 15, 25, and 32). The attenuation coefficients are random and time varying, but the delays are assumed to remain stable. The output of the first sensor in the cluster is shown in Figure 7.7b. The delays estimated using the algorithm described above are 0, 14.99, 25, and 31.99, and the estimated waveform is shown in Figure 7.7c. The mse is 0.1244. Note that the final output of the algorithm was quantized to binary level (±1). Since the aim was verification of the algorithm, we have not considered any noise in this simulation study.

7.2 Discrete Channel

Multipaths are caused by local scattering due to scattering objects in the neighborhood of the source or sensor, or reflections from the physical channel boundary as in propagation through a duct, and also due to large objects. In this section, we shall look at the multipaths caused by reflections from large

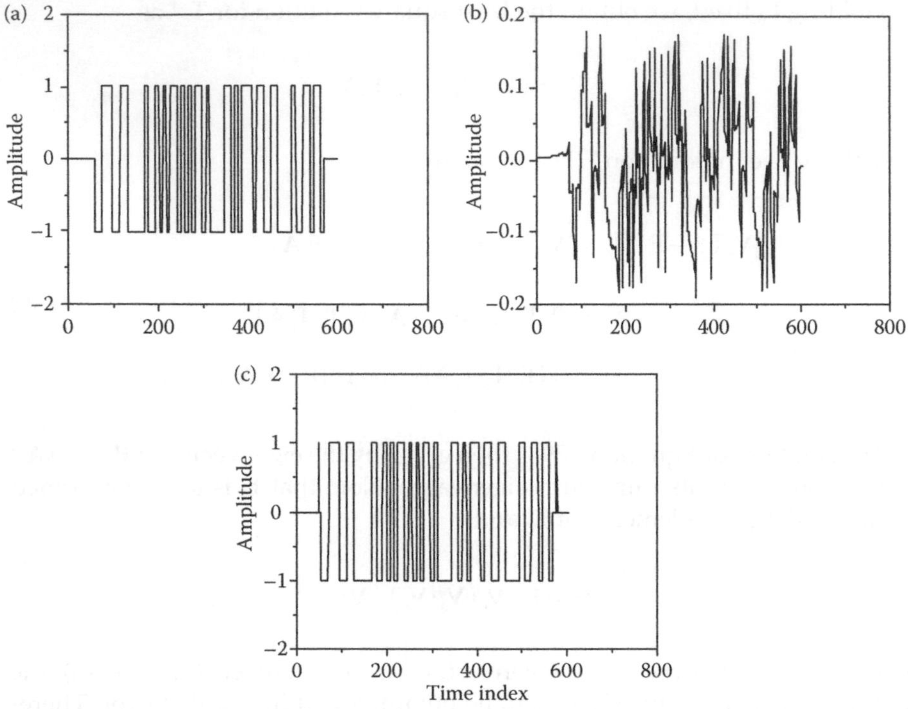

FIGURE 7.7
Sensor cluster; (a) transmitted signal, (b) received signal at the first sensor, (c) Processed waveform after binary quantization.

obstacles. Their number is usually small and they possess additional information, namely, direction of propagation, whose measurement will require the use of an array of sensors. In all, for each multipath, there are three parameters, namely, attenuation, delay, and direction of propagation. The number of multipaths is often unknown and difficult to estimate. Furthermore, there may be more than one source, resulting in co-channel interference. The problem of estimating all multipath parameters of all sources is indeed a daunting task. Fortunately, often it is not of practical interest to estimate all parameters of the multipaths, for example, in wireless communication, quantities of interest are delays and spatial response.

7.2.1 Reflections from Large Obstacles

Reflections from mountains, skyscrapers, steel towers used in power transmission, etc., are the typical multipaths encountered in a rural or semiurban environment in radio communication. The multipath signals are likely to arrive from different directions, but all paths are assumed to lie in a plane passing through the receiver, transmitter, and obstacles. Such multipaths are generally few in number (three to four) and their directions and delay parameters remain unchanged over many slots of data transmission (GSM mode), but the attenuation is likely to vary randomly from slot to slot. Our aim is to jointly estimate the DOA and delays. Using estimated DOA and delay, we go on to estimate the transmitted signal.

7.2.1.1 Uncorrelated Multipaths

We consider a source emitting a signal $f_0(t)$. The receiver is an M element uniformly spaced (ULA) sensor array. The signal travels via d paths and reaches the sensor array. The output of the array $\mathbf{f}(t)$ will be given by

$$\mathbf{f}(t)=\sum_{i=1}^{d}\mathbf{a}(\theta_i)m_i f_0(t-\tau_i), \tag{7.41}$$

where $\mathbf{a}(\theta_i)$ is the array response vector in the direction θ_i of ith path, m_i is the complex attenuation (fading) suffered by ith path, and τ_l is propagation delay of ith path. Thus, there are three unknown parameters (θ_i, m_i, τ_i) associated with ith path. (θ_i, τ_i) are assumed to be constant over many slots of data transmission, but the fading parameter is rapidly and randomly varying. Equation 7.41 may be expressed in a compact form as

$$\mathbf{f}(t)=\left[\mathbf{a}(\theta_1)\,(\theta_2)\ldots\mathbf{a}(\theta_d)\right]\mathrm{diag}\left\{m_1,m_2,\ldots,m_d\right\}\begin{bmatrix} f_0(t-\tau_1)\\ f_0(t-\tau_2)\\ \vdots\\ f_0(t-\tau_d)\end{bmatrix}. \tag{7.42}$$

We collect N snapshots at time instants $t_1, t_2, t_3, \ldots, t_N$, and form a data matrix $\mathbf{F}$,

$$\begin{aligned}\mathbf{F}&=[\mathbf{f}(t_1)\ \mathbf{f}(t_2)\ \ldots\ \mathbf{f}(t_N)]\\ &=\mathbf{A}\ \mathrm{diag}\ \{m_1, m_2, \ldots, m_d\}\mathbf{F}_0. \qquad (7.43)\\ &=\mathbf{AMF}_0,\end{aligned}$$

where

$$\mathbf{F}_0=\begin{bmatrix} f_0(t_1-\tau_1) & f_0(t_2-\tau_1) & \cdots & f_0(t_N-\tau_1)\\ f_0(t_1-\tau_2) & f_0(t_2-\tau_2) & \cdots & f_0(t_N-\tau_2)\\ \vdots & \vdots & & \vdots\\ f_0(t_1-\tau_d) & f_0(t_2-\tau_d) & \cdots & f_0(t_N-\tau_d)\end{bmatrix}.$$

Alternatively, we form a data vector by stacking one below the other the transpose of each row of **F**. Then the right-hand side of Equation 7.43 can be written in terms of Khatri–Rao (columnwise Kronecker) matrix product.

$$\mathbf{g}=(\mathbf{A}\circ\mathbf{F}_0)\mathbf{m}=\mathbf{G}(\theta,\tau)\mathbf{m}, \tag{7.44}$$

where $\theta=[\theta_1, \theta_2,..., \theta_d]$ and $\tau=[\tau_1, \tau_2,..., \tau_d]$ are parameter vectors; **g** is a vector of size $(MN\times 1)$, and **G** is a matrix of size $(MN\times d)$. It is called a space-time matrix. We assume that **m** is an uncorrelated random vector; the rationale being that the medium through which the signal propagates is fluctuating with time as in wireless communication or repeated independent experiments produce uncorrelated multipaths.

Next we compute the correlation matrix defined as

$$\begin{aligned}\mathbf{C}&=\sum_k \mathbf{g}_k\mathbf{g}_k^T=\mathbf{G}(\theta,\tau)\sum_k \mathbf{m}_k\mathbf{m}_k^T\mathbf{G}^T(\theta,\tau)\\&=\sigma_m^2\mathbf{G}(\theta,\tau)\mathbf{G}^T(\theta,\tau),\end{aligned} \tag{7.45}$$

where the summation over k refers to different data slots or repeated independent experiments. The size of **C** matrix is $MN\times MN$. Assuming **G** is full column rank, the rank of **C** will be equal to d $(d<MN)$. The nonzero eigenvectors of **C** will lie in space-time array manifold, that is, the space spanned by a space-time response vector

$$\mathbf{u}(\theta,\tau)=\mathbf{a}(\theta)\otimes\mathbf{f}_0^T(\tau),$$

as θ varies over a range of angles and τ over a range of delays. True space-time response vector will be orthogonal to the space spanned by the null eigenvectors of **C**, that is,

$$\mathbf{u}^H(\theta,\tau)\mathbf{E}_n\mathbf{E}_n^H\mathbf{u}(\theta,\tau)=0. \tag{7.46a}$$

In the presence of noise, the null eigenvalues will have a finite value depending upon the noise power. When the SNR is sufficiently large, the signal eigenvalues will stand out, making it possible to estimate the signal subspace and the number of multipaths. The quantity defined in Equation 7.46a will not be equal to zero but a finite value, though quite small. It is more convenient to define an inverse quantity, often known as MUSIC spectrum,

$$S(\theta,\tau)q=\frac{1}{\mathbf{u}^H(\theta,\tau)\mathbf{E}_n\mathbf{E}_n^H\mathbf{u}(\theta,\tau)}. \tag{7.46b}$$

We consider a multipath channel where there are two multipaths and a direct path. The direct path is incident at DOA of 10° and the two multipaths are incident at 15° and 25°. With respect to the direct path, the delays are 0.01 and 0.006 sec. The transmitted signal is a random narrowband signal. The center frequency is 990 Hz and the bandwidth is 20 Hz. An eight-sensor ULA is assumed. The sensor spacing is $\delta=\lambda/2$ ($\lambda=1.5$ m). The array output is sampled at intervals of 0.0005 sec.

Thirty-two time samples from each sensor were used to compute the spatio-temporal covariance matrix. Further, 64 independent experiments with the same input signal but with independent fading were simulated. A small amount background white noise was added to the signal (SNR≈60 db). A contour plot of MUSIC spectrum is shown in Figure 7.8. The DOA and delay parameters are seen to be correctly estimated.

7.2.1.2 Correlated Multipaths

The assumption of uncorrelated multipaths becomes untenable particularly when we have finite data. To overcome this problem we shall exploit the shift invariance structure of a data matrix instead of the structure of the covariance matrix. This leads us to an ESPRIT type of algorithm in place of the MUSIC algorithm. We begin with Equation 7.43, which is reproduced here for convenience

$$\mathbf{F}=\mathbf{AMF}_0.$$

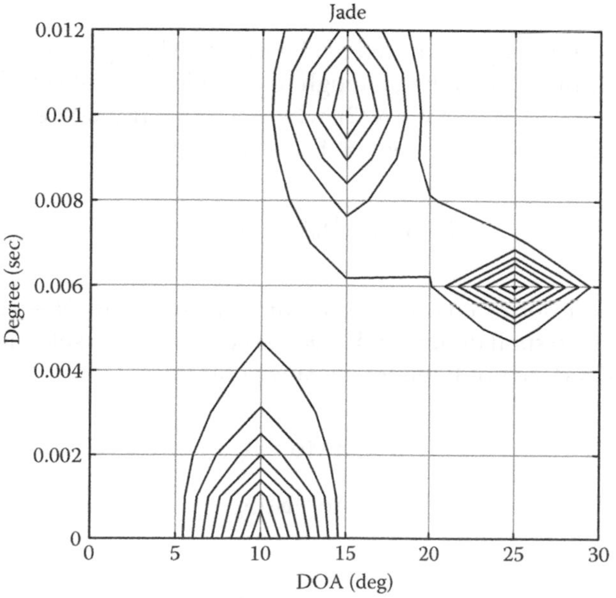

FIGURE 7.8
Contour plot of MUSIC spectrum. The DOA of the direct path and the DOA and relative delay of multipaths are correctly estimated.

Let **D** be the DFT matrix, which transforms a sequence into a DFT sequence. The DFT matrix is given by

$$\mathbf{D} = \begin{bmatrix} 1 & 1 & \cdots & 1 \\ 1 & \phi & \cdots & \phi^{N-1} \\ \vdots & \vdots & & \vdots \\ 1 & \phi^{N-1} & \cdots & \phi^{(N-1)^2} \end{bmatrix}, \qquad \phi = e^{-j(2\pi/N)}.$$

Now we right-multiply on both sides of Equation 7.43 and evaluate the product. We obtain

$$\begin{aligned} \mathbf{FD} &= \mathbf{AMF_0D} \\ \bar{\mathbf{F}} &= \mathbf{AMD_\phi} \operatorname{diag}\{\bar{\mathbf{f}}_0\}, \end{aligned} \tag{7.47}$$

where $\bar{\mathbf{F}}$ is a matrix whose rows are DFT *s* of the rows of **F**, and D_ϕ is given by

$$\mathbf{D}_\phi = \begin{bmatrix} 1 & \phi_1 & \phi_1^2 & \cdots & \phi_1^{N-1} \\ 1 & \phi_2 & \phi_2^2 & \cdots & \phi_2^{N-1} \\ \vdots & \vdots & \vdots & & \vdots \\ 1 & \phi_d & \phi_d^2 & \cdots & \phi_d^{N-1} \end{bmatrix}, \qquad \phi_i = e^{-j(2\pi/N)\tau_i}, \tag{7.48}$$

and $\bar{\mathbf{f}}_0$ is DFT of the signal emitted by the source. The signal is assumed to be real and bandlimited. It is then enough to consider a finite set of coefficients numbered from 0 to $N'-1$ ($N'<N/2$), where the spectral power is significant. It is now possible to factor $\bar{\mathbf{f}}_0$ out of $\bar{\mathbf{F}}$ and obtain

$$\tilde{\mathbf{F}} = \mathbf{F} \operatorname{diag}\{\mathbf{f}_0\}^{-1} = \mathbf{AMD}_\phi. \tag{7.49a}$$

Consider a subset of columns of $\tilde{\mathbf{F}}$ starting from i to $N'-p+i$ ($p \geq 1$ and $1 \leq i \leq p$), which we shall denote by $\tilde{\mathbf{F}}^i$. On carrying out this selection operation on the right-hand side of Equation 7.49a we obtain

$$\begin{aligned} \tilde{\mathbf{F}}^i &= \mathbf{AM}\phi^{i-1}\mathbf{D}_\phi \\ &= \mathbf{A}\phi^{i-1}\mathbf{MD}_\phi, \end{aligned} \tag{7.49b}$$

where

$$\phi = \operatorname{diag}\{\phi_1, \phi_2, \ldots, \phi_d\},$$

and D_ϕ now has dimensions of $d \times (N' - p + 1)$. We now define a new matrix $\mathcal{F}$

$$\mathcal{F} = \begin{bmatrix} \tilde{\mathbf{F}}^1 \\ \tilde{\mathbf{F}}^2 \\ \vdots \\ \tilde{\mathbf{F}}^p \end{bmatrix} = \begin{bmatrix} \mathbf{A} \\ \mathbf{A}\phi \\ \vdots \\ \mathbf{A}\phi^{p-1} \end{bmatrix} \mathbf{MD}_\phi, \tag{7.50}$$

$\mathcal{F}$ is a Hankel matrix of dimension $pM \times (N' - p + 1)$. Define $\mathcal{F}_u$ and $\mathcal{F}_1$ matrices as below

$$\mathcal{F}_u = \begin{bmatrix} \tilde{\mathbf{F}}^1 \\ \tilde{\mathbf{F}}^2 \\ \vdots \\ \tilde{\mathbf{F}}^{p-1} \end{bmatrix} \quad \mathcal{F}_1 = \begin{bmatrix} \tilde{\mathbf{F}}^2 \\ \tilde{\mathbf{F}}^3 \\ \vdots \\ \tilde{\mathbf{F}}^p \end{bmatrix}. \tag{7.51}$$

Using Equation 7.51 in Equation 7.50 we obtain the following relation

$$\mathcal{F}_u = \mathbf{A}_1\mathbf{MD}_\phi \quad \mathcal{F}_1 = \mathbf{A}_1\phi\mathbf{MD}_\phi,$$

where

$$\mathbf{A}_1 = \begin{bmatrix} \mathbf{A} \\ \mathbf{A}\phi \\ \vdots \\ \mathbf{A}\phi^{p-2} \end{bmatrix}.$$

Consider a pencil matrix

$$\begin{aligned} \mathcal{F}_1 - \gamma\mathcal{F}_u &= \mathbf{A}_1\phi\mathbf{MD}_\phi - \gamma\mathbf{A}_1\mathbf{MD}_\phi \\ &= \mathbf{A}_1\phi\mathbf{MD}_\phi - \mathbf{A}_1\gamma\mathbf{IMD}_\phi \\ &= \mathbf{A}_1[\phi - \gamma\mathbf{I}]\mathbf{MD}_\phi. \end{aligned}$$

The ϕ_I are given by the rank-reducing numbers of the pencil $(\mathcal{F}_1 - \gamma\mathcal{F}_u)$. From ϕ_I and Equation 7.48, we can compute τ_I for $1 \le i \le d$.

$$\tau_i = -\frac{N}{2\pi}\angle\gamma_i,$$

where $\angle\gamma_i$ is the phase angle of eigenvalue γ_i. The rank-reducing numbers are the same as the nonzero eigenvalues of

$$\mathbf{E}_1 = [\mathcal{F}_u^H \mathcal{F}_u]^{-1} \mathcal{F}_u^H \mathcal{F}_l. \tag{7.52}$$

The next task is to estimate θ, DOA. We do this independent of delays.

From Equation 7.50, we define two data matrices by selecting upper M–1 rows and lower $M-1$ rows from each element (matrix of dimension $M \times (N'-p+1)$) in $\mathcal{F}$. For example, the ith element in $\mathcal{F}$ is factored as follows:

$$\tilde{\mathbf{F}}_u^i = \begin{bmatrix} \tilde{\mathbf{f}}_1^i \\ \tilde{\mathbf{f}}_2^i \\ \vdots \\ \tilde{\mathbf{f}}_{M-1}^i \end{bmatrix} = \mathbf{A}_{M-1}\Phi^{i-1}\mathbf{MD}_\phi \quad \tilde{\mathbf{F}}_l^i = \begin{bmatrix} \tilde{\mathbf{f}}_2^i \\ \tilde{\mathbf{f}}_3^1 \\ \vdots \\ \tilde{\mathbf{f}}_M^i \end{bmatrix} = \mathbf{A}_{M-1}\Theta\Phi^{i-1}\mathbf{MD}_\phi,$$

where $\tilde{\mathbf{f}}_1^i$ is the 1st row of $\tilde{\mathbf{F}}^i$ and so on,

$$\Theta = \text{diag}\left\{\frac{2\pi}{\lambda}\delta\sin\theta_1, \frac{2\pi}{\lambda}\delta\sin\theta_2, \ldots, \frac{2\pi}{\lambda}\delta\sin\theta_d\right\},$$

and

$$\mathbf{A}_{M-1} = \begin{bmatrix} 1 & 1 & \cdots & 1 \\ \theta_1 & \theta_2 & \cdots & \theta_d \\ \vdots & \vdots & & \vdots \\ \theta_1^{M-2} & \theta_2^{M-2} & \cdots & \theta_d^{M-2} \end{bmatrix}.$$

The data matrices thus obtained are

$$\mathcal{F}_u = \begin{bmatrix} \mathbf{A}_{M-1} \\ \mathbf{A}_{M-1}\Phi \\ \vdots \\ \mathbf{A}_{M-1}\Phi^{p-1} \end{bmatrix} \mathbf{MD}_\phi \quad \mathcal{F}_l = \begin{bmatrix} \mathbf{A}_{M-1} \\ \mathbf{A}_{M-1}\Phi \\ \vdots \\ \mathbf{A}_{M-1}\Phi^{p-1} \end{bmatrix} \Theta\mathbf{MD}_\phi. \tag{7.53}$$

Consider a pencil matrix

$$\mathcal{F}_l - \gamma\mathcal{F}_u = \begin{bmatrix} \mathbf{A}_{M-1} \\ \mathbf{A}_{M-1}\Phi \\ \vdots \\ \mathbf{A}_{M-1}\Phi^{p-1} \end{bmatrix} [\Theta - \gamma\mathbf{I}_d]\mathbf{MD}_\phi.$$

The unknowns θ_i, $i=1, 2,\ldots, d$, are given by the rank-reducing numbers of the pencil,

$$\theta_i = \sin^{-1}\left(\frac{\angle\gamma_i}{2\pi}\frac{\lambda}{\delta}\right),$$

where $\angle\gamma_i$ is the phase angle of eigenvalue γ_i. The rank-reducing numbers are the same as the nonzero eigenvalues of

$$\mathbf{E}_2 = [\mathcal{F}_u^H \mathcal{F}_u]^{-1} \mathcal{F}_u^H \mathcal{F}_l. \tag{7.54}$$

The delay and DOA estimates are required to be correctly paired, that is, a given delay must be associated with a correct DOA so that they both correspond to a real wavefront. This is possible since $\mathbf{E}_1$ (Equation 7.52) and $\mathbf{E}_2$ commute, that is, $\mathbf{E}_1\mathbf{E}_2=\mathbf{E}_2\mathbf{E}_1$, they possess the same eigenvectors. A simple method for joint diagonalization is to first triangularize $\mathbf{E}_1$ (Schur decomposition),

$$\mathbf{U}^H \mathbf{E}_1 \mathbf{U} = \mathbf{T}_1,$$

where $\mathbf{T}_1$ is an upper triangular matrix whose diagonal entries are the eigen values of $\mathbf{E}_1$. Now apply this similarity transform to $\mathbf{E}_2$, yielding

$$\mathbf{U}^H \mathbf{E}_2 \mathbf{U} = \mathbf{T}_2,$$

where $\mathbf{T}_2$ is an upper triangular matrix (in the noise-free case) whose diagonal elements are correctly paired with those in $\mathbf{T}_1$ [26].

To illustrate the above algorithm we consider a four-element sensor array (ULA). A finite random bandlimited signal, having a bandwidth of 100 Hz, is incident on the array via five correlated paths. The path attenuation is assumed to be random, given by a zero mean unit variance uniformly distributed complex random variable in the range ±0.5. We have used an array output of 256 points (only one segments) for estimation of delay and DOA. The SNR was 60 dB. Other parameters used are $N'=54$ (Equation 7.49a) and $p=30$ (Equation 7.50). The results are summarized in Table 7.2. The algorithm is found to be quite sensitive to noise.

7.2.2 Duct Propagation

In Chapter 5, in the context of source localization in a bounded medium, we have modeled the array signal as

$$\mathbf{F}(\omega) = \mathbf{A}\mathbf{w}F_0(\omega) + \mathbf{N}(\omega), \tag{7.55}$$

where $\mathbf{A}=[\mathbf{a}_0, \mathbf{a}_1,\ldots, \mathbf{a}_{q-1}]$ and $\mathbf{a}_0, \mathbf{a}_1,\ldots, \mathbf{a}_{q-1}$ are direction vectors pointing to q paths ($\mathbf{a}_0$ for direct path), and $\mathbf{w}$ stands for the complex weight vector,

TABLE 7.2

Estimation of parameters of correlated multipaths

Actual Parameters		Estimated Parameters	
Delay	DOA (deg)	Delay (sec)	DOA (deg)
8	50	7.99	50.07
12	40	11.99	40.00
4	20	3.98	20.11
16	10	16.00	9.86
0	0	0.003	0.09

Note: The delays are in units sampling interval, 0.004 sec.

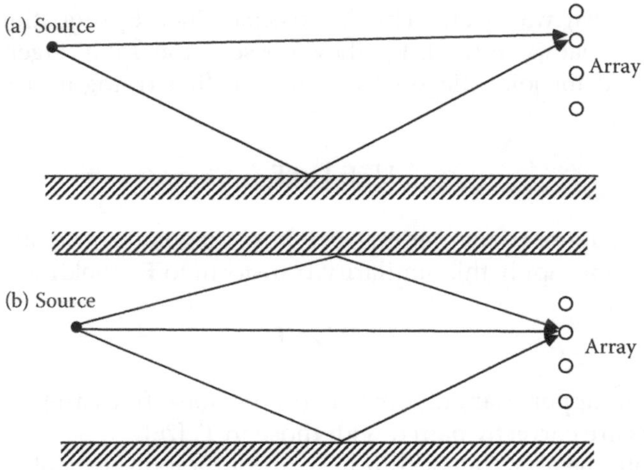

FIGURE 7.9
Two types of simple channels for which the vector may be estimated from a knowledge of source location and channel characteristics. (a) Single reflector and (b) two reflector channel.

$\mathbf{w} = [w_0, w_1, \ldots, w_{q-1}]$, where $w_0, w_1, \ldots, w_{q-1}$ are complex weighting coefficients applied to the temporal Fourier coefficients of the array signal. They represent attenuation and phase change due to propagation delays. If the channel is well characterized, both **A** and **w** can be estimated as a part of the source localization step, which, we will assume, has been carried out prior to waveform estimation. Thus, as a first step, assuming that **Aw** is known, let us explore how well we can recover the signal in the presence of background noise and interference.

7.2.2.1 Known Channel

A few examples of simple channels, whose characteristics are known or can be modeled reasonably well, are sketched in Figure 7.9. The channel shown in Figure 7.9a is a good approximation of surface-reflected radar signal, surface-reflected seismic signal, etc. The reflection coefficient of the surface is the unknown quantity, but the DOA of the direct wavefront as well as the reflected wavefront can be easily computed given the source location. For example, let the source be located at (l, h_s), where l is range and h_s is height above the (reflecting) surface. The DOA of the direct wavefront is given by

$$\tan\theta_0 = \frac{h_s - h_a}{l} \quad \text{and} \quad \tan\theta_1 = -\frac{h_s + h_a}{l}, \tag{7.56}$$

where h_a is the height of the array (midpoint) from the surface. In deriving Equation 7.56, we have used the method of images as outlined in Chapter 1. For this channel, the direction vectors and weight vector, respectively, are

$$\mathbf{A} = [\mathbf{a}_0, \mathbf{a}_1],$$

where

$$\mathbf{a}_0 = \text{col}\left\{1, e^{-j(2\pi/\lambda)d\sin(\theta_0)}, e^{-j(2\pi/\lambda)2d\sin(\theta_0)}, \ldots, e^{-j(2\pi/\lambda)(M-1)d\sin(\theta_0)}\right\}$$

$$\mathbf{a}_1 = \text{col}\left\{1, e^{-j(2\pi/\lambda)d\sin(\theta_1)}, e^{-j(2\pi/\lambda)2d\sin(\theta_1)}, \ldots, e^{-j(2\pi/\lambda)(M-1)d\sin(\theta_1)}\right\},$$

and

$$\mathbf{w} = \text{col}\,\{1, r\},$$

where r is the coefficient of reflection that we shall assume for simplicity is independent of the angle of incidence. We can now express **Aw** vector in terms of the direction vectors of the direct wavefront from the source and the one from the image of the source.

$$\mathbf{Aw} = [\mathbf{a}_0 + r\mathbf{a}_1]. \tag{7.57}$$

First, let us try the linear least squares estimate (LLSE) of $F_0(\omega)$. We shall convert Equation 6.11 into a frequency domain expression, leading to

$$\begin{aligned}\hat{F}_0(\omega) &= \left[\mathbf{w}^H\mathbf{A}^H\mathbf{Aw}\right]^{-1}\mathbf{w}^H\mathbf{A}^H\mathbf{F}(\omega)\\ &= F_0(\omega) + \left[\mathbf{w}^H\mathbf{A}^H\mathbf{Aw}\right]^{-1}\mathbf{w}^H\mathbf{A}^H\mathbf{N}(\omega).\end{aligned} \tag{7.58}$$

The quantity inside the square brackets in Equation 7.58 may be evaluated using Equation 7.57. It is given by

$$\begin{aligned}\mathbf{w}^H\mathbf{A}^H\mathbf{Aw} &= \mathbf{a}_0^H\mathbf{a}_0 + r^H\mathbf{a}_1^H\mathbf{a}_0 + r\mathbf{a}_0^H\mathbf{a}_1 + |r|^2\,\mathbf{a}_1^H\mathbf{a}_1\\ &= M\left(1 + |r|^2 + \frac{r^H\mathbf{a}_1^H\mathbf{a}_0 + r\mathbf{a}_0^H\mathbf{a}_1}{M}\right).\end{aligned} \tag{7.59}$$

The variance of the noise power in the filtered array output is given by

$$\sigma_{\hat{\eta}}^2 = \frac{\sigma_\eta^2}{\left[\mathbf{w}^H \mathbf{A}^H \mathbf{A} \mathbf{w}\right]}$$

$$= \frac{\sigma_\eta^2}{M\left(1+|r|^2+\dfrac{r^H \mathbf{a}_1^H \mathbf{a}_0 + r\mathbf{a}_0^H \mathbf{a}_1}{M}\right)}. \tag{7.60}$$

Compare Equation 7.60 with Equation 6.14, which was derived for two uncorrelated sources. In the present case, both sources are correlated (the second source is an image of the primary source). The variance of the noise is reduced by a factor of four, when $r=1$ and $\mathbf{a}_1 \approx \mathbf{a}_0$ (when $h_a \gg hs$). The multipath propagation has indeed helped to improve signal estimation.

Next, we shall try the Capon filter to estimate the waveform in the presence of interference, another source at known location, and the usual background white noise. The position of the sources and the receiving array are shown in Figure 7.10. We shall assume that the sources are uncorrelated. The DOA of the direct and the reflected wavefronts are given by

$$\tan\theta_{00} = \frac{h_\text{s} - h_\text{a}}{l}, \quad \tan\theta_{10} = -\frac{h_\text{s} + h_\text{a}}{l},$$

$$\tan\theta_{01} = \frac{h_\text{s} + \Delta h - h_\text{a}}{l}, \quad \tan\theta_{11} = \frac{h_\text{s} + \Delta h + h_\text{a}}{l}$$

For simplicity, we assume that the coefficient of reflection r is the same for both sources. The direction vectors are given by

$$\mathbf{A}_0\mathbf{w} = [\mathbf{a}_0 + r\mathbf{a}_1],$$

$$\mathbf{A}_1\mathbf{w} = [\mathbf{a}_{01} + r\mathbf{a}_{11}].$$

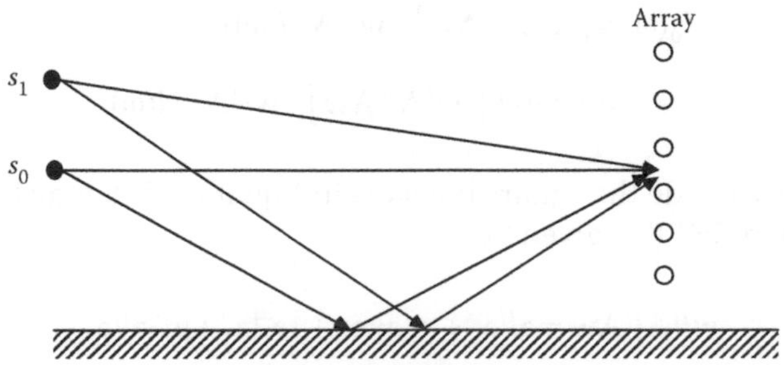

FIGURE 7.10
Two sources at the same range but at different heights. A vertical array of sensors is assumed.

The array output is modeled as

$$\mathbf{F}(\omega)=\mathbf{A}_0\mathbf{w}F_0(\omega)+\mathbf{A}_1\mathbf{w}F_1(\omega)+\mathbf{N}(\omega). \tag{7.61}$$

To compute the Capon filter given by Equation 6.26 we need the spectral matrix of the array output. Since the sources are assumed uncorrelated and the background noise is white, the spectral matrix may expressed as

$$\mathbf{S}(\omega)=\mathbf{A}_0\mathbf{w}\mathbf{w}^H\mathbf{A}_0^H S_0(\omega)+\mathbf{A}_1\mathbf{w}\mathbf{w}^H\mathbf{A}_1^H S_1(\omega)+\sigma_\eta^2\mathbf{I}. \tag{7.62}$$

The Capon filter is specified by Equation 6.26. To estimate the waveform emitted by the zeroth source the required filter is given by

$$\mathbf{H}_{\text{Cap}}=\frac{\mathbf{S}_f^{-1}\mathbf{A}_0\mathbf{w}}{\mathbf{w}^H\mathbf{A}_0^H(\omega,\theta)\mathbf{S}_f^{-1}\mathbf{A}_0\mathbf{w}}. \tag{7.63}$$

Applying the filter given in Equation 7.63 to the array output, we obtain

$$\mathbf{H}_{\text{Cap}}^H\mathbf{F}(\omega)=F_0(\omega)+\frac{\mathbf{w}^H\mathbf{A}_0^H\mathbf{S}_f^{-1}\mathbf{A}_1\mathbf{w}}{\mathbf{w}^H\mathbf{A}_0^H\mathbf{S}_f^{-1}\mathbf{A}_0\mathbf{w}}F_1(\omega)+\frac{\mathbf{w}^H\mathbf{A}_0^H\mathbf{S}_f^{-1}\mathbf{N}(\omega)}{\mathbf{w}^H\mathbf{A}_0^H\mathbf{S}_f^{-1}\mathbf{A}_0\mathbf{w}}. \tag{7.64}$$

While the signal from the zeroth source has been fully extracted, there are two terms in Equation 7.64 interfering with the desired signal. Here we shall consider the term representing the interference from the source #1. The second term, that is, the noise term, will not be considered as it follows the same approach used previously in connection with the single source case. The interference due to the second source will be measured in terms of cross-talk, as defined previously. In the present case, the cross-talk is given by

$$\text{Cross talk}=\frac{(\mathbf{w}^H\mathbf{A}_1^H\mathbf{S}_f^{-1}\mathbf{A}_0\mathbf{w})(\mathbf{w}^H\mathbf{A}_0^H\mathbf{S}_f^{-1}\mathbf{A}_1\mathbf{w})}{\left|\mathbf{w}^H\mathbf{A}_0^H\mathbf{S}_f^{-1}\mathbf{A}_0\mathbf{w}\right|^2}. \tag{7.65}$$

The cross-talk as a function of source separation for two uncorrelated sources is plotted in Figure 7.11. A 16-sensor vertical ULA was assumed. Notice that for a short range, the second source has little influence on the first source. But this influence grows rapidly as the range increases.

7.2.2.2 Partially Known Channel

In many real-life problems, the channel characteristics are never fully known as it is impossible to measure the microlevel variations causing path length variations on the order of a fraction of a wavelength. Such variations are

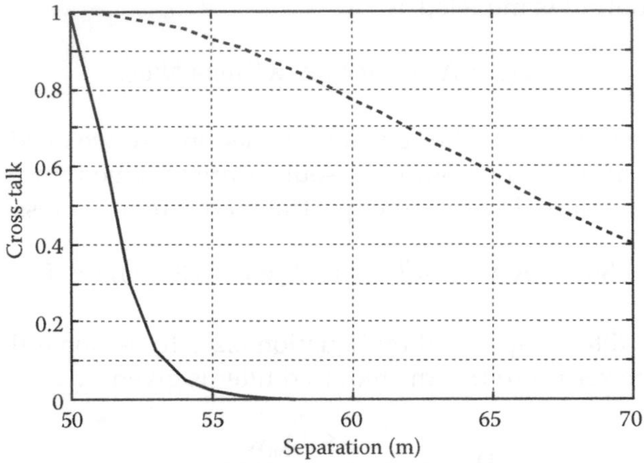

FIGURE 7.11
Cross-talk as a function of separation between two sources. The range is the same for both sources. Solid curve: range=1,000 m and dashed curve=5,000 m. The first source is 50 m above the surface.

known to affect the performance of source localization algorithms, particularly those belonging to a high-resolution class [8]. On the other hand, we may have fairly good knowledge about the general features of a channel but not enough to characterize at microlevel. One such example is a shallow water sea where the sea bottom variability is high. While the general geometry of the shallow water channel, broad undulations of the sea bottom, water temperature variations, etc., are available from the actual measurements, the factors which affect the performance of waveform estimation, such as the details of sea bottom, particularly sediment distribution, and sea surface undulation, are unknown. In Chapter 5, we have shown how source localization can be carried out with a full knowledge of a shallow water channel. Here we shall describe an approach capable of estimating the source location as well as signal waveform emitted by the source, given only partial knowledge.

Consider a single source in a shallow water channel and a vertical array of sensors. We have shown in Chapter 5 that the eigenvector corresponding to the largest eigenvalue is related to **Aw**,

$$\mathbf{E}_s = \frac{\mathbf{Aw}}{|\mathbf{Aw}|}, \quad |\mathbf{Aw}| \neq 0. \tag{7.66}$$

In Equation 7.66, the weight vector **w** is dependent on the channel parameters and the columns of **A** matrix on the source position. Notice that the weight vector **w** in Equation 7.66 occurs linearly, while the source parameters in **A** occur nonlinearly. A least squares method of estimating the nonlinear parameters by first eliminating the linear parameters followed by minimizing the norm of the error vector was first suggested by Guttman et al. [9] and applied to a signal-processing problem by Tufts and Kumaresan [10].

Here, we shall exploit this approach. In Equation 7.66, we assume that the source position is approximately known and write the equation in terms of the unknown **w** vector,

$$\mathbf{E}_s = \mathbf{A}\frac{\mathbf{w}}{|\mathbf{A}\mathbf{w}|} = \mathbf{A}\tilde{\mathbf{w}}. \tag{7.67}$$

Let $A^{\#}$ be the pseudoinverse of **A**. The least squares estimate of $\tilde{\mathbf{w}}$ will be given by

$$\hat{\tilde{\mathbf{w}}} = \mathbf{A}^{\#}\mathbf{E}_s.$$

Substitute back into Equation 7.66 and obtain an estimate of $\mathbf{E}_s$. The mse is given by

$$\|\text{error}\|_2 = \|(\mathbf{I} - \mathbf{A}\mathbf{A}^{\#})\mathbf{E}_v\|_2 = \|\mathbf{P}_A^{\perp}\mathbf{E}_s\|_2,$$

where $\mathbf{P}_A^{\perp}$ is the orthogonal projection complement of matrix **A**. The mse is now minimized with respect to the source location parameters. We define a parametric spectrum as,

$$S(R_0, Z_0)\frac{1}{\mathbf{E}_s^H\mathbf{P}_A^{\perp}\mathbf{E}_s}. \tag{7.68}$$

In practical implementation, the parametric spectrum is computed over a dense grid in the range-depth space. At each grid point the projection matrix $\mathbf{P}_A^{\perp}$ is computed and Equation 7.68 is evaluated. Thus, in the proposed method, the subspace spanned by the columns of the **A** matrix (range space) is steered until $\mathbf{P}_A^{\perp}$ coincides with $\mathbf{E}_\eta\mathbf{E}_\eta^H$, that is, when the parametric spectrum becomes very large (ideally infinite). After obtaining the source location parameters, we use them to estimate the weight vector,

$$\hat{\tilde{\mathbf{w}}} = \mathbf{A}_{\max}^{\#}\mathbf{E}_s, \tag{7.69}$$

where $\mathbf{A}_{\max}^{\#}$ is $A^{\#}$ evaluated where the parametric spectrum is maximum. Thus, $\hat{\tilde{\mathbf{w}}}$ is the least mean square estimate of $\tilde{\mathbf{w}}$.

An example of parametric spectrum is shown in Figure 7.12. A low power source (–10 dB) is assumed at range 4,600 m and depth 50 m in a Pekeris channel of depth 200 m. Pekeris channel is a constant depth water layer resting on a soft bottom (speed: 1,600 m/sec and relative density: 2.0). A vertical ULA consisting of 40 sensors is placed at a depth of 70 m. For the same channel, the reflection coefficients were computed from the eigenvector corresponding to the largest eigenvalue (Equation 7.69). The reflection coefficients are normalized with respect to $|\mathbf{A}\mathbf{w}|$, which may be obtained from the fact that the

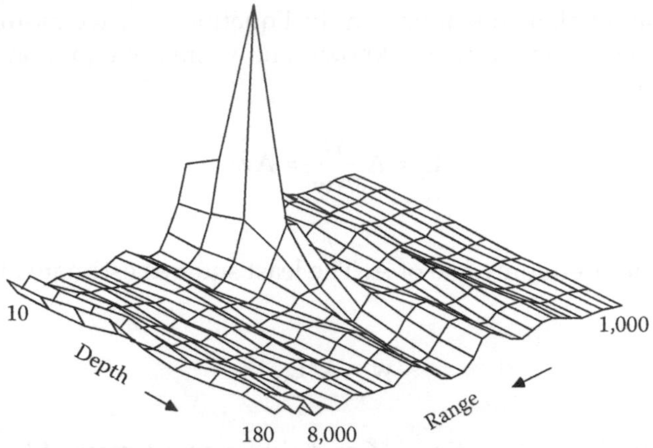

FIGURE 7.12
The parametric spectrum for a single source located at range 4,600 m and depth 50 m. Range scan is in steps of 200 m. Depth scan is in steps of 20 m.

TABLE 7.3
A comparison of estimated reflection coefficients (first eight coefficients) with true reflection coefficients, computed for a pekeris channel described on Chapter 1, Section 1.3

	True Reflection Coefficients		Estimated Reflection Coefficients			
	$M=40$		$M=40$		$M=60$	
Image #	Real	Imag	Real	Imag	Real	Imag
1	1.0	0.0	1.0	0.0	1.0	0.0
2	−0.861	−0.506	−0.847	−0.718	−0.870	−0.490
3	0.467	0.882	0.612	0.829	0.416	0.886
4	0.995	−0.064	1.075	−0.260	0.998	−0.091
5	0.981	0.167	1.395	−0.053	0.925	0.205
6	0.905	0.407	1.040	0.251	0.858	0.459
7	0.626	−0.768	0.705	−0.783	0.641	−0.752
8	0.107	0.981	−0.013	0.964	0.107	0.986

weighting coefficient corresponding to the direct path is by definition equal to one; hence its actual observed value must be equal to $|\mathbf{Aw}|$. In Table 7.3, the estimated and the actual reflection coefficients for the first eight images out of 20 multipaths used in computation are listed for two different array lengths.

In computer simulation, we have found that, for good results, the number of sensors must be many more than the lower limit given in Ref. [11]. The least mse in the estimated reflection coefficients for different number of sensors is shown in Figure 7.13. Here the number of significant images is 20 ($P=20$) and hence, according to the lower limit ($>2P+2$), the minimum number of sensors ought to be more than 42. We observe that an array of 60 sensors appears to be optimum. In addition to the unknown reflection coefficients

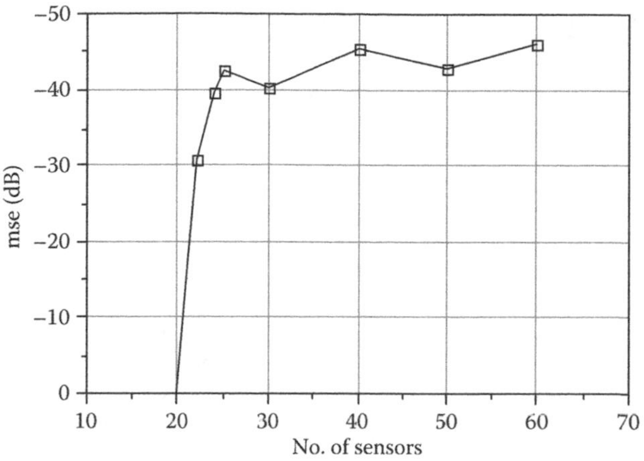

FIGURE 7.13
The role of array size on the mean square error in the estimated reflection coefficients is shown. Twenty multipaths were assumed.

we have background noise, which is likely to be both spatially and temporally correlated. But, since it is uncorrelated with the signal, it occurs in a linear combination; as such, it may be estimated using the approach used for the estimation of the reflection coefficients. Indeed, such an approach has been used by Boehme [12] for the estimation of the noise spectrum first using an approximate knowledge of the channel. In the next step, the previously estimated noise spectrum is used in the expression for the likelihood ratio, which is then maximized with respect to the unknown channel parameters and the source position. The maximization of the likelihood ratio is, however, highly computation intensive [13].

7.3 Scatter Channel

A large cluster of multipaths is created when a transmitter or receiver is surrounded by scatterers. They are characterized by delay and angle spreads, that is, delay and DOA are random variables. If the transmitter, or receiver, or both are moving, there is Doppler shift depending upon speed in addition to the delay and angle spreads. The delay spread introduces spectral variation or frequency-selective spreading, angle spread introduces space-selective spreading, and Doppler spread introduces time-selective spreading.

7.3.1 Geometrical Model

Wave scattering is caused by sharp pointed objects, corners of a building, embedded steel structure of modern RCC buildings, etc. Scattering point acts as a point source radiating in all directions with strength α, where α

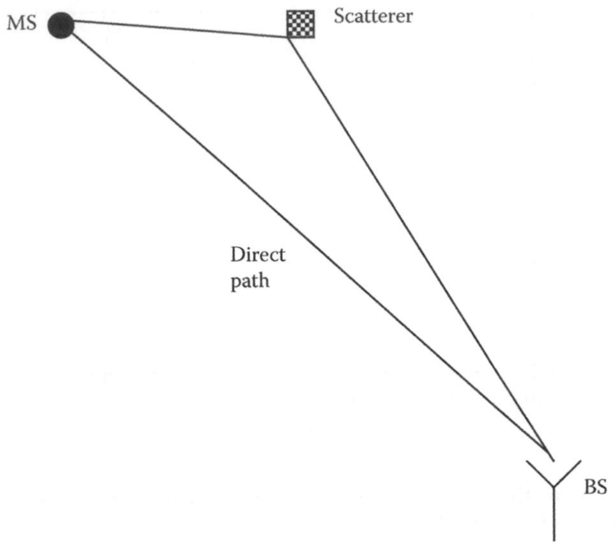

FIGURE 7.14
Scattered ray will always travel a longer path. Therefore, all scattered signals will reach the receiver base station (BS) after the direct signal. MS: mobile station.

is a complex number. When the incident wave is of unit strength, α is equal to a scattering coefficient and is $|\alpha| \le 1$. A scattered ray will always travel a longer path than the direct ray whatever the position of the scatterer (see Figure 7.14).

It is common practice to ignore multiple bounces or multiple scattering, which results in reduced amplitude. Scatterers are also found in the neighborhood of a receiver, such as a mobile phone or GPS receiver. In mobile communication, a base station carrying both transmitter and receiver is often placed on a tall building well above scatterers. But a mobile phone in the hands of a user is more likely to be surrounded by scatterers. A mobile phone is both a transmitter and receiver (transceiver). Sometimes, a base station is placed at a low level, as in indoors or in a large shopping mall, for communication. Thus, we have broadly two situations. Of the two transceivers, one is above the plane of scatterers and the other is surrounded by scatterer. The other possibility is where both transceivers are at low level, hence surrounded by scatterers. These two possibilities give rise to different distributions of angle of arrival. Models are illustrated in Figure 7.15.

In the first model, all scatterers lie within a circle centered at a mobile station. The scatterers are uniformly distributed inside the circle (see Figure 7.15a) whose radius is determined by the maximum permissible path loss due to excess path length. We notice that at the mobile station the scattered rays are uniformly distributed in all directions (360°), but at the base station the scattered rays are bundled around the direct ray. The angular spread of the scattered rays is within a few tens of degrees. The delay spread is also within a few tens of microseconds [14].

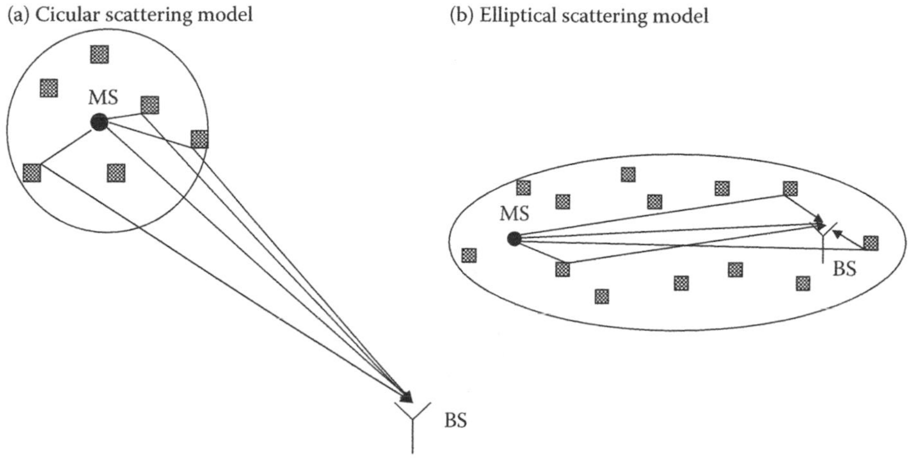

FIGURE 7.15
Scatterers around base station (BS) and mobile station (MS). (a) Circular scattering model; here only MS is surrounded by scatterers. (b) Elliptical scattering model; both BS and MS are surrounded by scatterers.

In the second model, the scatterers are present throughout the space (see Figure 7.15b). Consider an ellipse with transceivers as focii A scattered ray coming from any point on the ellipse will have same excess path that is greater than any other ray scattered from inside the ellipse. Depending upon the maximum permissible path loss, we can draw an elliptical boundary enclosing all significant scatterers. This is the elliptical scattering model studied in Refs. [15, 16]. The scattered rays are uniformly distributed in all directions at both transceivers. However, the delay spread is insignificant, less than 1 msec [14].

7.3.2 Scatterers around Mobile

A signal emitted by a transceiver is often scattered by point scatterers in the immediate neighborhood of the transceiver. What reaches a distant array is a collection of plane wavefronts, differing in phase, angle of arrival, and amplitude, but all wavefronts remain correlated. Let $(\theta_0+\delta\theta_k)$ be the DOA of the scattered wavefront from the kth scatterer and θ_0 is the nominal DOA of the direct wavefront from the source (see Figure 7.16). We assume that there are L scatterers in the vicinity of the source. The array output may be expressed as

$$\mathbf{f}(t)=\left[\sum_{k=0}^{L-1}\alpha_k e^{-j\omega_c\delta t_k}\mathbf{a}(\theta_0+\delta\theta_k)\right]f_0(t-\tau_0-\delta t_k), \tag{7.70}$$

where $\mathbf{a}(\theta_0+\delta\theta_k)$ is the array response vector defined as

$$\mathbf{a}(\theta_0+\delta\theta_k)$$
$$=[1, e^{-j(\omega_c d/c)\sin(\theta_0+\delta\theta_k)}, e^{-j(\omega_c 2d/c)\sin(\theta_0+\delta\theta_k)},\ldots, e^{-j(\omega_c (M-1)d/c)\sin(\theta_0+\delta\theta_k)}],$$

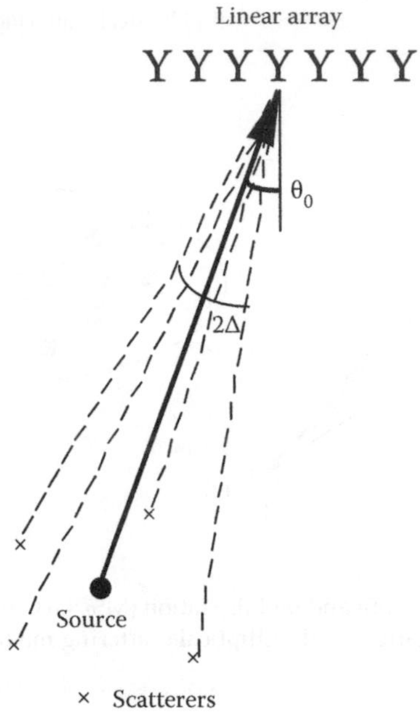

FIGURE 7.16
Local scattering model. Mobile station (MS) is surrounded by scatterers. The signal reaching *a* distant array consists of a suite of plane waves arriving with different DOAs and delays.

for *a* equispaced (spacing $=d$) linear array (see Chapter 2 for more details on the array response), α_k is the coefficient of scattering (complex), and δt_k is the delay (with respect to direct wavefront) of a wavefront from the *k*th scatterer; τ_0 is the propagation delay assumed to be known or estimated independently. Note that $k=0$ refers to the direct wavefront, for which $\alpha_0=1$ and $\delta t_0=0$. It is assumed that the source emits a narrowband signal, that is, bandwidth $\ll (2\pi/\delta t_k)$. We shall express $\mathbf{a}\,(\theta_0+\delta\theta_k)$ in Taylor's series expansion,

$$\mathbf{a}(\theta_0+\delta\theta_k)=\mathbf{a}(\theta_0)+\delta\theta_k\left.\frac{\partial\mathbf{a}(\theta)}{\partial\theta}\right|_{\theta=\theta_0}+\frac{(\delta\theta_k)^2}{2!}\left.\frac{\partial^2\mathbf{a}(\theta)}{\partial\theta^2}\right|_{\theta=\theta_0}+\ldots \tag{7.71}$$

$$\approx\mathbf{a}(\theta_0)+\delta\theta_k\mathbf{a}'(\theta_0).$$

Using Equation 7.71 with the first two terms only in Equation 7.70 we obtain the following approximate result [17] after correcting for the propagation delay

$$\mathbf{f}(t)\approx[\phi_0\mathbf{a}(\theta_0)+\phi_1\mathbf{a}'(\theta_0)]\,f_0(t), \tag{7.72a}$$

which for a ULA becomes

$$\mathbf{f}(t) \approx \phi_0 \begin{bmatrix} 1, (1 - j\frac{\phi_1}{\phi_0}\frac{2\pi d}{\lambda}\cos\theta_0)e^{-j\frac{2\pi d}{\lambda}\sin\theta_0}, \ldots, \\ (1 - j\frac{\phi_1}{\phi_0}\frac{2\pi d(M-1)}{\lambda}\cos\theta_0)e^{-j\frac{2\pi d(M-1)}{\lambda}\sin\theta_0} \end{bmatrix} f_0(t) \tag{7.72b}$$

where

$$\phi_0 = \sum_{k=0}^{L-1} \alpha_k e^{-j\omega_c \delta t_k},$$

and

$$\phi_1 = \sum_{k=0}^{L-1} \alpha_k \delta\theta_k e^{-j\omega_c \delta t_k}.$$

For DOA estimation the covariance matrix of the array output is of interest. Using the first order approximation in Equation 7.72a, we obtain the covariance matrix

$$\mathbf{C}_f \approx L\sigma_\alpha^2 \sigma_{f_0}^2 \left[\mathbf{a}(\theta_0)\mathbf{a}^H(\theta_0) + \sigma_\theta^2 \mathbf{a}'(\theta_0)\mathbf{a}'^H(\theta_0)\right], \tag{7.73}$$

where we have assumed that α_k and $\delta\theta_k$ are independent random variables whose variances are σ_α^2 and σ_θ^2, respectively, and $\sigma_{f_0}^2$ is variance of the source signal. Note $\mathbf{c}_f$ is a sum of two rank-one matrices; hence its maximum rank will be two.

For a ULA we can derive a more specific result. For small $\delta\theta$, $\sin\delta\theta \approx \delta\theta$ and $\cos\delta\theta \approx 1$ we have

$$\left[\mathbf{a}(\theta_0 + \delta\theta)\right]_m = e^{-j(2\pi d/\lambda)m(\sin\theta_0 + \delta\theta\cos\theta_0)}.$$

The coefficient of scattering is assumed to be uncorrelated. Then the covariance matrix simplifies to

$$\begin{aligned} \{\mathbf{C}_f\}_{mn} &= L\sigma_{f_0}^2 \sigma_\alpha^2 E\{\mathbf{a}(\theta_0 + \delta\theta)\mathbf{a}^H(\theta_0 + \delta\theta)\}_{mn} \\ &= L\sigma_{f_0}^2 \sigma_\alpha^2 e^{-j(2\pi d/\lambda)(m-n)\sin\theta_0} E\{e^{-j(2\pi d/\lambda)(m-n)\delta\theta\cos\theta_0}\}, \end{aligned} \tag{7.74a}$$

where $\sigma_{f_0}^2$ is the variance of the source signal and σ_α^2 is the variance of the coefficient of scattering. Assuming $\delta\theta$ is uniformly distributed over a range $\pm\Delta$, the expected value in Equation 7.74a may be shown to be

$$E\{e^{-j(2\pi d/\lambda)(m-n)\delta\theta\cos\theta_0}\} = \frac{\sin 2\pi\frac{d}{\lambda}\Delta(m-n)\cos\theta_0}{2\pi\frac{d}{\lambda}\Delta(m-n)\cos\theta_0}. \tag{7.74b}$$

Using Equation 7.74b in Equation 7.74a we obtain

$$\mathbf{C}_f \approx L\sigma_{f0}^2\sigma_\alpha^2\mathbf{D}(\theta_0)\mathbf{Q}\mathbf{D}^H(\theta_0), \tag{7.75}$$

where

$$\{\mathbf{Q}\}_{mn} = \frac{\sin 2\pi\frac{d}{\lambda}\Delta(m-n)\cos\theta_0}{2\pi\frac{d}{\lambda}\Delta(m-n)\cos\theta_0},$$

and

$$\mathbf{D} = \text{diag}\,[\mathbf{a}(\theta_0)].$$

Note that **Q** is a symmetric Toeplitz matrix, hence it has real eigenvalues. Let

$$\mathbf{Q} = \sum_{i=1}^{M} \lambda_i \mathbf{v}_i \mathbf{v}_i^H,$$

be the eigen-decomposition of **Q** matrix. It is known that $\lambda_1 \approx \lambda_2 \approx \lambda_3 \approx \ldots \lambda_r \approx 1$, where r is the rank of **Q** and the remaining eigenvalues are insignificant. The rank is approximately given by $r \approx [(Md/\lambda)\, 2\Delta\cos\theta_0]$, where $[x]$ stands for the largest integer greater than x. The eigenvectors corresponding to the significant eigenvalues are known as discrete prolate spheroidal sequences (DSSP) [18]. We have computed the rank of the matrix **Q** as shown in Table 7.4.

Using the eigen-decomposition of matrix **Q** in Equation 7.75 we obtain

$$\begin{aligned}\mathbf{C}_f &\approx L\sigma_{f0}^2\sigma_\alpha^2 \sum_{i=1}^{r} \lambda_i \mathbf{D}\mathbf{v}_i \mathbf{v}_i^H \mathbf{D}^H \\ &= L\sigma_{f0}^2\sigma_\alpha^2 \sum_{i=1}^{r} \lambda_i \mathbf{v}_i . \times \mathbf{a}(\theta_0)\mathbf{v}_i^H . \times \mathbf{a}^H(\theta_0),\end{aligned} \tag{7.76}$$

where $.\times$ stands for element by element multiplication (MATLAB® convention). The structure of the covariance matrix suggests that there are r uncorrelated sources whose response vectors are given by $\mathrm{v}_i . \times \mathrm{a}(\theta_0)$, $i = 1, 2, \ldots, r$.

TABLE 7.4

Rank of **Q** Matrix for Different Values of $\Delta\cos\theta_0$ and $M = 64$ and $d = \lambda/2$

$\Delta\cos\theta_0$	Rank
0.03 rad	2
0.1	7
0.15	10

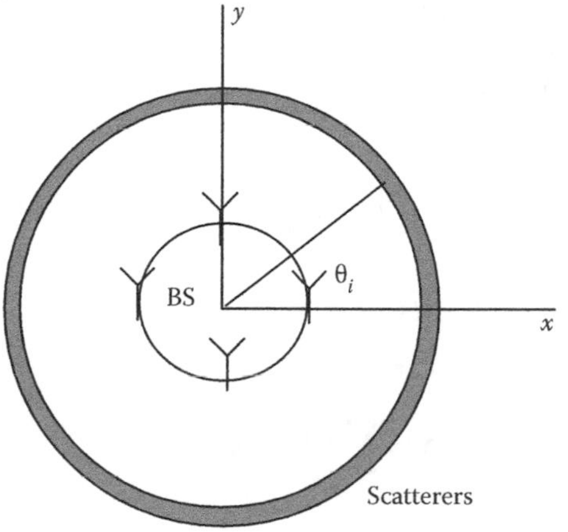

FIGURE 7.17
An idealized model of scatterers around a base station. The scatterers are in far-field region. The transmitter is also in the far field.

7.3.3 Scatterers around Base Station

We now consider another case where a base station is surrounded by scatterers, a situation commonly encountered when the base station is located indoors or at relatively low level in a city center environment. The base station array will receive the scattered waves from all directions, ideally 360°. Assume that we have a circular array of M elements at the base station and scatterers are in far-field region, which implies plane wave incidence on the array.

Consider a point scatterer (see Figure 7.17) at ith position at an angle θ_i measured with respect to the y-axis. The relative time delay of a wavefront from a scatterer with respect to a dummy receiver at the center of the circular array is given by

$$\delta\tau_i = \frac{R}{c}\cos(\theta_0 - \theta_i),$$

where R stands for the radius of a ring of scatterers (see Figure 7.17). The response of a circular array is given by

$$\mathbf{a}(\theta_i) = \left[e^{-j(\omega_c a/c)\cos(\theta_i)}, e^{-j(\omega_c a/c)\cos((2\pi/M)-\theta_i)}, \ldots, e^{-j(\omega_c a/c)\cos((2\pi(M-1)/M)-\theta_i)}\right]^T,$$

where a stands for the radius of circular array. The array output may be written as

$$\mathbf{f}(t) = \sum_i \alpha_i e^{-j\omega_c(R/c)\cos(\theta_0-\theta_i)}\mathbf{a}(\theta_i)f_0(t), \tag{7.77}$$

where α_i is the scattering coefficient, which is assumed to be an uncorrelated complex random variable with zero mean and variance equal to σ_s^2. Further, the scatterers are uniformly distributed over the scattering ring. The covariance function of the array output is given by

$$\mathbf{C}_f = E\{\mathbf{f}(t)\mathbf{f}^H(t)\} = \sigma_s^2 \sum_i E\{\mathbf{a}(\theta_i)\mathbf{a}^H(\theta_i)\}|f_0(t)|^2 \\ = \sigma_s^2 \mathbf{A}|f_0(t)|^2, \tag{7.78}$$

where

$$\mathbf{A} = \sum_i E\{\mathbf{a}(\theta_i)\mathbf{a}^H(\theta_i)\},$$

whose (m,n)th element is given by

$$\{\mathbf{A}\}_{m,n} = J_0\left(\frac{2\omega_c a}{c}\sin\frac{\pi}{M}(m-n)\right). \tag{7.79}$$

It may be observed that the covariance function of the array output does not depend on R, radius of the ring of scatterers. The normalized correlation is minimum between the diametrically opposite antennas, which is equal to $J_0(2\pi(2a/\lambda))$. The correlation between opposite antennas is zero whenever $2\pi(2a/\lambda) = z_i$, $i = 1, 2, \ldots$, where z_i is ith zero of Bessel function of order zero, for example for $i = 1$, $z_1 = 2.4048$.

7.4 Channel Estimation

A linear channel is where the input and output are linearly related through an impulse response function of the channel. In many practical problems, for example, in communication a linear channel model is extensively used. There are four types of linear channels, depending on the type of impulse response function [19].

7.4.1 Time and Frequency Selective

When an impulse response function is time varying, the output is related to the input through a convolution sum

$$y(n) = \sum_k h(n,k)x(n-k), \tag{7.80a}$$

where $h(n, k)$ is the impulse response of the channel at time n to a unit response at time $n-k$.

7.4.2 Frequency Selective (Time-Nonselective)

The impulse response function is time invariant. The input and output are related as

$$y(n) = \sum_k h(k)x(n-k). \quad (7.80b)$$

7.4.3 Time Selective (Frequency Nonselective)

When the impulse response function in Equation 7.80a is of the type $h(n, k)=h(n)\delta(k)$, where $\delta(k)$ is the Kronecker delta function, $\delta(k)=1$ when $k=0$, and $\delta(k)=0$ for all $k\neq 0$. The input and output relation becomes

$$y(n)=h(n)x(n). \quad (7.80c)$$

7.4.4 Constant

Finally, the impulse response function is simply a constant, h_0. The output is simply a scaled version of the input

$$y(n)=h_0x(n). \quad (7.80d)$$

Such a channel is also known as time-nonselective and frequency-nonselective. Our interest will be in the frequency selective channel (Equation 7.80b), which has been widely studied. A practical example of a frequency selective channel is radio wave propagation through air or acoustic wave propagation underwater. A transmitted signal reaches the receiver via different paths with different delays.

The impulse response function (in continuous time) may be modeled as follows:

$$h(t) = h_0(t) * \sum_{i=0}^{d-1} m_i\delta(t-\tau_i), \quad (7.81)$$

where $h_0(t)$ is a filter that accounts for propagation effects, pulse-shaping waveform, and filters used in transmitter and receiver, and m_i and τ_I, $i=0$, 1, $d-1$ are the attenuation and delay suffered by different multipaths. The length of the impulse response function is an important parameter, which we shall denote by L. Approximately, L is equal to τ_{max}+length of $h_0(t)$ ($h(t)=0$ $t\notin 0, L-1$.

7.4.5 Single Input Single Output

We shall express Equation 7.80b in a matrix form. We define the following vectors and a matrix:

(1) Output data vector consisting of N consecutive samples at time instant n

$$\underset{(N-L+1)\times 1}{\mathbf{y}=[y(L-1), y(L), \ldots, y(N-1)]^T}.$$

(2) Impulse response coefficient vector

$$\underset{L\times 1}{\mathbf{h}=[h(0), h(1), h(2),\ldots, h(L-1)]^T}.$$

where the length of the impulse response function is L.

(3) Input data matrix

$$\underset{(N-L+1)\times L}{\mathbf{X}=\begin{bmatrix} x(L-1) & x(L-2) & x(L-3) & \cdots & x(0) \\ x(L) & x(L-1) & x(L-2) & \cdots & x(1) \\ \vdots & \vdots & \vdots & & \vdots \\ x(N-1) & x(N-2) & x(N-3) & \cdots & x(N-L) \end{bmatrix}}.$$

In terms of these quantities, Equation 7.80b may be expressed as

$$\mathbf{y}=\mathbf{Xh}. \tag{7.82}$$

When the input data matrix $\mathbf{X}$ is known and is full column rank, it is straightforward to estimate the channel impulse response vector,

$$\mathbf{h}=[\mathbf{X}^H\mathbf{X}]^{-1}\,\mathbf{X}^H\mathbf{y}. \tag{7.83}$$

This method of channel estimation works well when the input signal, such as the training sequence in communication or the probing signal in system identification, is long compared to the length of the impulse response function ($N \gg L$).

There is an alternate method of expressing Equation 7.80b in matrix form. Instead of the data matrix, we define an impulse response matrix, also known as a filter matrix, as follows:

$$\mathbf{H} = \begin{bmatrix} h(0), h(1), h(2)\cdots h(L-1) & & & \\ & h(0), h(1), h(2)\cdots h(L-1) & & \\ & \ddots & \ddots & \\ & & & h(0), h(1), h(2)\cdots h(L-1) \end{bmatrix},$$
$$N \times (N+L-1)$$

an input data vector

$$\mathbf{x} = [x(n), x(n-1), x(n-2), \ldots, x(n-N-L+1)]^T.$$
$$(N+L-2)\times 1$$

and output data vector

$$\mathbf{y} = [y(n), y(n-1), y(n-2), \ldots, y(n-N+1)]^T.$$

The alternate representation of Equation 7.80b is

$$\mathbf{y} = \mathbf{H}\mathbf{x}. \tag{7.84}$$

Unfortunately, since there are more unknowns ($N+L-1$) than knowns (N), it is not possible to recover the input from the output even though the impulse response function is known. In a trivial case of $L=1$, when the channel is time-nonselective and frequency-nonselective, it is possible to estimate the input up to a scalar constant ($1/h_0$).

7.4.6 Single Input Multiple Outputs (SIMOs) Channel

The above limitation may be overcome by increasing the number of equations or knowns. There are two equivalent approaches to achieve the desired goal. In the first approach, we sample the output at a higher than the Nyquist rate, as in communication, the sampling rate is an integer multiple of the symbol rate [20, 25]. In the second approach, we use multiple receiving antennas with the assumption that the channel impulse response functions to different sensors are different, in the sense that they have different zeros, or in other words they are coprime. This is known as spatial diversity [20]. In this work, we shall emphasize the second approach. Consider an M sensor antenna array but single input. We define an array output data vector (dim: $MN\times 1$) by stacking the outputs of all individual receivers.

$$\mathbf{y} = \left[\mathbf{y}_0^T, \mathbf{y}_1^T, \ldots, \mathbf{y}_{M-1}^T\right]^T,$$

where

$$\mathbf{y}_i=[y_i\,(n), y_i\,(n-1), y_i\,(n-2),\ldots, y_i\,(n-\mathrm{N}+1)]^T,$$

is the output of the ith sensor. The array filter matrix is also defined by stacking the filter matrices of individual receivers.

$$\mathcal{H}=\begin{bmatrix}\mathbf{H}_0\\ \mathbf{H}_1\\ \vdots\\ \mathbf{H}_{M-1}\end{bmatrix}, \tag{7.85}$$

where H_i is the filter matrix of ith channel. The input data vector, however, remains the same. We use the above array output data vector and array filter matrix in Equation 7.84 to obtain the input/output relation of M sensor receiver array

$$\mathbf{y}=\mathcal{H}\mathbf{x}. \tag{7.86}$$

The size of $\mathcal{H}$ is $MN\times(N+L-1)$. For large M (≥ 2) and diversity in the individual receiver impulse response functions, the column rank of $\mathcal{H}$ will be equal to $(N+L-1)$. Next, we state the condition for $\mathcal{H}$ to be full column rank. Let $H_i(z)=h_0^i+h_1^i z^{-1}+\cdots+h_{L-1}^i z^{-(L-1)}$ be the z-transform of the impulse response of ith channel, that is, the transfer function. We assume that all impulse response functions are of the same length. The condition for full column rank of $\mathcal{H}$ is that the transfer functions, $H_i\,(z)$, $i=0, 1,\ldots, M-1$, do not possess a common factor, in other words, $H_i\,(z)$, $i=0, 1,\ldots, M-1$ are coprime [20]. If the transfer functions are not coprime there exists a common factor $c(z)$, we can then express the transfer function of ith sensor as $H_i(z)=c(z)\hat{H}_i(z)$ and the output of the ith sensor as

$$\begin{aligned}Y_i(z)&=c(z)\hat{H}_i(z)X(z)\\ &=\hat{H}_i(z)[c(z)X(z)].\end{aligned}$$

The common factor may also be combined with the input signal, then it would be difficult to distinguish from the output alone whether $c(z)$ is part of the signal or part of the transfer function of the channel. The full rank property of $\mathcal{H}$ enables us to estimate the input vector, given the output vector and the array filter matrix. But unfortunately, the filter matrix is not known and it must be estimated either with the help of a training signal or blindly.

7.4.7 Blind Channel Estimation

Estimation of a channel impulse response function without any knowledge of the input is known as blind estimation. We exploit the statistical properties of the input signal, in particular the rank of its covariance matrix. We compute the covariance matrix of the output data,

$$\begin{aligned}\mathbf{C}_y &= E\{\mathbf{y}\mathbf{y}^H\} = \mathcal{H}E\{\mathbf{x}\mathbf{x}^H\}\mathcal{H}^H \\ &= \mathcal{H}\mathbf{C}_x\mathcal{H}^H,\end{aligned} \tag{7.87}$$

where $\mathbf{C}_y$ is the output covariance matrix of size $(MN \times MN)$ and $\mathbf{C}_x$ is the input covariance matrix of size $(N+L-1)(N+L-1)$. We shall assume that $\mathbf{C}_x$ is full rank $(N+L-1)$. There are $(N+L-1)$ nonzero eigenvalues and $MN-N-L+1$ null eigenvalues of the output covariance matrix. Let $\mathbf{U}_0$ represent the collection of eigenvectors corresponding to the null eigenvalues (noise space),

$$\mathbf{U}_0 = \{\mathbf{u}_{N+L}, \mathbf{u}_{N+L+1}, \ldots, \mathbf{u}_{MN}\}.$$

Since the columns of $\mathcal{H}$ span the signal subspace of $\mathbf{C}_y$, we have an important result, reminiscent of the MUSIC algorithm,

$$\mathbf{U}_0^H\mathcal{H} = 0. \tag{7.88}$$

The columns of $\mathcal{H}$ are orthogonal to any vector in the noise space. Hence,

$$\mathbf{u}_i^H\mathcal{H} = 0, \quad i = N+L, \quad N+L+1, \ldots, MN. \tag{7.89}$$

We shall now divide vector $\mathbf{u}_i$ into M equal subvectors each of length equal to N, as shown below,

$$\mathbf{u}_i = [\mathbf{u}_i^{0T}, \mathbf{u}_i^{1T}, \ldots, \mathbf{u}_i^{M-1T}]^T,$$

where $\mathbf{u}_i^m$ $(N \times 1)$ is mth subvector. Equation 7.88, using Equation 7.85, may be written as

$$\sum_{m=0}^{M-1} \mathbf{u}_i^{mH}\mathbf{H}_m = \mathbf{0}. \tag{7.90a}$$

Note that $\mathbf{u}_i^{mH}\mathbf{H}_m$ is a convolution between a sequence defining the subvector $\mathbf{u}_i^m$ and impulse response sequence $h_m(0), h_m(l), \ldots, h_m(L-1)$. Further, note that the order in the convolution expression is reversible, that is,

$$y(n) = \sum_{k=0}^{L-1} h(k)x(n-k) \text{ or } y(n) = \sum_{k=n}^{n-L+1} h(n-k)x(k),$$

will yield the same result. Hence, Equation 7.90a may be written as

$$\sum_{m=0}^{M-1} \mathbf{h}_m^H \mathbf{u}_i^m = 0, \tag{7.90b}$$

where $\mathbf{u}_i^m$ L×($N+L-1$)) is now a filter matrix defined using the subvector $\mathbf{u}_i^m$. Stacking all M factors on the left-hand side of Equation 7.90b we obtain

$$\mathbf{h}^H \mathcal{U}_i = 0, \tag{7.91a}$$

where

$$\mathbf{h} = [\mathbf{h}_0^T, \mathbf{h}_1^T, \ldots, \mathbf{h}_{M-1}^T]^T,$$

is the composite impulse response vector ($LM \times 1$) and

$$\mathcal{U}_i = [\mathbf{u}_i^{0T}, \mathbf{u}_i^{1T}, \ldots, \mathbf{u}_i^{M-1T}]^T.$$

Equation 7.91a is more convenient than Equation 7.89 for subspace estimation of the composite impulse response vector. We shall express Equation 7.91a in terms of its squared norm,

$$\mathbf{h}^H \mathcal{U}_i \mathcal{U}_i^H \mathbf{h} = 0, \tag{7.91b}$$

which holds good for all eigenvectors spanning the noise space. Hence,

$$\mathbf{h}^H \sum_{i=MN-N-L}^{MN} \mathcal{U}_i \mathcal{U}_i^H \mathbf{h} = 0. \tag{7.92a}$$

We have assumed from the beginning a noise-free situation, but in practice there will always be some noise, which we model as a white uncorrelated (sensor to sensor) stochastic process. Then, the right-hand side of Equation 7.92a will represent noise power, and we select $\mathbf{h}$ that minimizes the noise power,

$$\mathbf{h}^H \sum_{i=MN-N-L}^{MN} \mathcal{U}_i \mathcal{U}_i^H \mathbf{h} = \min, \tag{7.92b}$$

subject to the constraint that $|\mathbf{h}| = 1$. The solution is given by the eigenvector corresponding to the smallest eigenvalue of $\mathbf{Q}$ where

$$\mathbf{Q} = \sum_{i=MN-N-L}^{MN} \mathcal{U}_i \mathcal{U}_i^H. \tag{7.92c}$$

The impulse response vector is thus obtained, up to a scale factor, without any knowledge of the input signal, except that its covariance matrix is full rank. Having estimated the impulse response matrix, we go on to the final step of estimating the input signal. Refer to Equation 7.86, where $\mathcal{H}$ has now been estimated. Since $\mathcal{H}$ is a full column rank matrix, we can compute its inverse and thus estimate the input signal:

$$\mathbf{x}=[\mathcal{H}^H\mathcal{H}]^{-1}\mathcal{H}^H\mathbf{y}. \tag{7.93}$$

Moulines [21] first proposed the above method.

We give an example to demonstrate the application of the above algorithm. A transmitter broadcasts a narrowband stochastic signal (center frequency=1,500 Hz and bandwidth=8 Hz). The signal arrives at a four-antenna array (ULA) with eight multipaths, which are incident at random directions in the range $\pm\pi/12$ and arrive with random delays in the range 0–6 units of sampling interval, here 0.025 sec. The signal duration is 48 time samples. The output of each antenna (base band signal only) is corrupted with independent random Gaussian noise with zero mean and 0.25 variance. The system filter is an FIR filter with coefficients (arbitrarily chosen) as 1, 0.4, 0.3, 0.2, 0.0, −0.3, −0.1, 0.0. We assumed 64 independent transmissions for computing the ensemble averaged output covariance matrix. The channel is assumed to be stationary over the duration of the experiment. First, we computed the channel impulse response function using Equation 7.92b. The estimated channel response function is then used to estimate, using Equation 7.93, the input signal or transmitted signal. The results are shown in Figure 7.18. The top panel shows the transmitted signal and the bottom panel shows the estimated signal. The channel impulse response function is not shown.

7.4.8 Multiple Input Multiple Output Channel

In single input multiple output (SIMO) channel, we have exploited spatial diversity to achieve a full rank channel matrix and thereby recover the input bit stream from the received signal. The multipath structure of the channel was responsible for this happening. Now this feature is further exploited by having multipath phenomenon both at receiver as well as at transmitter, as proposed in Ref. [27], or in the entire space enclosing the receiver and transmitter, as proposed in Ref. [28] (see elliptical mode illustrated in Figure 7.15 of scatterers). First, we would like to show how the channel matrix can be made full rank without any multipath structure but with very long transmit and receive array.

We assume that the channel is memoryless, that is, only current output is a linear combination of current input. The receive array response to a signal transmitted by m_t th transmitter is given by

$$\mathbf{h}_{m_t}=e^{-j\varphi_{m_t}}[1\,e^{-2\pi j(d_r/\lambda)\sin\theta_{m_t}}\cdots e^{-2\pi j(d_r/\lambda)(M_r-1)\sin\theta_{m_t}}]^T, \tag{7.94a}$$

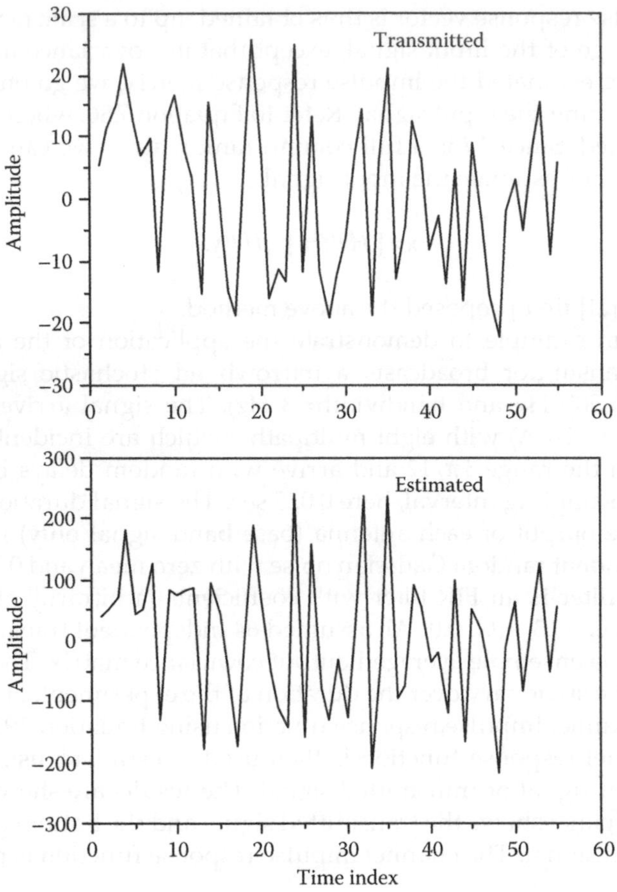

FIGURE 7.18
Blind channel and input signal estimation. Top panel shows transmitted signal and the bottom panel shows estimated signal.

where θ_{m_t} is a DOA from m_t th transmitter (see Figure 7.19) and $e^{-j\varphi_{m_t}}$ is a common phase term. The horizontal separation between receive and transmit arrays X is sufficiently large compared to the array aperture, that a plane wave approximation holds good. We can now write the channel matrix in terms of receive array responses (Equation 7.94a),

$$\mathbf{H} = \Gamma[\mathbf{h}_0\ \mathbf{h}_1 \cdots \mathbf{h}_{M_r-1}], \tag{7.94b}$$

where

$$\Gamma = \text{diag}\{e^{-j\varphi_0}\ e^{-j\varphi_1} \cdots e^{-j\varphi_{(M_t-1)}}\}.$$

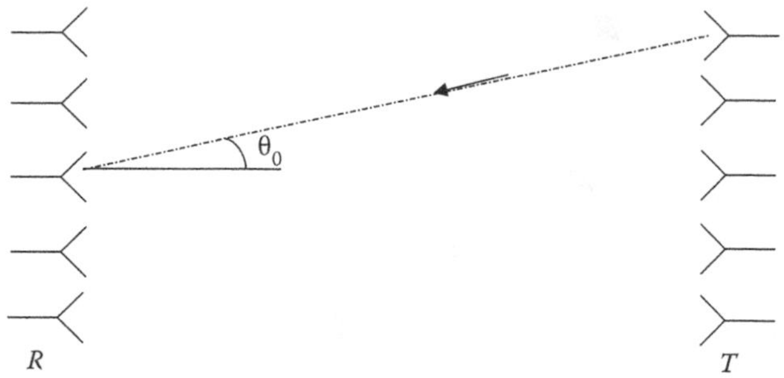

FIGURE 7.19
Transmit (*T*) and receive (*R*) antenna arrays separated by *X* units of distance. Each is a ULA with spacing d_t and d_r for transmit and receive array, respectively.

The channel matrix will be full rank when the columns of the matrix are uncorrelated. Let us compute the correlation between two adjacent columns from Equation 7.94b

$$\mathbf{h}_{m_t}^H \mathbf{h}_{m_t+1} = \sum_{m_r=0}^{M_r-1} e^{-2\pi j m_r (d_r/\lambda)(\sin\theta_{m_t+1} - \sin\theta_{m_t})}. \tag{7.95a}$$

We shall assume that $(M_t-1)\, d_t \ll X$, which will enable us to approximate $\sin\theta_{m_t} \approx \theta_{m_t} \approx (m_t d_t/X)$. Introducing this approximation into Equation 7.95a we obtain,

$$\frac{1}{M_r}\mathbf{h}_{m_t}^H \mathbf{h}_{m_t+1} \approx \frac{1}{M_r}\sum_{m_r=0}^{M_r-1} e^{-2\pi j m_r (d_r d_t/\lambda X)}$$

$$= \frac{\sin\left(2\pi \frac{d_r d_t}{\lambda X} M_r/2\right)}{M_r \sin\left(\pi \frac{d_r d_t}{\lambda X}\right)} e^{-\pi j (d_r d_t/\lambda X)(M_r-1)}. \tag{7.95b}$$

This function is shown in Figure 2.2. The correlation is small when

$$M_r \frac{d_r d_t}{\lambda X} \geq 1,$$

or

$$\frac{d_t}{X} \geq \frac{\lambda}{d_r M_r}. \tag{7.95c}$$

The right-hand side of Equation 7.95c is Rayleigh resolution (Chapter 2, Section 2.4) of a ULA and the left-hand side is the angular separation between two neighboring transmitters. Equation 7.95c implies that the

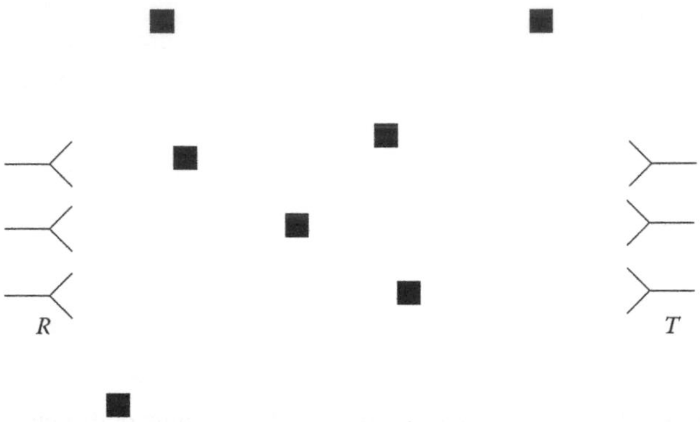

FIGURE 7.20
Scattering model of communication between transmit and receive arrays. There is no direct path.

Rayleigh resolution of the receive array must be less than the angular separation between two neighboring transmitters. Let $d_r = (\lambda/2)$ then Equation 7.95c implies $M_r \geq 2(X/dt)$. Evidently in scatter-free channel, the channel matrix will be full rank only when the receive array is very long compared to separation between the arrays. Note that it was necessary to assume that the size of a transmit array is much smaller than the separation so that a small angle approximation could be used.

In the presence of scatterers, full rank property of the channel matrix can be achieved even with a finite length of receive channel. We shall now consider a channel with a large number of scatterers located between the receive and transmit arrays, as illustrated in Figure 7.20.

It is assumed that delays of all scattering objects can be expressed in discrete time k (sampling interval). Multiple scattered paths, being too weak, are ignored. There are L discrete singly scattered paths. The signal received at m_rth antenna may be expressed as

$$y_{m_r}(k) = \sum_{l=0}^{L-1} \sum_{k=1}^{N_l} \alpha_{k,l} e^{j\varphi_{k,l,m_r}} \sum_{m_t=0}^{M_t-1} e^{j\vartheta_{k,l,m_r}} x_{m_t}(k-l). \tag{7.96}$$

There are N_l paths each having a delay of l time units and the maximum delay is L units; $\alpha_{k,l}$ is a complex scattering coefficient at kth scatterer with$l$ units of delay; φ_{k,l,m_r} is a carrier phase at m_r th receiver and ϑ_{k,l,m_t} at m_t th transmit antenna. We shall rewrite Equation 7.96, bringing out the role of impulse response function

$$\begin{aligned} y_{m_r}(k) &= \sum_{m_t=0}^{M_t-1} \sum_{l=0}^{L-1} \left[\sum_{k=1}^{N_l} \alpha_{k,l} e^{j(\varphi_{k,l,m_r} + \vartheta_{k,l,m_t})} \right] x_{m_t}(k-l) \\ &= \sum_{m_t=0}^{M_t-1} \sum_{l=0}^{L-1} h_{m_r,m_t}(l) x_{m_t}(k-l), \end{aligned} \tag{7.97a}$$

where

$$h_{m_r,m_t}(l)=\sum_{k=1}^{N_l}\alpha_{k,l}e^{j(\varphi_{k,l,m_r}+\vartheta_{k,l,m_t})}, \tag{7.97b}$$

is the impulse response function connecting m_t transmitter with m_r receiver. Let us introduce input data vector $\mathbf{x}(\cdot)=[x_0(\cdot)x_1(\cdot)\cdots x_{M_t-1}(\cdot)]^T$, output vector $\mathbf{y}(\cdot)=[y_0(\cdot)y_1(\cdot)\cdots y_{M_r-1}(\cdot)]^T$, and channel matrix $\mathbf{H}_l$ whose (m_r, m_t) th element is given by Equation 7.97b. Using these quantities we can express Equation 7.97a in a matrix form

$$\mathbf{y}(k)=\sum_{l=0}^{L-1}\mathbf{H}_l\mathbf{x}(k-l). \tag{7.98}$$

The above vector convolution sum can be expressed as a matrix equation in the same manner as we have done for scalar convolution sum (see Equation 7.84).

$$\begin{bmatrix}\mathbf{y}(n)\\ \mathbf{y}(n-1)\\ \mathbf{y}(n-2)\\ \vdots\end{bmatrix}=\begin{bmatrix}\mathbf{H}_0\ \mathbf{H}_1\cdots\mathbf{H}_{L-1} & & \\ & \mathbf{H}_0\ \mathbf{H}_1\cdots\mathbf{H}_{L-1} & \\ & \ddots \qquad \ddots & \\ & & \mathbf{H}_0\ \mathbf{H}_1\cdots\mathbf{H}_{L-1}\end{bmatrix}\begin{bmatrix}\mathbf{x}(n)\\ \mathbf{x}(n-1)\\ \mathbf{x}(n-2)\\ \vdots\end{bmatrix}, \tag{7.99a}$$

or

$$\bar{\mathbf{y}}=\mathcal{H}\bar{\mathbf{x}}. \tag{7.99b}$$

The eigenvalues of $\mathcal{H}^H\mathcal{H}$ are random numbers whose distribution has been analytically evaluated in Ref. [29]. The analysis shows $\mathcal{H}^H\mathcal{H}$ will be full rank for long antenna arrays with $M_t/M_r>1$ and rich scattering, that is, $N_l/M_r>1$.

7.5 Exercises

(1) Rewrite Equation 7.1 for two separate sources that are transmitting two different signals. Assume that multipaths follow different paths. Adopt the algorithm described in Section 7.1.1 for the two-source case. Suggest a method for matching a delay with the corresponding source.

(2) Consider the noise-free situation in Equation 7.31. Show that **Z** matrix is of rank d, the number of multipaths. In order to estimate d multipaths, it is necessary that the number of frequency samples must be greater than the number of multipaths, that is, $L>d$.

(3) Show that if second order approximation is used in Equation 7.71, an additional term will appear in Equation 7.73, namely, $(L/4)\sigma_\alpha^2 \nu_\theta^4 \mathbf{a}''(\theta_0)\mathbf{a}''^H(\theta_0)$, where

$$\nu_\theta^4 = E\{\delta\theta^4\}.$$

The pdf of $\delta\theta$ must be symmetric.

(4) Consider a multichannel FIR filter with a common input (see Figure 7.21). Show that any two outputs are related as

$$h^p * y^q = h^q * y^p,\ p, q = 1,\ldots, d.$$

This is known as cross-relation used in blind identification of FIR channels [22–24].

(5) In Exercise 7.4, the cross-relation can be expressed in a matrix form as in Equation 7.82. It is easy to show that the matrix relation is given by

$$\mathbf{Y}^q\mathbf{h}^p - \mathbf{Y}^p\mathbf{h}^q = 0, \quad p,q = 1,\ldots, d.$$

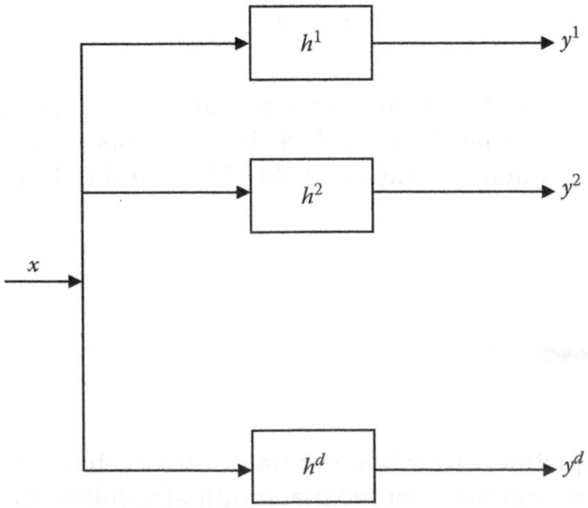

FIGURE 7.21
Multichannel FIR filter with a common input.

For $d=2$ we have just one equation

$$\mathbf{Y}^2\mathbf{h}^1 - \mathbf{Y}^1\mathbf{h}^2 = 0,$$

$$\left[\mathbf{Y}^2 \; -\mathbf{Y}^1\right]\begin{bmatrix}\mathbf{h}^1\\ \mathbf{h}^2\end{bmatrix} = 0.$$

When there are three sensors ($d=3$), we get two additional equations,

$$\mathbf{Y}^3\mathbf{h}^1 - \mathbf{Y}^1\mathbf{h}^3 = 0,$$

$$\mathbf{Y}^3\mathbf{h}^2 - \mathbf{Y}^2\mathbf{h}^3 = 0,$$

$$\begin{bmatrix}\mathbf{Y}^3 & 0 & -\mathbf{Y}\\ 0 & \mathbf{Y}^3 & -\mathbf{Y}^2\end{bmatrix}\begin{bmatrix}\mathbf{h}^1\\ \mathbf{h}^2\\ \mathbf{h}^3\end{bmatrix} = 0.$$

Continuing this process, show that additional equations available when Mth sensor is introduced to an existing M–1 sensor array are given in matrix form

$$\begin{bmatrix}\mathbf{Y}^M & & & -\mathbf{Y}^1\\ & \mathbf{Y}^M & & -\mathbf{Y}^2\\ & & \ddots & \vdots\\ & & \mathbf{Y}^M & -\mathbf{Y}^{M-1}\end{bmatrix}\begin{bmatrix}\mathbf{h}^1\\ \mathbf{h}^2\\ \vdots\\ \mathbf{h}^M\end{bmatrix} = 0.$$

By combining all these equations into one large composite matrix equation, it has been possible to estimate the channel impulse response functions (see Ref. [22] for details).

References

1. J. Vidal, M. Najar, and R. Jativa, High resolution time-of-arrival detection for wireless positioning systems, *IEEE Conf. Proc.*, pp. 2283–2287, 2002.
2. R. J. Vaccaro, C. S. Ramalingam, and D. W. Tufts, Least-squares time-delay estimation for transient signals in a multipath environment, *J. Acoust. Soc. Am.*, vol. 92, pp. 210–218, 1992.
3. P. S. Naidu and H. U. Shankar, Broadband source localization in shallow water, *Signal Processing*, vol. 72, pp. 107–116, 1999.

4. Y. T. Chan, J. M. Riley, and J. B. Plant, A parameter estimation approach to time-delay estimation and signal detection, *Trans. IEEE*, ASSP-28, pp. 8–16, 1980.
5. J.-J. Fuchs, Multipath time delay detection and estimation, *Trans. Signal Processing*, vol. 47, pp. 237–243, 1999.
6. Y. Bresler and A. H. Delaney, Resolution of overlapping echoes of unknown shape, *Conf. ICASSP*, 2657–2660, 1989.
7. A. L. Swindlehurst and J. G. Gunther, Methods for blind equalization and resolution of overlapping echoes of unknown shape, *Trans. Signal Processing*, vol. 47, pp. 1245–1254, 1999.
8. J. R. Daugherty and J. F. Lynch, Surface wave, internal wave, and source motion effects on matched field processing in a shallow water waveguide, *J. Acoust. Soc. Am.*, vol. 87, pp. 2503–2526, 1990.
9. I. Guttman, V. Peereyra, and H. D. Scholnik, Least squares estimation for a class of nonlinear models, *Technometrics*, vol. 15, pp. 209–218, 1973.
10. D. W. Tufts and R. Kumaresan, Improved spectral resolution II, *Proc. ICASSP*, pp. 392–397, 1980.
11. P. S. Naidu and T. Ganesan, Source localization in a partially known shallow water, *J. Acoust. Soc. Am.*, vol. 98, Pt. 1, pp. 2554–2559, 1995.
12. J. F. Boehme, Array processing. In *Advances in Spectrum Analysis and Array Processing*, S. Haykin (Ed.), pp. 1–63, Prentice-Hall, Englewood Cliffs, NJ, 1991.
13. C. F. Mecklenbraeuker, A. B. Gershman, and J. F. Boehme, ML estimation of environmental parameters in shallow ocean using unknown broadband sources, *Proc. IEEE*, ICNNSP-95, 1995.
14. A. Paulraj and C. B. Papadias, Space-time processing for wireless communications, *IEEE Signal Process. Mag.*, Vol. 11, pp. 49–83, 1997.
15. J. C. Liberti and T. S. Rappaport, A geometrically based model for line-of-sight multipath radio channels, *IEEE Conf. Proc.*, pp. 844–848, 1996.
16. R. B. Ertel and J. H. Reed, Angle and time of arrival statistics for circular and elliptical scattering models, *IEEE J. Select. Areas Comm.*, vol. 17, pp. 1829–1840, 1999.
17. D. Asztely, B. Ottersten, and A. L. Swindlehurst, Generalized array manifold model for wireless communication channels with local scattering, *IEE Proc. Radar Sonar, Navigat.*, vol. 145, pp. 51–57, 1998.
18. D. Slepian, H. O. Pollack, and H. J. Landau, Prolate spheroidal wave functions, Fourier analysis and uncertainty, *Bell Sys. Tech. J.*, vol. 40, pp. 43–84, 1961.
19. J. K. Tugnait, L. Tong, and Z. Ding, Single-user channel estimation and equalization, *IEEE Signal Process. Mag.*, Vol. 5, pp. 17–28, 2000.
20. L. Tong and S. Perreau, Multichannel blind identification: From subspace to maximum likelihood methods, *Proc. IEEE*, vol. 86, pp. 1951–1968, 1998.
21. E. Moulines, P. Duhamrl, J.-F. Cardoso, and S. Mayrargue, Subspace methods for the blind identification of multichannel FIR filters, *IEEE Trans. Signal Processing*, vol. 43, pp. 516–525, 1995.
22. Y. Hua, K. Abed-Merium, and M. Wax, Blind system identification using minimum noise subspace, *IEEE Trans. Signal Process.*, vol. 45, pp. 770–772, 1997.
23. W. Liu, S. Weiss and L. Hanzo, Subband adaptive generalized sidelobe canceller for broadband beamforming, *IEEE Conf. Proc.*, pp. 591–594, 2001.
24. G. Xu, H. Lui, L. Tong, and T. Kailath, A least-squares approach to blind channel identification, *IEEE Trans. Signal Process.*, vol. 43, pp. 2982–2993, 1995.
25. G. Ungerbroeck, Fractional tap-spacing equalizer and consequences for clock recovery in data modems, *IEEE Trans. Comm.*, COM-24, pp. 856–864, 1976.

26. A.-J. van der Veen, S. Talear, and A. Paulraj, A subspace approach to blind space-time signal processing for wireless communication systems, *IEEE Trans. Signal Process.*, vol. 45, pp. 173–190, 1997.
27. D. Gesbert, H. Boleski, D. A. Gore, and A. Paulraj, Outdoor MIMO wireless channels: Models and performance prediction, *IEEE Trans. Comm.*, vol. 50, pp. 1926–1934, 2002.
28. P. F. Driessen and G. J. Foschini, On the capacity formula for multiple input-multiple output wireless channels: A geometric interpretation, *IEEE Trans. Comm.*, vol. 47, pp. 173–176, 1999.
29. R. R. Muller, A random matrix model of communication via antenna arrays, *IEEE Trans. Inform. Theory*, vol. 48, pp. 2495–2506, 2002.

8

Wireless Communications

In modern communication systems, coded signals are used for transmitting information, which consists of a bit stream of ones and zeros. The bits are suitably coded into identifiable waveforms; for example "1" may be coded into a sinusoid of frequency f_1 and "0" is coded into another sinusoid of frequency f_0, as in frequency shift keying (FSK) modulation. The bits thus coded after mixing with a carrier are sequentially transmitted. Since the physical channel is a shielded cable (including optical cable), there is much less cross-channel interference. But, in a radio communication scenario, both transmitter and receiver are in the open space. Naturally, a sensor will receive signals from more than one source. It is therefore of great interest to minimize this co-channel interference, a problem unique to radio communication. To overcome the problem of co-channel interference, modern cellular radio communication has been devised. A user needs to communicate to the nearest base station, which, in turn, is connected to a central exchange. Thus, it is possible to communicate with a distant user without having to radiate a lot of power causing a drain on the battery (in case of a mobile transmitter) and creating interference to other users. Even this system seems to fail when many users in the same cell are trying to reach the base station. To further mitigate the problem of co-channel interference, it is proposed to use a highly directional antenna at the base station. This would enable the base station to separate the users having different bearings and reduce the co-channel interference among them.

In an urban environment there is the additional problem of multipath propagation due to scattering. Since the multipaths are likely to arrive from different directions, the use of a directional array, it is hoped, will help to alleviate the problem of fading, loss of bits, etc. We consider two types of environments. Local scattering caused by scatterers around a mobile station (MS) and scattering caused by distant large obstacles, like mountains, tall buildings, etc. In the first case the delay spread is over a fraction of a symbol duration, but in the second case it is over several symbol periods. Naturally, the two different situations call for different processing strategies.

8.1 Beamformation

The essential step in beamformation is estimation of the direction vector of each source. Given the direction vector, a beam may be formed in that direction using one of the beamforming methods described in Sections 6.1 and 6.2. When there are a large number of sources (users), ordinarily we need to have a large sensor array (more sensors than the number of users) for the estimation of direction vectors. The advantage of using coded signals is that this limitation no longer exists. We can have more number of users than the number of sensors. We shall illustrate this possibility for code division multiple access (CDMA) system of mobile communication.

8.1.1 Co-Channel Interference

In Section 5.3, we have derived a postcorrelation covariance matrix (see Equation 5.68) and expressed the same in a form (Equation 5.69) suitable for application of the subspace algorithm. The direction vectors to all users may be estimated provided the array has more sensors than the number of users. When this is not satisfied, the presence of users in excess of the number of sensors will only introduce interference, known as co-channel interference. We shall show how, by employing extra information that is available but not used, we can overcome the problem of co-channel interference.

At the base station we will assume an array of sensors (EM dipoles). The preferred array shape is a uniform circular array (UCA) with uniform response in all directions as the users are likely to be all around the base station. A uniform linear array (ULA) may also be used, particularly when most users are on the broadside where the array has the best possible response. The postcorrelation covariance matrix (Equation 5.68) is reproduced here for convenience.

$$\mathbf{C}_{g_0 g_0} = p_0^2 \mathbf{a}(\theta_0)\mathbf{a}(\theta_0)^H + \frac{2}{3L}\sum_{k=1}^{Q} p_k^2 \mathbf{a}(\theta_k)\mathbf{a}(\theta_k)^H + \frac{\sigma_\eta^2}{L}\mathbf{I}.$$

The first term on the right-hand side is signal of interest, as we would like to estimate the direction vector of the user of interest. The second term represents the co-channel interference from all other sources. Notice that this term will be small for large L (code length).

8.1.2 Estimation of all Direction Vectors

We like to estimate the direction vectors of all users in the same cell. For this we shall compute the postcorrelation covariance matrices for all users. This would require a knowledge of the codes used by all users present within the cell. We will now have Q equations of the type given by Equation 5.68

$$\mathbf{C}_{g_k g_k} = p_k^2 \mathbf{a}(\theta_k)\mathbf{a}(\theta_k)^H + \frac{2}{3L}\sum_{\substack{i=0 \\ k\neq i}}^{Q} p_i^2 \mathbf{a}(\theta_i)\mathbf{a}(\theta_i)^H + \frac{\sigma_\eta^2}{L}\mathbf{I} \tag{8.1}$$

$$k = 0, 1, \ldots, Q-1.$$

In Equation 8.1, we have Q matrix equations and Q matrix unknowns, $k=0,1,\ldots,Q-1$. We shall express Equation 8.1 in a matrix form.

$$\underset{Q\times 1}{\begin{bmatrix} \mathbf{C}_{g_0g_0} \\ \mathbf{C}_{g_1g_1} \\ \vdots \\ \mathbf{C}_{g_{Q-1}g_{Q-1}0} \end{bmatrix}} = \underset{Q\times Q}{\begin{bmatrix} 1 & \frac{2}{3}L & \cdots & \frac{2}{3}L \\ \frac{2}{3}L & 1 & \cdots & \frac{2}{3}L \\ \vdots & \vdots & & \vdots \\ \frac{2}{3}L & \frac{2}{3}L & \cdots & 1 \end{bmatrix}} \underset{Q\times 1}{\begin{bmatrix} p_0^2\mathbf{a}(\theta_0)\mathbf{a}(\theta_0)^H \\ p_1^2\mathbf{a}(\theta_0)\mathbf{a}(\theta_0)^H \\ \vdots \\ p_{Q-1}^2\mathbf{a}(\theta_0)\mathbf{a}(\theta_0)^H \end{bmatrix}} + \frac{\sigma_\eta^2}{T_s}\underset{Q\times 1}{\begin{bmatrix} \mathbf{I} \\ \mathbf{I} \\ \vdots \\ \mathbf{I} \end{bmatrix}}. \tag{8.2}$$

Note that each element of columns and matrices in Equation 8.2 is an $M\times M$ matrix. In a compact form

$$\mathbf{C} = \pounds\Theta + \frac{\sigma_\eta^2}{L}\mathbf{II}, \tag{8.3}$$

where $\mathbf{C}$, Θ, $\mathbf{II}$ are all matrices of size $MQ\times M$, but $\pounds$ is a full rank square matrix of size $MQ\times MQ$. Multiplying by $\pounds^{-1}$ on both sides of Equation 8.3, we can express it as

$$\Theta = \pounds^{-1}\mathbf{C} - \frac{\sigma_\eta^2}{L}\pounds^{-1}\mathbf{II}. \tag{8.4}$$

The error term in Equation 8.4 may be expressed as a product of a diagonal matrix and a column unit matrix. The elements of the diagonal matrix are equal to row sums of $\pounds^{-1}$. Thus, the noise covariance matrix in the estimated direction matrix remains diagonal. The variance of the noise may be estimated from the eigenvalues of the direction matrix. The power, transmitted by each user, that is, p_k^2, can also be estimated from the largest eigenvalue of the direction matrix.

8.1.3 Simulation Results

The estimated direction vector of a user is compared with the known direction vector. A dot product between the two vectors is computed as a measure of similarity,

$$\varepsilon_l = \frac{\hat{\mathbf{a}}^H(\theta_l)\mathbf{a}(\theta_l)}{|\hat{\mathbf{a}}^H(\theta_l)||\mathbf{a}(\theta_l)|}.$$

Note that $0 \le |\varepsilon_l| \le 1$, the lower limit represents the worst estimate and the upper limit represents the best estimate. We have computed the mean and the variance of ε_l as a measure of quality of estimate. The results are shown in Table 8.1. The postcorrelation matrix approach for the estimation of direction vectors as described here is not limited by the requirement that the number of sensors must be greater than the number of users, as in the approach described in Ref. [1] using both pre- and postcorrelation matrices. In fact, to verify this claim, the above simulation was repeated with no noise for a different number of users. The results are shown in Table 8.2. There is, however, a slight decrease in the quality of estimate.

8.1.4 Beamforming with Cyclostationary Signals

We consider P sources emitting cyclostationary signals with different but known cyclic frequencies. We like to find a set of weight coefficients that forms a beam in the direction of a single source having a specified cyclic frequency. The array output is governed by the signal model given in Equation 2.17e. The noise is stationary but not necessarily white. We use the frequency shifted version of $\mathbf{f}(t)$, which we defined in Equation 5.70. Let $\mathbf{w}_+$ and $\mathbf{w}_-$ be the beamforming weight coefficient vectors for $\mathbf{f}_+(t)$ and $\mathbf{f}_-(t)$, respectively. The cross-correlation (for lag τ) of the outputs is given by

TABLE 8.1

Error in the estimation of direction vector for different SNRs

SNR (dB)	Mean	Variance
−20	0.6029	0.1059
−10	0.9746	3.8480e−04
−5	0.9945	8.8470e−06
0	0.9979	8.6691e−07
No noise	0.9983	1.0934e−06

Note: Ten sensor ULA, randomly distributed 10 users, 100 snapshots, code length=63 chips.

TABLE 8.2

Error in the estimation of direction vector with increasing number of users

No. of Users	Mean	Variance
10	0.9983	1.0934e−06
20	0.9949	6.7374e−06
40	0.9930	1.5846e−05
50	0.9909	2.3899e−05
60	0.9878	3.6955e−05

Note: Ten sensor ULA, randomly distributed users, 100 snapshots, no noise, and code length=63 chips.

$$\rho = \frac{1}{T}\sum_{t=-T/2}^{T/2} E\left\{\mathbf{w}_{-}^{H}\mathbf{f}_{-}\left(t+\frac{\tau}{2}\right)\mathbf{f}_{+}^{H}\left(t-\frac{\tau}{2}\right)\mathbf{w}_{+}\right\}$$
$$T \to \infty \tag{8.5}$$
$$= \mathbf{w}_{-}^{H}\mathbf{c}_{f}^{\alpha}(\tau)\mathbf{w}_{+},$$

where $\mathbf{c}_f^\alpha$ is the cyclic covariance matrix defined in Equation 5.71a for lag τ. The filter coefficients are chosen to maximize the cross-correlation (Equation 8.5) or its magnitude square. Further, we require that $\mathbf{w}_+$ and $\mathbf{w}_-$ are unit norm vectors.

$$\left|\mathbf{w}_{-}^{H}\mathbf{c}_{f}^{\alpha}\mathbf{w}_{+}\right|^{2} = \max \qquad \mathbf{w}_{+}^{H}\mathbf{w}_{+} = \mathbf{w}_{-}^{H}\mathbf{w}_{-} = 1. \tag{8.6a}$$

The solution of Equation 8.6a is given by left and right singular vectors corresponding to the largest singular value of $\mathbf{c}_f^\alpha$. Also $\mathbf{w}_+$ and $\mathbf{w}_-$ are, respectively, the eigenvectors corresponding to the largest eigenvalues of $\mathbf{c}_f^\alpha\mathbf{c}_f^{\alpha H}$ and $\mathbf{c}_f^{\alpha H}\cdot\mathbf{c}_f^\alpha$ [2]. The cyclic covariance matrix appearing in Equation 8.6a is, in practice, replaced by a cyclic autocorrelation function defined in terms of time average

$$\hat{\mathbf{c}}_{f}^{\alpha} = \frac{1}{T}\sum_{-T/2}^{T/w}\mathbf{f}_{-}(t)\mathbf{f}_{+}^{H}(t). \tag{8.6b}$$

When the carrier frequencies of different sources are sufficiently apart and the signal duration T is large, the cyclic autocorrelation matrix given by Equation 8.6b approaches the cyclic covariance matrix of a single source, which is a rank one matrix, as shown in Equation 5.71b whose left and right singular vectors corresponding to the largest singular value are equal to $\mathbf{a}_0/\sqrt{M}$ (for ULA). Thus, we have an interesting result

$$\mathbf{w}_{+} = \mathbf{w}_{-} = \frac{\mathbf{a}_{0}}{\sqrt{M}}. \tag{8.6c}$$

8.1.5 Cyclic adaptive Beamforming

We shall consider a simple signal model given by Equation 2.17c, which has been considered in the previous subsection. Let $\mathbf{w}$ be a vector of weight coefficients for beamforming [3]. The output is given by

$$\hat{f}(t) = \mathbf{w}^{H}\mathbf{f}(t).$$

We choose the weight vector to minimize the mean square difference between the output, $\hat{f}(t)$, and the incident signal, $f_c(t)$ where the subscript c stands for complex,

$$\frac{1}{T}\sum_{t=-T/2}^{T/2}\left|\mathbf{w}^H\mathbf{f}(t)-f_c(t)\right|^2=\min. \quad T\to\infty \tag{8.7}$$

Minimization of Equation 8.7 leads to the following equation

$$\mathbf{w}=\mathbf{C}_f^{-1}\mathbf{c}_{ff_c}=\mathbf{C}_f^{-1}\mathbf{a}(\theta_0)\sigma_{f_c}^2, \tag{8.8}$$

where

$$\mathbf{C}_f=\frac{1}{T}\sum_{t=-T/2}^{T/2}\mathbf{f}(t)\mathbf{f}^H(t),$$

is the covariance matrix of the array output. The important point to note in Equation 8.8 is that to obtain the weight vector it is necessary to know the correct array response vector. In practice this is not possible without precise array calibration.

This difficulty is overcome by exploiting the property of a cyclostationary signal [5], namely,

$$\mathbf{f}(t-\tau)e^{-2j\pi\alpha t}=\rho_{f_cf_c^*}^{\alpha}(\tau)\mathbf{a}(\theta_0)f_c(t)+\eta_0(t), \tag{8.9a}$$

where $\rho_{f_c}^{\alpha}(\tau)$ is a cross-cyclic correlation coefficient defined as

$$\rho_{f_c}^{\alpha}(\tau)=\frac{\frac{1}{T}\sum_{t=-T/2}^{T/2}f_c(t+\tau/2)f_c^*(t-\tau/2)e^{-j2\pi\alpha t}}{\sqrt{\frac{1}{T}\sum_{t=-T/2}^{T/2}\left|f_c(t)\right|^2\frac{1}{T}\sum_{t=-T/2}^{t=+T/2}\left|f_c(t-\tau/2)e^{-j2\pi\alpha t}\right|^2}}. \quad T\to\infty$$

In Equation 8.9a, $\eta_0(t)$ consists of two parts, namely, a signal orthogonal to $f_c(t)$ and the background noise. It is expressed as

$$\eta_0(t)=\sqrt{1-\left|\rho_{f_c}^{\alpha}(\tau)\right|^2}\,\mathbf{a}(\theta_0)f_c^{\perp}(t)+\eta(t-\tau)e^{-j2\pi\alpha t}, \tag{8.9b}$$

where $f_c^{\perp}(t)$ is orthogonal to $f_c(t)$ (i.e., $(1/T)\sum_{t=-T/2}^{T/2}f_c(t)f_c^{\perp}(t)=0$). We seek a weighting coefficient vector that minimizes the mean square difference,

$$\left\{\frac{1}{T}\sum_{t=-T/2}^{T/2}\left|(\mathbf{w}^H\mathbf{f}(t)-r(t)\right|^2\right\}_{T\to\infty}=\min, \tag{8.10}$$

where $r(t)$ is a new reference function given by

$$r(t)=\mathbf{c}^H\mathbf{f}(t-\tau)e^{-j2\pi\alpha t},$$

and $\mathbf{c}$ is an arbitrary vector not orthogonal to $\mathbf{a}(\theta_0)$. It is often known as a control vector. Because of Equation 8.9a, $r(t)$ may be considered as a scaled and corrupted version of the unknown signal. Hence, it is fit to be a reference signal. Minimization of Equation 8.10 with respect to $\mathbf{w}$ (keeping τ and α constant) leads to

$$\mathbf{w}=\mathbf{C}_f^{-1}\mathbf{c}_{fr}, \tag{8.11a}$$

where

$$\mathbf{c}_{fr}=\left\{\frac{1}{T}\sum_{t=-T/2}^{T/2}\mathbf{f}(t)\mathbf{c}^H\mathbf{f}(t-\tau)e^{-j2\pi\alpha t}\right\}_{T\to\infty}. \tag{8.11b}$$

Using Equation 8.9a in Equation 8.11b, we obtain

$$\begin{aligned}\mathbf{c}_{fr}&=\rho^{\alpha}_{f_cf_c^*}(\tau)\mathbf{c}^H\mathbf{a}(\theta_0)\frac{1}{T}\sum_{t=-T/2}^{T/2}\mathbf{f}(t)f_c(t)\\&=[\rho^{\alpha}_{f_cf_c^*}(\tau)\mathbf{c}^H\mathbf{a}(\theta_0)]\mathbf{a}(\theta_0)\sigma^2_{f_c},\end{aligned} \tag{8.11c}$$

where we have used the fact that $f(t)$ is uncorrelated with $\eta_0(t)$. Substituting Equation 8.11c in Equation 8.11a, we obtain

$$\mathbf{w}=[\rho^{\alpha}_{f_cf_c^*}(\tau)\mathbf{c}^H\mathbf{a}(\theta_0)]\mathbf{C}_f^{-1}\mathbf{a}(\theta_0)\sigma^2_{f_c}. \tag{8.12}$$

This is the least squares (LS) self-coherence restoral (SCORE) algorithm proposed by Agee [4, 5]. We have derived the algorithm for a single source. For multiple sources, if all signals have distinct cyclic frequency, each signal may be treated as a single signal.

Further improvement in the LS-SCORE may be achieved by selecting $\mathbf{w}$ and $\mathbf{c}$ to maximize the correlation between $\hat{f}(t)$ and the reference signal, $r(t)$.

$$\left|\frac{1}{T}\sum_{t=-T/2}^{T/2}\hat{f}(t)r(t)\right|^2 = \left|\mathbf{w}^H\frac{1}{T}\sum_{t=-T/2}^{T/2}\mathbf{f}(t)\mathbf{f}^H(t-\tau)e^{j2\pi\alpha t}\mathbf{c}\right|^2$$
$$= \left|\mathbf{w}^H\mathbf{C}_f^{\alpha}(\tau)\mathbf{c}\right|^2 = \max. \tag{8.13a}$$

The weight vectors **w** and **c** are constrained to unit norm

$$\mathbf{w}^H\mathbf{w} = \mathbf{c}^H\,\mathbf{c} = 1. \tag{8.13b}$$

The solution of Equation 8.13a subject to constraint Equation 8.13b is obtained utilizing the Lagrange multiplier method [3],

$$\mathbf{C}_f^{\alpha}(\tau)\mathbf{c}\mathbf{c}^H\mathbf{C}_f^{\alpha}(\tau)^H\,\mathbf{w} = \lambda_1\mathbf{w}, \tag{8.14a}$$

$$\mathbf{C}_f^{\alpha}(\tau)^H\,\mathbf{w}\mathbf{w}^H\mathbf{C}_f^{\alpha}(\tau)\mathbf{c} = \lambda_2\mathbf{c}. \tag{8.14b}$$

Note that in Equation 8.14a, $\mathbf{C}_f^{\alpha}(\tau)\mathbf{c}\mathbf{c}^H\mathbf{C}_f^{\alpha}(\tau)^H$ is rank one matrix, hence its eigenvector corresponding to the only nonzero eigenvalue is proportional to $\mathbf{C}_f^{\alpha}(\tau)\mathbf{c}$. Therefore,

$$\mathbf{w} = \frac{1}{\sqrt{\lambda_1}}\mathbf{C}_f^{\alpha}(\tau)\mathbf{c}, \tag{8.15a}$$

and similarly from Equation 8.14b we get

$$\mathbf{c} = \frac{1}{\sqrt{\lambda_2}}\mathbf{C}_f^{\alpha}(\tau)^H\,\mathbf{w}. \tag{8.15b}$$

Combining Equation 8.15a and b, we obtain the following equations

$$\sqrt{\lambda_1\lambda_2}\,\mathbf{w} = \mathbf{C}_f^{\alpha}(\tau)\mathbf{C}_f^{\alpha}(\tau)^H\,\mathbf{w}, \tag{8.16a}$$

and

$$\sqrt{\lambda_1\lambda_2}\,\mathbf{c} = \mathbf{C}_f^{\alpha}(\tau)^H\mathbf{C}_f^{\alpha}(\tau)\mathbf{c}. \tag{8.16b}$$

Thus, the weight vectors **w** and **c** are left and right singular vectors corresponding the largest singular value of $\mathbf{C}_f^{\alpha}(\tau)$, which will naturally satisfy Equation 8.13a. Further, since $\mathbf{C}_f^{\alpha}(\tau)^H\mathbf{C}_f^{\alpha}(\tau)$ is Hermitian positive semi-definite (being covariance matrices), the eigenvalues are real positive. The maximum correlation between $\hat{f}(t)$ and $r(t)$ is equal to $\sqrt{\lambda_1\lambda_2}$. Equations 8.16a and 8.16b form the basis of the cyclic adaptive beamforming (CAB) algorithm described in Ref. [3]. In Ref. [4] a different approach is taken, where normalized correlation between $\hat{f}(t)$ and the reference signal $r(t)$ is maximized with respect to **w** and **c**. In this approach (known as cross-SCORE), a more complex result is derived, where one needs to compute generalized eigenvalue decomposition.

For the simple signal model given by Equation 2.17e, we can easily compute,

$$\begin{aligned}\mathbf{C}_f^{\alpha}(\tau) &= \frac{1}{T}\sum_{t=-T/2}^{T/2} \mathbf{f}(t)\mathbf{f}^H(t-\tau)e^{j2\pi\alpha t} \\ &= \mathbf{a}(\theta_0)\frac{1}{T}\sum_{t=-T/2}^{T/2} f_c(t)f_c^H(t-\tau)e^{j2\pi\alpha t}\mathbf{a}^H(\theta_0) \\ T &\to \infty \\ &= \mathbf{a}(\theta_0)c_{f_c}^{\alpha}(\tau)\mathbf{a}^H(\theta_0).\end{aligned} \quad (8.17)$$

From Equation 8.17, it is clear that $\mathbf{C}_f^{\alpha}(\tau)$ is a rank one matrix. This property holds good even in the presence of other signals, but with a different cyclic frequency and uncorrelated noise. Using Equation 8.17 in Equation 8.16a we obtain

$$\mathbf{C}_f^{\alpha}(\tau)\mathbf{C}_f^{\alpha}(\tau)^H = \mathbf{a}(\theta_0)\left|c_{f_c}^{\alpha}(\tau)\right|^2 \left|\mathbf{a}(\theta_0)\right|^2 \mathbf{a}^H(\theta_0),$$

also a rank one matrix. Hence, its eigenvector corresponding to the largest eigenvalue is simply proportional to $\mathbf{a}(\theta_0)$,

$$\mathbf{w} \propto \mathbf{a}(\theta_0). \quad (8.18)$$

We derived a similar result in Equation 8.6b, where we simply maximized the filter power output under unit norm constraint on the filter.

8.2 Multipath Communication Channel

A wireless communication channel may be classified into two groups according to the delay spread: (1) Small delay spread less than one symbol duration typically caused by local scattering. Such a channel is memory-less, where only the current input gives rise to current output. (2) Large delay spread over several symbol duration typically caused by reflections from obstacles and often accompanied by local scattering. Current output is a linear combination of the current and past inputs. A finite impulse response (FIR) filter model is appropriate to describe such a channel.

8.2.1 Small Delay Spread Channel

A transmitter is surrounded by many local scatterers, which results in a collection of rays, a ray tube (Figure 8.1), being incident on the array from the same transmitter. The rays within a tube travel closely parallel to the axis of the ray tube, that is, the line of sight (LOS). The local scatterers are the metallic point scatters, such as buildings, other vehicles in the neighborhood. A ray tube will be represented by a composite direction vector defined as a linear combination of direction vectors of all subrays, including their relative delays. Here a method is proposed to estimate the composite direction vector without requiring to know the number, direction, and delays of the subrays constituting a ray tube.

Within a ray tube there are many rays caused by reflections at the scattering objects, such as buildings around a source (Figure 8.2). We shall assume R discrete scatterers; therefore R multipaths all lie within a narrow angular range. The signal transmitted by a source propagates via different multipaths and reaches the receiver array with random delays with respect to the direct arrival. The relative delays are assumed to be uniformly distributed over several chip periods, but well within one symbol duration. Then, we say that there is no intersymbol interference (ISI). The scattering coefficient

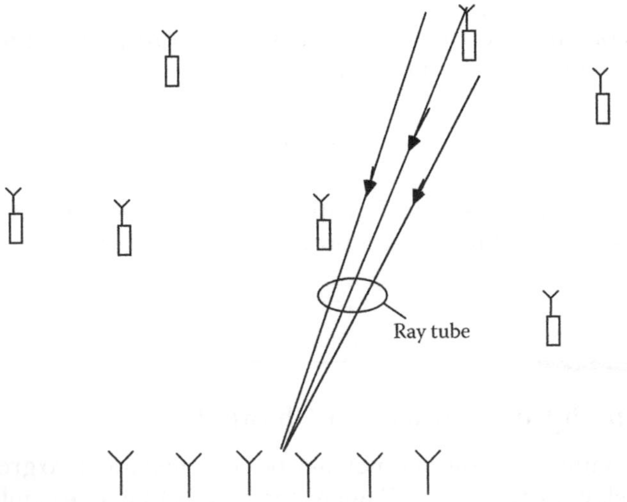

FIGURE 8.1
Base station array and many co-channel users.

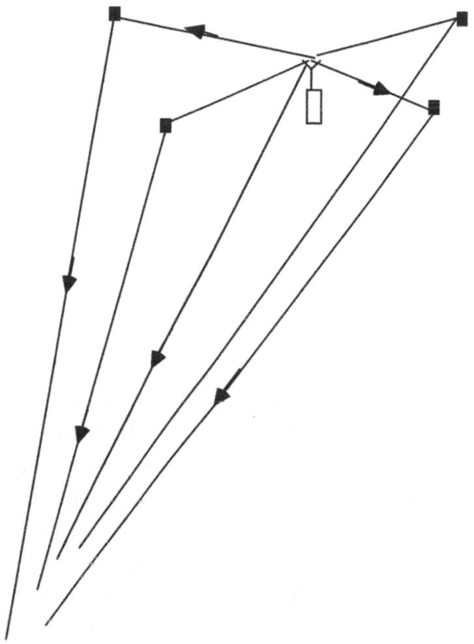

FIGURE 8.2
A transmitter is surrounded by many local scatterers, which result in many rays, a ray tube, being incident on the array from the same transmitter. The rays within a tube travel closely parallel to the axis of the ray tube, that is, line of sight (LOS).

is also assumed to be a complex random variable whose amplitude is uniformly distributed in the range of 0–1 and its phase in the range of $\pm\pi$. The polarization properties of a scattered signal are likely to be different from those of the actual transmitted signal. Finally, since the base station is normally located at a high point, on top of a tall building, we shall assume there are no scatterers present around it. The array output from kth transmitter can be written as

$$\mathbf{f}_k(t) = p_k e^{j\omega_c(t-\tau_k)} \mathbf{a}_1(\theta_k) f_0(t-\tau_k), \tag{8.19a}$$

where

$$\mathbf{a}_1(\theta_k) = \left[\sum_{r=0}^{R-1} \alpha_r e^{-j\omega_c \delta t_r} \mathbf{a}(\theta_k + \delta\theta_r) \right]. \tag{8.19b}$$

The propagation delay of direct path is τ_k and the excess delay is δt_r. It will be called here a composite direction vector as $\mathbf{a}_1$ represents a weighted average of R scattered paths. The power output of the array may be obtained from Equation 8.19a as follows

$$\begin{aligned} E\left\{\|\mathbf{f}_k(t)\|^2\right\} &= p_k^2 \|\mathbf{a}_1(\theta_k)\|^2 \\ &= p_k^2 \sum_r E\{\alpha_r^H \alpha_r\} E\{a^H(\theta_k+\delta\theta_r) a(\theta_k+\delta\theta_r)\} \\ &= p_k^2 \left(1+\frac{R-1}{3}\right) M, \end{aligned} \tag{8.20}$$

where we assumed that the scattering coefficients α_r and the scattering angles $\delta\theta_r$ are independent and uncorrelated random variables. For the purpose of evaluating the average power, we have assumed that the scattering coefficients are uniformly distributed random variables (the magnitude in the range of 0–1 and phase in the range $\pm\pi$), the angular deviations of the multipaths from the dominant path are random variables that are uniformly distributed in the range $\pm\Delta/2$. Note that the power increases linearly with the number of multipaths. Note that for direct path, $\alpha_0 = 1$.

For Q independent transmitters, each having its own suit of multipaths, the array output is given by

$$\mathbf{f}(t) = \sum_{k=0}^{Q-1} e^{j\omega_c(t-\tau_k)} \mathbf{a}_1(\theta_k) f_k(t-\tau_k).$$

And in place of Equation 8.2, we have the matrix equation (Equation 8.21). The composite direction vectors can be estimated as outlined in Section 8.1.2 for normal direction vectors, that is, having only direct paths.

$$\begin{bmatrix} \mathbf{C}_{z_0 z_0} \\ \mathbf{C}_{z_1 z_1} \\ \vdots \\ \mathbf{C}_{z_{Q-1} z_{Q-1}} \end{bmatrix} = \begin{bmatrix} 1 & \frac{2}{3L} & \cdots & \frac{2}{3L} \\ \frac{2}{3L} & 1 & \cdots & \frac{2}{3L} \\ \vdots & \vdots & & \vdots \\ \frac{2}{3L} & \frac{2}{3L} & \cdots & 1 \end{bmatrix} \begin{bmatrix} p_0^2 \mathbf{a}_1(\theta_0)\mathbf{a}_1^H(\theta_0) \\ p_1^2 \mathbf{a}_1(\theta_1)\mathbf{a}_1^H(\theta_1) \\ \cdots \\ \cdots \\ \cdots \\ p_{Q-1}^2 \mathbf{a}_1(\theta_{Q-1})\mathbf{a}_1{}^H(\theta_{Q-1}) \end{bmatrix} + \frac{\sigma_\eta^2}{L} \begin{bmatrix} \mathbf{I} \\ \mathbf{I} \\ \vdots \\ \mathbf{I} \end{bmatrix}. \tag{8.21}$$

The only difference between Equations 8.2 and 8.21 is in the direction vectors. In a multipath channel model, the direction vector (Equation 8.19b) is a linear combination of the direction vectors to all multipaths suitably weighted by the scattering coefficients and the excess delays. It must be emphasized that we are really not interested in the direction of arrival to different users or in the number of multipaths, which may not be the same for all transmitters. Our aim is only in the estimation of the direction vectors and possibly the radiated power for $k=0, 1,\ldots, Q-1$. The algorithm developed in Section 8.1 can readily be applied to the present problem.

As an illustration we quote the results of simulation of the above algorithm. A four-sensor array was assumed and eight users spread uniformly over an angular sector $\pm 60°$. All users are assumed to radiate the same power. A local scattering channel is assumed. Each source generates eight multipaths having random amplitudes, uniformly distributed in the range of 0–1 and random phase, uniformly distributed in the range of $\pm\pi$. The angular deviation of the multipaths about the direct or the dominant path is $\pm 2.5°$ and the delay spread is random in the range of 1–3 chips. There are 32 chips in the spreading sequence (BPSK sequence). The propagation delays are assumed to be known. The signal-to-noise ratio (SNR) (at the first sensor) was −6.67 dB. The postcorrelation matrix was computed after matched filtering with the individual spreading sequence (assumed to be known) and averaged over 1,024 data bits. The data stream estimated and the actual data bits transmitted for the second source are shown in Figure 8.3. There were only 77 error bits out of 1,024 transmitted bits. This figure naturally varied from run to run.

8.2.2 Large Delay Spread Channel

We encounter another type of channel where the delay spread of different multipaths is over several symbol periods. There is a strong ISI. As many symbols are involved in defining the output at any time instant, it is best described through an impulse response function

$$x(t) = \sum_{k=0}^{\infty} h(t-kT)s(k). \tag{8.22a}$$

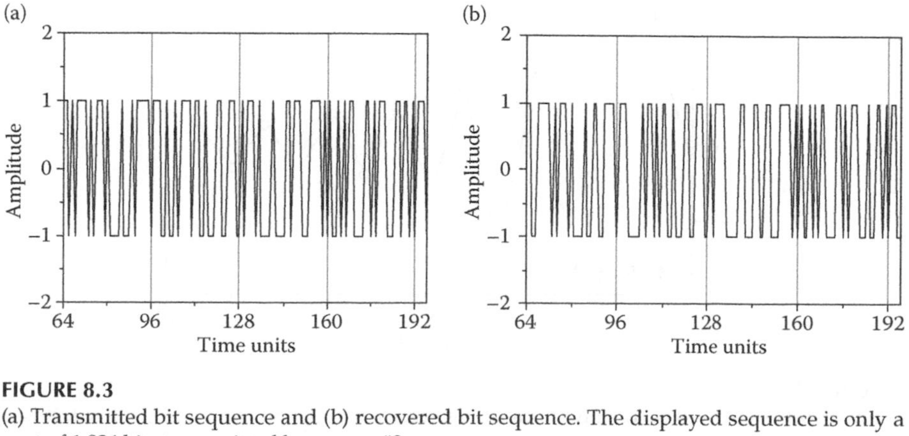

FIGURE 8.3
(a) Transmitted bit sequence and (b) recovered bit sequence. The displayed sequence is only a part of 1,024 bits transmitted by source #2.

The impulse response function may typically consist of many spikes spread over several symbols. An example of spikes spread over four symbols is given in Figure 8.4. The actual impulse response function is a convolution of the spike train with the symbol-shaping function and transmitter and receiver filter response functions (shown in the inset of Figure 8.4).

The impulse response function can be highly variable within a symbol duration, which may require sampling at a rate higher than the symbol rate. Let the sampling rate be P/T or the sampling interval be $T_s = T/P$ [14].

In Equation 8.22a, let $t = (iT_s + nT)$ where T_s is fractional sampling interval, and T is symbol duration.

$$\begin{aligned} x(iT_s + nT) &= \sum_{q=1}^{Q} \sum_{l=0}^{\infty} h(iT_s + nT - lT) s_q(l) \\ &= \sum_{q=1}^{Q} \sum_{l=0}^{L-1} h(iT_s + lT) s_q(n-l) \quad i = 0, 1, \ldots, P-1. \end{aligned} \tag{8.22b}$$

Fractionally spaced sampling effectively generates P *virtual channels.* Define a vector $\mathbf{x}_n^i$ consisting of N received samples from ith virtual channel.

$$\mathbf{x}_n^i = [x(iT_s + nT), x(iT_s + (n-1)T), \ldots, x(iT_s + (n-N+1)T)]^T.$$

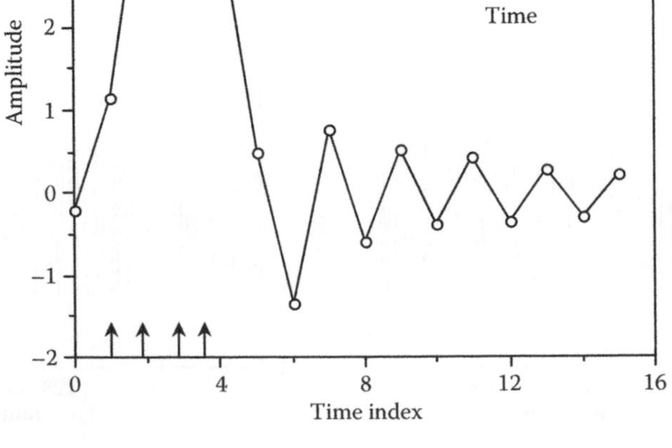

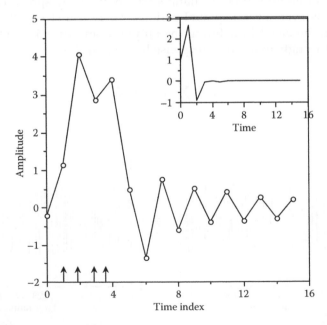

FIGURE 8.4
Train of spikes (arrows) is convolved with filter response function shown in the inset. It is a raised cosine sinc function [10].

Similarly define a stacked vector out of P virtual channels (but a single sensor)

$$\mathbf{x}_n = [\mathbf{x}_n^{0T}, \mathbf{x}_n^{1T}, \ldots, \mathbf{x}_n^{(P-1)T}]^T.$$

$$PN \times 1$$

We define a symbol vector

$$\mathbf{s}_{n,q} = [s_q(n), s_q(n-1), \ldots, s_q(n-N+1), \ldots, s_q(n-N-L+2)]^T,$$

$$(N+L-1) \times 1$$

and all Q vectors into one long stacked vector.

$$\mathbf{s}_n = [\mathbf{s}_{n,0}^T, \mathbf{s}_{n,1}^T, \ldots, \mathbf{s}_{n,Q-1}^T]^T.$$

$$Q(N+L-1) \times 1$$

Next, we define a filter matrix (see Chapter 7, Section 7.4)

$$\mathbf{H}_{q,N}^i = \begin{bmatrix} h_q(iT_s), \ldots, h_q(iT_s + (L-1)T), \ldots & \ldots & 0 \\ 0 \quad h_q(iT_s), \ldots, h_q(iT_s + (L-1)T) & \ldots & 0 \\ \vdots \quad \ddots & & \vdots \\ 0 \quad \ldots \quad 0 \quad h_q(iT_s), \ldots, h_q(iT_s + (L-1)T) & & \end{bmatrix}$$

$$N \times (N+L-1)$$

$$\mathbf{H}_{q,N} = [\mathbf{H}_{q,N}^{0}{}^T, \mathbf{H}_{q,N}^{1}{}^T, \ldots, \mathbf{H}_{q,N}^{P-1T}]^T.$$

$$PN \times (N+L-1)$$

Finally,

$$\mathcal{H} = [\mathbf{H}_{0,N}, \mathbf{H}_{1,N}, \ldots, \mathbf{H}_{(Q-1),N}],$$

$$PN \times Q(N+L-1)$$

$$\mathbf{x}_n = \mathcal{H}\mathbf{s}_n. \tag{8.23}$$

$\mathcal{H}$ is full (column) rank if $PN > Q(N+L-1)$, that is, it is a tall matrix. Equation 8.23 is easily adopted for multiple antenna receivers. Assume that there are M receivers, not necessarily in any particular geometry. We have multichannel output that is then sampled at a fractionally spaced sampling interval. We have thus P virtual and M real channels. The dimension of the output vector $\mathbf{x}_n$ will be $PMN\times 1$ and that of composite filter matrix $\mathcal{H}$ will be $PMN\times Q(N+L-1)$. Estimation of the composite filter matrix follows the procedure of Moulines, which we have described in Chapter 7. This is a stochastic algorithm requiring computation of covariance matrix. In the next section, we shall describe a deterministic approach that directly uses the array output, presumed to be noise free.

8.3 Symbol Estimation

The ultimate goal, naturally, is to extract the bit stream transmitted by different users. The presence of multipath propagation in a real channel is the main culprit, over and above the system and measurement noise. There has been vast research effort to mitigate the problems imposed by real channels. It is outside the scope of this work to summarize the current research results. Instead, to give a flavor, we shall consider two types of commonly encountered channels, namely, the local scattering channel with small delay spread, and the FIR filter channel with large delay spread and elaborate one algorithm for each type of channel. This is by no means a complete review of the present state of technology.

8.3.1 Small Delay Spread

The main object of processing is to estimate the data bits, $k=0, 1,\ldots, Q-1$; $n=0, 1,\ldots, N-1$ from the array output. In the previous section, we have outlined a method for estimating the composite direction vectors and the transmitter power. Using this information, we go on to estimate the data bits or the bit stream from each user. For this purpose we refer to the basic signal model given in Equation 5.60a. Define the following matrices

$$\hat{\mathbf{A}}=\left[\mathbf{a}_1(\theta_0), \mathbf{a}_1(\theta_1),\ldots, \mathbf{a}_1(\theta_{Q-1})\right]\mathrm{diag}\left\{p_0 e^{-j\omega_c\tau_0}, p_1 e^{-j\omega_c\tau_1},\ldots, p_{Q-1} e^{-j\omega_c\tau_{Q-1}}\right\},$$

$$\mathbf{b}_k=\mathrm{col}\{b_{k,0}, b_{k,1},\ldots, b_{k,N-1}\},$$

$$\mathbf{s}_k(t-\tau_k)=\mathrm{col}\{s_k(t-\tau_k), s_k(t-T_s-\tau_k),\ldots, s_k(t-(N-1)T_s-\tau k)\},$$

$$k=0, 1,\ldots, N-1.$$

Equation 5.60a may be expressed in a compact form as

$$\mathbf{f}(t) = \hat{\mathbf{A}}\ \mathrm{col}\left\{\mathbf{b}_0^H \mathbf{s}_0(t-\tau_0),\ \mathbf{b}_1^H \mathbf{s}_1(t-\tau_1),\ldots,\mathbf{b}_{Q-1}^H \mathbf{s}_{Q-1}(t-\tau_{Q-1})\right\} + \eta(t). \quad (8.24a)$$

In the previous section, we have shown how to estimate the composite direction vectors and the transmitted power. Given the propagation delays we can define $\hat{\mathbf{A}}$. Let $\hat{\mathbf{A}}^{\#}$ be the pseudoinverse of $\hat{\mathbf{A}}$ given by

$$\hat{\mathbf{A}}^{\#} = (\hat{\mathbf{A}}^H \hat{\mathbf{A}})^{-1} \hat{\mathbf{A}}^H.$$

To obtain the minimum mean square error (MMSE) solution we premultiply on both sides of Equation 8.24a with $\hat{\mathbf{A}}^{\#}$. The result is

$$\begin{aligned} &\mathrm{col}\left\{\mathbf{b}_0^H \mathbf{s}_0(t-\tau_0),\ \mathbf{b}_1^H \mathbf{s}_1(t-\tau_1),\ldots,\mathbf{b}_{Q-1}^H \mathbf{s}_{Q-1}(t-\tau_{Q-1})\right\} \\ &= \hat{\mathbf{A}}^{\#}\mathbf{f}(t) - \hat{\mathbf{A}}^{\#}\eta(t). \end{aligned} \quad (8.24b)$$

Next, to extract the bit stream of kth user, we multiply on both sides of Equation 8.24b by $\mathbf{s}_k(t-\tau_k)^H$ and integrate the resulting scalar quantity over the bit duration, nT_s to $(n+1)T_s$. This will yield nth bit of kth user; of course, there will be some error on account of background noise and also due to the fact that finite duration codes are only approximately orthogonal. Notice that a knowledge of the *propagation delays* τ_k, $k=0, 1,\ldots, N-1$, is essential for a successful bit recovery.

8.3.2 Bit Error Rate (BER)

The error in the transmission occurs largely on account of invalidity of the assumptions made in the model. For example, the codes used by different users are assumed to be uncorrelated, which indeed is true only when the code length is infinite. For finite code length there will be co-user interference, which increases with the increasing number of users. A numerical experiment was performed in which we increased the number of users from 8 to 40 in a single sector. The multipath propagation was assumed to be absent. The results are shown in Figure 8.5. The bit error rate (BER) is significantly lower when the number of users is less than 32, that is, less than the code length.

Next we study the effect of increasing the number of multipaths. The BER monotonically increases with an increasing number of multipaths, as illustrated in Figure 8.6. Letting the maximum BER (before decoding) be 0.05, up to nine multipaths are tolerable. Even though the multipaths contribute more power from each transmitter, as shown in Equation 8.18, hence higher SNR, the net BER increases with the increasing number of paths, which creates higher co-channel interference. To further decrease BER, we have

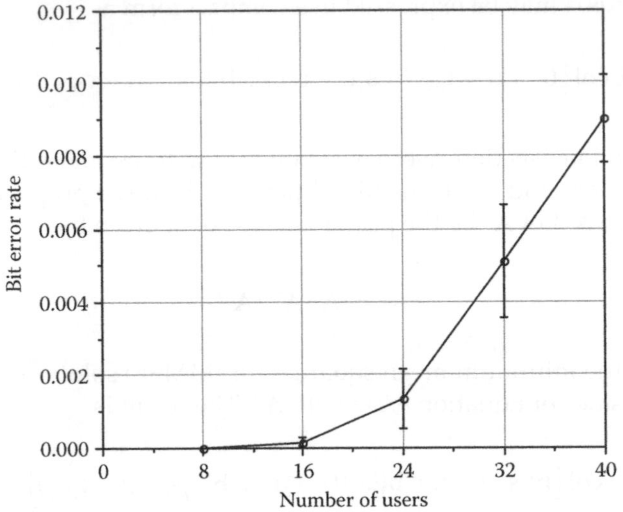

FIGURE 8.5
The bit error rate (BER) as a function of the number of users. Four antenna ULA, no multipaths, code length=32 chips. Two sigma error bars are shown. Average of 100 trials.

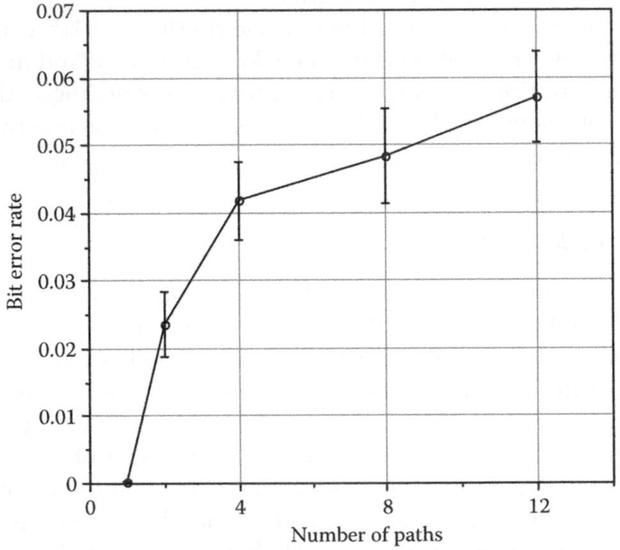

FIGURE 8.6
The bit error rate as a function of the number of paths. Four antenna ULA, code length=32, number of users=16, averaged over 100 trials, two sigma error bars are shown.

experimented with a larger array. An eight-antenna array significantly lowers BER, as illustrated in Figure 8.7. But the decrease of BER from that for a single antenna to that for an array of four antennas is the most remarkable. The presence of multipaths significantly increases BER. At 5% BER, the channel

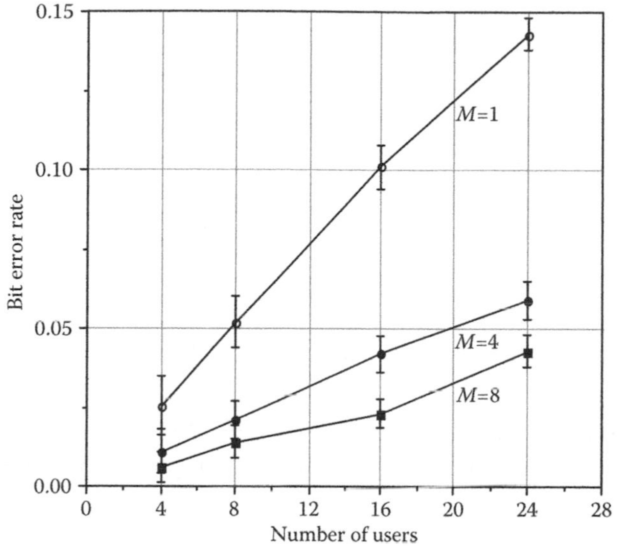

FIGURE 8.7
The bit error rate as a function of the number of users for different array sizes. Multipaths=4, code length=32, averaged over 100 trials, two sigma error bars are shown.

may not have more than nine multipaths (Figure 8.6). There is a considerable capacity increase (two and half times) from a single antenna to a mere four-antenna array (Figure 8.7). Increase in the capacity with the use of antenna array is well documented [6, 7].

8.3.3 Large Delay Spread

We have M receivers and Q transmitters; and connecting Q transmitters with M receivers is a matrix impulse response function

$$\mathbf{H}(t)=[h_{m,q}(t), \quad m=0,1,\ldots,M-1, \quad q=0,1,\ldots,Q-1],$$

where $h_{m,q}(t)$ is modeled as in Equation 7.81. The discrete output of the ith receiver at time instant $(n+\tau)$, where n is an integer and τ is a fraction $0\le\tau\le1$, is given by

$$x_i(n+\tau)=\sum_{k=0}^{L-1} h_{i,0}(k+\tau)s_0(n-k)+\cdots+\sum_{k=0}^{L-1} h_{i,Q-1}(k+\tau)s_{Q-1}(n-k) \quad i=0,1,\ldots,M-1. \tag{8.25}$$

We assume that the over-sampling factor is P, therefore,

$$\tau=\frac{1}{P},\frac{2}{P},\ldots,\frac{P-1}{P}.$$

We start sampling at $t=0$ and collect samples from N symbol periods; then construct a data matrix $\mathbf{X}$ as,

$$\mathbf{X} = [\mathbf{x}_0, \ldots, \mathbf{x}_{N-1}]$$

$$= \begin{bmatrix} \mathbf{x}(0) & \mathbf{x}(1) & \cdots & \mathbf{x}(N-1) \\ \mathbf{x}\left(\frac{1}{P}\right) & \mathbf{x}\left(1+\frac{1}{P}\right) & \cdots & \\ \vdots & \vdots & & \vdots \\ \mathbf{x}\left(\frac{P-1}{P}\right) & \mathbf{x}\left(1+\frac{P-1}{P}\right) & \cdots & \mathbf{x}\left(N-1+\frac{P-1}{P}\right) \end{bmatrix}, \tag{8.26}$$

where each entry is a column vector of size $M \times 1$, therefore the size of $\mathbf{X}$ matrix is $MP \times N$. The data matrix as defined in Equation 8.26 has a factorization **[8–10]**

$$\mathbf{X} = \mathbf{HS}, \tag{8.27}$$

where

$$\mathbf{H} = \begin{bmatrix} \mathbf{H}(0) & \mathbf{H}(1) & \cdots & \mathbf{H}(L-1) \\ \mathbf{H}\left(\frac{1}{P}\right) & & \cdots & \\ \vdots & & & \vdots \\ \mathbf{H}\left(\frac{P-1}{P}\right) & & \cdots & \mathbf{H}\left(L-1+\frac{P-1}{P}\right) \end{bmatrix}. \tag{8.28}$$

Each element in the above matrix (Equation 8.28) is itself a matrix of size $M \times Q$, therefore, the size of $\mathbf{H}$ matrix is $MP \times QL$. To see more clearly the structure of $\mathbf{H}$ matrix, let us consider a simple example. Let $M=4$, $Q=2$, $P=2$, and $L=2$. The matrix takes the following form

$$\mathbf{H} = \begin{bmatrix} h_{00}(0) & h_{01}(0) & h_{00}(1) & h_{01}(1) \\ h_{10}(0) & h_{11}(0) & h_{10}(1) & h_{11}(1) \\ h_{20}(0) & h_{21}(0) & h_{20}(1) & h_{21}(1) \\ h_{30}(0) & h_{31}(0) & h_{30}(1) & h_{31}(1) \\ h_{00}(\frac{1}{2}) & h_{01}(\frac{1}{2}) & h_{00}(\frac{3}{2}) & h_{01}(\frac{3}{2}) \\ h_{10}(\frac{1}{2}) & h_{11}(\frac{1}{2}) & h_{10}(\frac{3}{2}) & h_{11}(\frac{3}{2}) \\ h_{20}(\frac{1}{2}) & h_{21}(\frac{1}{2}) & h_{20}(\frac{3}{2}) & h_{21}(\frac{3}{2}) \\ h_{30}(\frac{1}{2}) & h_{31}(\frac{1}{2}) & h_{30}(\frac{3}{2}) & h_{31}(\frac{3}{2}) \end{bmatrix}.$$

$$(8 \times 4)$$

The symbol matrix **S** is given by

$$\mathbf{S} = \begin{bmatrix} \mathbf{s}_0 & \ddots & \mathbf{s}_{N-2} & \mathbf{s}_{N-1} \\ \mathbf{s}_{-1} & \ddots & \ddots & \mathbf{s}_{N-2} \\ \ddots & \ddots & \ddots & \ddots \\ \mathbf{s}_{-L+2} & \mathbf{s}_{-L+3} & \ddots & \ddots \\ \mathbf{s}_{-L+1} & \mathbf{s}_{-L+2} & \ddots & \mathbf{s}_{N-L} \end{bmatrix}. \tag{8.29}$$

Each entry in the symbol matrix is a vector of size $Q\times 1$, thus making it a block Toeplitz matrix. For example, for $N=4$ the symbol matrix is given by

$$\mathbf{S} = \begin{bmatrix} \mathbf{s}_0 & \mathbf{s}_1 & \mathbf{s}_2 & \mathbf{s}_3 \\ \mathbf{s}_{-1} & \mathbf{s}_0 & \mathbf{s}_1 & \mathbf{s}_2 \\ \mathbf{s}_{-2} & \mathbf{s}_{-1} & \mathbf{s}_0 & \mathbf{s}_1 \\ \mathbf{s}_{-3} & \mathbf{s}_{-2} & \mathbf{s}_{-1} & \mathbf{s}_0 \end{bmatrix}.$$

Clearly the symbol matrix has a block Toeplitz structure.

The blind identification problem is to estimate **H** and **S** from **X.** For such factorization to be unique, it is necessary that **H** and **S** are full column rank and row rank, respectively. This requires that $MP>QL$ and $N>QL$. We shall assume, for the sake of simplicity, that this condition is satisfied. (Veen [8–10] suggests extending the **X** matrix by left-shifting and stacking when the above assumption is not satisfied.) To factor **X** into **HS**, the strategy is to find either **S,** whose row span is equal to the row span of **X,** or **H,** whose column span is equal to the column span of **X.** We shall take the latter approach, as the dimension of **H** remains unchanged with the increasing number of symbols. In the first approach, the dimension of **S** increases with increasing N.

We form the following matrix product

$$\mathbf{C}_x=\mathbf{X}\mathbf{X}^H=\mathbf{H}\mathbf{S}\mathbf{S}^H\mathbf{H}^H, \tag{8.30}$$

where $\mathbf{C}_x$ is a matrix of size $(MP\times MP)$. Here we assume that $\mathbf{S}\mathbf{S}^H$ is a full rank matrix (rank$=QL$), which requires that symbols are spatially and temporally uncorrelated and **H** is a full column rank $(=QL)$. The rank of $\mathbf{C}_x$ will be QL. Let $\mathbf{U}_0$ be the null space spanned by the null eigenvectors of $\mathbf{C}_x$,

$$\underset{(QL\times N)}{\mathbf{U}_0^H\mathbf{C}_x\mathbf{U}_0 = \mathbf{U}_0^H\mathbf{H}\mathbf{S}\mathbf{S}^H\mathbf{H}^H\mathbf{U}_0 = 0.} \tag{8.31}$$

Since $\mathbf{S}\mathbf{S}^H$ is full rank, from Equation 8.31 it follows that

$$\mathbf{U}_0^H\mathbf{H} = 0. \tag{8.32}$$

Note that the rank of $\mathbf{U}_0$ is $(MP-r_c)$, where r_c stands for the rank of $\mathbf{C}_x$. Therefore, the null space of $\mathbf{U}_0$ will be of dimension r_c. It is not possible to uniquely estimate $\mathbf{H}$ as there is an unknown nonsingular $(QL \times QL)$ matrix, which may be used to postmultiply Equation 8.32 without altering the equality. In Refs. [8–10], the problem of nonuniqueness was resolved by resorting to a clever technique of augmenting the data matrix and exploiting the constant modulus property of symbol sequence (see Chapter 5).

8.3.4 Augmentation of Data Matrix

The data matrix is left shifted and then stacked. As an example, consider a one-step shifting and stacking. The data matrix given in Equation 8.26 becomes

$$\mathcal{X}_2 = \begin{bmatrix} \mathbf{x}_0, \ldots, \mathbf{x}_{N-2} \\ \mathbf{x}_1, \ldots, \mathbf{x}_{N-1} \end{bmatrix}.$$

Two-step shifting and stacking transforms the data matrix as

$$\mathcal{X}_3 = \begin{bmatrix} \mathbf{x}_0, \ldots, \mathbf{x}_{N-3} \\ \mathbf{x}_1, \ldots, \mathbf{x}_{N-2} \\ \mathbf{x}_2, \ldots, \mathbf{x}_{N-1} \end{bmatrix}.$$

Likewise in m–1 (m=2, 3,…) steps the data matrix is transformed into

$$\mathcal{X}_m = \begin{bmatrix} \mathbf{x}_0, \ldots, \mathbf{x}_{N-m} \\ \mathbf{x}_1, \ldots, \mathbf{x}_{N-m+1} \\ \vdots \quad\quad \vdots \\ \mathbf{x}_{m-1}, \ldots, \mathbf{x}_{N-1} \end{bmatrix}. \tag{8.33}$$

Note that the matrix $\mathbf{X}_m$ is of size $mMP \times (N-m+1)$. The augmented data matrix has a factorization

$$\mathcal{X}_m = \mathcal{H}_m \mathcal{S}_m, \tag{8.34}$$

where

$$\mathcal{H}_m = \underbrace{\begin{bmatrix} 0 & & \mathbf{H} \\ & \mathbf{H} \;\; \cdot^{\cdot^{\cdot}} & \\ \mathbf{H} & & \end{bmatrix}}_{mMP \times Q(L+m-1)},$$

and

$$S_m = \begin{bmatrix} \mathbf{s}_{m-1} & \ddots & \mathbf{s}_{N-2} & \mathbf{s}_{N-1} \\ \mathbf{s}_{m-2} & \ddots & \ddots & \mathbf{s}_{N-2} \\ \ddots & \ddots & \ddots & \ddots \\ \mathbf{s}_{-L+2} & \mathbf{s}_{-L+3} & \ddots & \ddots \\ \mathbf{s}_{-L+1} & \mathbf{s}_{-L+2} & \ddots & \mathbf{s}_{N-L-m+1} \end{bmatrix}.$$

$$Q(L+m-1) \times (N-m+1)$$

In practice, we do not know in advance Q (number of sources) and L (length of channel response). These quantities are required to achieve factorization of the data matrix, as in Equation 8.34. An ingenious method to estimate Q and L is suggested in Refs. [8–10]. If $\mathcal{H}_m$ and S_m have full column rank and row rank, respectively, then the rank of $\mathcal{X}_m$ will be $r_{\mathcal{X}_m} = Q(L+m-1)$. Now, increase m by 1, that is, m+1 left shifts. It is easy to show that

$$r_{\mathcal{X}_{m+1}} - r_{\mathcal{X}_m} = Q. \tag{8.35}$$

Thus, the number of sources can be estimated from the difference in the rank of augmented data matrices. Having estimated Q, it is straightforward to estimate the value of L, given by

$$L = \frac{r_{\mathcal{X}_m}}{Q} - m + 1. \tag{8.36}$$

8.3.5 Estimation of Filter Matrix

We shall now work with data matrix instead of correlation matrix (Equation 8.30), which shall make it convenient to compare with the results given in Refs. [8–10]. Both approaches are, however, completely equivalent. Let $\mathcal{U}_m$ be the null space of $\mathcal{X}_m$. Analogous to Equation 8.32, we can obtain the following result

$$\mathcal{U}_m^H \mathcal{H}_m = 0. \tag{8.37}$$

In Refs. [8–10] it is shown that Equation 8.37 is equivalent to

$$\hat{\mathcal{U}}_m^H \hat{\mathbf{H}} = 0, \tag{8.38}$$

where $\hat{\mathbf{H}}$ is a matrix formed by stacking the columns of the filter matrix Equation 8.28, that is,

$$\hat{\mathbf{H}} = \begin{bmatrix} \mathbf{H}_0 \\ \mathbf{H}_1 \\ \vdots \\ \mathbf{H}_{L-1} \end{bmatrix}, \tag{8.39a}$$

and

$$\hat{\mathcal{U}}_m^H = \begin{bmatrix} \mathcal{U}_m^H & & 0 \\ \vdots & \ddots & \\ \mathcal{U}_1^H & & \mathcal{U}_m^H \\ & \ddots & \vdots \\ 0 & & \mathcal{U}_1^H \end{bmatrix}, \tag{8.39b}$$

where $\mathcal{U}_m^H = [\mathcal{U}_1^H \ \mathcal{U}_2^H \cdots \mathcal{U}_m^H]$. Note that the size of $\mathcal{U}_m^H$ is $(mMP - r_{\mathcal{X}_m}) \times MP$, where $r_{\mathcal{X}_m}$ is the rank of $\mathcal{X}_m$. Each column of $\hat{\mathcal{U}}_m^H$ is vertically displaced by an amount equal to MP.

From Equation 8.38, we can conclude that $\hat{\mathbf{H}}$ spans the null space of $\hat{\mathcal{U}}_m^H \hat{\mathcal{U}}_m$. Estimation of $\hat{\mathbf{H}}$, however, is still nonunique, to the extent that there exists an unknown matrix of size $Q \times Q$ (not $QL \times QL$) in the estimated $\hat{\mathbf{H}}$. The non-uniqueness can, however, be removed by exploiting the constant modulus property of the transmitted symbols. This step becomes particularly trivial when Q=1, as demonstrated in the numerical example given below. Finally, from the estimated $\hat{\mathbf{H}}$ matrix, we reconstruct $\mathbf{H}$ matrix as in Equation 8.28. Using Equation 8.27, the symbol matrix is then computed as

$$\hat{\mathbf{S}} = (\mathbf{H}^H \mathbf{H})^{-1} \mathbf{H}^H \mathbf{x}. \tag{8.40}$$

We shall now give a numerical example. Consider a four element ULA, and a single source with four multipaths, which are randomly incident on the array. A small amount of noise is also added to the signal, keeping SNR = 0 dB. Also, the relative propagation delays of the multipaths are randomized. The fractional sampling rate is four times the Nyquist rate. The channel filter is assumed to be a random filter of length eight. Then, the size of channel filter will be (16×8). The data matrix is augmented with three left shifts (m=3). Next, we have computed the correlation matrix and its null space. The null space of the correlation matrix was used to compute $\hat{\mathcal{U}}_m^H \hat{\mathcal{U}}_m$, whose minimum eigenvector yields an estimate of $\hat{\mathbf{H}}$ within an unknown complex constant. The filter matrix $\mathbf{H}$ is derived from $\hat{\mathbf{H}}$ and used in Equation 8.40. Since the symbol sequence is ±1, it is enough to consider the sign of the filter output of Equation 8.40. The result, based on a single simulation, is 217 correct bits out of 256 transmitted bits. A

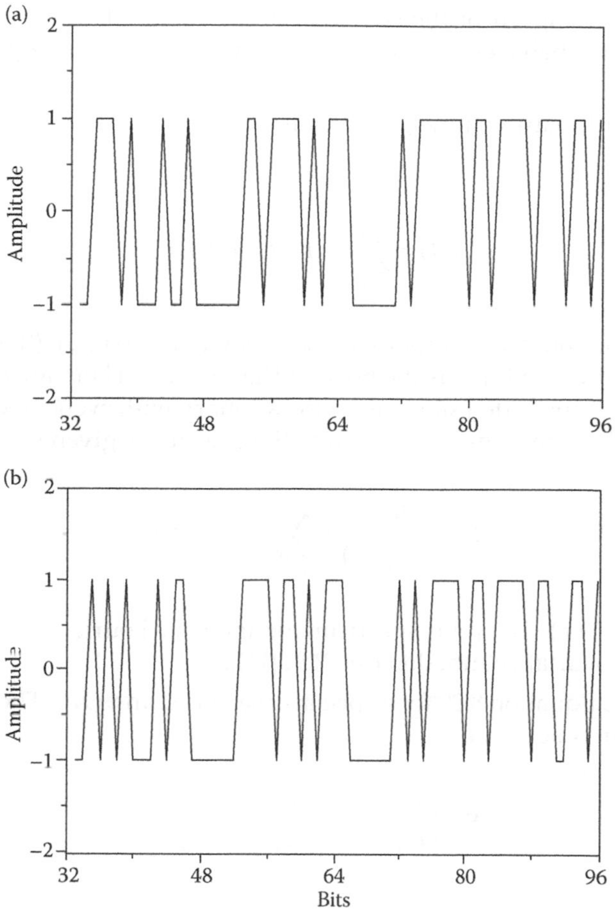

FIGURE 8.8
Blind identification. (a) Estimated bits and (b) transmitted bits. 217 bits out of 256 transmitted bits were correctly identified at 0 dB SNR.

sample of the output is shown in Figure 8.8. Under noise-free condition, however, all bits were correctly recovered without any error.

8.4 Exercises

(1) A cyclostationary source with cycle frequency α is moving away from a receiver at an angle θ with respect to the line joining the source and the receiver. Let

$$\gamma = \frac{v\cos(\theta)}{c},$$

where v is speed of the source and c is speed of wave. Show that the source behaves as if its cyclic frequency is equal to $\alpha/(1-\gamma)$. (See [11])

(2) In a multipath channel, both noisy input $x(t)=s(t)+\eta_t(t)$ and noisy output,

$$y(t) = \sum_{k=1}^{d} a_k s(t-\tau_k) + \eta_1(t),$$

where a_k and τ_k are amplitude and delay of kth path; $\eta_t(t)$ is noise at transmitter, and $\eta_r(t)$ is the noise at the receiver. There are d paths in all. The transmitted signal is cyclostationary with cyclic frequency α. Show that the transfer function of the channel is given by

$$H(f) = \frac{S^{\alpha}_{yx}(f)}{S^{\alpha}_{xx}(f)} = \sum_{k=1}^{d} a_k e^{-j2\pi(f+\alpha)\tau_k},$$

where $S^{\alpha}_{xx}(f)$ is cyclic spectrum defined as Fourier transform of cyclic covariance function (see Ref. [12]).

(3) Express Equation 8.29 for a special case of L=2 and Q=1. The **S** matrix now turns out to be

$$\mathbf{S} = \underset{(2\times N)}{\begin{bmatrix} \mathbf{s}_0 & \cdots & \mathbf{s}_{N-2} & \mathbf{s}_{N-1} \\ \mathbf{s}_{-1} & \cdots & \mathbf{s}_{N-3} & \mathbf{s}_{N-2} \end{bmatrix}}.$$

Define an equalizing filter

$$\mathbf{W} = [\mathbf{w}_0 \ \ \mathbf{w}_1] = \mathbf{H}[\mathbf{H}^H \ \mathbf{H}]^{-1}.$$

Apply the equalizing filter to a block N symbols, which yields

$$\mathbf{W}^H\mathbf{X} = \begin{bmatrix} \mathbf{s}_0 & \cdots & \mathbf{s}_{N-2} & \mathbf{s}_{N-1} \\ \mathbf{s}_{-1} & \cdots & \mathbf{s}_{N-3} & \mathbf{s}_{N-2} \end{bmatrix}.$$

Apply the same filter to one unit delayed block of N symbols and show that [13]

$$\mathbf{w}_0^H\mathbf{X}^0 = \mathbf{w}_1^H\mathbf{X}^1 = [s_0, s_1, \ldots, s_{N-1}],$$

where $\mathbf{X}^0$ and $\mathbf{X}^1$ data matrices formed with a block of data and delayed block of data, respectively.

(4) Take the example of **H** matrix given on page 428 (after Equation 8.28). Construct a two-step augmented filter matrix $\mathcal{H}_2$, as in Equation 8.34. What limits the maximum value of m, the number of shifts?

(5) Derive the result given in Equations 8.14a and 8.14b by maximizing Equation 8.13a under the constraints given by Equation 8.13b. What is the significance of constants λ_1 and λ_2?

References

1. H. Liu and M. D. Zoltowski, Blind equalization in antenna array CDMA systems, *IEEE Trans. Signal Processing*, vol. 45, pp. 161–172, 1997.
2. K. -L. Du and M. N. S. Swamy, An iterative blind cyclostationary beamforming algorithm, Proc. IEEE Conf. on Communications, pp. 145–148, April 2002.
3. Q. Wu and K. M. Wong, Blind adaptive beamforming for cyclostationary signals, *IEEE Trans. Signal Processing*, vol. 44, pp. 2757–2767, 1996.
4. B. G. Agee2, S. V. Schell, and W. A. Gardner, Spectral self-coherence restoral: A new approach to blind adaptive signal extraction using antenna arrays, *Proc. IEEE*, vol. 78, pp. 753–767, 1990.
5. B. G. Agee3, S. V. Schell, and W. A. Gardner, The SCORE approach to blind adaptive signal extraction: An application of the theory of spectral correlation, *IEEE Proc. Fourth Workshop on Spectrum Estimation and Modelling*, Minneapolis, MN, pp. 277–282, 1988.
6. D. Falconer, Spatial-temporal signal processing for wideband wireless systems in wireless communication in the 21st Century, in *Wireless Communication in the 21st Century*, Eds. M. Shafi, J. Mizusawa, T. Hattori, and S. Ogose, IEEE Press, New York, pp. 1–45, 2000.
7. A. F. Naguib, A. Paulraj, and T. Kailath, Capacity improvement with base station antenna array in cellular CDMA, *IEEE Trans. Vehicular Tech.*, vol. 43, pp. 691–698, 1994.
8. A.-J. van der Veen, S. Talwar, and A. Paulraj, A subspace approach to blind space-time signal processing for wireless communication systems, *IEEE Trans. Signal Processing*, vol. 45, pp. 173–190, 1997.
9. A.-J. van der Veen, S. Talwar, and A. Paulraj, Blind estimation of multiple digital signals transmitted over FIR channels, *IEEE Signal Processing Lett.*, vol. 2, pp. 99–102, 1995.
10. A.-J. van der Veen, S. Talwar, and A. Paulraj, Blind identification of FIR channels carrying multiple finite alphabet signals, *IEEE Proc. ICASSP*, pp. 1213–1216, 1995.
11. H. H. Fan and H. Yan, Instantaneous DOA estimation for moving wideband cyclostationary sources under multipath, *IEEE Proc. ICASSP*, IV-949–952, 2005.
12. G. Gelli, L. Izzo, A. Napolitano, and L. Paura, Multipath-channel identification by an improved Prony algorithm based on spectral correlation measurements, *Signal Processing*, vol. 31, pp. 17–29, 1993.

13. A.-J. van der Veen and A. Trindade, Combining blind equalization with constant modulus properties, *Proc. Asilomar Conf. Sig. Sys. Comput.*, pp. 1568–1572, 2000.
14. V. Zarzoso, A. K. Nandi, J. L. Garcia, and L. V. Dominguez, Blind identification and equalization of MIMO FIR channels based on second-order statistics and blind source separation, *IEEE Conf.*, DSP-2002, pp. 135–138, 2002.

9

Tomographic Imaging

As a wavefield propagates through a medium, it is subjected to time delays and loss of power. The wavefield is reflected from interfaces separating media of different impedances, and is scattered by inhomogeneities present in the medium. By observing these effects, it is possible to study the characteristics of the medium through which the field has propagated. Seismic exploration, on which depends the future discoveries of the petroleum deposits, exploits these effects of propagation to produce a detailed image of the subsurface geologic structure, which may be conducive to the accumulation of the hydrocarbon deposits. Likewise, the ultrasonic imaging used in medical diagnoses and in nondestructive testing also exploits the propagation effects of the wavefield. In this chapter, we shall study these effects of propagation for the purpose of constructing a three-dimensional image of the medium. Tomography refers to cross-sectional imaging of objects from transmitted, reflected, or diffracted wavefields. Accordingly, there are three different types of tomographic-imaging methods. One or more effects of propagation, such as accumulated attenuation, travel time, wavefield produced by diffraction or scattering, are observed in all directions (360° for 3D imaging). The observations such as travel time delays or accumulated attenuation are inverted by solving a system of linear equations. Where the medium is a diffracting type, that is, the size of inhomogeneities is comparable to the wavelength of the illuminating wavefield, the preferred approach is Fourier inversion. The subject of tomographic imaging is covered in the next four sections. In the last section, we investigate how to estimate the shape of an object from its scattered field.

9.1 Radiation

When the wavelength of illuminating radiation (e.g., x-rays, ultrasound) is much smaller than the dimensions of the inhomogeneities in the propagating medium, the concept of ray propagation becomes useful. The rays may travel in a straight line or along a curved line, depending upon the average wave speed: a straight line when the average speed is constant or curved path when the average speed is spatially variable as in a layered medium. The local speed variation is assumed to have a negligible effect on the ray paths.

The propagation has two effects on the wave, namely, wave attenuation and delay, both of which are of great significance from the point of tomographic imaging. Typically, in x-ray tomography, wave attenuation is used, and in ultrasound and seismic tomography, total time delay is used.

9.1.1 Absorption

Consider for example an object cross section represented by a function $f(x, y)$. A straight-line ray intersects the object and suffers a certain amount of attenuation depending upon the length of the ray path lying inside the object (see Figure 9.1). Let N_{in} be the number of photons incident on the object at point A and N_d be the number of photons coming out at point B within the time interval of measurement. N_{in} and N_d are related as below

$$N_d = N_{in} \exp\left[-\int_A^B f(x,y)ds\right]. \tag{9.1a}$$

Define attenuation as negative of log N_d/N_{in}, which turns out to be equal to the integral of $f(x, y)$ along path AB,

$$p = -\log_e \frac{N_d}{N_{in}} = \int_A^B f(x,y)ds, \tag{9.1b}$$

where p is also known as projection. The equation of line AB may be expressed as $x\cos(\theta)+y\sin(\theta)=t$ where θ is the slope of the line and t stands for the

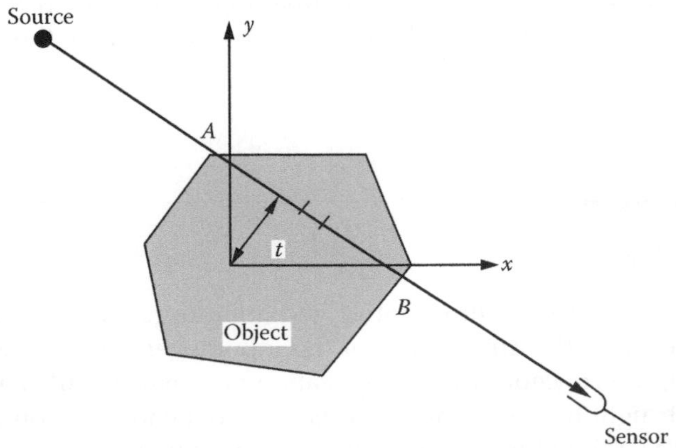

FIGURE 9.1
A ray passing through an absorbing object suffers an attenuation proportional to the integral over the path lying within the object.

perpendicular distance to the line from the origin (Figure 9.1). Equation 9.1b may be expressed as follows:

$$
\begin{aligned}
p_\theta(t) &= -\log_e \frac{N_d}{N_{\text{in}}} = \int_A^B f(x,y)ds \\
&= \int_{-\infty}^{\infty}\int_{-\infty}^{\infty} f(x,y)\delta(x\cos(\theta)+y\sin(\theta)-t)dxdy.
\end{aligned} \tag{9.2}
$$

$p_\theta(t)$ is known as Radon transform (Chapter 1, Section 1.2). For a fixed θ and variable t, we obtain $p_\theta(t)$, a continuous function of t, known as parallel projection, which may be generated by illuminating an object with a parallel beam and scanning the output with a receiver (see Figure 9.2a).

Taking Fourier transform on both sides of Equation 9.2, we obtain

$$
\begin{aligned}
P_\theta(\omega) &= \int_{-\infty}^{\infty} p_\theta(t)\exp(-j\omega t)dt \\
&= \int_{-\infty}^{\infty}\int_{-\infty}^{\infty} f(x,y)dxdy \int_{-\infty}^{\infty} \delta(x\cos(\theta)+y\sin(\theta)-t)\exp(-j\omega t)dt \\
&= \int_{-\infty}^{\infty}\int_{-\infty}^{\infty} f(x,y)\exp(j\omega(x\cos(\theta)+y\sin(\theta))dxdy \\
&= F(\omega\cos(\theta), \omega\sin(\theta)),
\end{aligned} \tag{9.3}
$$

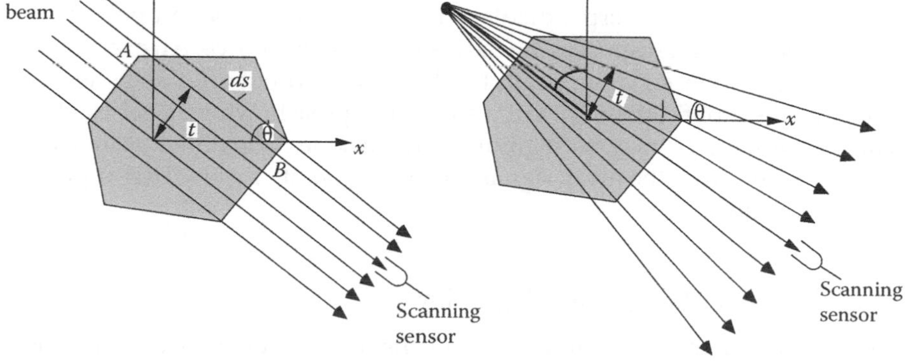

FIGURE 9.2
(a) A parallel projection is obtained by illuminating an object with a parallel beam. (b) A fan projection is obtained by illuminating an object with a fan beam generated by a point source at a radial distance d and angular distance $(90+\alpha)°$.

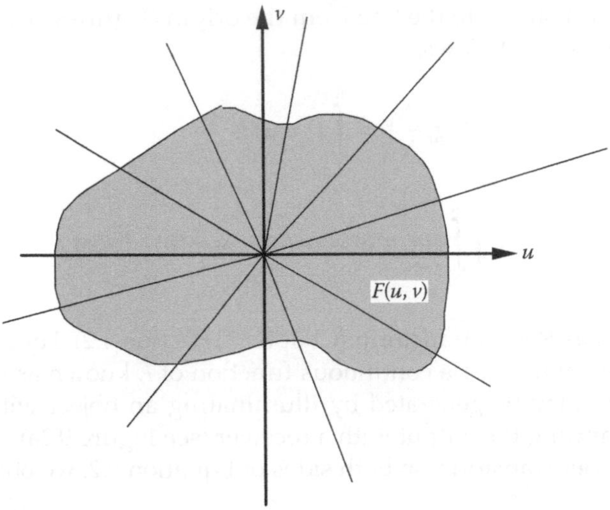

FIGURE 9.3
The Fourier transform of the object function is scanned along a series of radial lines, one for each projection.

where $F(\omega\cos(\theta), \omega\sin(\theta))$ is the 2D Fourier transform of the object function $f(x, y)$ evaluated on $u=\omega\cos(\theta)$ and $v=\omega\sin(\theta)$. The Fourier transform of a projection of an object function taken at an angle θ is equal to the slice of 2D Fourier transform of the object function evaluated on a radial line at an angle θ, as shown in Figure 9.3. This is known as Fourier slice theorem [1, p. 372]. By changing θ continuously, the 2D Fourier transform of the object function is evaluated over a series of radial lines as shown in Figure 9.3.

A fan beam is more appropriate, as a point source at finite distance emits a spherical beam which, when used to illuminate a finite target, may be considered as a conical beam in three dimensions or a fan beam in two dimensions. Both the slope and the perpendicular distance of a ray depend upon the ray angle measured with respect to the radius vector of the source. An exact reconstruction algorithm for a fan beam has been worked out after some tedious geometrical simplifications [1], but a simpler approach, where a series of fan beams covering 360° may be regrouped into a series of parallel beams, is of greater interest from the array signal processing point of view. This approach is outlined below.

9.1.2 Filtered Backprojection Algorithm

The reconstruction algorithm consists of the inverse Fourier transformation of a radially sampled object Fourier transform, which in turn is obtained from parallel projection data. Let us express the 2D inverse Fourier transform in polar coordinates.

$$f(x,y) = \frac{1}{4\pi^2} \int\int_{-\infty}^{+\infty} F(u,v) e^{j(ux+vy)} du dv$$
$$= \frac{1}{4\pi^2} \int_0^{\infty} s ds \int_0^{2\pi} F(s,\theta) e^{js(x\cos\theta + y\sin\theta)} d\theta. \tag{9.4}$$

The inner integral may be expressed as a sum of two integrals,

$$\int_0^{2\pi} F(s,\theta) e^{js(x\cos\theta+y\sin\theta)} d\theta$$
$$= \int_0^{\pi} F(s,\theta) e^{js(x\cos\theta+y\sin\theta)} d\theta + \int_{\pi}^{2\pi} F(s,\theta) e^{js(x\cos\theta+y\sin\theta)} d\theta$$
$$= \int_0^{\pi} F(s,\theta) e^{js(x\cos\theta+y\sin\theta)} d\theta$$
$$+ \int_0^{\pi} F(s,\pi+\theta) e^{js(x\cos(\pi+\theta)+y\sin(\pi+\theta))} d\theta.$$

Using the mapping

$$F(s,\ \pi+\theta) = F(-s,\ \theta),$$

Equation 9.4 may be written as

$$f(x,y) = \frac{1}{2\pi} \int_0^{\pi} \left\{ \frac{1}{2\pi} \int_{-\infty}^{\infty} F(s,\theta) |s| e^{js(x\cos\theta+y\sin\theta)} ds \right\} d\theta$$
$$= \frac{1}{2\pi} \int_0^{\pi} \left\{ \frac{1}{2\pi} \int_{-\infty}^{\infty} F(s,\theta) |s| e^{jst} ds \right\} d\theta. \tag{9.5}$$

Note that, for a fixed θ, $F(s, \theta)$ is equal to $P_\theta(\omega)$ given in Equation 9.3. Thus, the quantity inside the curly brackets in Equation 9.5 may be obtained from the parallel projection by simply filtering it with a filter having a transfer function, $H(s) = |s|$.

$$f(x,\ y) = \frac{1}{2\pi} \int_0^{\pi} \tilde{f}(x\cos\theta + y\sin\theta) d\theta, \tag{9.6}$$

where

$$\tilde{f}(x\cos\theta + y\sin\theta) = \left\{ \frac{1}{2\pi} \int_{-\infty}^{\infty} P_\theta(\omega)|\omega| e^{j\omega(x\cos\theta + y\sin\theta)} d\omega \right\}, \tag{9.7}$$

is often known as a filtered projection. In Equation 9.7, the projection data is projected back onto a section $t=x\cos\theta+y\sin\theta$ for a fixed θ and hence the reconstruction procedure is called the backprojection algorithm. The process of reconstruction consists of filtering each projection with a filter whose transfer function is $H(\omega)=|\omega|$ and then backprojecting according to Equation 9.6.

9.1.3 Algebraic Reconstruction

The backprojection algorithm requires a uniform linear array (ULA) capable of going round the target object. Such an idealistic experimental setup cannot be achieved at least in one important area of application, namely, exploration for earth resources. The sensor array tends to be nonuniform and generally distributed over a large area. Furthermore, since experimental observations are necessarily confined to the earth's surface or a few deep borewells, it is practically impossible to go round the target object to get a 4π solid angle coverage. The backprojection algorithm cannot be applied in most real situations except perhaps in seismic exploration for oil where a near ideal experimental setup can be realized. It is, therefore, necessary to devise an alternate approach, albeit less accurate and of lower resolution. The basic idea in this alternate approach is to divide the target object into as many homogeneous cells as possible. The cell size is small enough to allow the assumption of no variation in physical parameters (e.g., wave speed) over the size of a cell, and large enough to allow the validity of ray propagation. As a ray propagates through the target object, it passes through many cells lying in its path. The ray, as it reaches a sensor, carries the cumulative effect of all cells, for example, sum of all delays introduced by all those cells lying in its path. It is further assumed that a ray does not suffer any refraction (or reflection) on the boundary of a cell and hence it travels in a straight line joining source and sensor. For this assumption to hold good, the change in the acoustic impedance must be small (on the order of 10%). It is, however, possible to relax this assumption, but only at the cost of increased computation as the ray paths have to be traced numerically. The reconstruction process becomes iterative where, starting from an initial estimate of the speed variation, successive corrections are introduced consistent with the available time delay information. The ray paths will have to be traced afresh at the beginning of each iteration. Each cell would introduce a delay proportional to the path length inside the cell and the unknown wave speed in the cell. Let δ represent the length of each side of a cell and c_m be the wave speed in mth cell. The maximum delay introduced by mth cell will be $\tau_m = \sqrt{2}\delta/c_m$. The contribution of a cell toward the total delay observed at a sensor would depend upon the sensor position.

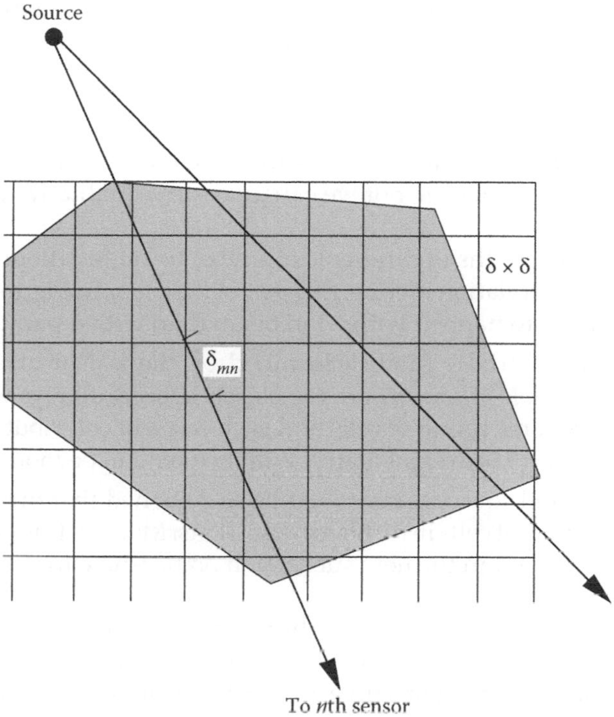

FIGURE 9.4
A target is divided into many square cells (or cubes in 3D). The wave speed is assumed to be constant in a cell. The path delay introduced by *m*th cell is given by (δ_{nm}/c_m), where δ_{nm} is path length in *m*th cell of a ray going to *n*th sensor.

It is possible that some cells do not contribute at all. Indeed, in any typical sensor array distribution there are many cells through which a ray joining the source to sensor will not pass at all; hence, there can be no contribution from such cells. This may be easily seen in Figure 9.4, where out of 56 cells a ray passes through less than 10 cells. Let $w_{m,n}$ represent a weighting coefficient, which when used along with the maximum delay gives a delay contributed by *m*th cell to *n*th sensor. In terms of path length of a ray in the *m*th cell, the weighting coefficient is given by $w_{m,n}=\delta_{mn}/\sqrt{2}\delta$ (see Figure 9.4). Thus, the total delay observed at the *n*th sensor is given by

$$t_n = \sum_{m=0}^{M-1} w_{m,n}\tau_m \quad n=0,1,\ldots,N-1, \tag{9.8a}$$

where we have assumed that there are *M* cells and *N* sensors. In matrix form, Equation 9.8a may be expressed as

$$\mathbf{t}=\mathbf{w}\boldsymbol{\tau}, \tag{9.8b}$$

where $\mathbf{t}=\text{col}\{t_0, t_1, \ldots, t_N\}$, $\tau=\text{col}\{\tau_0, \tau_1, \ldots, \tau_M\}$, and $\mathbf{w}$ is a $N\times M$ matrix of weight coefficients. When $N \geq M$, a unique solution of Equation 9.8b may be given by

$$\tau=(\mathbf{w}^T\mathbf{w})^{-1}\,\mathbf{w}^T\mathbf{t}, \tag{9.9}$$

provided $\mathbf{w}^T\mathbf{w}$ is nonsingular. The question of the rank of the weight matrix $\mathbf{w}$ has no quantitative answer, but we can give some qualitative guidelines:

(1) The ray path lengths in different cells must be quite different so that there is no correlation among weight coefficients. In Figure 9.4, this is more likely to happen with a fan beam than with a parallel beam.
(2) The weight matrix is fully determined by the sensor array geometry. For example, if sensors are too closely spaced, all rays will travel through the same group of cells and each cell will contribute roughly the same delay. The weight matrix will then tend to be more singular.
(3) The sensor and source arrays must be so designed that the rays pass through different cells in different directions. More on this possibility will be discussed in the next subsection on borehole tomography.

It may be noted that since the weight matrix is generally a large, sparse matrix, efficient techniques have been designed for fast and economical (in terms of memory requirements) inversion of the weight matrix. This is, however, beyond the scope of this book. The interested reader may look to a review article by Ivansson [2].

9.1.3.1 Borehole Tomography

The use of a source array often improves the rank condition of the weight matrix. Consider a P source array arranged in some unspecified form. Equation 9.8b may be used to express the output as

$$\mathbf{t}_p=\mathbf{w}_p\tau, \quad p=0,1, \ldots, P-1, \tag{9.10}$$

where $\mathbf{t}_p$ is the array output due to pth source and $\mathbf{w}_p$ is the weight matrix corresponding to the position of pth source. Next, we stack up all array vectors into a single vector. Note that τ is independent of the source position. Equation 9.10 reduces to

$$\tilde{\mathbf{t}} = \tilde{\mathbf{w}}\tau, \tag{9.11}$$

where $\tilde{\mathbf{t}} =\{\mathbf{t}_0, \mathbf{t}_1, \ldots, \mathbf{t}_{P-1}\}$ and $\tilde{\mathbf{w}} =\{\mathbf{w}_0, \mathbf{w}_1, \ldots, \mathbf{w}_{P-1}\}$. The solution of Equation 9.11 may be expressed as

$$\tau = (\tilde{\mathbf{w}}^T\tilde{\mathbf{w}})^{-1}\tilde{\mathbf{w}}^T\tilde{\mathbf{t}}. \tag{9.12}$$

We shall now consider a specific example of source and sensor arrays used in borehole tomography in geophysical exploration [3]. A typical arrangement

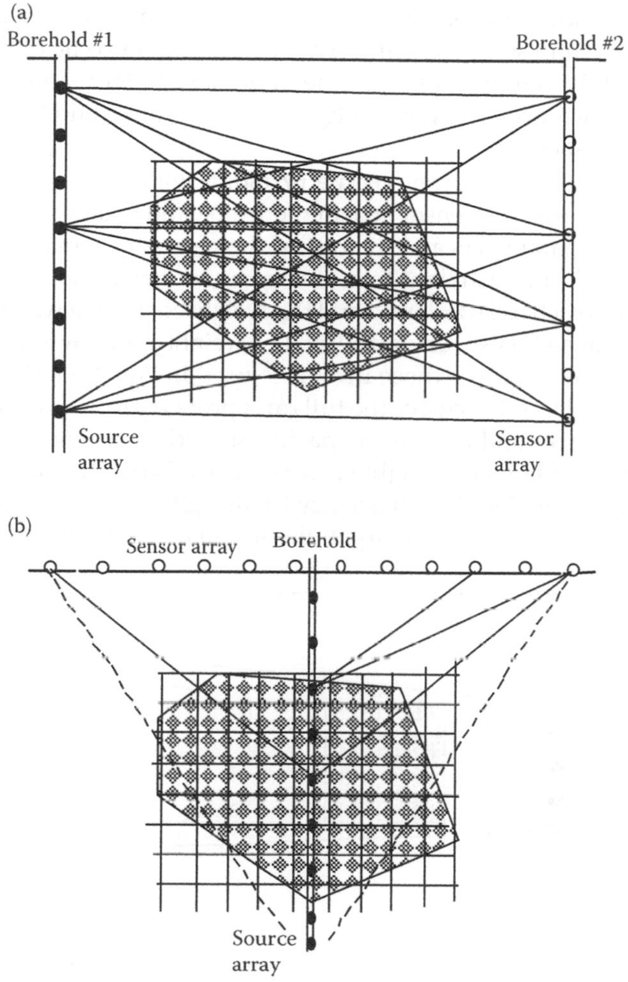

FIGURE 9.5
Source and sensor arrays in borehole tomography. (a) A target lies between two boreholes. In borehole #1 sources are arranged as a ULA of sources and in borehole #2 sensors are arranged as ULA of sensors. (b) In another arrangement the sensor array is on the surface and the source array is in the borehole.

in borehole tomography is shown in Figure 9.5. The sources are fired sequentially and the transmitted signals are recorded for later processing.

9.1.3.2 Source/Sensor Array Design

For successful reconstruction of wave speed variations, the primary requirement is that $\tilde{\mathbf{w}}^T\tilde{\mathbf{w}}$ in Equation 9.12 must be invertible. Since $\tilde{\mathbf{w}}$ is entirely determined by the source and sensor array geometry, it is possible to come up with a proper design for the source and sensor arrays that would make the rank of $\tilde{\mathbf{w}}$ equal to the number of cells. A simple numerical example is worked out to show how the rank of $\tilde{\mathbf{w}}^T\tilde{\mathbf{w}}$ depends on the number of sources, source spacing, and sensor spacing. The interesting outcome of this exercise is the fact that the number of sensors need not be greater than the number of cells. We may achieve by using multiple sources what could be achieved by using more sensors.

The source and sensor array geometry along with the target location are shown in Figure 9.6. The sensor spacing (d units) and source spacing (s units) are the variable parameters and the rest of the geometry remains fixed. From each source, eight rays (straight lines) were drawn toward eight sensors. The line intercept in each cell was found and the weight coefficient was computed as described on page 441 (also see Figure 9.4). The weight matrix $\tilde{\mathbf{w}}$ is first computed and then the rank of $\tilde{\mathbf{w}}^T\tilde{\mathbf{w}}$, whose inverse is used in Equation 9.12. The results are shown in Table 9.1. To achieve the full rank property for $\tilde{\mathbf{w}}^T\tilde{\mathbf{w}}$, we must have three to five sources and the sensor spacing should be around 1.5 units. Note that we have considered only eight sensors, which is half the number of cells. We have compensated for this deficiency by using three to five sources. When the source array is close to the target, the angular width of the illuminating

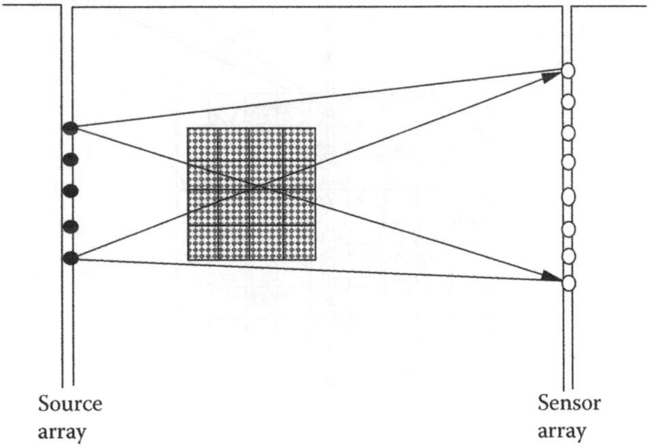

FIGURE 9.6
A target consisting of 16 cells (unit cells) lies between two boreholes separated by 16 units. An eight-sensor array is located in the right hole and the source array (up to five sources) is in the left hole. The center of sensor array and source array is at the average depth of the target.

TABLE 9.1
Rank of $\tilde{\mathbf{w}}^T\tilde{\mathbf{w}}$ as a function of source and sensor spacing

	d=1			d=1.5	
Number of Sources	s=0.5	s=1.0	s=1.5	s=1.0	s=1.5
1	6			7	
3	13	14	15	16 (1,350)	16 (1,850)
5	15	15	16 (2,200)	16 (2,200)	16 (835)

Note: The bracketed quantity represents the eigenvalue spread.

beam becomes large, which in turn requires a large aperture sensor array to capture the wide illuminating beam. However, an indefinite increase of the sensor spacing will not help. There exists a range of sensor separation over which not only $\tilde{\mathbf{w}}^T\tilde{\mathbf{w}}$ is full rank, but it is also stable, as shown in Figure 9.7, whereas for sensor separation between 1.5 and 3.0 units, the eigenvalue spread of $\tilde{\mathbf{w}}^T\tilde{\mathbf{w}}$ is low and the matrix becomes singular outside the range 1.0–4.0. The above findings are specific to the geometry of source and sensor arrays and the target; nevertheless, similar behavior is expected in other situations.

9.2 Diffracting Radiation

Diffraction becomes important whenever the inhomogeneities in an object are comparable in size to the wavelength of the wavefield used for imaging. In tomographic imaging, an object is illuminated from many different directions, either sequentially or simultaneously, and the image is reconstructed from the scattered wavefield collected by an array of sensors, usually a linear array. Early workers who attacked the problem of deriving the inversion algorithm for tomography with diffracting wavefields were Iwata and Nagata [4] and Mueller et al. [5], who based their work on Wolf's work [6] on the inverse scattering problem assuming the first order Born approximation. A good review of diffraction tomography may be found in Refs. [1, 7].

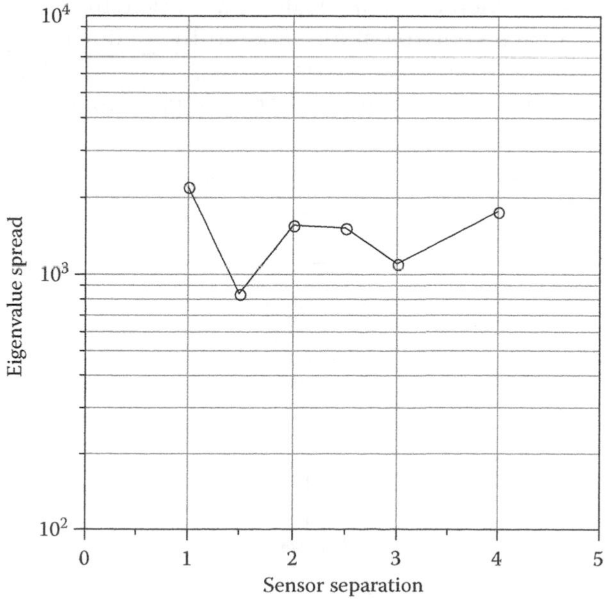

FIGURE 9.7
Eigenvalue spread as a function of sensor separation. Five sources are spaced at intervals of 1.5 units. The target used is the same as in Figure 9.6.

9.2.1 Linear Array

An object is illuminated from various directions with a diffracting source of radiation, such as acoustic waves whose wavelength is comparable to the scale of inhomogeneities. The incident wave energy is scattered in all directions by the diffraction process within the object. A long linear array of sensors facing the incident wavefield is used to record the forward scatter (see Figure 9.8). In Section 1.6, we derived an expression for the scattered field in the x-y plane due to a plane wave traveling in z direction and illuminating a three-dimensional object (spherical). A similar result for a two-dimensional object (cylindrical) was also given. For the sake of simplicity, we shall talk about tomographic imaging of a two-dimensional object. Consider an arrangement wherein a cylindrical object is illuminated with a plane wave traveling at right angle to the axis and a linear array of sensors located on the opposite side, as shown in Figure 9.8. The Fourier transform of the scattered field, which is measured at a set of discrete points by the sensor array, may be obtained from Equation 1.82, where set $u_0=0$ and $v_0=k_0$,

$$P_0(u) = \frac{jk_0^2}{2} \frac{e^{j\sqrt{k_0^2-u^2}l}}{\sqrt{k_0^2-u^2}} \Delta\tilde{c}(u, \sqrt{k_0^2-u^2}-k_0) \qquad |u| \le k_0, \tag{9.13a}$$

where $\Delta\tilde{c}(u,v)$ is a 2D Fourier transform of $\delta c(x, y)$. As u varies from $-k_0$ to $+k_0$, $P_0(u)$ traces a cross section of $\Delta\tilde{c}(u,v)$ along a semicircular arc, as shown in Figure 9.9. The circle is centered at $(0, -k_0)$ and the radius is equal to k_0. The circular arc is described by an equation $v = \sqrt{k_0^2-u^2}-k_0$, $|u| \le k_0$. The entire $\Delta\tilde{c}(u, v)$ may be sampled over a series of arcs either by rotating the object but

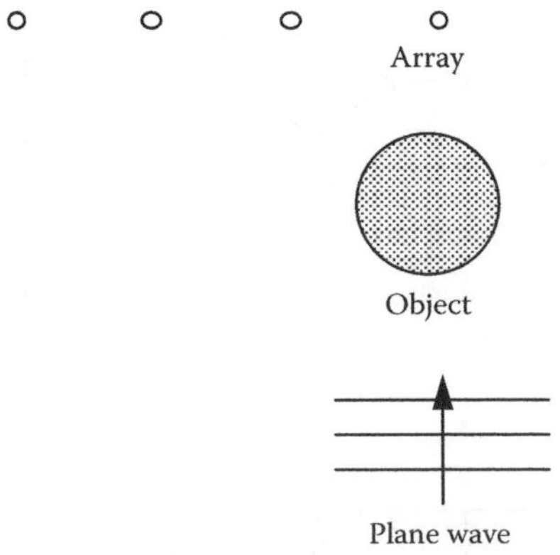

FIGURE 9.8
A basic experimental setup for tomographic imaging. A linear array, ideally of infinite aperture, is used to receive the forward scatter from the object.

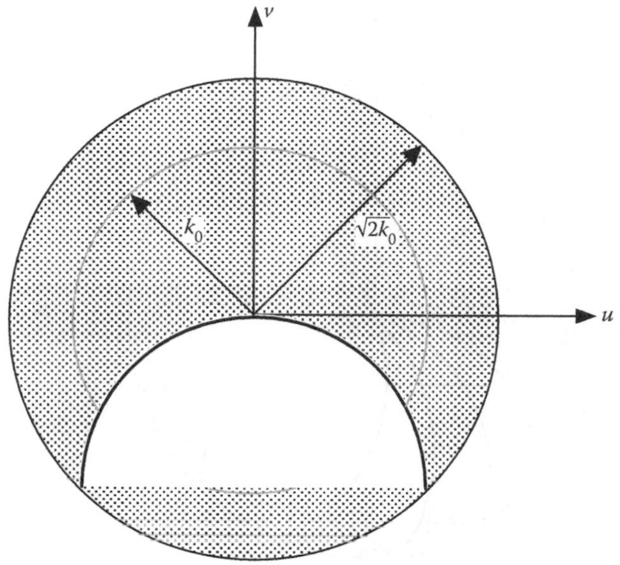

FIGURE 9.9
As u varies from $-k_0$ to $+k_0$, $P_0(u)$ traces a cross section of $\Delta\tilde{c}(u, v)$ along a semicircular arc (thick curve).

keeping the direction of illumination and the position of array fixed or vice versa. We consider the first case. When the object is rotated so is its Fourier transform through the same angle. Hence, for an object, which is rotated through an angle φ, the scattered field, analogous to Equation 9.13a, is given by the following equation:

$$P_\varphi(u) = \frac{jk_0^2}{2}\frac{e^{j\sqrt{k_0^2-u^2}l}}{\sqrt{k_0^2-u^2}}\Delta\tilde{c}$$
$$\times\left(u\cos\varphi + (\sqrt{k_0^2-u^2}-k_0)\sin\varphi, -u\sin\varphi + (\sqrt{k_0^2-u^2}-k_0)\cos\varphi\right) \tag{9.13b}$$
$$|u| \le k_0 \text{ and } -\frac{\pi}{2} \le \varphi \le \frac{\pi}{2}.$$

The center of the circular arcs will all lie on a circle of radius k_0 centered at $(-k_0 \sin\varphi, -k_0 \cos\varphi)$ (see Figure 9.10).

9.2.2 Filtered Backpropagation Algorithm

This is an adaptation of the filtered backprojection algorithm developed for nondiffracting radiation to diffracting radiation. The essential difference is that the sampling paths are now arcs of a circle instead of radial lines in

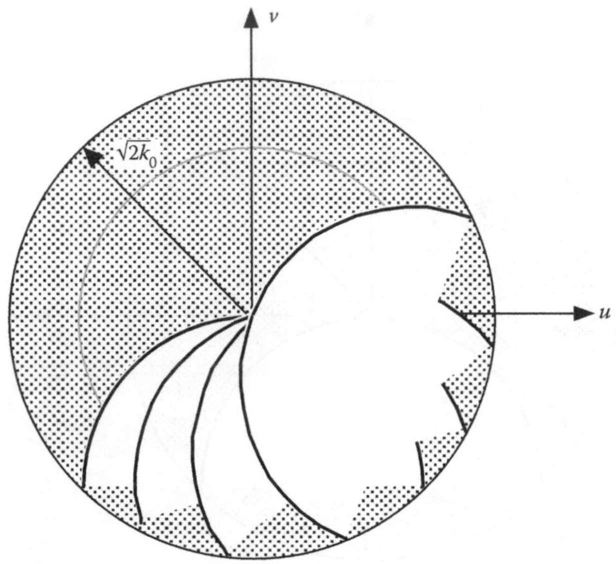

FIGURE 9.10
The entire 2D Fourier transform, $\Delta\tilde{c}(u, v)$, is sampled along a series of semicircular arcs by rotating the object, keeping the transmitter and array fixed. Because u must lie within $-k_0$ to $+k_0$, the radius of the disc spanned by the semicircular sampling arcs is equal to $\sqrt{2k_0}$.

nondiffracting radiation. We start by expressing the Fourier integral in polar coordinates (see Equations 9.5–7.7)

$$\begin{aligned} f(x,y) &= \frac{1}{2\pi}\int_0^{\pi}\left\{\frac{1}{2\pi}\int_{-\infty}^{\infty} F(s,\varphi)|s|e^{js(x\cos\varphi+y\sin\varphi)}ds\right\}d\varphi \\ &= \frac{1}{2\pi}\int_0^{\pi}\left\{\frac{1}{2\pi}\int_{-\infty}^{\infty} F(s,\varphi)|s|e^{jst}ds\right\}d\varphi \\ &= \frac{1}{2\pi}\int_0^{\pi}\tilde{f}(x\cos\varphi+y\sin\varphi)d\varphi, \end{aligned}$$

where

$$\tilde{f}(x\cos\varphi+y\sin\varphi) = \left\{\frac{1}{2\pi}\int_{-\infty}^{\infty} F(s,\varphi)|s|e^{js(x\cos\varphi+y\sin\varphi)}ds\right\}.$$

We shall use the above representation of 2D Fourier transform in polar coordinates. Note that in place of $f(x,y)$ we have $\delta\tilde{c}$ and in place of $F(s, \varphi)$ we have $\Delta\tilde{c}$. Equation 9.13b relates the Fourier transform of the scattered field to the object Fourier transform, that is,

$$\Delta\tilde{c}(u\cos\varphi + u'\sin\varphi, -u\sin\varphi + u'\cos\varphi)$$

$$= \frac{2}{jk_0^2}\sqrt{k_0^2 - u^2}\, e^{-j\sqrt{k_0^2-u^2}l} P_\varphi(u),$$

where $|u| \le k_0$ and $u' = (\sqrt{k_0^2 - u^2} - k_0)$. Using the above relation, we obtain the reconstruction

$$\delta\tilde{c}(x,y) = \frac{1}{2\pi}\int_0^\pi \delta\tilde{\tilde{c}}(x\cos\varphi + y\sin\varphi)\,d\varphi, \qquad (9.14a)$$

where

$$\delta\tilde{\tilde{c}}(x\cos\varphi + y\sin\varphi)$$

$$= \left\{\frac{1}{2\pi}\int_{-k_0}^{k_0} \frac{2}{jk_0^2}\sqrt{k_0^2 - u^2}\, e^{-j\sqrt{k_0^2-u^2}l} P_\varphi(u)|u| e^{ju(x\cos\varphi + y\sin\varphi)}\,du\right\}. \qquad (9.14b)$$

According to Equation 9.14b, the scattered field measured by the linear array is filtered with a filter whose transfer function is given by

$$H(u) = \sqrt{k_0^2 - u^2}\, e^{-j\sqrt{k_0^2-u^2}l}\,|u|. \qquad (9.14c)$$

The term of interest in the filter transfer function is $e^{-j\sqrt{k_0^2-u^2}l}$, which represents backward propagation of the wavefield from the plane of observation to the target (see Chapter 1, Section 1.2). For this reason the reconstruction algorithm described above is called the filtered backpropagation algorithm. Except for this difference, the algorithm is quite similar to the filtered backprojection algorithm.

9.2.3 Multisource Illumination

There are many situations where it is not possible to turn an object around for multiple illuminations nor is it possible to turn around the source-array configuration, keeping the object fixed as the space around the object may not be accessible as is the case in geophysical imaging, nondestructive testing, remote monitoring, etc. In such a situation, it is recommended to employ an array of sources, often arranged as a linear equispaced array. A typical example is borehole tomography, which we have already considered in the previous section in the context of algebraic reconstruction. In this section, we

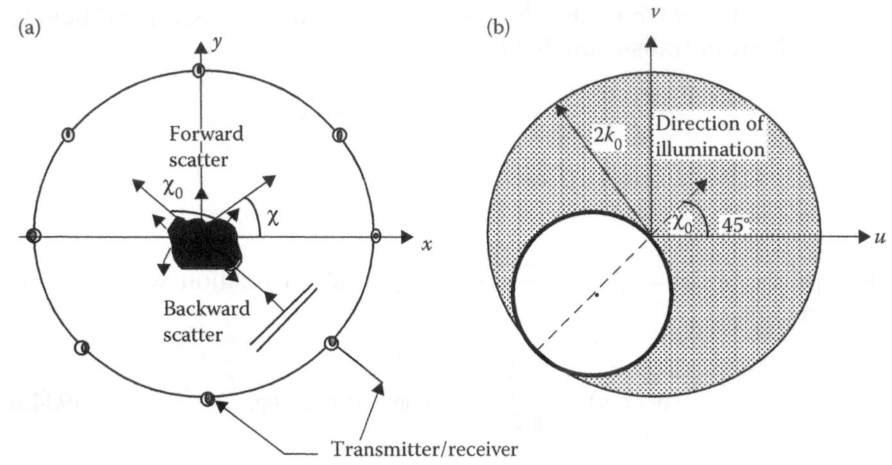

FIGURE 9.11
(a) An experimental setup of transceivers for tomographic imaging. (b) $\Delta c(u, v)$ is now sampled over a circumference of a circle of radius k_0, as shown. The angle of incidence of the plane wave illumination is 45°.

shall reconsider the same in the context of the backpropagation algorithm [8]. But, first, let us look at a simpler system, a circular array of transceivers of interest in medical imaging (see Figure 9.11).

9.2.3.1 Circular Array

The backscatter is lost in a linear array tomographic system; naturally, some potentially useful information is also lost, in the sense that only a part of the object spectrum lying within a disc of radius equal to $\sqrt{2k_0}$ is utilized. An alternate tomographic system consisting of a circular array of sensors (transceivers) encircling the object is proposed [9]. In this configuration, both forward scatter and backward scatter are captured. This results in the doubling of the area of the spectrum coverage, a disc of radius $2k_0$.

Arbitrarily shaped measurement boundaries were suggested in Refs. [10, 11], whose authors have shown that on a straight-line boundary or on a circular boundary it is enough to measure either the diffracted field or its normal derivative. For a completely arbitrary boundary, we need both types of measurements [11]. A circular transducer array was used to illuminate an object with a pulse (broadband) from different directions [12, 13], and the backscatter alone, measured as a function of time, was employed for the purpose of object reconstruction. A circular array for ultrasound holographic imaging was used by Qin et al. [14], but they have approximated a circular array by a series of linear arrays and then applied the backpropagation algorithm. A circular array of transceivers was suggested for microwave diffraction tomography [15], where a near-field diffraction phenomenon was used. The object Fourier transform was related to the scattered field through a two-dimensional convolution relation. The scattered field measured with a large circular array surrounding the object (see Figure 9.11a) is proportional to the Fourier transform of the object profile taken on the circumference of a circle of radius equal to the wavenumber and centered at ($-k_0\cos$, $-k_0\sin$). This result is called here a Fourier diffraction theorem (FDT) for a circular array [9].

$$P_s(R,\chi,\chi_0)=\frac{k_0^2}{4}e^{j(k_0R+(\pi/4))}\sqrt{\frac{2}{\pi k_0 R}}\Delta C(k_0(\cos\chi-\cos\chi_0),k_0(\sin\chi-\sin\chi_0)). \quad (9.15)$$

The left-hand side is simply the observed scattered field on a large circle. The right-hand side is a Fourier transform of the object function, which is evaluated on a circle of radius k_0 and centered at $k_x=-k_0\cos\chi_0$ and $k_y=-k_0\sin\chi_0$ (see Figure 9.11).

By changing the angle of incidence of the wavefront, χ_0, it is possible to cover the Fourier plane with a series of circles spanning a disc of radius equal to $2k_0$ (see Figure 9.12). The increased coverage has been possible because we captured the backscatter as well. Note the important difference is that the scattered field measured with a circular array, being in the far-field region, directly yields the object Fourier transform. On the contrary, with a linear array we need to Fourier transform the observed field. This important result is first verified against the measured scattered field.

9.2.3.2 Verification of FDT

First, we shall verify the FDT through an example where an exact scattered field as well as its object Fourier transform are known. Consider a liquid

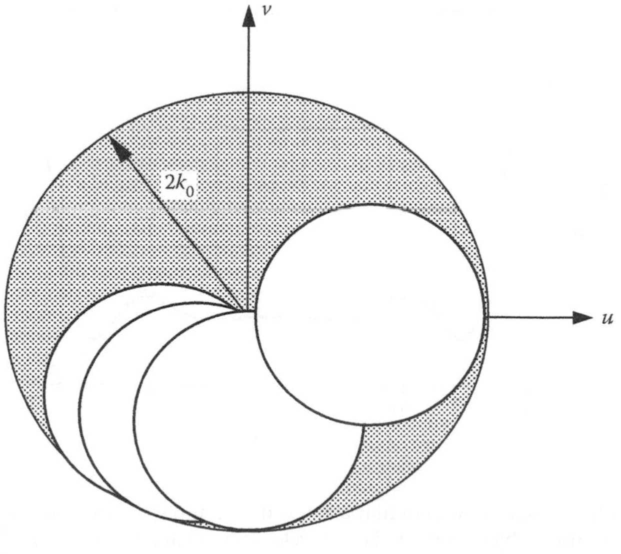

FIGURE 9.12
By changing the angle of incidence of the wavefront, χ_0, it is possible to cover the Fourier plane with a series of circles spanning a disc of radius equal to $2k_0$.

cylinder in water and assume that its refractive index is slightly above that of the water, $\delta n \le \lambda/4a$, where δn is the change in the refractive index, a is the radius of the cylinder, and λ is the wavelength of illuminating wave (Born approximation). The scattered field due to a liquid cylinder, when it is illuminated by a plane wave, was experimentally studied in Ref. [16] and theoretically in Ref. [17]. We have computed the scattered field using the derivation given in Ref. [17]. Next, we evaluate the object profile Fourier transform

$$\tilde{O}(k_0(\sin\chi - \sin\lambda_0), k_0(\cos\lambda - \cos\lambda_0)) = 2\pi\delta_n a \frac{J_1(ka)}{k},$$

where

$$k = \sqrt{(\sin\chi - \sin\chi_0)^2 + (\cos\chi - \cos\chi_0)^2}\, k_0,$$

as a function of χ for a fixed direction of illumination; in the present case, $\chi_0 = 0°$. The scattered field measured by a circular array is now compared with the Fourier transform of the object profile evaluated on a circle of radius k_0 centered at $(-k_0/\sqrt{2}, -k_0/\sqrt{2})$ (see Figure 9.13).

9.2.4 Bilinear Interpolation Algorithm

As the Fourier transform of sound speed fluctuations is obtained from the scattered field, either from a linear array or circular array, in principle it is possible, by inverse Fourier transformation, to estimate the sound speed fluctuations. In practice, however, this is not a trivial step. As the object Fourier transform is measured over a series of arcs, it will be necessary to interpolate

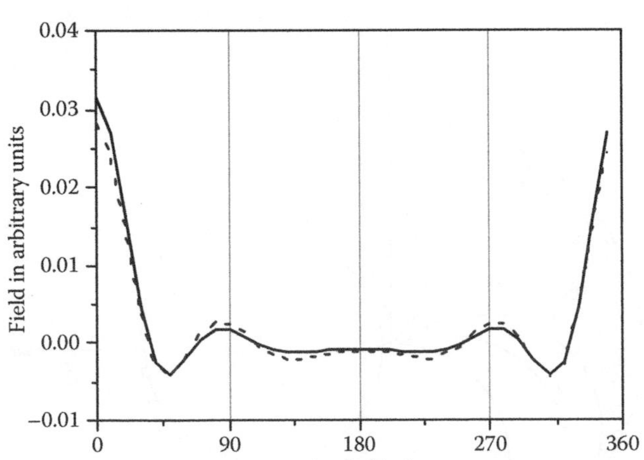

FIGURE 9.13
The scattered field measured by a circular array is now compared with the Fourier transform of a uniform cylindrical object evaluated on a circle centered at $(-k_0/\sqrt{2}, -k_0/\sqrt{2})$, as shown in Figure 9.11. The mean square error is 1.759×10^{-5}. (From P. S. Naidu, A. Vasuki, P. Satyamurthy and L. Anand, Diffraction tomographic imaging with circular array, *IEEE Trans.*, UFFC-42, pp. 787–789, 1995. With permission.)

to the nearest square grid point and thus create a discrete Fourier transform matrix for inversion. Alternatively, the backpropagation method originally developed for nondiffracting radiation may also be used. We shall use the interpolation approach.

Consider a circular array of transceivers. The angle of illumination χ_0 is varied over 360° by switching on a sensor to transmit mode, one at a time, and keeping the remaining sensors in the receive mode. The received field can be expressed as a function of two parameters, namely, the angle of illumination and the angular coordinate of each sensor, that is, (χ, χ_0). We must then map every point in (χ, χ_0) space onto (k_x, k_y) space. The reverse mapping, i.e., from (k_x, k_y) space onto (χ, χ_0) space, is, however, more convenient to use. The to-and-fro mapping functions are as follows:

$$\cos\chi - \cos\chi_0 = \frac{k_x}{k_0}, \ \sin\chi - \sin\chi_0 = \frac{k_y}{k_0}. \tag{9.16}$$

Solving the above equations for χ_0, we get the following inverse-mapping functions:

$$\chi_0 = \tan^{-1}\left\{\frac{-k_y - k_x\sqrt{\frac{2}{p}-1}}{-k_x + k_y\sqrt{\frac{2}{p}-1}}\right\}$$

$$\chi = \tan^{-1}\left\{\frac{k_y - k_x\sqrt{\frac{2}{p}-1}}{k_x + k_y\sqrt{\frac{2}{p}-1}}\right\}, \tag{9.17}$$

where $p = (k_x^2 + k_y^2)/2k_0^2$. Equations 9.16 and 9.17 together give a set of transformation equations that can be used to map from the k-plane into the χ-plane. Every point in the k-plane is mapped onto the χ-plane. The values of (χ, χ_0) thus obtained may not correspond to any of those points where the scattered field is observed; then we must take recourse to some form of interpolation. For example, bilinear interpolation is given by

$$O(\chi, \chi_0) = \sum_i \sum_j O(\chi_i, \chi_j) h_1(\chi - \chi_i) h_2(\chi_0 - \chi_j),$$

where $h_1(\chi) = 1 - |\chi|/\Delta\chi$, for $|\chi| \leq \Delta\chi$ otherwise $=0$ and $h_2(\chi_0) = 1 - (|\chi_0|/\Delta\chi_0)$ for $|\chi_0| \leq \Delta\chi_0$ otherwise $=0$. Here $\Delta\chi$ and $\Delta\chi_0$ are the sampling intervals. Once the values of the Fourier transform are obtained over a rectangular grid in (k_x, k_y) space, the inverse two-dimensional Fourier transform can be computed to obtain the object profile. The above algorithm is essentially an adaptation of the frequency domain interpolation algorithm, which is known to be very fast [1].

9.2.5 Imaging with a Circular Array

Since a circular array captures the entire diffracted energy, that is, both forward and backward scattered energy, a greater part of the object spectrum is utilized, indeed twice that of forward-scatter-only (linear array) setup. Consequently, we expect a better resolution of small inhomogeneities. To demonstrate this, we have carried out the following numerical experiment [18]. A small inhomogeneity of radius 0.25λ is embedded in a larger cylinder of radius 2λ having a refractive index contrasts with respect to the surrounding medium of 0.01 and 0.005, respectively (Figure 9.14b). The target is insonified with a narrowband plane wave radiation of wavelength 1.0λ. A circular array of 64 transceivers is assumed. For comparison, we have also considered a linear array of the same length and one transmitter located on the broadside, and the scattered field was calculated using the object Fourier transform over semicircular arcs. For a circular array, however, the scattered field was computed using the exact solution given in Ref. [16]. The reconstruction (a central slice) of the target is shown in Figure 9.14. The reconstruction obtained using a linear array is shown in Figure 9.14a and that obtained using a circular array is shown in Figure 9.14b. Clearly, the circular array outperforms the equivalent linear array as the small inhomogeneity is more accurately located. Next, we would like to emphasize the role of the array size on object reconstruction. When using a linear array, it is necessary that the array output be Fourier transformed before it is used for reconstruction. Consequently, the errors in the Fourier transformation due to the finite size of the array will degrade the reconstruction. This effect is demonstrated in Figure 9.15. The first three figures (Figure 9.15a, b, and c) were obtained using a linear array of three different sizes, namely, 64, 128, and 512 receivers spaced at $\lambda/2$ and 100λ away from the object, and a cylinder of radius 1λ with a refractive index contrast of 0.01. The scattered field was computed using the exact solution given in Ref. [16]. The reconstruction shown in Figure 9.15d was obtained using a 64-element circular array (radius $=100\lambda$). The reconstruction obtained with the circular array is superior to that obtained with a linear array of a much larger size (512 receivers). Notice that the sidelobes have practically disappeared.

9.3 Broadband Illumination

The object of interest is illuminated with a broadband signal, which results in better coverage of the spatial spectrum of the object with fewer illuminations. In this section, we shall examine the effectiveness of broadband illumination.

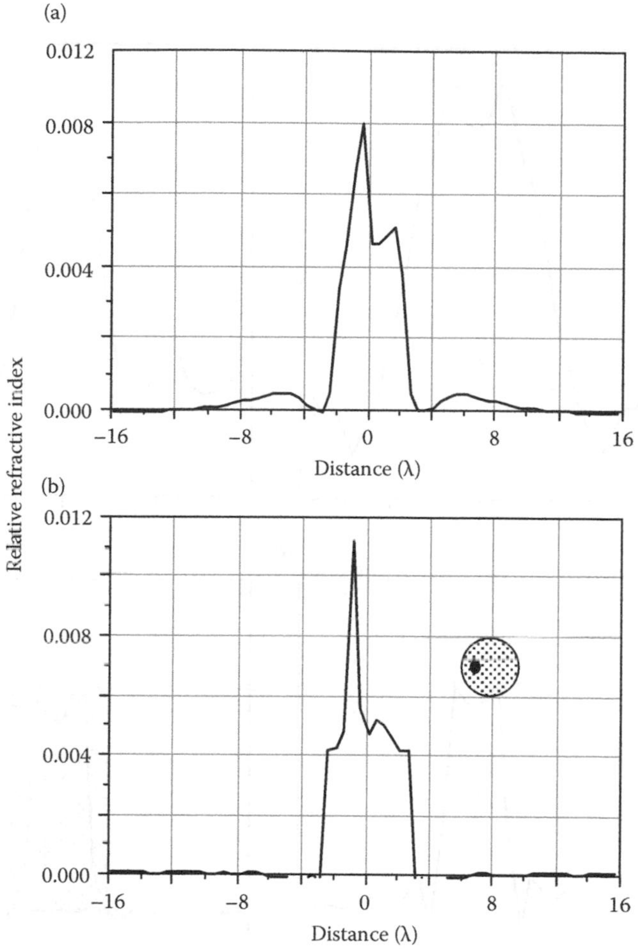

FIGURE 9.14
An example of improved performance of the circular array over linear array. A cylinder of radius 2λ with a small inhomogeneity of radius 0.25λ embedded in it (see inset in (b)) is used as a test target. (a) Reconstruction using a linear array and (b) using a circular array. (From A. Vasuki and P. S. Naidu, *Acoust. Lett.* (UK), vol. 21, pp. 144–150, 1998. With permission.)

9.3.1 Spectrum Coverage

If the object is illuminated from one direction with a set of different frequencies covering a spectral band, the scattered field will correspond to a set of circular arcs covering a crescent-shaped region in the object Fourier plane, as shown in Figure 9.16b. The radii of the inner and outer circles forming the crescent are related to the lower and upper cutoff frequencies. Let f_l and f_u be the lower and upper cutoff frequencies, respectively; the radii of the circles are $2\pi f_l/c_0$ and $2\pi f_u/c_0$, respectively. By suitably selecting the lower or upper cutoff frequencies, it is possible to emphasize the low or high frequency spatial

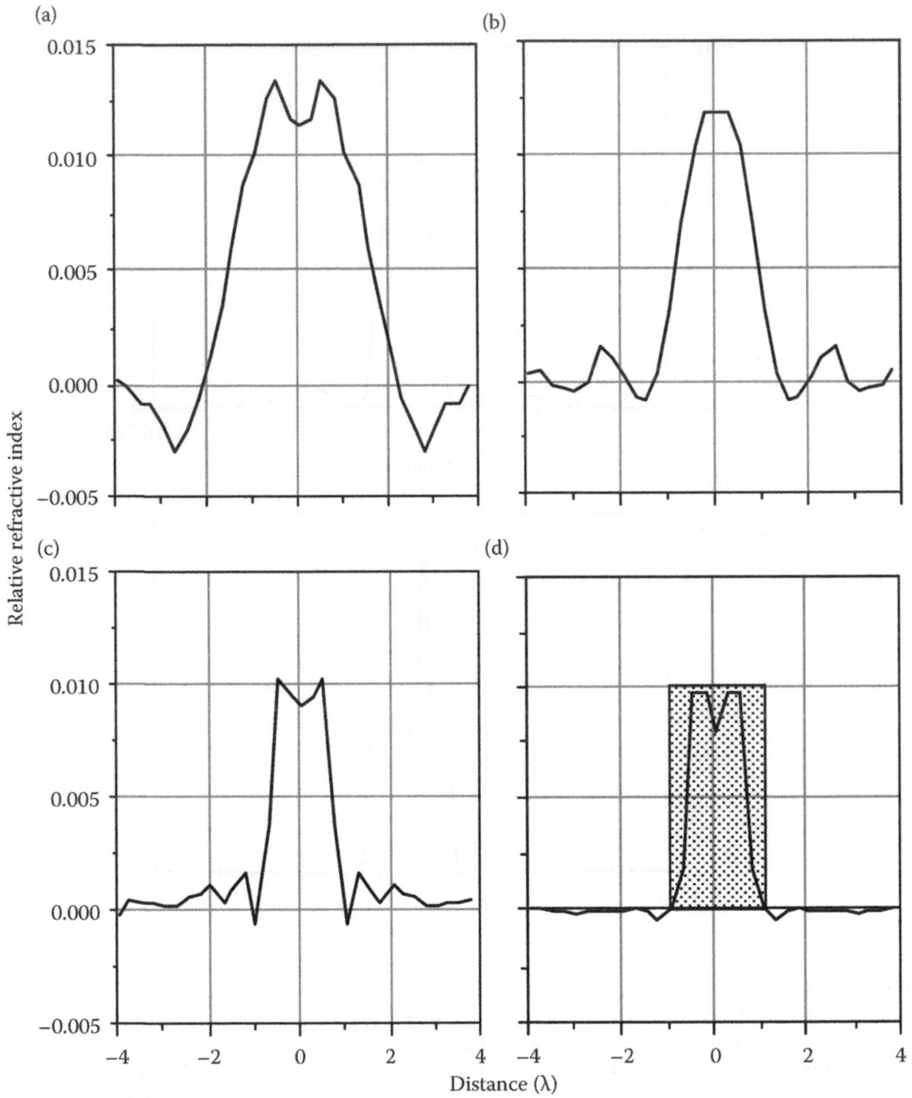

FIGURE 9.15
A comparison of performance of a linear array of finite size with that of a circular array. The number of illuminations in *all* cases was 64. (From A. Vasuki and P. S. Naidu, *Acoust. Lett.* (UK), vol. 21, pp. 144–150, 1998. With permission.)

spectrum of the object. With a single illumination it is thus possible to cover a fraction r of the disc of maximum size of radius $2f_u$, where

$$r = \frac{f_u^2 - f_l^2}{4f_u^2}. \tag{9.18}$$

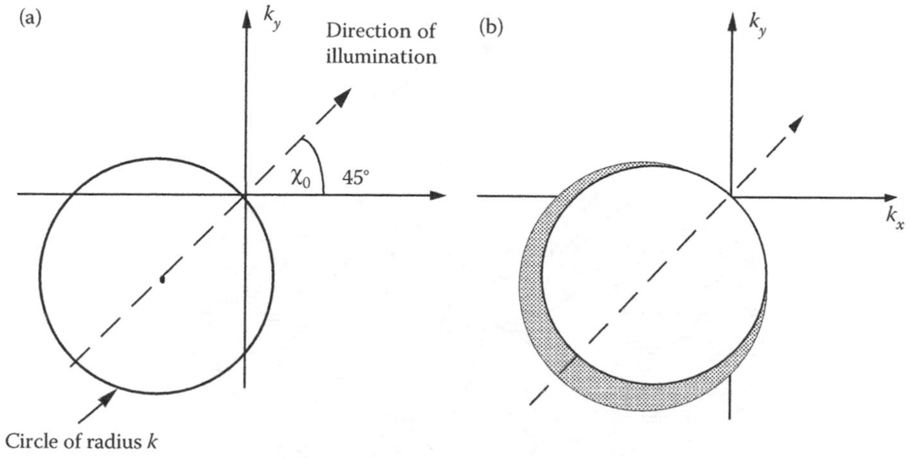

FIGURE 9.16
(a) For narrowband the scattered field is proportional to the object Fourier transform evaluated on a circle. (b) For finite band, the scattered field is proportional to the object Fourier transform evaluated inside a crescent-shaped region. (From A. Vasuki and P. S. Naidu, *Acoust. Lett.* (UK), vol. 21, pp. 144–150, 1998. With permission.)

Note that $r \leq 0.25$, where the maximum is achieved when $f_l = 0$. To cover the entire disc we need four or more illuminations, but certainly far fewer than the number of illuminations for narrowband illumination. With more than two illuminations, certain regions are likely to be covered more than once; thus the fraction of covered area will be less than that predicted by Equation 9.18. For example, for four illuminations the area covered is shown in Figure 9.17, where we have assumed that $f_l = 1.5$ kHz and $f_u = 3.0$ kHz. The fraction of the Fourier plane covered is 75%. It is further possible to increase this fraction, in particular, by decreasing the lower cutoff frequency. It is straightforward to analytically compute the area covered by one or more crescents, although it becomes difficult with an increasing number of illuminations. In Table 9.2, we list the fraction of area covered as a function of lower cutoff, but keep the upper cutoff frequency fixed at 3 kHz. Note that 100% coverage is not possible even when $f_l = 0$.

Further, to demonstrate the effect of the lower cutoff, we have carried out computer reconstruction using four illuminations and 64 sensors (circular array). The results are shown in Figure 9.18. There is an overall improvement in the reconstruction with decreasing lower cutoff frequency. Evidently, increasing the number of illuminations will also increase the coverage of the Fourier plane. For example, in Table 9.3 we show how the coverage increases as we increase the number of illuminations. Here the lower and upper cutoff frequencies are held fixed at 250 and 3,000 Hz, respectively. Four to eight illuminations seem to be ideal as not much is gained by going beyond eight illuminations, which gives almost 95% coverage.

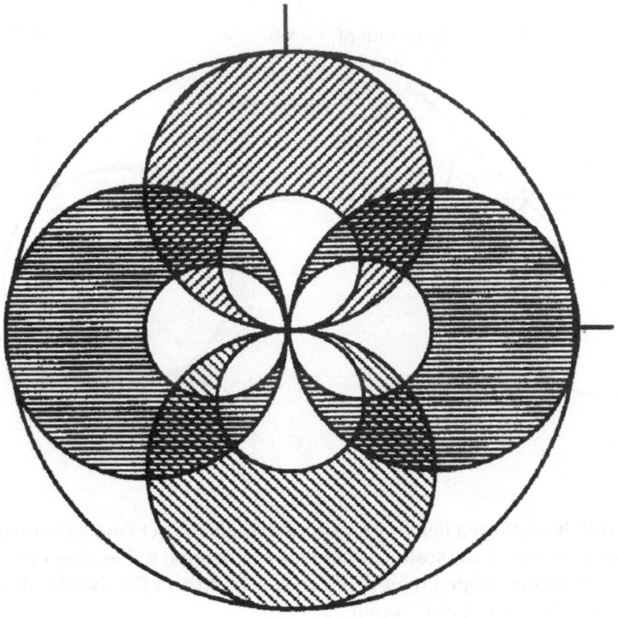

FIGURE 9.17
Broadband illumination provides a better coverage of the object Fourier transform. Just with four illuminations, it is possible to get a 75% coverage. The lower and upper cutoff frequencies are 1.5 and 3.0 kHz, respectively. (From A. Vasuki and P. S. Naidu, *Acoust. Lett.* (UK), vol. 21, pp. 144–150, 1998. With permission.)

TABLE 9.2
The fraction of the disc area covered with a broadband signal whose lower cutoff is varied and upper cutoff is held fixed (3.0 kHz)

Lower Cutoff Frequency (Hz)	Area Covered (%)
2,000	34.35
1,500	75.00
750	80.72
500	81.37
250	81.71
0	82.00

Source: A. Vasuki and P. S. Naidu, *Acoust. Lett.* (UK), vol. 21, pp. 144–150, 1998. With permission.
Note: We have assumed four illuminations.

9.3.2 Signal Processing Issues

The major issue in tomographic imaging from the point of signal processing relates to the fact that the object Fourier transform is sampled on a nonstandard grid, such as polar rastor in nondiffracting tomography or circular arcs

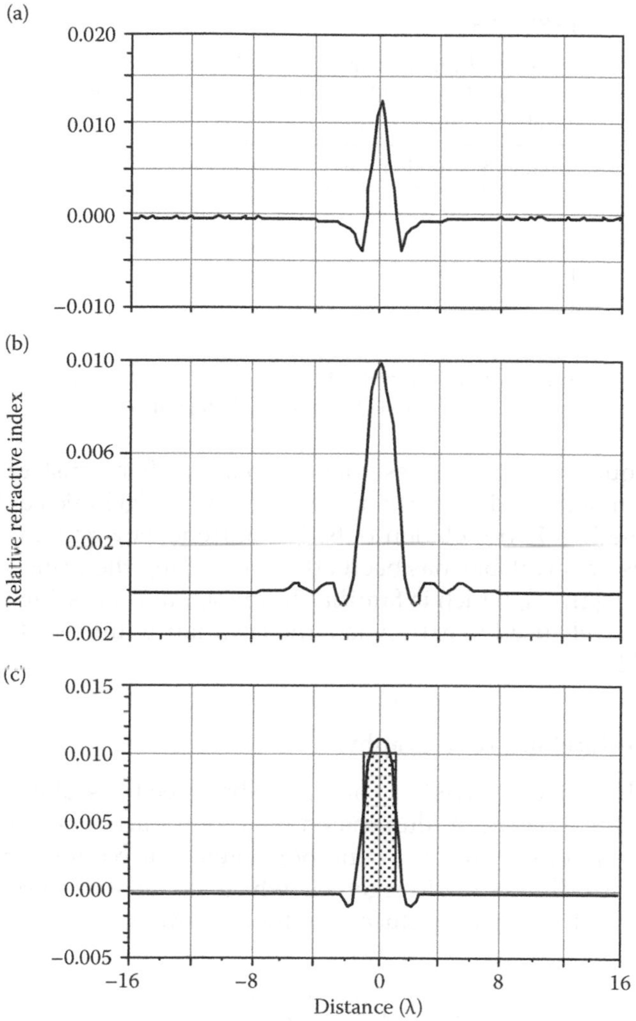

FIGURE 9.18
The effect of the lower cutoff frequency on the reconstruction. (a) 1,500 Hz, (b) 500 Hz, and (c) 250 Hz. The upper cutoff frequency is 3,000 Hz. (From A. Vasuki and P. S. Naidu, *Acoust. Lett.* (UK), vol. 21, pp. 144–150, 1998. With permission.)

in diffraction tomography. All existing methods of reconstruction require interpolation to convert from nonstandard grid to standard square grid. Another issue relates to the fact that in any practical implementation of the tomographic-imaging scheme only a limited number of views, often covering a finite angular interval, are likely to be available, leaving large gaps in the Fourier plane. As shown earlier, broadband illumination can help to reduce the gaps. Signal-processing tools have also been suggested for extrapolation of the observed Fourier transform into the missing gaps. *A priori* information about the object, such as a limit on the support of the object function either in

TABLE 9.3

Fraction of the disc area covered with a broadband signal (250–3,000 Hz) for a different number of illuminations

Number of Illuminations	Area Covered (%)
1	24.83
2	49.66
4	81.71
8	94.9
12	97.2

Source: A. Vasuki and P. S. Naidu, *Acoust. Lett.* (UK), vol. 21, pp. 144–150, 1998. With permission.

space or frequency domain, does not help to uniquely reconstruct the object function from a limited number of samples on an algebraic contour in the Fourier plane [19]. Extrapolation outside the frequency domain, where the observations are available, has been attempted using the principle of maximum entropy [21, 22], which is familiar to the signal-processing community as it is extensively used to extrapolate the covariance function for spectrum estimation [20].

9.3.3 Cross-Borehole Tomography

In Chapter 1, we have derived a relationship between the scattered wavefield from a target that has been illuminated by a source array positioned in one borehole and a receiver array in another borehole (Chapter 1, Section 1.2). Consider a two-dimensional target for which the relationship is given by Equation 1.91, which we reproduce here for convenience,

$$F_1(u_1,u_2) = 2k_0^2\Delta\tilde{c}\left(\frac{u_1}{d}+\frac{u_2}{d}, \sqrt{k_0^2-\left(\frac{u_1}{d}\right)^2}-\sqrt{k_0^2-\left(\frac{u_2}{d}\right)^2}\right)$$

$$\times\frac{e^{-j\sqrt{k_0^2-\left(\frac{u_1}{d}\right)^2}L_a}}{\sqrt{k_0^2-\left(\frac{u_1}{d}\right)^2}}\frac{e^{-j\sqrt{k_0^2-(u_2/d)^2}L_b}}{\sqrt{k_0^2-\left(\frac{u_2}{d}\right)^2}}.$$

Let $d=\lambda_0/2$, $u_1'=2u_1/\lambda_0$, and $u_2'=2u_2/\lambda_0$. For $\pi \le u_1, u_2 \le -\pi$ it turns out that $k_0 \le u_1', u_2' \le -k_0$. A point in the Fourier plane (u_1',u_2') would correspond to a point (u, v) in the Fourier plane of $\Delta\tilde{c}$ where

$$\begin{aligned} u &= u_1' + u_2' \\ v &= \pm\left\{\sqrt{k_0^2-(u_1')^2}-\sqrt{k_0^2-(u_2')^2}\right\}. \end{aligned} \tag{9.19}$$

Eliminating u_2' from two equations in Equation 9.19 we obtain

$$\left[v \pm \sqrt{k_0^2 - (u_1')^2}\,\right]^2 + (u - u_1')^2 = k_0^2. \tag{9.20}$$

For a given value of u_1', Equation 9.20 describes a circle with radius k_0 and centered at $(u_1', \pm\sqrt{k_0^2 - (u_1')^2})$, for example; for $u_1' = 0$ the two circles are centered on the y-axis at $\pm k_0$. In Figure 9.19, we show different circles (arcs) for different values of u_1'; in particular, the thick arcs are for $u_1' = 0$ and $u_1' = -k_0$, and the thin arcs are for $u_1' = \pm(k_0/\sqrt{2}\,)$. Similarly, by eliminating u_1' from Equation 9.19 we obtain

$$\left[v \pm \sqrt{k_0^2 - (u_2')^2}\,\right]^2 + (u - u_2')^2 = k_0^2, \tag{9.21}$$

which describes a circle with radius k_0 and centered at $(u_2', \pm\sqrt{k_0^2 - (u_2')^2}\,)$. For a fixed u_2', the object Fourier transform is scanned over a circle, for example, A in Figure 9.20.

Let the receiving array (ULA) be beamed to receive the wavefield in some particular direction, that is, for some fixed u_2'. For $u_2' = 0$, the circle described by Equation 9.21 will intersect circle A at two points, namely, O' and O (see Figure 9.20). The object Fourier transform is sensed only at these points of

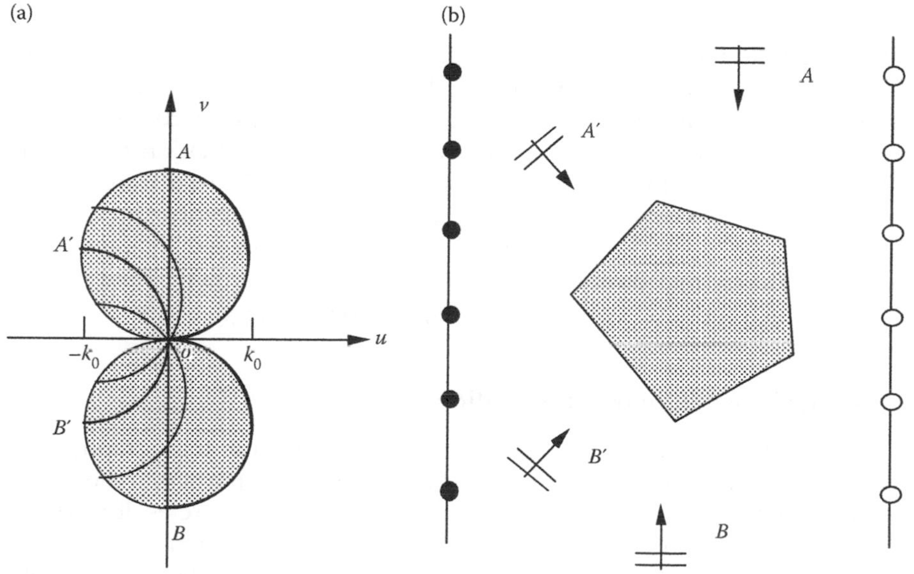

FIGURE 9.19
(a) The object Fourier transform lying inside two circular disks is scanned along a series of semicircular arcs. (b) As an example, consider four plane wavefronts A, A', B, B' emitted by distant sources in the source array. Corresponding circular arcs are shown in (a). The object Fourier transform is scanned on semicircle AO by wavefront A and on OB by wavefront B. Wavefronts A' and B' scan the object Fourier transform on $A'OB'$.

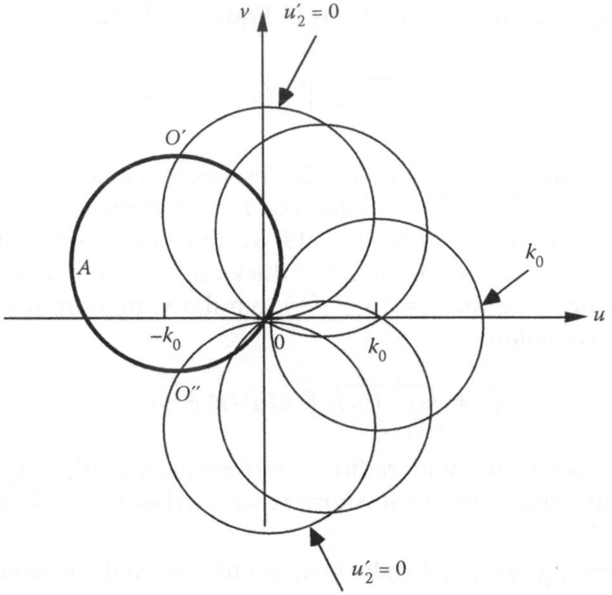

FIGURE 9.20
For a fixed u_1', the object Fourier transform is scanned over a circle *A*. For different values of u_2', that is, for different directions of the beam, a series of circles will intersect the circle *A* over an arc *O′′OO′*. The object Fourier transform will be sensed over this arc only.

intersection. For different values of u_2', that is, for different directions of the beam, we can draw a series of circles, which will intersect the circle *A* over an arc *O″OO′*. Some of these circles are shown in Figure 9.20. There will be as many arcs as the number of sources. In the ideal situation of an infinitely long source array, the arcs over which the object Fourier transform is sensed will fill two circles shown in Figure 9.19. A few arcs corresponding to $u_2'=\pm k_0, \pm(k_0/\sqrt{2})$ are also shown in this figure. In summary, in cross-borehole tomography the object Fourier transform can be scanned over a pair of circles along a series of semicircular arcs, as shown in Figure 9.19.

9.3.4 Maximum Entropy Extrapolation

The object Fourier transform scanned in cross-borehole tomography (see Figure 9.19) skips a good portion of the Fourier plane, in particular along the *u*-axis. Consider a square, $4k_0 \times 4k_0$, superscribing the two circles. It is found that just about 39% of the square is included within the two circles. Consequently, the resolution particularly along the *x*-axis will be considerably deteriorated. This is true only for infinitely long source and sensor arrays. Additionally, since in any borehole tomographic experiment the array size is bound to be finite, there will be a further reduction of the scanned area [21]. To overcome the effect of undercoverage of the object Fourier transform, it has been suggested to extrapolate the measured Fourier transform into an area where no measurements were possible, using the principle of maximum entropy [21, 22]. We shall briefly review this approach.

The aim of tomographic imaging is to determine the wave speed function $1+\delta\tilde{c}(x, y)$. We have already shown how, from the scattered field, we can obtain the Fourier transform of the wave speed variations over a limited domain (Figure 9.19), hereafter referred to as Ω, which is completely determined by the geometry of experiment. Nothing is, however, known of the wave speed outside this domain. Of the many possible functions, we choose one, which is consistent with the observed Fourier transform in the specified domain and is maximally noncommittal with regard to unavailable data. This is the principle of maximum entropy founded by Burg [23] and widely used in spectrum analysis [20]. The constrained optimization problem may be formulated as follows:

$$\text{minimize: } H = \iint_{\Gamma} (1+\delta\tilde{c}(x,y))\ln(1+\delta\tilde{c}(x,y))dxdy, \tag{9.22}$$

$$\text{subject to: } \Delta\tilde{c}(u,v) = \Delta\tilde{c}(u,v)\big|_{\text{measured in } \Omega} \quad (u,v)\in\Omega, \tag{9.23}$$

$$\begin{aligned}\partial H &= \iint_{\Gamma} \partial\delta\tilde{c}(x,y)[1+\ln(1+\delta\tilde{c}(x,y))]dxdy \\ &= \frac{1}{4\pi^2}\int_{-k_0}^{k_0}\int_{-k_0}^{k_0} \partial\Delta\tilde{c}(u,v)\iint_{\Gamma}[1+\ln(1+\delta\tilde{c}(x,y))]e^{j(ux+vy)}dxdydudv.\end{aligned}$$

H will be minimized when $\partial H=0$. Since $\Delta\tilde{c}\ (u, v)$ is already specified in $(u, v)\in\Omega$, $\partial\Delta\tilde{c}\ (u, v)=0$ in Ω. Therefore, maximization of entropy requires

$$\iint_{\Gamma}[1+\ln(1+\delta\tilde{c}(x,y))]e^{j(ux+vy)}dxdy = 0 \qquad \text{in } (u,v)\notin\Omega. \tag{9.24}$$

The solution is obtained by requiring to alternatively satisfy Equations 9.23 and 9.24. The algorithm for the constrained optimization has the following steps:

(1) Compute $q(u,v) = \iint_{\Gamma}[1+\ln(1+\delta\tilde{c}(x,y))]e^{j(ux+vy)}dxdy.$

(2) Set $q(u,v)=0$ for $(u, v)\notin\Omega$.

(3) Compute FT $\left\{\delta\tilde{c}(x, y) = e^{\text{IFT}\{q(u,v)\}-1} - 1\right\}$.

(4) Set $\Delta\tilde{c}(u, v)=\Delta\tilde{c}(u, v)\big|_{\text{measured}}$ for $(u,v)\in\Omega$.

(5) Compute $\delta\tilde{c}(x, y)=\text{IFT}\{\Delta\tilde{c}(u, v)\}$ and go to step 1.

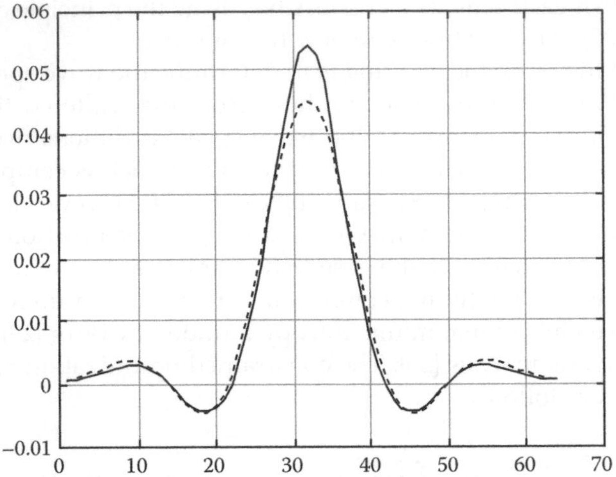

FIGURE 9.21
Maximum entropy reconstruction, dashed line (--) filtered object and solid line (__) maximum entropy reconstruction. There is only a marginal improvement in the maximum entropy reconstructed object.

The procedure is terminated when the reconstructed function meets some criterion of convergence. The algorithm is demonstrated through an example. A square object of size (5×5) with wave speed $c(=1+0.1)$ is embedded in a background with wave speed, $c_0=1.0$. The Fourier transform of the object was computed over a grid of 64×64. Let us assume that cross-borehole geometry permits measuring the object Fourier transform over a figure of 8 (see Figure 9.19a), where the radius of the pass disc is four. The filtered object Fourier transform was then used in the maximum entropy reconstruction algorithm described above. The reconstructed object (a horizontal cross section) after four iterations is shown in Figure 9.21, but it did not change much even after 40 iterations. Much of the spectrum lost during filtering remains unrecoverable except what lies between the upper and lower discs.

9.4 Reflection Tomography

The wavefield returned by an object may be considered either as a reflection at the surface of discontinuity in physical parameters (wave speed, density, etc.) or backscattering from deep inside the object due to variable physical parameters. In this section, we shall deal with the latter situation and reserve the former to be dealt with in Chapter 10. The backscatter from an inhomogeneous object may be related to the inhomogeneity inside the object under the usual assumption of weak scattering (Born approximation). Indeed, the

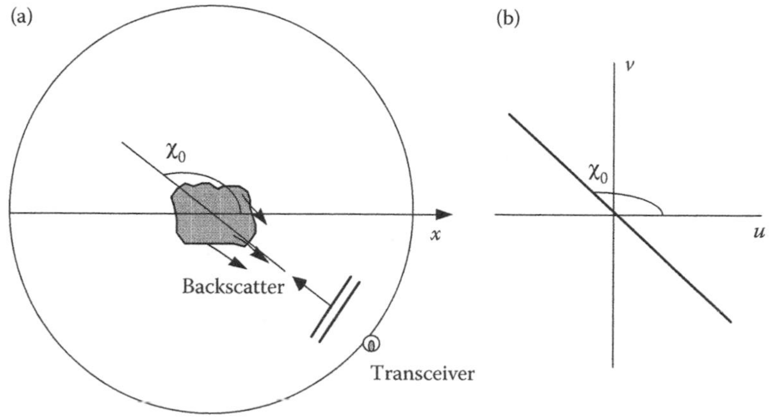

FIGURE 9.22
(a) A single transceiver is used to illuminate an object and receive the backscatter. A plane wavefront is assumed to be incident at an angle χ_0. (b) Backscatter coefficients are proportional to the object Fourier transform over a radial line as shown.

backscattering coefficient in the frequency domain may be easily derived from Equation 9.15 by setting $\chi=-(180-\chi_0)$. We obtain

$$P_s(R,\chi_0)=\frac{k_0^2}{4}e^{j(k_0R+(\pi/4))}\sqrt{\frac{2}{\pi k_0 R}}\Delta C(-2k_0\cos\chi_0,-2k_0\sin\chi_0), \qquad (9.25)$$

where R now stands for distance to the center of the object whose size is assumed to be much smaller than R, so that the far-field approximation hold good. The experimental setup is shown in Figure 9.22a. The backscatter coefficient at a fixed frequency, $P_s(R,\chi_0)$, is proportional to the object Fourier transform at spatial frequency $(-2k_0\cos\chi_0, -2k_0\sin\chi_0)$, where χ_0 is the angle of illumination (see Figure 9.22b). If we now use a broadband signal for illuminating the object (keeping χ_0 fixed) and Fourier decompose the received signal, we shall obtain the object Fourier transform over a radial line at an angle χ_0 (see Figure 9.22b). By illuminating the object repeatedly over 360°, either by physically taking the transceiver around the object or rotating the object around its axis, we can cover the entire Fourier plane. This commonly used experimental setup was suggested by Ref. [13]. The above result is akin to the Fourier slice theorem of nondiffracting tomography (Section 7.1). Naturally, many of the reconstruction methods developed for transmission (nondiffracting) tomography, in particular, backprojection, backpropagation, and interpolation methods, can be used in the present case of reflection tomography. Additionally, there is further similarity between the transmission tomography and reflection tomography. As noted in Section 7.1, the projection of an object is equal to the line integral of some physical quantity (e.g., absorption in x-ray tomography) over the ray path. A similar physical insight can be given to reflection tomography.

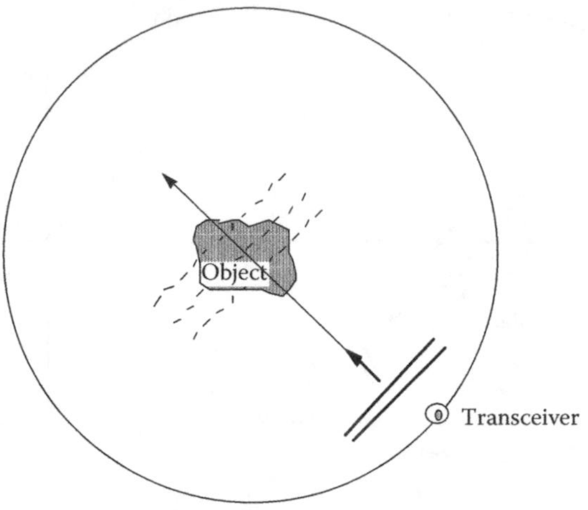

FIGURE 9.23
At any time instant the receiver will receive echoes from all scattering elements that lie on a surface, *s*.

9.4.1 Line Integral

Let a broadband plane wavefront be incident on a scattering object. A series of echoes will be emitted as the wavefront penetrates the object. At any time instant, the receiver will receive echoes from all scattering elements that lie on a surface (see Figure 9.23). Let $f(x, y)$ be the reflectivity function

$$\begin{aligned} p_s(\rho, \chi_0) &= \int_s f(x,y)ds \\ &= \int_0^\infty \int_0^{2\pi} f(r,\theta)\delta\left(\sqrt{r^2+R^2+2rR\cos(\theta-\chi_0)}-\rho\right)rdrd\theta. \end{aligned} \tag{9.26}$$

Different variables appearing in Equation 9.26 are illustrated in Figure 9.24. The limits on integrals suggest that the object is of infinite size. But in practice, the object is finite and the transceiver is placed outside the object. To overcome this difficulty we shall assume that the reflectivity is zero outside the domain of the object.

For fixed χ_0 we have a waveform that is a function of time or ρ ($=ct$). We then compute a Fourier transform of the waveform $p_s(\rho/c, \chi_0)$. After simplification we obtain

$$P_s(k,\chi_0) = \int_0^\infty \int_0^{2\pi} f(r,\theta)\exp(-jk\left(\sqrt{r^2+R^2+2rR\cos(\theta-\chi_0)}\right)rdrd\theta. \tag{9.27}$$

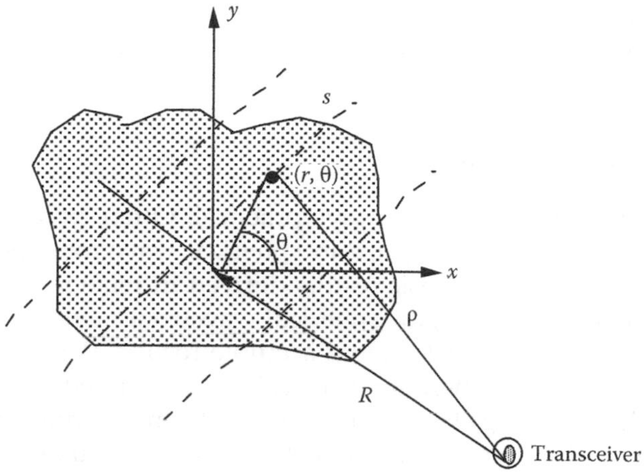

FIGURE 9.24
Symbols used in Equation 9.26 are explained in the figure.

Under the assumption $R \gg r$, the exponent in Equation 9.27 may be expanded in binomial series. Retaining the first two terms in the binomial expansion

$$\sqrt{r^2+R^2+2rR\cos(\theta-\chi_0)}$$
$$\approx R+r\cos\left((\theta-\chi_0)+\frac{r^2}{2R}(1-\cos^2(\theta-\chi_0))\right),$$

we obtain

$$P_s(k,\chi_0)=e^{-jkR}\int_0^{\infty}\int_0^{2\pi} f(r,\theta)e^{-jk((r^2/2R)(1-\cos^2(\theta-\chi_0)))}e^{-jkr\cos(\theta-\chi_0)}r\,dr\,d\theta. \tag{9.28}$$

It is easy to see that

$$r\cos(\theta-\chi_0)=x\cos(\chi_0)+y\sin(\chi_0),$$

and

$$r^2(1-\cos^2(\theta-\chi_0))=(x\sin(\chi_0)-y\cos(\chi_0))^2.$$

Substituting in Equation 9.28 we obtain

$$P_s(k,\chi_0)=e^{-j\omega R}$$
$$\times\int_{-\infty}^{\infty}\int_{-\infty}^{\infty} f(x,y)e^{-jk((x\sin(\chi_0)-y\cos(\chi_0))^2/2R)}e^{-jk(x\cos(\chi_0)+y\sin(\chi_0))}dx\,dy. \tag{9.29a}$$

If we assume that the object size is much smaller than R, the middle term in Equation 9.29a may be set to 1, leading to a simple Fourier transform relation,

$$P_s(k,\chi_0)=F(k\cos(\chi_0),k\sin(\chi_0)), \tag{9.29b}$$

where $F(.)$ is the Fourier transform of $f(x, y)$. When this assumption is not acceptable, the presence of a middle term can be accounted for through an iterative procedure described by Ref. [24].

9.4.2 Spherical Wavefronts

Plane wave illumination is not always practical. Often, point sources at a finite distance are used, giving rise to spherical wavefronts penetrating a scattering object. While the basic results derived in relation to a simplified plane wave model hold good, the spherical wavefront model, though mathematically more complex, yields more accurate results. We shall derive the exact result in one case where the transceivers are placed on a horizontal plane above the scattering object. Consider a uniform two-dimensional array of transceivers and a three-dimensional object with arbitrary speed variation (see Figure 9.25). The transceivers are sequentially fired (in any order) and all returned signals are arranged with common zero time, that is, as if all transceivers were fired at the same time instant. Interestingly, in this setup it is practical to think of using a single transceiver and move it from place to place. Consider a weakly inhomogeneous (speed fluctuations only) object illuminated by a point source with the scattered field being received by a sensor kept close to the source. From every scattering volume element, a backscatter radiation reaches the detector as modeled. The first order backscatter may be derived from Equation 1.77, which we reproduce here for convenience,

$$f_1(\mathbf{r},t) = \frac{1}{4\pi}\int\!\!\int\!\!\int_{-\infty}^{+\infty} \frac{2k_0^2\delta\tilde{c}e^{j(k_0|\mathbf{r}-\mathbf{r}'|)}}{|\mathbf{r}-\mathbf{r}'|} f_0(\mathbf{r}',t)dx'dy'dz',$$

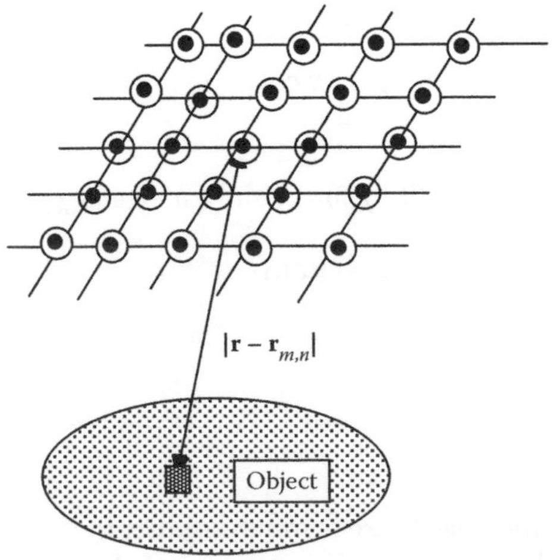

FIGURE 9.25
Uniform planar array of transceivers above a three-dimensional object.

where $f_0(\mathbf{r}', t)$ is an illuminating wavefield, which, for a point source emitting a sinusoid, is given by

$$f_0(\mathbf{r}',t) = \frac{e^{j(k_0|\mathbf{r}_s-\mathbf{r}'|)}}{|\mathbf{r}_s - \mathbf{r}'|} e^{-j\omega_0 t},$$

where $\mathbf{r}_s$ is source position vector. Since the source and detector are at the same location, $\mathbf{r}=\mathbf{r}_s$. Using the point source illumination expression given in Equation 1.77, we obtain

$$f_1(\mathbf{r},t) = \frac{1}{4\pi} e^{-j\omega_0 t} \iiint_{-\infty}^{+\infty} \frac{2k_0^2 \delta\tilde{c} e^{j2(k_0|\mathbf{r}-\mathbf{r}'|)}}{|\mathbf{r}-\mathbf{r}'|^2} dx'dy'dz'. \tag{9.30a}$$

The scattered field on a $z=0$ surface in rectangular coordinates may be written as

$$f_1(x,y,z=0,\omega_0) = \frac{1}{4\pi} \iiint_{-\infty}^{+\infty} \frac{2k_0^2 \delta\tilde{c}(x',y',z') e^{j2(k_0[(x-x')^2+(y-y')^2+(z-z')^2]^{1/2})}}{\left[(x-x')^2+(y-y')^2+(z')^2\right]} dx'dy'dz', \tag{9.30b}$$

which we shall express in a form that enables us to use a result (Equation 1.78), derived in Chapter 1, Section 1.6,

$$\frac{\partial}{\partial\omega_0}\left(\frac{f_1(x,y,z=0,\omega_0)}{\omega_0^2}\right) = \frac{j}{\pi c^3} \iiint_{-\infty}^{+\infty} \frac{\delta\tilde{c}(x',y',z') e^{j2(k_0[(x-x')^2+(y-y')^2+(z-z')^2]^{1/2})}}{\left[(x-x')^2+(y-y')^2+(z')^2\right]^{1/2}} dx'dy'dz'. \tag{9.31}$$

Now using the result (Equation 1.78) in Equation 9.31 we go into the frequency domain,

$$\frac{\partial}{\partial\omega_0}\left(\frac{f_1(x,y,z=0,\omega_0)}{\omega_0^2}\right) = \frac{1}{2\pi^2 c^3} \iint_{-\infty}^{+\infty} \frac{\Delta\tilde{c}(u,v,\sqrt{4k_0^2-u^2-v^2})}{\sqrt{4k_0^2-u^2-v^2}} e^{j(ux+vy)} du dv. \tag{9.32}$$

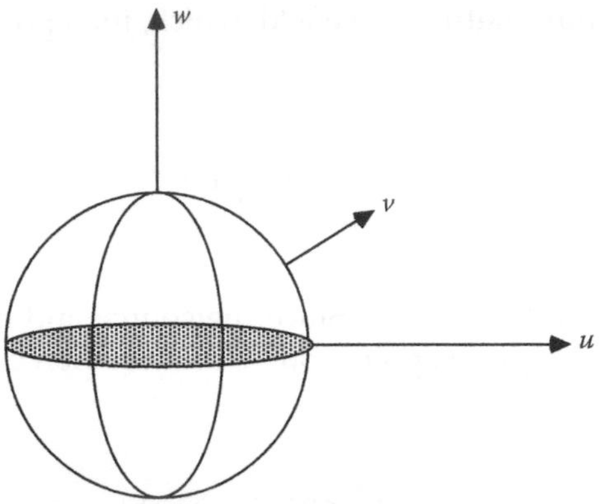

FIGURE 9.26
The Fourier transform of a reflected signal (echo) corresponds to the Fourier transform of the object on a sphere centered at the origin and radius equal to $2k_0$.

From Equation 9.32 we can get the Fourier transform of the speed fluctuations in terms of the Fourier transform of the wavefield observed on the surface,

$$\begin{aligned}&\Delta\tilde{c}(u,v,\sqrt{4k_0^2-u^2-v^2})\\&=\frac{c^3}{2}\sqrt{4k_0^2-u^2-v^2}\,\frac{\partial}{\partial\omega_0}\left(\frac{F_1(u,v,\omega_0)}{\omega_0^2}\right).\end{aligned} \tag{9.33}$$

Thus, the Fourier transform of the speed variations is derived from the Fourier transform of the backscatter measured on a plane surface. It is interesting to observe that the Fourier transform thus computed actually corresponds to the Fourier transform of the object on a sphere centered at the origin and with radius equal to $2k_0$ (see Figure 9.26). A broadband signal will be necessary to cover the entire Fourier transform of the speed variation function.

9.5 Object Shape Estimation

If the boundary of an object is piecewise linear, the corner points are sufficient for pattern recognition, image compression and coding, shape analysis, etc., [25]. The corner detection algorithms work on spatial image data in the form of a photograph. The sensor arrays are used for corner detection from the scattered wavefield (acoustic or electromagnetic). When an object, whose refractive index is slightly different with respect to that of the surrounding medium, is illuminated with a plane wave, the scattered field measured around the object is proportional to the Fourier transform of the object. Thus, the shape information is buried in the scattered field. It is of some interest in medical diagnosis, in subsurface imaging, and in nondestructive testing to be able to recognize the shape of the buried object from the scattered acoustic or electromagnetic field, particularly when only a few limited views are permitted. We shall show that when the object is binary, convex, and having a nondegenerate polygonal cross section, the scattered field is a sum of sinusoids, a function of wavenumber, and corners of the polygon. The object is illuminated with a broadband plane wave and the scattered field is measured as a function of wavenumber. The frequencies of the sinusoids are estimated from the scattered field using an algorithm described in Ref. [20].

It is shown in Section 7.2 that the scattered field measured by a circular array is proportional to the 2D Fourier transform of the object profile taken on the circumference of a circle of radius equal to the wavenumber and centered at $(-k_0 \cos \chi_0, -k_0 \sin \chi_0)$, where χ_0 is angle illumination (see Figure 9.11). By changing the direction of illumination (0–360°), the object Fourier transform is scanned over a disk of radius k_0. When the object is binary (i.e., refractive index is constant throughout the object) the interest is in the shape of the object. The shape information may be directly obtained from the scattered field or 2D Fourier transform of the object. This approach was taken by Milanfar and co-workers [26] in the context of ray tomography, where the input data are projections of object.

9.5.1 Fourier Transform of Binary Convex Polygonal Object

Consider the evaluation of the 2D Fourier transform over a p-sided binary and convex polygonal domain (see Figure 9.27). Take any point inside the

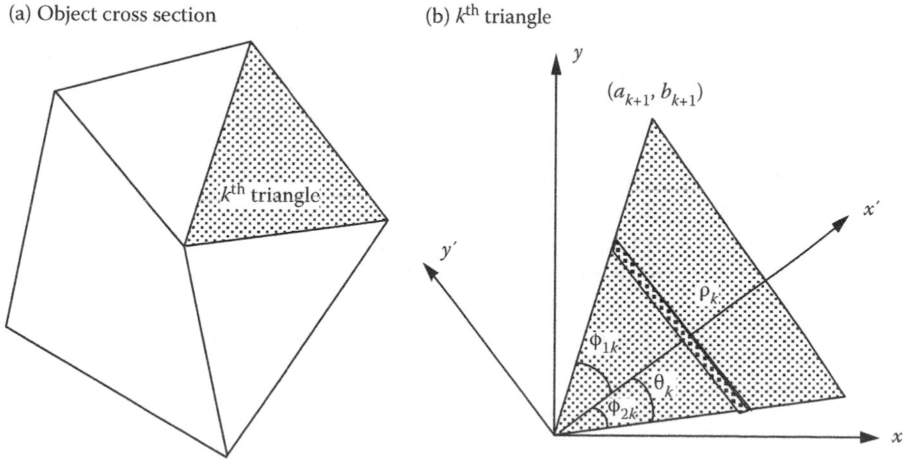

FIGURE 9.27
To evaluate the Fourier transform of a polygonal object we consider each triangle. (From A. Buvaneswari and P. S. Naidu, *IEEE Trans. Med. Imaging*, vol. 7, pp. 253–257, 1998. With permission.)

polygon and join it to all corners forming p triangles that lie entirely inside the polygon and make this the origin of the coordinate system

$$
\begin{aligned}
F(u,v) &= \iint\limits_{\text{over polygon}} e^{j(ux+vy)}dxdy \\
&= \sum_{n=1}^{p} \iint\limits_{\text{over } n\text{th triangle}} e^{j(ux+vy)}dxdy.
\end{aligned}
\tag{9.34}
$$

To evaluate the integral over nth triangle refer to Figure 9.27b, where we show the integration along a narrow strip under the rotated coordinate system such that the new x-axis is perpendicular to nth side. Note that this is valid only for convex objects (for nonconvex objects, it's not possible to drop a perpendicular from the origin to at least one edge, such that the perpendicular lies entirely within the object) that are nondegenerate. The triangle is then covered by a series of strips. Equation 9.34 reduces to

$$
F(u,v) = \sum_{n=1}^{p} \int_{0}^{\rho_n} \int_{-x'\tan\phi_{2n}}^{x'\tan\phi_{1n}} e^{\{-j((u\cos\theta_n+v\sin\theta_n)x'+(v\cos\theta_n-u\sin\theta_n)y')\}}dx'dy', \tag{9.35}
$$

where $x'=(x\cos\theta_k+y\sin\theta_k)$ and $y'=(y\cos\theta_k-x\sin\theta_k)$. Evaluate the integral in Equation 9.35 first with respect to y' followed by integration with respect to x'. We obtain

$$
F(u,v) = \sum_{n=1}^{p} \frac{[e^{-j(u'-v'\tan\phi_{1n})\rho_n}-1]}{v'(u'-v'\tan\phi_{1n})} - \frac{[e^{-j(u'+v'\tan\phi_{2n})\rho_n}-1]}{v'(u'+v'\tan\phi_{2n})}, \tag{9.36}
$$

where $u'=(u\cos\theta_n+v\sin\theta_n)$ and $v'=(v\cos\theta_n-u\sin\theta_n)$. We shall now rewrite Equation 9.36 by replacing θ_n, ϕ_{1n}, and ϕ_{2n} in terms of the coordinates of the two corners corresponding to the nth side, namely, (a_n, b_n) and (a_{n+1}, b_{n+1}). The following relations are used for this purpose:

$$\rho_n = a_n\cos\theta_n + b_n\sin\theta_n = a_{n+1}\cos\theta_n + b_{n+1}\sin\theta_n$$

$$a_n = \rho_n(\cos\theta_n + \sin\theta_n\tan\phi_{1n}),$$

$$b_n = \rho_n(\sin\theta_n - \cos\theta_n\tan\phi_{1n})$$

$$a_{n+1} = \rho_n(\cos\theta_n - \sin\theta_n\tan\phi_{2n})$$

$$b_{n+1} = \rho_n(\sin\theta_n + \cos\theta_n\tan\phi_{2n}).$$

We obtain

$$F(u,v)=\sum_{n=1}^{p}\rho_n\left[\frac{[e^{-j(ua_n+vb_n)}-1]}{v'(ua_n+vb_n)}-\frac{[e^{-j(ua_{n+1}+vb_{n+1})}-1]}{v'(ua_{n+1}+vb_{n+1})}\right]. \tag{9.37}$$

Our goal is to determine (a_n, b_n) and (a_{n+1}, b_{n+1}) from Equation 9.37. This may be achieved by expressing Equation 9.37 on the $k_y=0$ and $k_x=0$ axes. We get the following equations

$$\begin{aligned} u^2F(u,v=0) &= -\sum_{n=1}^{p}\rho_n\left\{\frac{e^{-jua_n}-1}{a_n\sin\theta_n}-\frac{e^{-jua_{n+1}}-1}{a_{n+1}\sin\theta_n}\right\} \\ v^2F(u=0,v) &= \sum_{n=1}^{p}\rho_n\left\{\frac{e^{-jvb_n}-1}{b_n\sin\theta_n}-\frac{e^{-jvb_{n+1}}-1}{b_{n+1}\sin\theta_n}\right\}. \end{aligned} \tag{9.38}$$

The above equations may be solved by modeling them as a sum of sinusoids and using the well-known Prony's algorithm or its more modern versions [20]. From the coefficients in the exponents of the complex sinusoids we obtain (a_n, a_{n+1}) and (b_n, b_{n+1}), but we are yet to pair them, that is, select the right x and y coordinate pair that will form a valid corner. We note that $F(u, v=0)$ and $F(u=0, v)$ represent backscatter due to a broadband illumination along the x- and y-axes, respectively (see Equation 9.29b).

9.5.2 Pairing Algorithm

In the previous section, we saw how to obtain the x- and y-coordinates of the corners of the polygon. This alone will not suffice to define a unique convex polygon. We need some additional information on how to pair a given x-coordinate with the right y-coordinate from the list of estimated y-coordinates. This problem is resolved by using an additional illumination at an angle θ

$$k^2F(k\cos\theta, k\sin\theta)$$

$$\sum_{n=1}^{p}\Gamma_n\left\{\frac{[e^{-jk(a_n\cos\theta+b_n\sin\theta)}-1]}{(a_n\cos\theta+b_n\sin\theta)}-\frac{[e^{-jk(a_{n+1}\cos\theta+b_{n+1}\sin\theta)}-1]}{(a_{n+1}\cos\theta+b_{n+1}\sin\theta)}\right\}, \tag{9.39}$$

where $\Gamma_n=(\rho_n/\sin(\theta-\theta_n))$. From the backscatter due to an illumination at angle θ ($\neq$ or $\pi/2$), we can estimate the coefficients in the exponents of the complex sinusoids as described in Refs. [20, 27]. Thus, we get the additional information in the form of linear combination of the x- and y-coordinates of the corners $(a_n \cos\theta+b_n \sin\theta)$ $n=1,2, \ldots, p$. The steps in the pairing algorithm are as follows:

(1) Generate a list of x-coordinates, y-coordinates, and the linear combination of the x- and y-coordinates. It is presumed that the list is not in the same order as the indexed corners.

(2) Take the first element from the x-coordinate list and any one element from the y-coordinate list and form a linear combination, $(a_1 \cos\theta+b_n \sin\theta)$ $n=1,2, \ldots, p$.

(3) Compare the result of the linear combination with those estimated with θ (≠ or π/2) illumination. The best match (within the limits of estimation error) will indicate the correct choice of b_n.

(4) Take the next element from the x-coordinate list and go to step (2).

For the purpose of illustration we consider a square object of size (6 m, 6 m), rotated by 30° and shifted away from the origin by (5 m, 5 m). It is illuminated from three directions, 0, π, and π/6. The x- and y-coordinates obtained from the noiseless scattered field in the first two directions and their linear combination ($\theta = 30°$) are shown in Table 9.4, and those estimated from the scattered field obtained in the third direction are shown in Table 9.5. The application of the pairing algorithm is illustrated in Table 9.6. The best match with the estimated coefficients is shown in column three in bold figures and the corresponding y-coordinate is shown in the last column.

We may encounter the problem of repeated x- or y-coordinates or their projections. The projections of two corners may overlap or come very close to each other, depending upon the orientation of the object. As shown in Figure 9.28, for a square object depending upon the orientation, the adjacent projections (e.g., x_1 and x_2) may come close to each other or overlap. The problem of repeated projection can be resolved by selecting another direction of illumination whenever the number of sinusoids estimated in x, y differs. The number of sinusoids observed in all three illuminations must be equal. In practice, it may be necessary to illuminate an unknown object along several directions and estimate the sinusoids along each direction. From this set, choose three directions, preferably two orthogonal directions, having an

TABLE 9.4

The x- and y-coordinates and their linear combination ($\theta = 30°$) are shown

x	y	$x \cos\theta + y \sin\theta$
9.0981	6.0981	10.9282
3.9019	9.0981	7.9282
0.9019	3.9019	2.7321
6.0981	0.9019	5.7321

Source: A. Buvaneswari and P. S. Naidu, *IEEE Trans. Med. Imaging*, vol. 7, pp. 253–257, 1998. With permission.

TABLE 9.5

The estimated projections from the scattered field (noise free) are shown

x	9.0981, 6.0981, 3.9019, 0.9019
y	9.0981, 6.0981, 3.9019, 0.9019
$x \cos\theta + y \sin\theta$	10.9282, 5.7321, 7.9282, 2.7321

Source: A. Buvaneswari and P. S. Naidu, *IEEE Trans. Med. Imaging*, vol. 7, pp. 253–257, 1998. With permission.

TABLE 9.6

A numerical illustration of the pairing algorithm ($\theta=30°$)

x	y	$x \cos\theta+y \sin\theta$	Best Match for x
9.0981	9.0981	12.4282	6.0981
	6.0981	**10.9282**	
	3.9019	9.8302	
	0.9019	8.3301	
6.0981	9.0981	9.8302	0.9019
	6.0981	8.3301	
	3.9019	7.2321	
	0.9019	**5.7321**	
3.9019	9.0981	**7.9282**	9.0981
	6.0981	6.4290	
	3.9019	5.3301	
	0.9019	3.8302	
0.9019	9.0981	5.3301	3.9019
	6.0981	3.8302	
	3.9019	**2.7321**	
	0.9019	0.5000	

Source: A. Buvaneswari and P. S. Naidu, *IEEE Trans. Med. Imaging*, vol. 7, pp. 253–257, 1998. With permission.

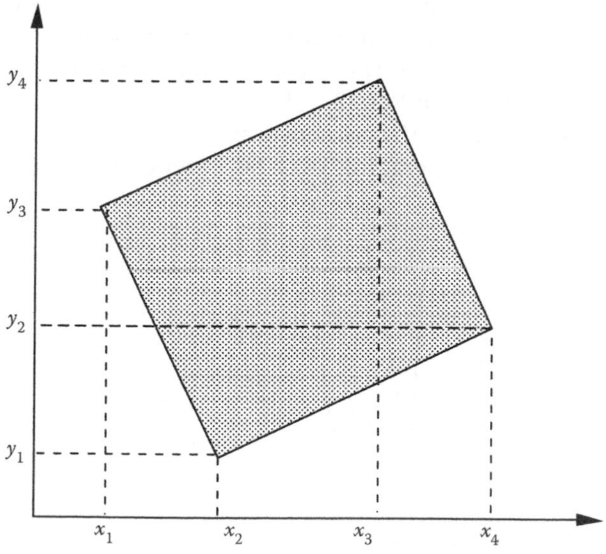

FIGURE 9.28
A square object and the projections of its corners on the x- and y-axes. Note that, depending upon the orientation, the adjacent projections (e.g., x_1 and x_2) may come close to each other. (From A. Buvaneswari and P. S. Naidu, *IEEE Trans. Med. Imaging*, vol. 7, pp. 253–257, 1998. With permission.)

equal number of sinusoids. The number of sinusoids that may be determined from a finite data in the presence of noise is indeed a complex problem and hence it is outside the scope of this book. The reader may like to refer to a book, e.g., Ref. [20], or current literature on this topic.

When there are two or more objects, as the center of coordinates would lie outside all but one object, it is necessary to modify Equation 9.31, which was derived under the assumption that the center lies inside the object. Also, there may be some ambiguity in the process of constructing the object shape even when all x- and y-coordinates are correctly paired. The amplitude of the sinusoid corresponding to a corner can then be used to resolve such an ambiguity in addition to the fact that the objects are convex and the number of objects is known (see Ref. [27] for more details).

9.5.3 Performance Analysis

The performance of the object shape estimation procedure has been investigated through numerical experiments [27]. For this we have considered a square object (refractive index contrast equal to 0.01) of size 8×8 m^2 and rotated by 9° with respect to the x-axis (see Figure 9.28). The object was illuminated with a broadband plane wavefront whose wave number varied from $\pi/64$ to π in steps of $\pi/64$ along three directions, namely, x-axis, y-axis, and a radial direction at an angle $\theta=30°$. The backscatter at each wavenumber was computed using the Fourier transform approach described in Ref. [1]. A typical example of backscatter caused by x-axis illumination is shown in Figure 9.29. To this scattered field, sufficient white Gaussian noise was added so that the SNR became equal to a specified figure. Here the SNR is defined as ten times the logarithm (base 10) of the ratio of the average scattered energy to noise variance. The corners and also the T_n' were estimated using the procedure given in Ref. [27]. The mean square error (MSE) in the estimation of coordinates of the corners was studied as a function of SNR. The results, obtained by averaging over 50 independent experiments, are

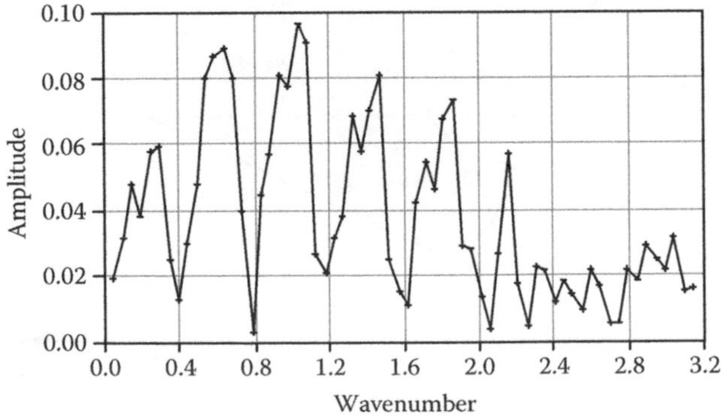

FIGURE 9.29
Computed backscatter from a weakly scattering square object (shown in Figure 7.28) illuminated by a broadband signal. (From A. Buvaneswari and P. S. Naidu, *IEEE Trans. Med. Imaging*, vol. 7, pp. 253–257, 1998. With permission.)

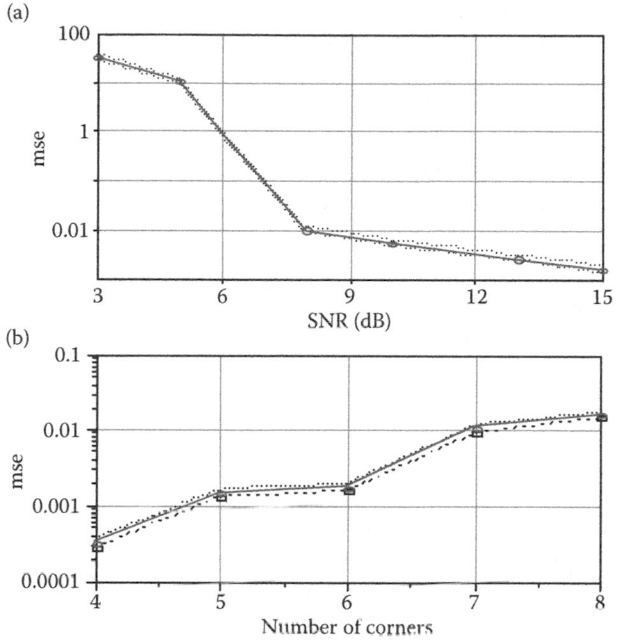

FIGURE 9.30
(a) Mean square error (m^2) in the estimation x- and y-coordinates as a function of SNR. (b) Mean square error (m^2) as a function of the number of corners (SNR=20 dB). The dotted lines show the error bounds for 95% confidence.(From A. Buvaneswari and P. S. Naidu, Estimation of shape of binary polygonal object from scattered field, *IEEE Trans.* on Med Imaging, vol. 7, pp. 253-257, 1998. With permission.)

shown in Figure 9.30a. Notice that mse rises very rapidly for SNR below 8 dB. This is largely on account of the fact that the projections of two adjacent corners (e.g., x_1 and x_4, and x_2 and x_3 in Figure 9.29) are close to each other; in this example they are 1.2515 m apart. For a different orientation, say, at 6° when the separation becomes 0.8362 m, the MSE rapidly rises for SNR below 15 dB. The estimation error (MSE) also depends upon the number of corners in a polygonal object. The numerical results are shown in Figure 7.30b.

9.6 Exercises

(1) What is the essential difference between an array of sensors used for DOA estimation (Chapters 2 and 5) or for signal waveform estimation (Chapters 3 and 6) and the array used in nondiffracting radiation tomography?

(2) Show that the filter function used in the backpropagation algorithm, (Equation 9.14c), reduces to that used in the backprojection algorithm (page 440) for the radiation of wavelength much smaller than the scale of speed/density fluctuations.

(3) A rectangular object (2D) is illuminated by a plane wavefront as shown in Figure 9.31. An array of sensors measures the relative delays. Assume that the speed of propagation inside the object is constant and that there are no diffraction effects. (a) Sketch the delay profile (delay as a function of sensor position) for different object orientations, $\theta=0°$ (as seen Figure 9.31), $\theta=45°$, and $\theta=90°$. (b) How would you estimate the size of the object?

(4) Consider a simple arrangement of sources and sensors in two boreholes, as shown in Figure 9.32. Compute the sampling points in the Fourier plane. See Ref. [28] on how to compute the sampling points in a more general case.

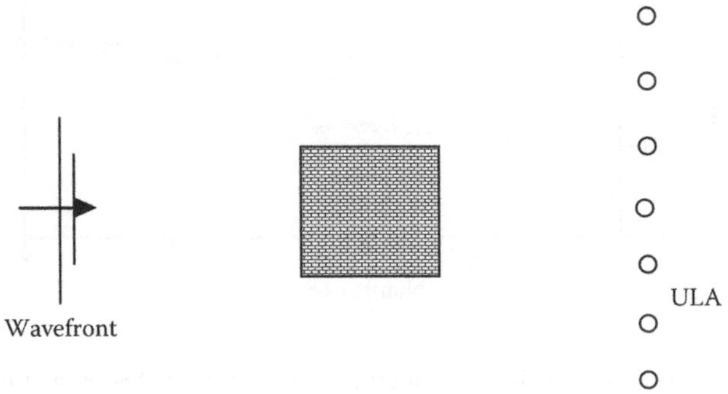

FIGURE 9.31
A rectangular object (2D) is illuminated by a plane wavefront.

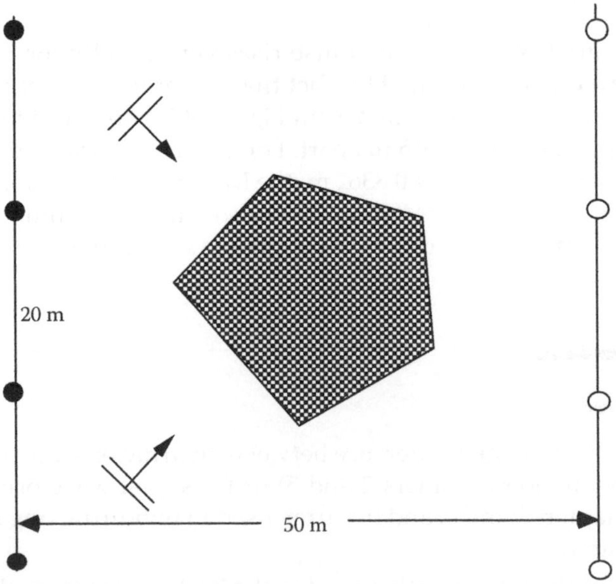

FIGURE 9.32
A cross-borehole tomography experimental setup.

References

1. A. C. Kak, Tomographic imaging with diffracting and non diffracting sources, In *Array Signal Processing*, Ed. S. Haykin, Prentice-Hall, Englewood Cliffs, NJ, pp. 351–428, 1985.
2. S. Ivansson, Seismic borehole tomography – theory and computational methods, *Proc. IEEE*, vol. 74, pp. 328–338, 1986.
3. M. Gustavsson, S. Ivansson, P. Moren and J. Pihl, Seismic borehole tomography – measurement system and field studies, *Proc. IEEE*, vol. 74, pp. 339–346, 1986.
4. K. Iwata and R. Nagata, Calculation of refractive index distribution from interferograms using the Born and Rytov's approximation, *Japan J. Appl. Phys.*, vol. 14, Suppl. 14-1, pp. 379–383, 1974.
5. R. K. Mueller, M. Kaveh and G. Wade, Reconstructive tomography and applications to ultrasonics, *Proc. IEEE*, vol. 67, pp. 567–587, 1979.
6. E. Wolf, Three dimensional structure determination of semi-transparent objects from holographic data, *Opt. Commun.*, vol. 1, pp. 153–156, 1969.
7. A. J. Devaney, Diffraction tomography, In *Inverse Methods in Electromagnetic Imaging*, Part 2, Eds W. M. Boerner et al., D. Reidel Pub. Co., pp. 1107–1135, 1985.
8. A. J. Devaney, Geophysical diffraction tomography, *IEEE Trans.*, vol. GE-22, pp. 3–13, 1984.
9. P. S. Naidu, A. Vasuki, P. Satyamurthy and L. Anand, Diffraction tomographic imaging with circular array, *IEEE Trans.*, UFFC-42, pp. 787–789, 1995.
10. R. Porter, Determination of structure of weak scatterers from holographic images, *Opt. Commun.*, vol. 39, pp. 362–364, 1981.
11. A. J. Devaney and G. Beylkin, Diffraction tomography using arbitrary transmitter and receiver surfaces, *Ultrasonic Imaging*, vol. 6, pp. 181–193, 1984.
12. S. J. Norton, Tomographic reconstruction of acoustic reflectivity using circular array: Exact solution, *J. Acoust. Soc. Am.*, vol. 67, pp. 1266–1273, 1980.
13. S. J. Norton and M. Linzer, Ultrasonic reflectivity tomography with circular transducer arrays, *Ultrasonic Imaging*, vol. 1, pp. 154–184, 1979.
14. Z. Qin et al., Circular array ultrasound holographic imaging using the linear array approach, *IEEE Trans.*, vol. UFFC-36, pp. 485–493, 1989.
15. J. M. Ruis, M. Ferrando, L. Jofre, E. De Los Reyes, A. Elias and A. Broquet, Microwave tomography: An algorithm for cylindrical geometries, *Electron. Lett.*, vol. 23, pp. 564–565, 1987.
16. P. Tamarkin, Scattering of an underwater ultrasonic beam from liquid cylindrical obstacles, *J. Acoust. Soc. Am.*, vol. 21, pp. 612–616, 1949.
17. S. J. Bezuszka, Scattering of underwater plane ultrasonic waves by liquid cylindrical obstacles, *J. Acoust. Soc. Am.*, vol. 25, pp. 1090–1095, 1953.
18. A. Vasuki and P. S. Naidu, Broadband tomographic imaging with circular array, *Acoust. Lett.* (UK), vol. 21, pp. 144–150, 1998.
19. J. L. Sanz, On the reconstruction of band-limited multidimensional signals from algebraic contours, IBM Research Report, RJ 4351, (47429), 1984.
20. P. S. Naidu, *Modern Spectrum Analysis of Time Series*, CRC Press, Boca Raton, FL, 1995.
21. T.-W. Lo, G. L. Duckworth and M. N. Toksoz, Minimum cross entropy seismic diffraction tomography, *J. Acoust. Soc. Am.*, vol. 87, pp. 748–756, 1990.

22. A. Mohammad-Djafari and G. Demoment, Maximum entropy Fourier synthesis with application to diffraction tomography, *Appl. Opt.*, vol. 26, pp. 1745–1754, 1987.
23. J. P. Burg, Maximum entropy spectrum analysis, Presented at the International Meeting of Exploration Geophysicists, Orlando, FL, 1967.
24. C. Q. Lang and W. Xiong, An iterative method of ultrasonic reflection mode tomography, *IEEE Trans. Med. Imaging*, vol. 13, pp. 419–425, 1994.
25. G. Medioni and Y. Yasumoto, Corner detection and curve representation using cubic B-spline, *Computer Vision Graphics and Image Processing*, vol. 39, pp. 267–278, 1987.
26. P. Milanfar, W. C. Karl, and A. S. Willsky, Reconstructing binary polygonal objects from projections: A statistical view, *Graphical Models and Image Processing*, vol. 56, pp. 371–391, 1994.
27. A. Buvaneswari and P. S. Naidu, Estimation of shape of binary polygonal object from scattered field, *IEEE Trans. Med. Imaging*, vol. 7, pp. 253–257, 1998.
28. D. Nahamoo, S. X. Pan, and A. C. Kak, Synthetic aperture diffraction tomography and its interpolation free computer implementation, *IEEE Trans.*, vol. SU-31, pp. 218–229.

10

Imaging by Wavefield Extrapolation

In Chapter 2, we have seen that the wavefield on a horizontal plane can be derived from the wavefield observed on another horizontal plane given that the space lying between the two planes is homogeneous and free from any sources. This operation is called extrapolation of wavefield; forward extrapolation is when we go away from the sources and backward extrapolation is when we go toward the sources (Figure 10.1). Wavefield extrapolation enables us to map the source distribution provided it is known that all sources are confined to a layer. This problem is known as an inverse source problem [1]. A slightly different situation arises in the scattering problem. An external source induces a field on the surface of a scatterer, which in turn will radiate wavefield, known as scattered field, back into the space. This scattered field contains information about the scatterers. The inverse scattering problem pertains to extraction of information about the scatterers from the scattered field. The tomographic imaging covered in Chapter 7 falls in the realm of the inverse scattering problem. In the present chapter, we seek a means of reconstructing a layered (but not necessarily horizontally layered) medium using the reflected wavefield. This problem is of great significance in seismic exploration, where it is commonly known as migration. An image of subsurface reflectors can also be achieved through focused beamformation, which gives an estimate of the reflected energy received from a subsurface point. The focused beamformation is based on a ray theoretic description of the wavefield, as in optics, but the migration is based on diffraction properties of the wavefield. Both approaches lead to similar results. For imaging, an essential input is the wave speed, which, fortunately, has to be estimated from the observed wavefield only.

10.1 Migration

The interface between two homogeneous layers may be considered as a thin layer of point sources (scatterers). This forms the basis of the exploding reflector model [2]. The wavefield observed on the surface of earth can be extrapolated downward into the earth. The interfaces separating homogeneous layers reflect or scatter wave energy. Such an interface may be modeled as a surface with point scatterers whose density is proportional to the

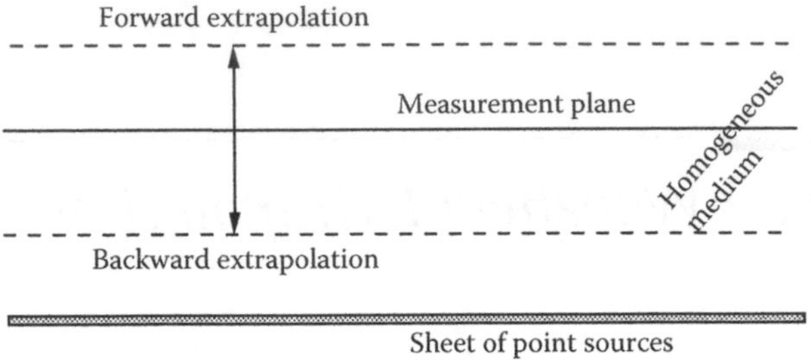

FIGURE 10.1
The wavefield measured on a horizontal plane can be extrapolated upward (forward extrapolation) or downward toward the source (backward extrapolation).

impedance discontinuity. To image an interface, that is, to map the impedance discontinuity, it is necessary to compute the distribution of the wave energy on an interface. This problem has been treated as a boundary value problem [3, 4] or an initial value problem [5]. As a boundary value problem, we solve the wave equation in homogeneous half space with a boundary condition that the wavefield is given on the surface of the earth ($z=0$).

The boundary value problem has been solved in the time domain by Claerbout [3] by solving a finite difference equation with a time varying boundary condition, and in the frequency domain by Stolt [4] by expressing the extrapolation as a filtering problem. Extrapolation in the z direction may also be expressed as propagation in backward time. Note that in the wave equation the double derivative with respect to z differs from the double derivative with respect to time only in a scale factor given by $-c^2$. Thus, starting from some time instant, the wavefield observed at the surface ($z=0$) is propagated backward in time, a process that is equivalent to extrapolation of the wavefield to a lower level ($z<0$). The boundary value problem may be reformulated as a source problem with zero boundary condition but driven by an external source that is given as a time-reversed output of each receiver [6]. In another approach, extrapolation is posed as an initial value problem but marching backward [5]. In this approach, the time axis is scaled by the wave speed that converts a recorded seismic section into a wavefield throughout the subspace as it might have appeared at the latest recording time. Thus, the converted wavefield is next propagated backward in time.

10.1.1 Imaging Conditions

Imaging requires two steps, namely, (1) extrapolation in space or reverse propagation in time; and (2) an imaging condition, that is, how to decide when an image has been formed. In optical imaging, convergence of all rays emerging from a point to another point (image point) is the imaging condition. In seismic or acoustic imaging, the imaging condition commonly used is when the depropagated field reaches the starting time, which is the time when the scatterer was illuminated or excited. This information can be found given the wave speed and the distance from the source (see Figure 10.2). It is also possible to set the excitation time to zero provided the scattered wavefront and illuminating wavefront travel along the same path but in opposite directions. This forms the basis of the popular migration principle called exploding reflector, which we shall describe in detail in the next section. Qualitatively speaking, imaging is focusing of wave energy. An imaging condition based on how well the wave energy is focused at a point is also a likely candidate as an imaging condition. Indeed, in seismic imaging, it has been suggested that when p-waves and s-waves are focused at the same point, an image of the point is obtained [7].

10.1.2 Downward Continuation of Sources and Sensors

The source and sensor arrays are normally placed on the same surface. It is possible to analytically compute the field when both source and sensor arrays are relocated onto another plane given the field on the observation plane. For simplicity, we shall assume that both the observation plane and the plane onto which the source and sensor arrays are relocated are horizontal. We are already familiar with continuation of wavefield from one plane to another (see Chapter 1 and also later in this chapter). Continuation of a source array requires an additional concept of reciprocity, which states that when the positions of an omnidirectional source and an omnidirectional sensor are interchanged, the observed field remains unchanged [8]. To extrapolate the source array keeping the sensor array fixed, we need to interchange the source and the sensor arrays and then apply the wavefield extrapolation algorithm. By virtue of the principle of reciprocity, the result of the above approach will be same as that of actual relocation of the source array. In actual application, the source and the sensor arrays are relocated alternatively in small steps. As the source and the sensor arrays are continued downwards toward a reflector, at some stage the two arrays will completely coincide when they reach the

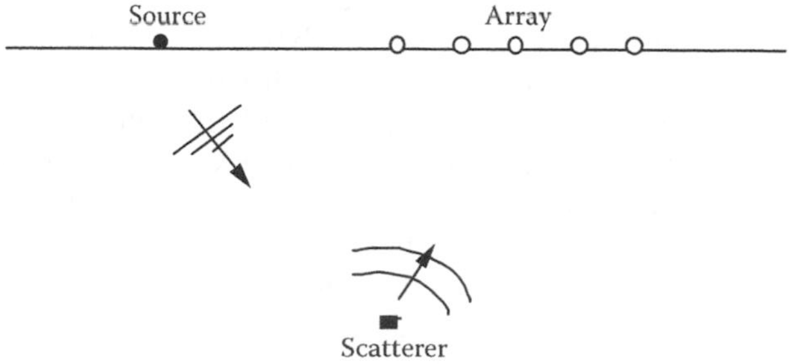

FIGURE 10.2
Depropagate a wavefront to a time instant when the illuminating wavefront hits the scatterer. When this happens, the scattered wavefield is found to be concentrated around the scattering point.

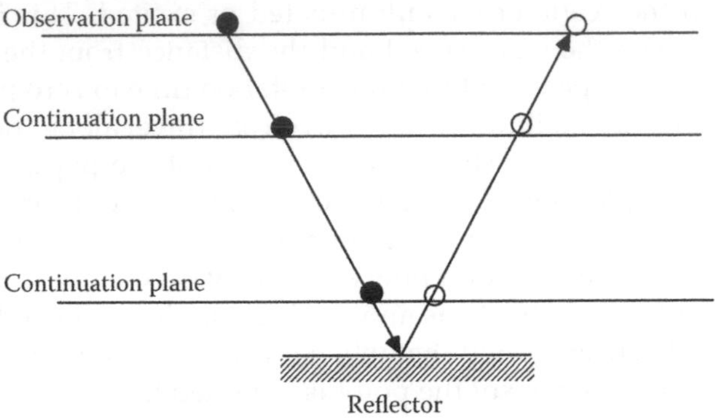

FIGURE 10.3
The wavefield measured in one plane is continued to another plane as if the source and the sensor arrays are located on that plane. It may be recalled that as the wavefield is continued the wave energy actually propagates along a ray.

reflector after a lapse of time equal to the one-way travel time. Occurrence of such a coincidence of source and sensor may be used as a condition for imaging. This phenomenon is illustrated in Figure 10.3.

10.2 Exploding Reflector Model

The wavefield, measured on a horizontal surface, is propagated backward in time in order to arrive at the reflector location, giving an estimate of the wavefield as existing at the reflector. The seismic traces are first stacked with proper delays so that the output corresponds to a hypothetical sensor kept close to the source. A set of such stacked seismic traces may be modeled as a wavefield observed in an imaginary experiment in which small charges are placed on a reflector and all of them are exploded at the *same* time instant. The wavefield is assumed to propagate upwards and reach the surface at time t_0. Conversely, the wavefield when propagated backwards will reach the reflector point after t_0 time units. It is assumed that there are no multiple reflections, surface waves, refractions, etc. Indeed, during the process of stacking, since the array is focused downward, much of the interference would be attenuated. The exploding reflector model, also known as the Loewenthal model [2], consists of tiny charges placed on a reflector and fired at the same time instant (Figure 10.4). The quantity of charge placed at a given point on the interface is proportional to the reflection coefficient at that point.

Let the interface be described by a function, $z=g(x,y)$, where the acoustic impedances above and below the interface are constant but different (see Figure 10.4). The reflection coefficient for vertical incidence is given by $r_0=((\rho_2c_2-\rho_1c_1)/(\rho_2c_2+\rho_1c_1))$. The reflectivity function may be written as

$$r(x,y,z)=r_0\delta(z-g(x,y)). \tag{10.1}$$

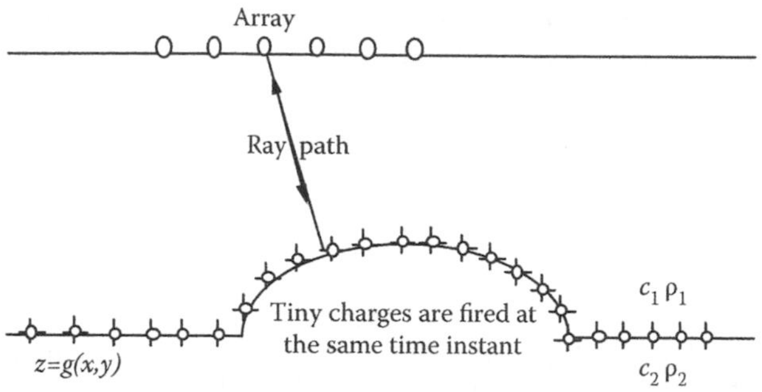

FIGURE 10.4
Exploding reflector model, also known as the Loewenthal model. Tiny charges are placed on a reflector and fired at the same time instant. The ray paths are perpendicular to the reflector. The wavefront, at the time of firing, that is, $t{=}0$, coincides with the reflector.

The wavefield in a medium bounded by the earth's surface above and the interface below will satisfy the following boundary condition and initial conditions:

Boundary condition: $f(x,y,z,t)|_{z=0}{=}f_0(x,y,t)$, that is, pressure field observed on the surface.

Initial condition: $f(x,y,z,t)|_{t=0}{=}r(x,y,z)\delta(t)$, that is, pressure field generated by the exploding charges on the interfaces.

The wavefront at time instant $t{=}0$ is the interface itself. As the time progresses, the wavefront travels upward toward the surface. The wavefield observed at the surface acts as a boundary condition, and the shape of the wavefront at $t{=}0$ is the initial condition which, in practice, is not known. The boundary condition (when given over an infinite plane) is enough to solve the wave equation. The wavefield thus obtained at all points within the space bounded from above by the observation plane and from below by the initial wavefront and for all t in the range $0{\le}t{\le}\infty$. Observe that the wavefield at $t{=}0$ is actually the pressure field generated by setting off charges on the interface and everywhere else it is zero. Conversely, if the wavefield observed on the surface is propagated backward in time until we reach the time instant $t{=}0$, we shall indeed reach the surface where the charges were set off. This is the rationale for imaging a reflector by propagation backward in time.

10.2.1 Initial Value Problem

Alternatively, an equivalent description of imaging is through a solution of initial value problem, which may also be expressed as an inverse source problem [9]. The wave equation with the right-hand side equal to a source creating a wavefield is

$$\nabla^2 f = \frac{1}{c_0^2}\frac{d^2 f}{dt^2} + r(x,y,z)\delta'(t),$$

where $r(x,y,z)$ as before stands for the reflectivity function and $\delta'(t)$ is the derivative of $\delta(t)$. At each point an explosive charge proportional to $r(x,y,z)$ is set off at $t=0$. The waves propagate unhindered by other reflecting interfaces (no multiple reflections). The solution of the inhomogeneous wave equation on the $z=0$ surface is given by

$$f(x,y,z=0,\omega)$$
$$=\frac{j\omega}{4\pi}\int\int\int_{-\infty}^{+\infty} r(x',y',z')\frac{e^{jk\sqrt{(x-x')^2+(y-y')^2+(z')^2}}}{\sqrt{(x-x')^2+(y-y')^2+(z')^2}}dx'dy'dz', \tag{10.2}$$

which may be further simplified following the procedure used in obtaining Equation 1.81

$$f(x,y,z=0,\omega)=\frac{\omega}{4\pi^2}\int\int_{-\infty}^{+\infty}\frac{R(u,v,w)}{w}e^{j(ux+vy)}dudv, \tag{10.3a}$$

$$F_0(u,v,\omega)=\omega\frac{R(u,v,w)}{w}, \tag{10.3b}$$

where $w=-\text{sgn}(\omega)\sqrt{k^2-u^2-v^2}$ for upgoing waves. From Equation 10.3b we can obtain the unknown reflectivity from the surface pressure field

$$R(u,v,w)=\frac{w}{\omega}F_0(u,v,\omega). \tag{10.4}$$

Next, we shall show how the same result (that is, Equation 10.4) can be obtained as a boundary value problem (see Equation 10.7).

10.3 Extrapolation in ω-k Plane

Recall the integral representation of a wavefield in a homogeneous medium (see Chapter 1; Equation 1.19), which we reproduce here for quick reference,

$$f(x,y,z,t)=\frac{1}{8\pi^3}\int\int\int_{-\infty}^{-\infty}F_0(u,v,\omega)e^{+j\sqrt{k^2-u^2-v^2}z}e^{-j(ux+vz-\omega t)}dudvd\omega.$$

We have chosen the positive sign in $e^{\pm j\sqrt{k^2-u^2-v^2}z}$, as the wavefield is propagated from the surface to the interface where charges are placed, that is, propagation is toward the source; hence, as per our convention (Chapter 1), +ve is chosen.

10.3.1 Downward Continuation

The wavefield measured on the surface may be continued downwards to any depth and for all times (see Equation 1.26). Using the initial condition in the exploding charge model, the wavefield at time $t=0$ is equal to the reflectivity function,

$$\begin{aligned} r(x,y,z) &= f(x,y,z,t)\big|_{t=0} \\ &= \frac{1}{8\pi^3}\iiint_{-\infty}^{\infty} F_0(u,v,\omega)e^{+j\sqrt{k^2-u^2-v^2}z}e^{-j(ux+vy)}dudvd\omega. \end{aligned} \tag{10.5}$$

Note that in Equation 10.2, $k=\omega/c$, where c is the wave speed in the medium above the interface and it is assumed to be known. Further, we relate the temporal frequency to the vertical spatial frequency, w. Since $k=\omega/c=-w\sqrt{1+(s^2/w^2)}$, where $s=\sqrt{u^2+v^2}$, we can express $w=-\omega/c\sqrt{1-(sc/w^2)}$ for an upgoing wave and $d\omega=-(c/\sqrt{1+(s^2/w^2)})$. Using these results in Equation 10.5 we obtain [10]

$$r(x,y,z)=\frac{c}{8\pi^3}\iiint_{-\infty}^{\infty}\frac{1}{\sqrt{1+\frac{s^2}{w^2}}}F_0\left(u,v,-cw\sqrt{1+\frac{s^2}{w^2}}\right)e^{-j(ux+vy+wz)}dudvdw. \tag{10.6a}$$

Equation 10.6a is applicable only when the sources actually replace the reflecting interfaces, but in practice the wave excitation is done on the surface and the wave propagates into the medium and it is then reflected at an interface back to the surface. In this process, since the wavefield travels first down and then up, the travel time is doubled; equivalently, the wave speed may be halved. Hence, we have

$$r(x,y,z)=\frac{c}{16\pi^3}\iiint_{-\infty}^{\infty}\frac{1}{\sqrt{1+\frac{s^2}{w^2}}}F_0\left(u,v,-\frac{c}{2}w\sqrt{1+\frac{s^2}{w^2}}\right)e^{-j(ux+vy+wz)}dudvdw, \tag{10.6b}$$

where $w=-(2\omega/c)\sqrt{1-(sc/2w)^2}$. This result agrees with that given in Ref. [9]. Computing the inverse Fourier transform on both sides of Equation 10.6b, we obtain

$$R(u,v,w)=\frac{c}{\sqrt{1+\frac{s^2}{w^2}}}F_0\left(u,v,-cw\sqrt{1+\frac{s^2}{w^2}}\right), \tag{10.7a}$$

or

$$R(u,v,w)=\frac{1}{\sqrt{1+\frac{s^2}{w^2}}}\frac{c}{2}F_0\left(u,v,-\frac{c}{2}w\sqrt{1+\frac{s^2}{w^2}}\right). \tag{10.7b}$$

The wavefield in a homogeneous (also in horizontally layered) medium has a radial symmetry in the (x,y) plane. For this case the appropriate extrapolation equation in polar coordinates is Equation 1.27b, reproduced here for convenience,

$$f(r,z,t)=\frac{1}{4\pi^2}\int_{-\infty}^{\infty}F_0(s,\omega)e^{j\omega t}d\omega\int_0^{\infty}sJ_0(sr)e^{\pm j\left(\sqrt{k^2-s^2}z\right)}ds,$$

where $F_0(s,\omega)$ is the Fourier transform of the surface wavefield having a radial symmetry. Let us rewrite Equation 1.27 in terms of plane wave deposition (PWD). We map s domain to γ domain where $s=k\sin(\gamma)$ and rewrite Equation 1.27b as

$$f(r,z,t)=\frac{1}{8\pi^2}\int_{-\infty}^{\infty}k^2e^{j\omega t}d\omega\int_0^{\infty}F_0(\sin(\gamma),\omega)e^{\pm j(kz\cos(\gamma))}\cos(2\gamma)J_0(k\sin(\gamma)r)d\gamma. \tag{10.8}$$

$F_0(\sin(\gamma),\omega)$ in Equation 10.8 or more specifically its inverse Fourier transform, $F_0(\sin(\gamma),t)$, may be obtained from Equation 1.32, that is, by slant stacking or Radon transform. Recall that $F_0(\sin(\gamma),t)$ is a plane wave incident at angle γ, a result of PWD of a point source. $F_0(\sin(\gamma),\omega)$ is also known as the angular spectrum of the wavefield on the surface. The angular spectrum at depth z is given by

$$F(\sin(\gamma),z,\omega)=F_0(\sin(\gamma),\omega)e^{\pm j(kz\cos(\gamma))}. \tag{10.9}$$

10.3.2 Layered Medium

We can extend the wavefield extrapolation problem from a single layer to a two-layer medium. Each layer is separated by a plane interface, either horizontal as in Figure 10.5a or inclined as in Figure 10.5b. In each layer, the density

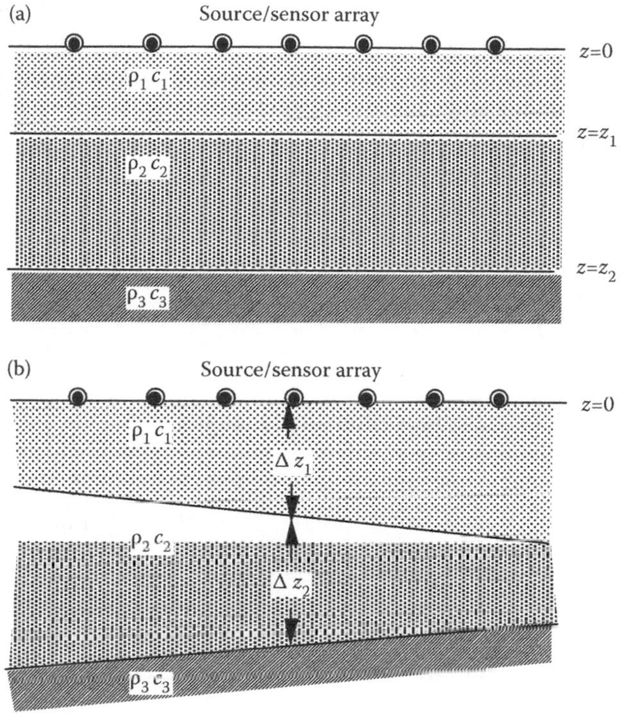

FIGURE 10.5
Wavefield extrapolation in two-layer medium. (a) Horizontal layers and (b) inclined layers. ◉ represents transceiver.

and wave speed are constant. We shall assume that zero-offset processing has removed all multiple reflected waves.

Let $f_1(x,y,z.t)$ and $f_2(x,y,z,t)$ be the wavefields produced by the exploding charges placed on the interface I and interface II, respectively. The total field is given by $f(x,y,z.t)=f_1(x,y,z.t)+f_2(x,y,z.t)$. The wavefield observed on the surface $z=0$ is given by $f_0(x,y,0,t)=[f_1(x,y,z.t)+f_2(x,y,z.t)]|_{z=0}$. The wavefield generated by exploding charges on two interfaces is given by

$$f(x,y,z,t)|_{t=0}=r_1\delta(z-g_1(x,y))+r_2\delta(z-g_2(x,y)), \tag{10.10}$$

where r_1 and r_2 are reflection coefficients and $g_1(x,y)$ and $g_1(x,y)$ are the surfaces separating the two layers. It may be noted that by removing the multiple reflections we have decoupled the two interfaces; in effect, we have linearized the propagation effects. Extension to the N-layer medium is straightforward when all layers are decoupled.

For extrapolation, it is necessary that the correct propagation speed of the material in each layer is used. First, extrapolation to the first interface is carried out using Equation 1.26. The Fourier transform of the wavefield observed on the surface is multiplied with the propagation filter function $\exp\left(-j\sqrt{k_1^2-u^2-v^2}z\right)$, $0 \le z \le z_1$, where $k_1=(\omega/c_1)$. Extrapolation to the second interface is obtained by multiplying with a filter function, $\exp\left(-j\sqrt{k_2^2-u^2-v^2z}\right)$, $z_1 \le z \le z_2$, where $k_2=(\omega/c_2)$.

$$
\begin{aligned}
&f_2(x,y,z,t)\big|_{t=0} \\
&= \frac{1}{8\pi^3}\iiint_{-\infty}^{-\infty} F_0(u,v,\omega)e^{-j\sqrt{k_1^2-u^2-v^2}z_1}e^{-j\sqrt{k_2^2-u^2-v^2}(z-z_1)}e^{-j(ux+vz)}dudvd\omega \\
&= \frac{1}{8\pi^3}\iiint_{-\infty}^{-\infty} F_0(u,v,\pm c_2\,\mathrm{sgn}(\omega)\sqrt{w^2+u^2+v^2}) \\
&\quad\times \frac{c_2 w}{\sqrt{w^2+u^2+v^2}}e^{-j\left\{\sqrt{k_1^2-u^2-v^2}-\sqrt{k_2^2-u^2-v^2}\right\}z_1}e^{-j(ux+vy-wz)}dudvdw \\
&= r_2(x,y,z).
\end{aligned}
\tag{10.11}
$$

Note that $r_1(x,y,z)$ is zero in the second layer.

10.3.3 Sloping Interface

We shall now look at an example of a single sloping reflector, as shown in Figure 10.6 [10]. We shall assume a uniform linear array (ULA) of transceivers oriented along the slope. For simplicity, let us assume that the transceiver radiates a spike and receives, after a delay, the same spike. The seismic data are assumed to have been stacked so that we have a zero-offset data corresponding to a field from hypothetical charges placed on the interface

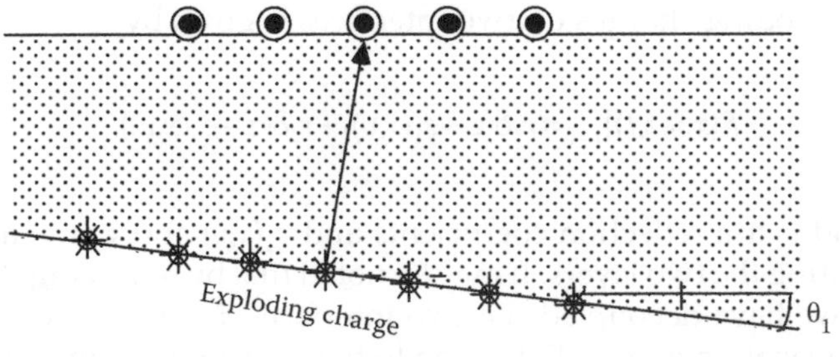

FIGURE 10.6
A sloping reflector and source/sensor (transceiver) array on the surface. By a process of stacking, the seismic field is reduced to a field generated by hypothetical exploding charges on the sloping interface.

(exploding reflector model described in Section 8.2). The zero-offset data may be expressed as

$$f(md,t)=\delta(t-m\tau_0-t_0), \quad m=0,1,\ldots,M-1,$$

where $\tau_0=(d/c)\sin\theta_0$, and $t_0=(z_0/c)\cos\theta_0$. Note that z_0 is a depth to the interface below sensor $m=0$. The Fourier transform of the zero-offset data may be obtained assuming the array size is very large,

$$F_0(u,\omega)=e^{-j\omega t_0}\delta(ud+\omega\tau_0). \tag{10.12}$$

Using Equation 10.12 in Equation 10.6a, that is, its 2D equivalent, we obtain after simplification

$$\begin{aligned} r(x,z) &= \frac{c}{4\pi^2}\int\int_{-\infty}^{-\infty}\frac{1}{\sqrt{1+\frac{s^2}{w^2}}}e^{jc\sqrt{w^2+u^2}t_0}\delta(ud+c\tau_0\sqrt{w^2+u^2})e^{-j(ux+wz)}du\,dw \\ &= \frac{c}{2\pi}\int_{-\infty}^{-\infty}\cos\theta_0 e^{-j(ct_0/\cos\theta_1)w}e^{-j(-\tan\theta_0 x+z)w}dw \\ &= c\cos\theta_0\delta(-\tan\theta_0 x - z_0 + z). \end{aligned} \tag{10.13}$$

From Equation 10.13 the reflector is given by

$$z=\tan\theta_0 x+z_0. \tag{10.14}$$

It is interesting to note that the slope of the interface is not equal to the slope of the line joining the spike arrivals in the zero-offset seismic data (see Section 10.6, Exercise 3).

10.3.4 Depropagation of Wavefield

Extrapolating a wavefield backward in time is known as depropagation; when continued it is possible to home onto the starting point. In the exploding reflector model, all scatter points are simultaneously fired. We shall now show that the field observed on the surface is equal to the field generated at scatter points and then propagated to the surface. By depropagating, we hope to obtain the field generated at the exploding reflectors. The zero-offset data provides only a blurred image of the reflectors, but by depropagating, the blurring can be reduced [5]. Consider a 2D model with uniform wave speed with reflecting facets. The wavefield in a uniform medium is given, in the frequency domain, by

$$f(x,z,t)=\frac{1}{4\pi^2}\int\int_{-\infty}^{+\infty} F(u,\omega)e^{j(-ux+\sqrt{k^2-u^2}z)}e^{j\omega t}dud\omega$$
$$=\frac{c}{4\pi^2}\int\int_{-\infty}^{+\infty}\frac{1}{\sqrt{1+\frac{u^2}{v^2}}}F(u,\omega)e^{-j(ux+vz)}e^{-j\sqrt{u^2+v^2}ct}dudv, \tag{10.15}$$

where $v=-\text{sgn}(\omega)\sqrt{k^2-u^2}$. The field on the surface, $z=0$, is given by

$$f(x,z=0,t)=\frac{c}{4\pi^2}\int\int_{-\infty}^{+\infty}\frac{1}{\sqrt{1+\frac{u^2}{v^2}}}F(u,\omega)e^{-j(ux+\sqrt{u^2+v^2}ct)}dudv. \tag{10.16}$$

The mathematical framework for imaging by wavefield extrapolation is covered in great detail in Ref. [11].

The wavefield at the exploding reflector at a depth z and backpropagated to the surface at time $t=z/c$ is given by

$$f\left(x,z,t=\frac{z}{c}\right)=\frac{c}{4\pi^2}\int\int_{-\infty}^{+\infty}\frac{1}{\sqrt{1+\frac{u^2}{v^2}}}F(u,\omega)e^{-j(ux+vz)}e^{-j\sqrt{u^2+v^2}z}dudv. \tag{10.17}$$

Since the wavefield is by and large vertically propagating, $u \ll v$ (Equation 10.17) may be approximated as

$$f\left(x,z,t=\frac{z}{c}\right)\approx\frac{1}{4\pi^2}\int\int_{-\infty}^{+\infty} cF(u,\omega)e^{-j(ux+2vz)}dudv. \tag{10.18}$$

The wavefield observed on the surface ($z=0$) is mapped into the (x,z) plane by substituting $(z/(c/2))$ for t. Note that the time axis of the recorded seismic data refers to two-way time and hence wave speed is halved as is common in seismic data processing. We obtain

$$f\left(x,z=0,t=\frac{z}{c/2}\right)\approx\frac{1}{4\pi^2}\int\int_{-\infty}^{+\infty} cF(u,\omega)e^{j(-ux-2vz)}dudv. \tag{10.19}$$

Thus, from Equations 10.18 and 10.19 we have an important approximate result that provides an initial wavefield for depropagation,

$$f\left(x,0,t=\frac{z}{c/2}\right)\approx f\left(x,z,t=\frac{z}{c}\right). \tag{10.20}$$

In words, the wavefield observed on the surface with its time axis mapped into a *z*-axis $t=(z/(c/2))$ is approximately equal to the wavefield at the reflector, which is then propagated to the surface.

Now given $f(x, 0, t=(z/(c/2)))$ or $f(x, z, t=(z/c))$, we need to depropagate until we reach the point of reflection. To depropagate by Δt time units we shall use Equation 10.15 with approximation $u \ll v$

$$\begin{aligned} f_d(x,z,t-\Delta t) &\approx \frac{1}{4\pi^2}\int\limits_{-\infty}^{+\infty}\int cF(u,\omega)e^{-j(ux+vz)}e^{-jvc\Delta t}dudv \\ &= \frac{c}{4\pi^2}\int\limits_{-\infty}^{+\infty}\int FT\left\{f\left(x,0,t=\frac{z}{c/2}\right)\right\}e^{-j(ux+vz)}e^{-jvc\Delta t}dudv, \end{aligned} \tag{10.21}$$

where subscript d stands for depropagated field (Figure 10.7).

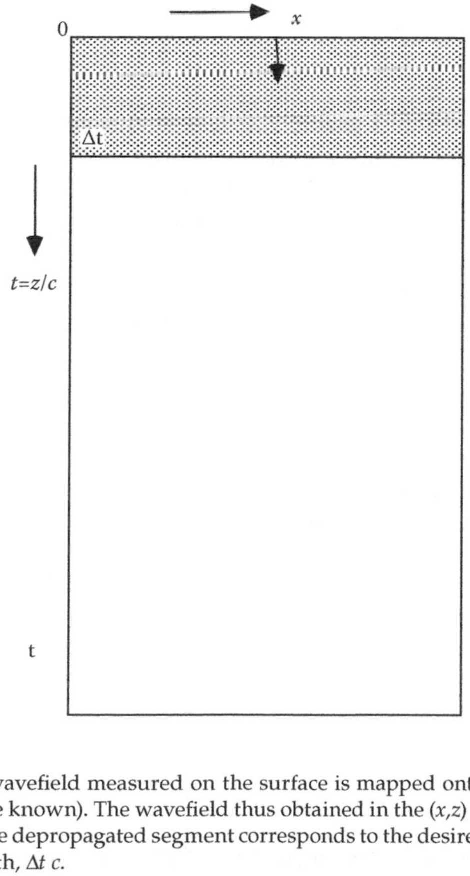

FIGURE 10.7
The time axis of the wavefield measured on the surface is mapped onto the *z*-axis (the wave speed is assumed to be known). The wavefield thus obtained in the (*x*,*z*) plane is now depropagated in steps of Δt. The depropagated segment corresponds to the desired reflectivity function over a segment of depth, $\Delta t\, c$.

There is an interesting application of seismic migration, in particular the above method of depropagation, to imaging of buried polyethylene pipes used in public utilities [22]. A ground-penetrating radar (GPR) is used to illuminate a target area and the reflected/scattered electromagnetic field is measured at a number of locations. Applying down continuation in time to $t=0$ gave the correct location of the buried pipes.

10.3.5 Relation to Diffraction Tomography

We shall now show that a close relation exists between imaging by extrapolation and diffraction tomography, which we have dealt with in Chapter 9. Consider a weakly inhomogeneous object (speed fluctuations only) illuminated by a point source and the scattered field being received by a sensor kept close to the source. We have shown in Chapter 9, Section 9.4 that the Fourier transform of the speed variations evaluated on a spherical surface of radius $2k$ is related to the Fourier transform of the field received by an array of transceivers (Figure 9.25). The relationship (Equation 9.33) is reproduced here for convenience:

$$\Delta\tilde{c}(u,v,\sqrt{4k_0^2-u^2-v^2})=\frac{c^3}{2}\sqrt{4k_0^2-u^2-v^2}\,\frac{\partial}{\partial\omega_0}\left(\frac{F_1(u,v,\omega_0)}{\omega_0^2}\right).$$

It may be observed that the zero-offset field measured in seismic exploration is essentially an output of an array of transceivers that may be considered as a mathematical model of the zero-offset field measurement setup.

To relate $\Delta\tilde{c}\left(u,v,\sqrt{4k_0^2-u^2-v^2}\right)$ to the reflectivity, we shall use Equation 10.7a, where we have shown the relationship between the Fourier transform of the reflectivity and the wavefield on the surface (zero-offset). It can be shown that the partial derivative appearing in Equation 9.33 may be given by

$$\frac{\partial}{\partial\omega}\left(\frac{F_1(u,v,\omega)}{\omega^2}\right)=-\frac{1}{2}\frac{\partial}{\partial w}\left(\frac{R(u,v,w)}{\sqrt{w^2+s^2}\,w}\right)\frac{\sqrt{w^2+s^2}}{w}, \tag{10.22}$$

where $R(u,v,w)$ is the spatial Fourier transform of $r(x,y,z)$. Substituting Equation 10.22 in Equation 9.33 we obtain

$$\Delta\tilde{c}\left(u,v,\sqrt{4k_0^2-u^2-v^2}\right)=-\frac{c^3}{4}\sqrt{w^2+s^2}\,\frac{\partial}{\partial w}\left(\frac{R(u,v,w)}{\sqrt{w^2+s^2}\,w}\right). \tag{10.23}$$

After carrying out the required differentiation, we obtain the following result:

$$\Delta\tilde{c}(u,v,w)=\frac{c^3}{4}\left(\frac{R(u,v,w)}{(w^2+s^2)}+\frac{R(u,v,w)}{w^2}-\frac{R_w(u,v,w)}{w}\right), \tag{10.24}$$

where $R_w(u,v,w)$ stands for the derivative of $R(u,v,w)$ with respect to w.

10.3.6 Continuation of Sources and Sensors

In Section 10.1, we mentioned that imaging can be realized by continuing both the source and sensor arrays down to the reflector. This is particularly useful for imaging when zero-offset data are not available [12]. We shall now derive filters for the downward continuation of the source and sensor arrays. The transfer function of a filter to continue the wavefield to a plane Δz below the observation plane may be obtained from Equation 1.18

$$H(u_r, v_r, \Delta z) = \exp\left(+j\sqrt{(k^2 - u_r^2 - v_r^2}\Delta z\right), \tag{10.25a}$$

where subscript r refers to the receiver coordinate. In the first step the sources are held fixed on the $z=0$ plane (observation plane) and the sensor array is displaced downward. Next, we interchange the source array with the displaced receiver array. By virtue of the reciprocity theorem, the wavefield at the new position of the sensor array due to the source array also at its new position will be equal to the previously downward continued field. In the second step, we downward continue the previously continued field, but the continuation is done in source coordinate space. The transfer function for downward continuation is given by

$$H(u_s, v_s, \Delta z) = \exp\left(+j\sqrt{(k^2 - u_s^2 - v_s^2}\Delta z\right), \tag{10.25b}$$

where subscript s refers to source coordinate. The downward continuation is alternatively done in the source and sensor spaces. In practice, the downward continuation is carried out as follows. Assume that we have two source and sensor ULAs each having M devices. There will be M^2 outputs, which we like to arrange on a square grid as shown in Figure 10.8. All outputs in a row (known as source gathers) are continued downwards in the sensor space and all outputs in a column (known as sensor gathers) are continued in the source space.

10.4 Focused Beam

When a source is in the far field, the directions of arrival (DOA), azimuth, and elevation are of interest. We have already seen how a beam is formed in a given direction (Chapters 4 and 5). On the other hand, when a source is in the near-field region, a beam may be formed to receive energy not only from a given direction, but also from a given point. This is akin to focusing in an optical system. In seismic exploration, the array size is of the same order as the depth of reflectors; therefore, it is often inappropriate to assume a far-field or plane wavefront condition, which is required for the purpose of imaging, that is, to form a focused beam to receive reflected energy from a given reflector.

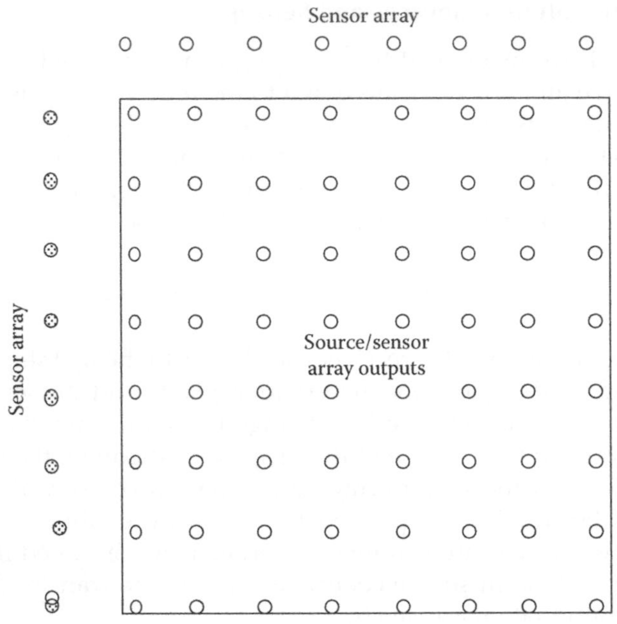

FIGURE 10.8
M^2 outputs source/receiver ULAs are arranged over a square grid. All outputs in a row are known as source gathers and all outputs in a column are known as sensor gathers.

10.4.1 Zero-Offset Wavefield

In reflection seismic exploration it is often desired to position both source and detector at the same location, although in practice this cannot be achieved and some amount of offset is always present. The main advantage of this arrangement is that the returned signal is a normally reflected signal; consequently, there is no conversion of wave energy into *s*-waves (see Chapter 1, Section 1.3). The zero-offset wavefield can, however, be obtained through the array-processing approach. For this we must focus a beam at the foot of a perpendicular drawn from the array midpoint to the reflector. A typical example of a source and receiver setup is shown in Figure 10.9. All signals reaching the sensor array emanate from the same reflecting element. In seismic parlance, these signals are called common depth point (CDP) gathers.

10.4.1.1 CDP

A linear array of sources and detectors are so arranged that the received signal always comes from the same reflecting element. This arrangement is illustrated in Figure 10.9a, and is commonly known as a CDP setup. The total time of travel from the source to the receiver is a function of separation or offset between the source and the receiver. When the reflecting element is horizontal, the travel time is given by

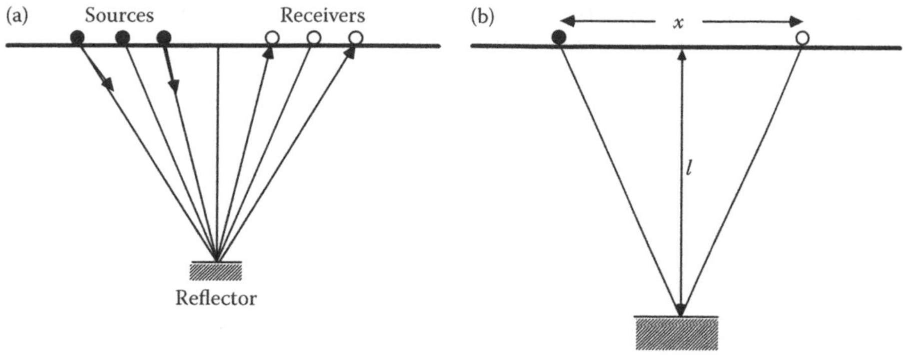

FIGURE 10.9
Common depth point (CDP) setup. (a) The source and receiver arrays are arranged in such a manner that a signal is always reflected from the same reflecting element. (b) Source-receiver pair.

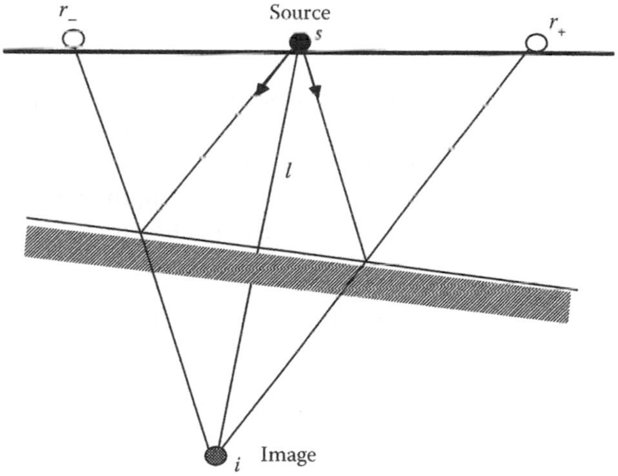

FIGURE 10.10
Inclined reflector. There are two sensors, one sensor, r_+, is downslope and another sensor, r_-, is upslope. To compute the travel time we consider an image of the source. The time of travel to the downslope sensor is equal to $(i\,r_+)/c$ and similarly for the upslope sensor.

$$T_x^2 = T_0^2 + \frac{x^2}{c^2}, \tag{10.26}$$

where x stands for separation between the source and receiver, $T_0=2l/c$, and l is the depth to the reflecting element (see Figure 10.9b). A plot of T_x vs. x is very useful because it enables us to estimate T_0 and the wave speed.

Let us now consider a sloping reflecting element, but the source and receiver arrays are, as before, on a horizontal surface (see Figure 10.10). Let the slope of the reflecting element be α. To compute the travel time to a downslope or an upslope sensor, we consider the image of the source and compute the distance from the image to the sensor. For a downslope sensor,

$$\begin{aligned} T_{x_+} &= \frac{(ir_+)}{c} \\ &= \frac{\sqrt{(x_+)^2 + 4l^2 - 4lx_+ \cos(\angle isr_+)}}{c}, \end{aligned} \tag{10.27}$$

and squaring on both sides of Equation 10.27 we obtain

$$T_{x_+}^2 = \left(\frac{x_+}{c}\right)^2 + T_0^2 + 2T_0 \frac{x_+}{c} \sin(\alpha). \tag{10.28}$$

Similarly, we have for an upslope receiver

$$T_{x_-}^2 = \left(\frac{x_-}{c}\right)^2 + T_0^2 - 2T_0 \frac{x_-}{c} \sin(\alpha). \tag{10.29}$$

The upslope and downslope travel times are shown in Figure 10.11, where $T_0 = 2l/c$ and l is the perpendicular depth to the reflector below the source position (see Figure 10.9). Note that Equations 10.28 and 10.29 reduce to Equation 10.26 for $\alpha = 0$ and for $x \to 0$, $T_x \to T_0$. Rearranging the terms in Equations 10.28 and 10.29, we can express them in a compact form

$$T_{x_\pm}^2 = \left(T_0 \pm \frac{x_\pm}{c} \sin(\alpha)\right)^2 \pm \left(\frac{x_\pm}{c}\right)^2 \cos(\alpha). \tag{10.30}$$

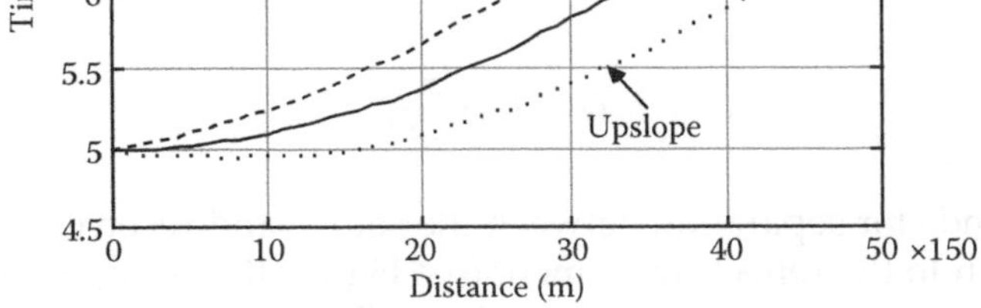

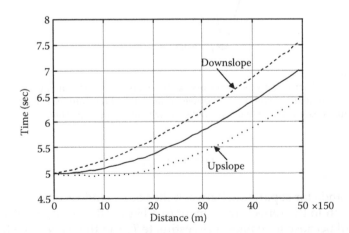

FIGURE 10.11
Downslope and upslope travel times (Equations 8.21 and 8.22) in seconds as a function of distance in meters. Wave speed = 1,500 m/sec, $T_0 = 5$ sec. Solid curve is for horizontal reflector (Equation 10.19).

Let the downslope and the upslope sensor be equidistant from the source, that is, $x_+=x_-=\bar{x}$. The average of the squares of downslope and upslope travel times turns out to be independent of the slope of the reflector,

$$\frac{T_{x_+}^2+T_{x_-}^2}{2}=\left(\frac{\bar{x}}{c}\right)^2+T_0^2. \tag{10.31}$$

Similarly, the difference of the squares of downslope and upslope travel times is given by

$$\frac{T_{x_+}^2-T_{x_-}^2}{4}=T_0\frac{\bar{x}}{c}\sin(\alpha). \tag{10.32}$$

10.4.2 Layered Medium

We now consider a layered medium overlying a reflector. The overlying medium is modeled as a stack of uniform layers (see Figure 10.12). The round-trip travel time T_x and the receiver position x are the two parameters of interest. They are given by

$$T_x=\sum_{k=1}^{N}\frac{2\Delta z_k}{c_k\sqrt{(1-p^2c_k^2)}}, \tag{10.33a}$$

$$x=p\sum_{k=1}^{N}\frac{2\Delta z_k c_k}{\sqrt{(1-p^2c_k^2)}}, \tag{10.33b}$$

where p is the ray parameter ($p=\sin(\theta_k)/c_k$), where θ_k is the angle of incidence in the kth layer (see Chapter 1, Section 1.2 for more information on the ray

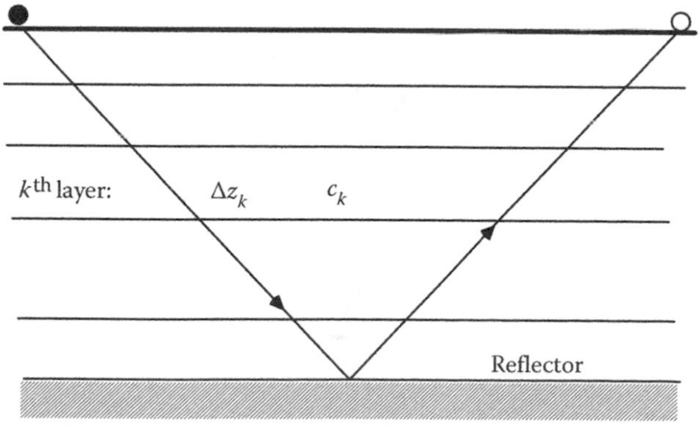

FIGURE 10.12
Layered medium overlying a reflector. Each layer is homogeneous with constant thickness and speed. The ray path consists of a series of linear segments. At an interface between two layers, the Snell's law (see Chapter 1, Section 1.3) must be satisfied.

parameter), and N stands for the number of layers. Let us expand T_x as a function of x in Taylor's series,

$$T_x = T_0 + \left.\frac{d^2T_x}{dx^2}\right|_{x=0} \frac{x^2}{2!} + \left.\frac{d^4T_x}{dx^4}\right|_{x=0} \frac{x^4}{4!} + \ldots \tag{10.34}$$

Since, for a horizontally layered medium, T_x is a symmetric function of x, only even terms are retained in the Taylor's series expansion (Equation 10.27). Further, the second derivative can be shown to be equal to

$$\left.\frac{d^2T_x}{dx^2}\right|_{x=0} = \frac{1}{\sum_{k=1}^{N} \Delta z_k c_k}.$$

Equation 10.34 simplifies to

$$T_x \approx T_0 + \frac{x^2}{2\sum_{k=1}^{N} \Delta z_k c_k}. \tag{10.35}$$

Upon squaring on both sides of Equation 10.35 and retaining only the second order terms, we obtain

$$T_x^2 \approx T_0^2 + \frac{x^2 T_0}{\sum_{k=1}^{N} \Delta z_k c_k},$$

which after rearranging reduces to a form identical to Equation 10.26,

$$T_x^2 \approx T_0^2 + \frac{x^2}{\frac{1}{T_0}\sum_{k=1}^{N} \frac{\Delta z_k}{c_k} c_k^2} = T_0^2 + \frac{x^2}{c_{\text{rms}}^2}, \tag{10.36}$$

where

$$c_{\text{rms}}^2 = \frac{1}{T_0}\sum_{k=1}^{N} \frac{\Delta z_k}{c_k} c_k^2,$$

is known as a root mean square (rms) speed of the layered medium. Indeed, we may replace a layered medium by a uniform medium having a speed equal to c_{rms}. For the purpose of focused beamformation, we may use the rms speed.

10.4.3 Focusing

To form a focused beam we must have many CDP gathers, which are obtained by means of a specially designed source-receiver array. For example, consider a linear array where every location is occupied either by a source or a sensor. The array is fired as many times as the number of locations. From every reflecting element we obtain a number of gathers equal to the number of receivers. This is illustrated in Figure 10.13. A ULA of four sensors is headed by a source. In position (A) the source is fired and the reflected signal is received by sensor #1. The entire receiver-source array is moved laterally by a half-sensor spacing

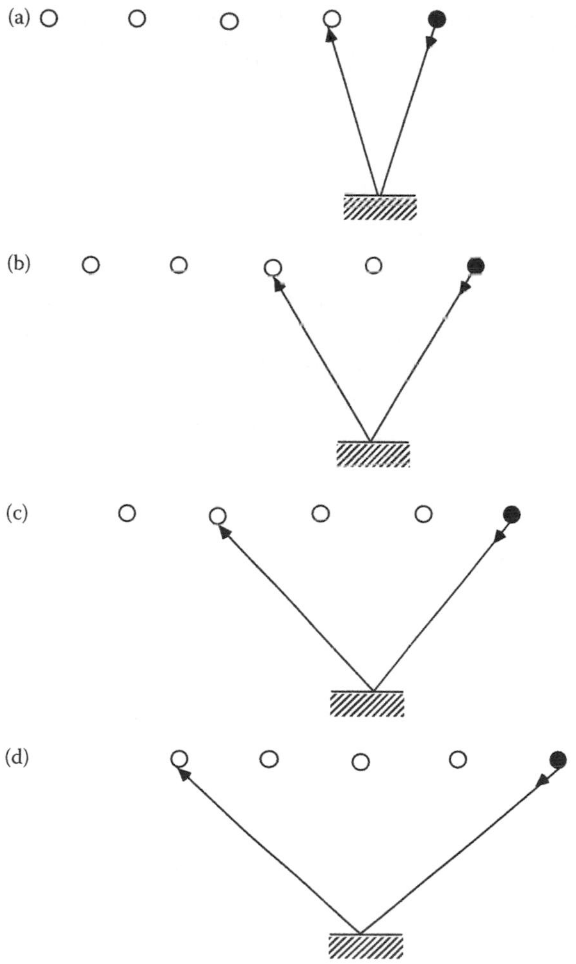

FIGURE 10.13
A reflecting element is illuminated by a source at different angles. The reflected signal is received by one of the sensors as shown. The entire receiver-source array is moved laterally by a half-sensor spacing.

as in position (B) and the source is fired once again. The reflected signal is received by sensor #2. This procedure is continued as in (C) and (D). Thus, we get four CDP gathers.

Let $T_1, T_2, \ldots, T_N$ be the round-trip travel time from the source to sensor #1, #2, …, #N, respectively. Let $f_0(t)$ be the signal transmitted by the source and $f_i(t), i=1,2,\ldots,N$ be the reflected signals received by four receivers. Since these are the delayed versions of $f_0(t)$, we can express them as

$$f_i(t)=f_0(t-T_i), \quad i=1,2,\ldots,N. \tag{10.37}$$

Focusing involves firstly correcting for delayed reception and secondly summing coherently after correction. Let the delay correction be given by $\hat{T}_1,\hat{T}_2,\ldots,\hat{T}_N$, which are computed using Equation 10.26 for an assumed depth to the reflector,

$$\hat{T}_n=\hat{T}_0\sqrt{\left(1+\left(\frac{\Delta tn}{\hat{T}_0}\right)^2\right)}, \tag{10.38}$$

where $\hat{T}_0=2\hat{l}/c$, $\hat{l}$ is assumed the depth to the reflector, and $\Delta t=d/c$. Recall that d stands for sensor spacing. We assume that the wave speed c is known. The CDP gathers are coherently summed after correction for the delays computed from Equation 10.38. As before, we shall assume that the source emits a broadband signal. The coherently summed output may be expressed as follows:

$$\begin{aligned} g(t) &= \frac{1}{2\pi}\int_{-\infty}^{\infty} F(\omega)\frac{1}{N}\sum_{n=0}^{N-1} e^{-j\omega\left[T_0\sqrt{(1+(\Delta tn/T_0)^2)}-\hat{T}_0\sqrt{(1+(\Delta tn/\hat{T}_0)^2)}\right]}e^{j\omega t}\,d\omega \\ &= \frac{1}{2\pi}\int_{-\infty}^{\infty} F(\omega)H_N(\omega)e^{j\omega t}\,d\omega, \end{aligned} \tag{10.39a}$$

where

$$H_N(\omega)=\frac{1}{N}\sum_{n=0}^{N-1} e^{-j\omega\left[T_0\sqrt{(1+(\Delta tn/T_0)^2)}-\hat{T}_0\sqrt{(1+(\Delta tn/\hat{T}_0)^2)}\right]}, \tag{10.39b}$$

is the filter transfer function. A numerical example of the transfer function is shown in Figure 10.14. A horizontal reflector is assumed at a depth corresponding to round-trip time equal to 5 sec. The sensor spacing, measured in units of propagation time, $\Delta t=d/c=0.1$ sec.

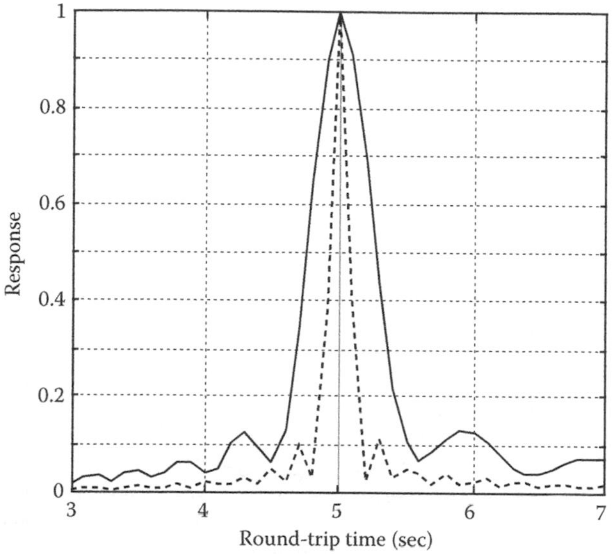

FIGURE 10.14
Response function of focused beamformation. The reflector is at a depth corresponding to 5 sec round-trip time. The array aperture is measured in propagation time (x/c). The solid line is for array aperture of 2.5 sec and dashed line for 5.0 sec. Further, the angular frequency is assumed to be 100 radians.

10.4.4 Depth of Focus

The response function has a finite width. A sharp reflector now appears as a diffused zone whose width is known as the depth of focus, analogous to that in optics. Ideally, one would like the depth of focus to be as narrow as possible. For the purpose of quantitative measure, we shall define the depth of focus as a distance between two 3 dB points on the response function. The depth of focus, measured in the units of round-trip propagation time ($=2\Delta l/c$, where Δl is depth of focus), and the array aperture, also measured in terms of the propagation time ($=x/c$), are shown in Figure 10.15. Notice that the minimum occurs when the aperture size is about four times the round-trip propagation time, in this case 5 sec. With a further increase in the aperture size, the depth of focus rapidly deteriorates. The minimum depth of focus appears to be independent of depth to the reflector; however, the required array aperture increases rapidly as the reflector depth increases. The dependence of the depth of focus on frequency is significant, as shown in Figure 10.16. Notice that the depth of focus becomes very narrow beyond about 50 Hz.

10.4.5 Inclined Reflector

We consider a sloping reflector. A small segment of the reflector is illuminated at different angles by means of a source-receiver array similar to the one used for a horizontal reflector. As the array is laterally shifted the point of reflection

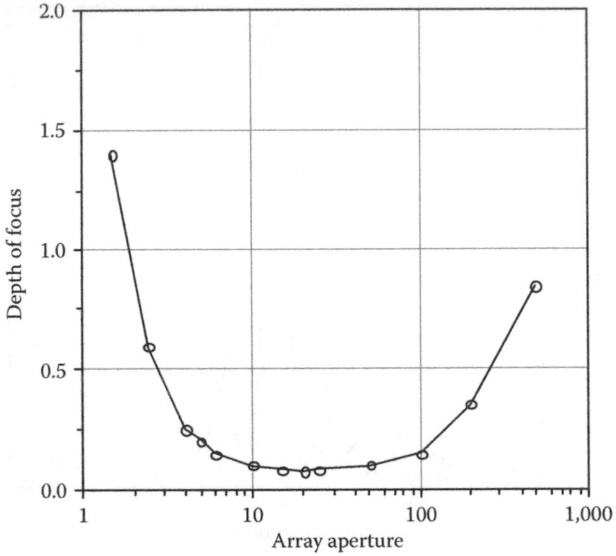

FIGURE 10.15
Depth of focus in units of round-trip propagation time (sec) as a function of array aperture, also measured in units of propagation time (sec).

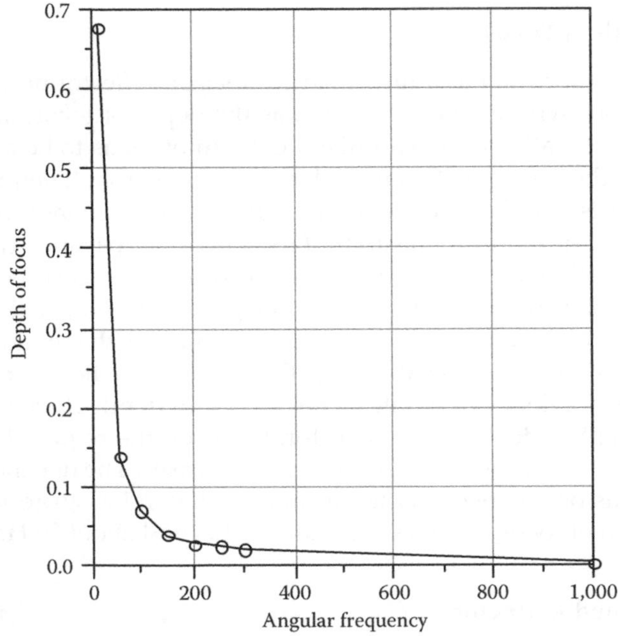

FIGURE 10.16
Depth of focus vs. angular frequency. The reflector is at a depth of 5 sec (round-trip travel time). The array aperture is held fixed at 20 sec (propagation time).

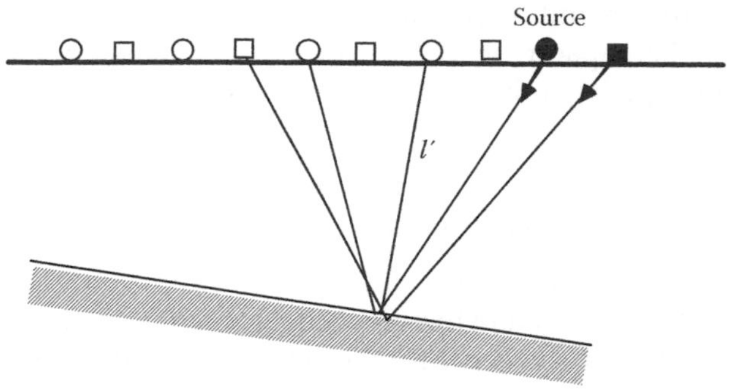

FIGURE 10.17
Common depth point (CDP) gathers from a sloping reflector. The reflecting element is at a depth of l' (round-trip travel time $=2l'/c$). Two positions of source-sensor array are shown; position #1: circles and position #2: squares. Notice the displacement of the reflecting point.

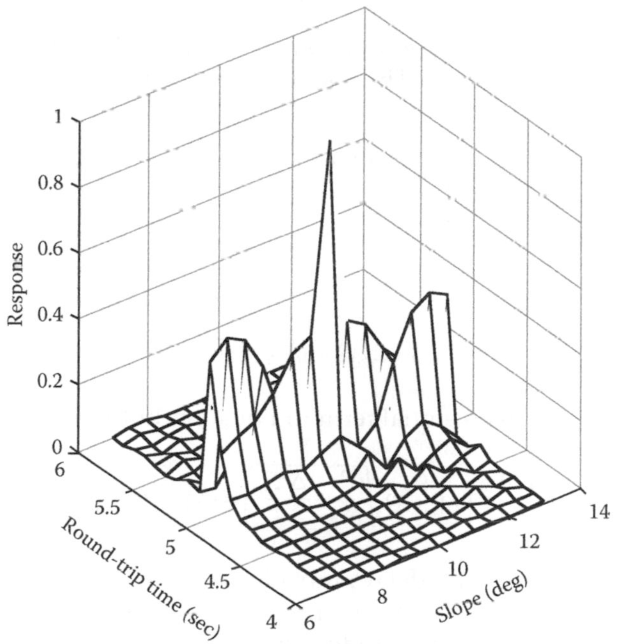

FIGURE 10.18
A mesh plot of the response function of the focused beamformer as a function of slope of the reflector and round-trip time. Array aperture is 20 sec (propagation time). Angular frequency is assumed equal to 100 radians.

changes, but only slightly, depending upon the slope (Figure 10.17). When the slope is zero, all CDP gathers emanate from the same point on the reflecting element. For gentle slope, the spread of the reflecting points is assumed to be small, small enough to allow the assumption of a constant slope.

The response of the focused beamformer as a function of the slope and the return-travel time is shown in Figure 10.18. We have assumed that $2l'/c=5$

and array aperture, in units of propagation time, is equal to $x/c=20$ sec. There are 200 CDP gathers spaced at an interval equal to 0.1 sec. The result shown in Figure 10.18 is perhaps the best one can expect as we have assumed the optimum array size.

10.4.6 Relation between Focusing and Downward Extrapolation

Focusing appears to achieve what downward extrapolation does in migration. Indeed, both are related in the sense that focusing is a simplified version of extrapolation. In order to see this relationship, let us examine the impulse response function of the downward extrapolation filter whose transfer function is given by

$$H(\omega,u,v)=\exp\left(j\Delta z\sqrt{\left(\left(\frac{\omega}{c}\right)^2-u^2-v^2\right)}\right). \tag{10.40a}$$

To get the impulse response function we compute the 2D inverse Fourier transform of Equation 10.40a. The result is given in Ref. [13]

$$h(r,\omega)=\frac{1}{2\pi}\frac{\partial}{\partial z}\left[\frac{\exp\left(j\frac{\omega}{c}r\right)}{r}\right]=\frac{1}{2\pi}\frac{j\frac{\omega}{c}r+1}{r^3}\exp\left(j\frac{\omega}{c}r\right), \tag{10.40b}$$

where

$$r=\sqrt{x^2+y^2+\Delta z^2}.$$

Let us rewrite the exponential term in Equation 10.40b as

$$\exp\left(j\frac{\omega}{c}r\right)=\exp\left(j\omega\sqrt{\frac{\rho^2}{c^2}+\frac{\Delta z^2}{c^2}}\right)=\exp\left(j\omega\sqrt{T_0^2+\frac{\rho^2}{c^2}}\right), \tag{10.40c}$$

where $\rho^2=x^2+y^2$. The phase delays introduced by the downward extrapolation filter are identical to those used in the focused beamformer. The difference, however, lies in the amplitude term. In place of a variable amplitude, we use a constant amplitude in the focused beamformer.

10.4.7 Focused Beamformation for Imaging

A focused beam enables us to estimate the amount of scattered or reflected power from a scattering volume element or a reflecting surface element located at a point in space. Consider a volume element illuminated by a point source with the scattered waves being received by a ULA (see Figure 10.19). To form a focused beam we need to compute the travel time from the source to the volume element and from the volume element to a sensor. For simplicity, let us assume that the background wave speed is constant so that ray paths

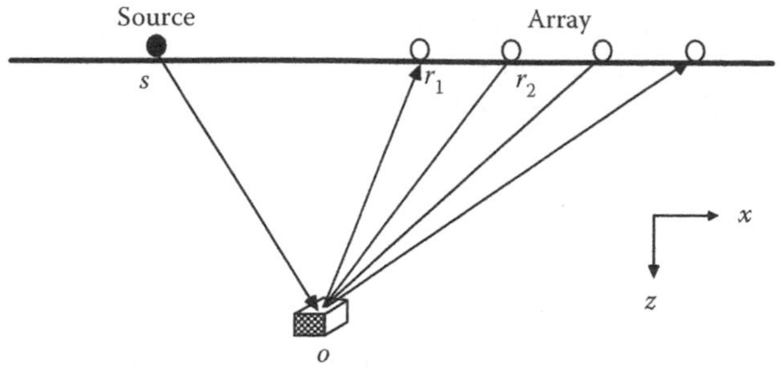

FIGURE 10.19
The array output is focused to a point o where a scattering volume element is presumed to be located. The power output is proportional to the scattering strength of the volume element.

are straight lines. In any realistic problem, however, the ray paths are more likely to be curved. It will be necessary to trace the rays before we are able to compute the travel times. Such a situation was considered in Ref. [14]. Let $f_n(t)$, $n=1,2,\ldots,N$ be the sensor outputs. The focused beam output is given by

$$g(t)=\frac{1}{N}\sum_{n=1}^{N} f_n(t-t(so)-t(or_n)),$$

where $t(so)$ stands for the travel time from the source to the volume element and $t(or_n)$ stands for the travel time from the volume element to the nth sensor. The scattered power from the volume element is given by $\int_{T_s}|g(t)|^2\,dt$ where T_s stands for signal duration. The process is repeated at every point in the space (x-z plane). The resulting map provides an image of the scattering strength.

The question of resolution needs to be answered. The size of the volume element is controlled, along the x-axis, by the array aperture (inversely proportional to array aperture), along the z-axis, by depth of focus, and finally along the y-axis, by the bandwidth of the signal (inversely proportional to bandwidth) [14]. An interesting application of the focusing by means of backpropagation of wavefield to detection of an acoustic source in a room without considering the reflections from the walls is reported in Ref. [15]. An array of microphones is deployed in a 3D space surrounding the source(s) of sound energy. The output from each microphone is backpropagated to a point in 3D space. Since we assume that no reflections from the walls or scattering exist, the backpropagation is simply equal to delaying the microphone output by an amount equal to the travel time along a straight-line path from the sensor to the assumed position of the source. The delayed output is cross-correlated with similarly delayed outputs from other sensors. An average of the cross-correlation is a measure of the acoustic energy present at the selected point. Ideally, the averaged cross-correlation function will have a sharp peak at the true sound source position. In practice, it is shown in Ref. [15] that a broad peak stands at the true sound source position. Another interesting application of the focused beamformation method is in medical ultrasonic imaging of cysts embedded in soft tissues below a bone [23]. A series of images are created with different assumed propagation speeds. All observed bright points are then combined to create an image of reflectors/scatterers.

10.5 Estimation of Wave Speed

We have assumed that the average background wave speed is known and this information was used for the purpose of imaging. We shall now look into how the wave speed can be estimated from the recorded wavefield itself. Firstly, the medium is assumed to be homogenous for which one needs to estimate just one unknown, namely, the wave speed of the homogenous medium. Secondly, the medium is a stack of layers for which we need to estimate the rms speed defined earlier (Equation 10.36). It turns out that the approach remains the same for both.

10.5.1 Wave Speed from CDP Gathers

In the context of CDP gathers, the round-trip travel time is given by Equation 10.26 for a homogeneous medium and by Equation 10.36 for a layered medium. The only difference is that in place of wave speed of a homogenous medium, we have rms speed, c_{rms}. To estimate the wave speed or rms wave speed, consider the travel time at two distinct sensor locations and compute the difference of the squares

$$T_{x_1}^2 - T_{x_2}^2 = \frac{x_1^2 - x_2^2}{c^2}, \tag{10.41}$$

from which the wave speed can be obtained as

$$c = \sqrt{\frac{x_1^2 - x_2^2}{T_{x_1}^2 - T_{x_2}^2}}. \tag{10.42}$$

For estimation of c_{rms} we use an identical approach. When the reflector is sloping, we use the relation between the round-trip travel time and the source sensor separation given by Equations 10.28 and 10.29. Consider two sensors, equidistant from the source, one located downslope and the other upslope. From Equation 10.31 it is clear that the slope of the reflector does not affect the average round-trip travel time. We can now use Equation 10.31 in place of Equation 10.26 for the estimation of wave speed.

10.5.1.1 Interval Speed

There is an interesting possibility of estimating the wave speed of a particular layer. Consider c_{rms} at two different depths having N and $N+1$ layers. From Equation 10.36 we have

$$c^2_{\text{rms}}\big|_N = \frac{1}{T_0\big|_N}\sum_{k=1}^{N}\frac{\Delta z_k}{c_k}c_k^2$$

$$c^2_{\text{rms}}\big|_{N+1} = \frac{1}{T_0\big|_{N+1}}\sum_{k=1}^{N+1}\frac{\Delta z_k}{c_k}c_k^2.$$

By subtraction we obtain an estimate of the wave speed in the $N+1$st layer

$$c_{N+1} = \frac{c^2_{\text{rms}}\big|_{N+1} T_0\big|_{N+1} - c^2_{\text{rms}}\big|_N T_0\big|_N}{\Delta z_{N+1}}. \tag{10.43}$$

Now consider an interval containing p layers. Again from Equation 10.36 we obtain after subtraction

$$\sum_{k-1}^{p} c_{N+k}\Delta z_{N+k} = c^2_{\text{rms}}\big|_{N+p} T_0\big|_{N+p} - c^2_{\text{rms}}\big|_N T_0\big|_N, \tag{10.44a}$$

which may be rewritten in a form,

$$\frac{1}{T_0\big|_{N+p} - T_0\big|_N}\sum_{k=1}^{p}\frac{\Delta z_{N+k}}{c_{N+k}}c^2_{N+k} = \frac{c^?_{\text{rms}}\big|_{N+p} T_0\big|_{N+p} - c^2_{\text{rms}}\big|_N T_0\big|_N}{T_0\big|_{N+p} - T_0\big|_N}.$$

By definition, the right-hand side is the interval rms wave speed in the interval containing p layers

$$\begin{aligned} c^2_{\text{rms}}\big|_p &= \frac{1}{T_0\big|_{N+p} - T_0\big|_N}\sum_{k=1}^{p}\frac{\Delta z_{N+k}}{c_{N+k}}c^2_{N+k} \\ &= \frac{c^2_{\text{rms}}\big|_{N+p} T_0\big|_{N+p} - c^2_{\text{rms}}\big|_N T_0\big|_N}{T_0\big|_{N+p} - T_0\big|_N}. \end{aligned} \tag{10.44b}$$

10.5.2 Estimation in the Presence of Errors

Accurate knowledge of the wave speed is essential not only for imaging, but also for characterization of the medium. The geophysical literature is proliferated with references on wave speed estimation. Several different approaches have been tried to estimate this important parameter. Briefly, correlation between the sensor outputs was used to compute the so-called velocity spectrum in Refs. [16, 17], minimum entropy approach was suggested in Ref. [18], wavefield extrapolation in Ref. [19], and maximum likelihood estimate of rms wave speed was suggested in Ref. [20]. Our interest shall be limited to just two approaches, namely, maximum likelihood and focusing with optimum wave speed.

10.5.2.1 Maximum Likelihood

We model the measurement error in T_x, as an additive random variable, that is,

$$\hat{T}_{x_i} = \sqrt{T_0^2 + \frac{x_i^2}{c_{\text{rms}}^2}} + \tau_i, \tag{10.45}$$

where τ_i is a measurement error in $\hat{T}_{x_i}$, assumed to be a zero mean Gaussian random variable. We introduce the following vectors:

$$\hat{\mathbf{T}}_x = \text{col}\{\hat{T}_{x_1}, \hat{T}_{x_2}, \ldots, \hat{T}_{x_N}\}$$

$$\mathbf{T}_m = \text{col}\left\{\sqrt{T_0^2 + \frac{x_1^2}{c_{\text{rms}}^2}}, \sqrt{T_0^2 + \frac{x_2^2}{c_{\text{rms}}^2}}, \ldots, \sqrt{T_0^2 + \frac{x_N^2}{c_{\text{rms}}^2}}\right\},$$

$$\tau = \text{col}\{\tau_i \; i = 1, 2, \ldots, N\}.$$

Further, let c_{rms}^2 be a random variable with probability density function (pdf) (c_{rms}^2). We would like to maximize the conditional pdf $(c_{\text{rms}}^2/\hat{\mathbf{T}}_x)$. Since

$$\text{pdf}(c_{\text{rms}}^2/\hat{\mathbf{T}}_x) = \frac{\text{pdf}(\hat{\mathbf{T}}_x/c_{\text{rms}}^2)\text{pdf}(c_{\text{rms}}^2)}{\text{pdf}(\hat{\mathbf{T}}_x)}, \tag{10.46}$$

it is sufficient to maximize the numerator in Equation 10.46. For this, note that

$$\begin{aligned}\text{pdf}(\hat{\mathbf{T}}_x/c_{\text{rms}}^2) &= \text{pdf}(\tau) \\ &= \frac{1}{(2\pi)^{N/2}\sqrt{\mathbf{C}}}\exp\left[-\frac{1}{2}(\hat{\mathbf{T}}_x - \mathbf{T}_m)^H \mathbf{C}^{-1}(\hat{\mathbf{T}}_x - \mathbf{T}_m)\right],\end{aligned} \tag{10.47a}$$

where $\mathbf{C}$ is covariance matrix of measurement errors. Further, c_{rms}^2 is also assumed to be a Gaussian random variable with mean equal to $\overline{c}_{\text{rms}}^2$ and standard deviation σ

$$\text{pdf}(c_{\text{rms}}^2) = \frac{1}{\sqrt{2\pi}\sigma}\exp\left(-\frac{1}{2}\left[\frac{(c_{\text{rms}}^2 - \overline{c}_{\text{rms}}^2)}{\sigma}\right]^2\right). \tag{10.47b}$$

Substitute Equations 10.47a and 10.47b in the numerator of Equation 10.46, which is then maximized with respect to c_{rms}^2. We shall maximize the logarithm (natural) of the numerator

$$\max\left\{(\hat{\mathbf{T}}_x - \mathbf{T}_m)^H \mathbf{C}^{-1}(\hat{\mathbf{T}}_x - \mathbf{T}_m) + \left[\frac{(c_{\text{rms}}^2 - \overline{c}_{\text{rms}}^2)}{\sigma}\right]^2\right\}_{w.r.t.\, c_{\text{rms}}^2}. \tag{10.48}$$

Differentiate Equation 10.48 with respect to c_{rms}^2 and set the derivative to zero to obtain

$$3\hat{\mathbf{T}}_x^H \mathbf{C}^{-1} \frac{\partial \mathbf{T}_m}{\partial (c_{\text{rms}}^2)} - 3\mathbf{T}_m^H \mathbf{C}^{-1} \frac{\partial \mathbf{T}_m}{\partial (c_{\text{rms}}^2)} - 2\frac{(c_{\text{rms}}^2 - \bar{c}_{\text{rms}}^2)}{\sigma^2} = 0, \tag{10.49}$$

where

$$\frac{\partial \mathbf{T}_m}{\partial (c_{\text{rms}}^2)} = -\frac{1}{2c_{\text{rms}}^4} \times \text{col}\left\{ \frac{x_1^2}{\sqrt{T_0^2 + \frac{x_1^2}{c_{\text{rms}}^2}}}, \frac{x_2^2}{\sqrt{T_0^2 + \frac{x_2^2}{c_{\text{rms}}^2}}}, \ldots, \frac{x_N^2}{\sqrt{T_0^2 + \frac{x_N^2}{c_{\text{rms}}^2}}} \right\} = \frac{-1}{2c_{\text{rms}}^4}\mathbf{x}. \tag{10.50}$$

Substituting for the derivative in Equation 10.50, we obtain

$$\frac{3}{4c_{\text{rms}}^4}\left[\mathbf{T}_m^H \mathbf{C}^{-1}\mathbf{x} - \hat{\mathbf{T}}_x^H \mathbf{C}^{-1}\mathbf{x}\right] = \frac{(c_{\text{rms}}^2 - \bar{c}_{\text{rms}}^2)}{\sigma^2}. \tag{10.51}$$

We shall now introduce a few approximations. Let the measurement errors be uncorrelated. Then, the covariance function becomes

$$\mathbf{C} = \text{diag}\left\{\sigma_{\tau_1}^2, \sigma_{\tau_2}^2, \ldots, \sigma_{\tau_N}^2\right\},$$

where $\sigma_{\tau_i}^2$ is the variance of the error in ith measurement. With $\sigma^2 \to \infty$ the right-hand side of Equation 10.51 becomes zero. Equation 10.51 may be simplified to

$$\sum_{i=1}^{N} \frac{x_i^2}{\sigma_{\tau_i}^2}\left[1 - \frac{\hat{T}_{x_i}}{\sqrt{T_0^2 + \frac{x_i^2}{c_{\text{rms}}^2}}}\right] = 0, \tag{10.52}$$

which is equal to the minimum mean square error estimate [20]. Numerical simulations presented in Ref. [20] indicate that the minimum mean square error estimate is very close to the maximum likelihood estimate, which is computationally far more involved. Since the unknown quantity c_{rms}^2 in Equation 10.52 occurs inside the square root term, it is not possible to explicitly solve for it. To overcome this, we shall introduce an approximation $T_0^2 \ll x^2/c_{\text{rms}}^2$, which

TABLE 10.1

Mean and Standard Deviation of Wave Speed Estimates given by Equation 10.46

Measurement Errors (std)	Mean Wave Speed	Wave Speed Estimate (std)
0.01 sec	1532.2 m/sec	16.3 m/sec
0.05 sec	1535.1 m/sec	102.8 m/sec

Note: The following parameters were assumed: $T_0=5.0$ sec, $c=1{,}500$ m/sec, maximum source-receiver separation = 1,500 m, number of source-receiver pairs = 100.

enables us to replace the square root term by $T_0(1+(1/2)(x_i^2/T_0^2c_{\text{rms}}^2))$ and then solve for the unknown,

$$\hat{c}_{\text{rms}}^2 \approx \frac{1}{2}\frac{\sum_{i=1}^{N}\frac{\hat{T}_{x_i}x_i^2}{T_0^3}\frac{x_i^2}{\sigma_{\tau_i}^2}}{\sum_{i=1}^{N}\frac{x_i^2}{\sigma_{\tau_i}^2}\left[\frac{\hat{T}_{x_i}}{T_0}-1\right]}. \tag{10.53}$$

The wave speed estimate given by Equation 10.53, on account of approximation $T_0^2 \ll x^2/c_{\text{rms}}^2$, suffers from a large error even for small measurement error. This is demonstrated in Table 10.1. The denominator in Equation 10.53 is strongly influenced by the measurement errors for limited array aperture. As noted in the context of focused beamformation, for best depth of focus the array aperture was required to be four times the depth to the reflector. It is understood that a similar requirement exists for wave speed estimation.

10.5.3 Focusing with Optimum Wave Speed

In Section 10.4, we saw how CDP gathers are focused at the point of reflection. We were required to know the correct wave speed. We shall now assume that the wave speed is unknown. In fact, we wish to find an optimum speed for which the focus is the best, in the sense that the magnitude of the focused signal is the highest possible. The response function of a focused beamformer, when the wave speed is unknown, may be obtained from Equation 10.32b

$$H_N(\omega)=\frac{1}{N}\sum_{n=0}^{N-1}e^{-j\omega\left[T_0\sqrt{(1+(\Delta tn/T_0)^2)}-\hat{T}_0\sqrt{(1+(\Delta\hat{t}n/\hat{T}_0)^2)}\right]}, \tag{10.54}$$

where $\Delta\hat{t}=\Delta x/\hat{c}$, $\hat{c}$ is the assumed speed and Δx is the basic unit of source and receiver separation. Note that $\Delta t=\Delta x/c$ is the actual propagation time. The maximum value of the response function (Equation 4.54) is one and it is achieved only when the exponent is equal to zero, which requires $\hat{T}_0=T_0$ and $\Delta\hat{t}=\Delta t$ or $\hat{c}=c$. For the present, we assume that $\hat{T}_0=T_0$ and evaluate $|H_N(\omega)|^2$

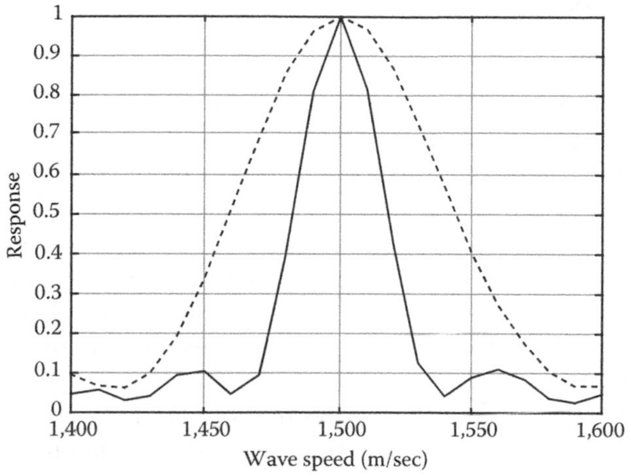

FIGURE 10.20
Response (magnitude square) of focused beamformer as a function of wave speed. Maximum source sensor separation: 8.5 km (solid line) and 4.5 km (dashed line). Angular frequency = 62.84.

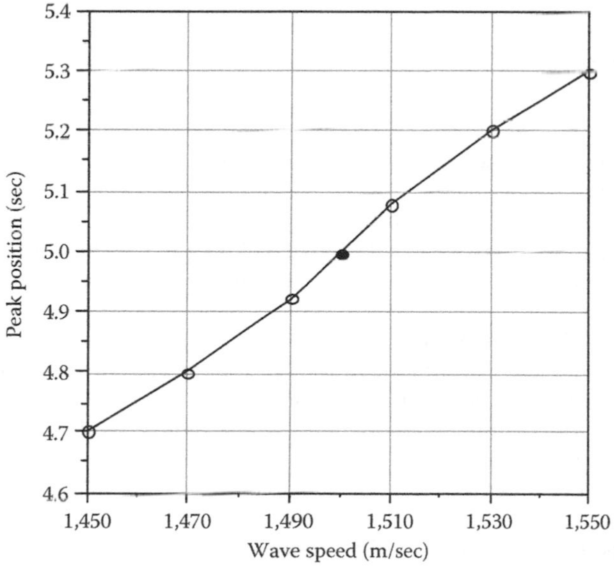

FIGURE 10.21
Focused beam position as a function of wave speed in m/sec. All other parameters are as in Figure 10.20. The maximum sensor separation is 4.5 km.

for a series of assumed wave speeds. Response function as a function of the assumed speed is shown in Figure 10.20. The position of the peak (point of reflection) changes with the assumed speed. The correct position is obtained only when the assumed wave speed is equal to the actual speed. The variation of the peak position is illustrated in Figure 10.21.

10.6 Exercises

(1) A point scatterer is located at a depth l units below the observation surface. A transceiver, placed on the surface, records the round-trip travel time. Show that this travel time is exactly the same as in the CDP experiment with a reflector replacing the scattering point.

(2) CDP seismic gathers containing an echo from a horizontal reflector are cross-correlated. Show that the maximum of the cross-correlation function lies on a hyperbolic time-distance curve. This property is the basis for a technique of seismic speed estimation known as velocity spectrum [21].

(3) Show that the slope of the line joining all reflected signals in a zero-offset seismic data over an incline reflector is related to the slope of the reflector through the following relation, $\tan\alpha=\sin\theta_0$, where α is the slope of the line joining the reflected signals and θ_0 is the slope of the reflector.

References

1. A. J. Devaney, Inverse source and scattering problem in ultrasonics, *IEEE Trans.*, SU-30, pp. 355–364, 1983.
2. D. Loewenthal, L. Lu, R. Robertson, and J. Sherwood, The wave propagation applied to migration, *Geophys. Prospect.*, vol. 24, pp. 380–399, 1976.
3. J. F. Claerbout, Toward unified theory of reflector mapping, *Geophysics*, vol. 31, pp. 467–481, 1971.
4. R. H. Stolt, Migration by Fourier transforms, *Geophysics*, vol. 43, pp. 23–48, 1978.
5. D. Lowenthal and I. D. Mufti, Reverse time migration in spatial frequency domain, *Geophysics*, vol. 48, pp. 627–635, 1983.
6. G. A. McMechan, Migration by extrapolation of time dependent boundary values, *Geophys. Prospect.*, vol. 31, pp. 413–420, 1983.
7. R. Sun and G. A. McMechan, Pre-stack reverse time migration, *Proc. IEEE*, vol. 74, pp. 457–467, 1989.
8. J. Claerbout, *Fundamentals of Geophysical Data Processing*, McGraw Hill, New York, 1972.
9. G. Cheng and S. Coen, The relationship between Born inversion and migration for common midpoint stack data, *Geophysics*, vol. 49, pp. 2117–2131, 1984.
10. E. A. Robinson, Frequency-domain analysis of multidimensional time series data, in *Handbook of Statistics*, Eds D. R. Brillinger and P. R. Krishnaiah, vol. 3, pp. 321–342, Elsevier Science, Amsterdam, 1983.
11. A. J. Berkhout, *Seismic Migration – Imaging of Acoustic Energy by Wave Field Extrapolation*, Elsevier, Amsterdam, 1980.
12. P. S. Schultz and J. W. C. Sherwood, Depth migration before stack, *Geophysics*, vol. 45, pp. 376–393, 1980.

13. W. A. Schneider, Integral formulation for migration in two and three dimensions, *Geophysics*, vol. 43, pp. 49–76, 1978.
14. R. A. Phinney and D. M. Jurdy, Seismic imaging of deep crust, *Geophysics*, vol. 44, pp. 1637–1660, 1979.
15. K. Kido, H. Noto, A. Shima, and M. Abe, Detection of acoustic signal by use of cross spectra between every pair of sensors distributed in space, *IEEE*, ICASSP-86, pp. 2507–2510, 1986.
16. M. T. Taner and F. Koehler, Velocity spectra digital computer derivation and applications of velocity functions, *Geophysics*, vol. 34, pp. 859–881, 1969.
17. W. A. Schneider and M. Backus, Dynamic correlation analysis, *Geophysics*, vol. 33, pp. 105–126, 1968.
18. D. De Vries and A. J. Berkhout, Velocity analysis based on minimum entropy, *Geophysics*, vol. 49, pp. 2132–2142, 1984.
19. O. Yilmaz and R. Chambers, Migration velocity analysis by wave field extrapolation, *Geophysics*, vol. 49, pp. 1664–1674, 1984.
20. R. L. Kirlin, L. A. Dewey, and J. N. Bradely, Optimum seismic velocity estimation, *Geophysics*, vol. 49, pp. 1861–1868, 1984.
21. L. C. Wood and S. Treitel, Seismic signal processing, *Proc. IEEE*, vol. 63, pp. 649–661, 1975.
22. R. T. Hussian, W. A. Sandham, and R. Chapman, Application of migration to ground probing radars, *IEEE Proc. ICASS*, pp. 1208–1211, 1988.
23. M. A. Haun, D. L. Jones, and W. D. O'Brien Jr, Adaptive focusing through layered media using the geophysical "Time migration" concept, *IEEE Ultrasonic Symp.*, pp. 1635–1638, 2002.

Index

C

D

E

G

H

N

T

U

V